EMERGENCE *and* COLLAPSE *of* EARLY VILLAGES

ORIGINS OF HUMAN BEHAVIOR AND CULTURE

Edited by Monique Borgerhoff Mulder and Joe Henrich

1. *Behavioral Ecology and the Transition to Agriculture*, edited by Douglas J. Kennett and Bruce Winterhalder
2. *Pattern and Process in Cultural Evolution*, edited by Stephen Shennan
3. *The Hadza: Hunter-Gatherers of Tanzania*, by Frank W. Marlowe
4. *Life Histories of the Dobe !Kung: Food, Fatness, and Well-being over the Life Span*, by Nancy Howell
5. *Friendship: Development, Ecology, and Evolution of a Relationship*, by Daniel J. Hruschka
6. *Emergence and Collapse of Early Villages: Models of Central Mesa Verde Archaeology*, edited by Timothy A. Kohler and Mark D. Varien

EMERGENCE *and* COLLAPSE *of* EARLY VILLAGES

Models of Central Mesa Verde Archaeology

Edited by

Timothy A. Kohler and Mark D. Varien

UNIVERSITY OF CALIFORNIA PRESS
Berkeley   Los Angeles   London

University of California Press, one of the most distinguished university presses in the United States, enriches lives around the world by advancing scholarship in the humanities, social sciences, and natural sciences. Its activities are supported by the UC Press Foundation and by philanthropic contributions from individuals and institutions. For more information, visit www.ucpress.edu.

Origins of Human Behavior and Culture, No. 6

University of California Press
Berkeley and Los Angeles, California

University of California Press, Ltd.
London, England

Library of Congress Cataloging-in-Publication Data

Emergence and collapse of early villages : models of central Mesa Verde archaeology / edited by Timothy A. Kohler and Mark D. Varien.
p. cm. — (Origins of human behavior and culture v.6)
Includes bibliographical references and index.
ISBN 978-0-520-27014-5 (cloth : alk. paper)
1. Pueblo Indians—Colorado—Mesa Verde National Park—History. 2. Pueblo Indians—Agriculture—Colorado—Mesa Verde National Park. 3. Pueblo Indians—Colorado—Mesa Verde National Park—Antiquities. 4. Mesa Verde National Park (Colo.) — Antiquities. I. Kohler, Timothy A. II. Varien, Mark D., 1954–
E99.P9E435 2012
978.8'27—dc23 2011028668

19 18 17 16 15 14 13 12
10 9 8 7 6 5 4 3 2 1

The paper used in this publication meets the minimum requirements of ANSI/NISO Z39.48-1992 (R 1997) (*Permanence of Paper*).∞

Cover illustration: Spruce Tree House, Mesa Verde National Park. Photo by Nate Crabtree.

The publishers gratefully acknowledge a production grant from the Bureau of Land Management Canyons of the Ancients National Monument and the National Landscape Conservation System. This material is based upon work supported by the National Science Foundation under grants BCS-0119981 and DEB-081640.

CONTENTS

PREFACE AND ACKNOWLEDGMENTS

When people began to domesticate plants and animals around 11,000 years ago, they unwittingly unleashed powerful forces of human population growth, the effects of which we are still dealing with today. The first people to do so, though, were the village dwellers of the early Neolithic, whose dates depend on when farming lifeways were adopted in different parts of the world.

Farming was rather late in arriving to the northern portions of the U.S. Southwest, but the special characteristics of the archaeological record in the Four Corners allow us to view and reconstruct it with a precision that cannot be achieved anywhere else in the world. In this book we examine the fortunes and fates of prehispanic Pueblo societies between A.D. 600 and A.D. 1300 in southwestern Colorado, a dramatic period marked by large swings in climate, population, and settlement size, as well as in amount and type of conflict.

What we know about these societies and the environments they inhabited is due to the accumulation of a century of archaeological and other scientific research—notably bolstered by increasing collaboration with the descendants of these societies over the last few years especially. It would really be impossible to give a full accounting of the intellectual debts incurred by the sort of synthetic effort this book represents.

All we can do is credit those people and institutions whose help has been most immediate. Blake Edgar encouraged us to submit this book to the University of California Press and saw it through the review and acceptance process. Mike Adler, Jim Allison, Bill Lipe, and Scott Ortman read all or major parts of it at some stage and provided thoughtful and useful comments. Kyle Bocinsky and Stefani Crabtree helped with nearly all aspects of production, especially the huge job of getting the figures and text into a standard format. Kyle helped with coding the simulations; Kyle and Ben Ford also helped run the simulations; and Stefani assembled the time-allocation data for small-scale societies used in several chapters. Jeff Dean and Carla Van West consulted on the use of paleoclimatic data and on generating the paleoproduction landscapes. Matthew Salzer allowed us to use some of his unpublished data. Kristie Arrington, Matthew Bailey, Kay Barnett, George Burr, Andrew Duff, Jerry Fetterman, Robert Gillson, Linda Honeycutt, Tim Hovezak, John Jones, Jim Judge, Tim Kearns, Laura Kochanski, Claire and Sander Kohler, Lee Lyman, John E. McCray, Larry Nordby, Ken Petersen, Doug Ramsey, Charles Reed, Bob

Reynolds, Hugh Robinson, John M. Shafer, Marilyn Von Seggern, Rich Wilshusen, Yuejun (Eugene) Yan, and Uncle Jimmy all provided valuable assistance. The survey crew for the Village Ecodynamics Project (VEP) Community Center Survey included Fumiyasu Arakawa, Donna Glowacki (director), Dave Johnson, and Hugh Robinson. We particularly appreciate the support and permissions for site visits granted by the Bureau of Land Management, Crow Canyon Archaeological Center, the National Park Service, the U.S. Forest Service, the Archaeological Conservancy, and 16 generous private landowners. To all of you—and to others too numerous to mention—thank you.

We want to make our basic data available to interested researchers. Appendix B presents key information about the largest settlements in the study area. We are depositing a similar data tabulation on all the habitations in this area—though necessarily without their specific locations—in the publicly accessible archives of Digital Antiquity (www.digitalantiquity.org/). The Swarm code generating the simulations reported here can be found at www.openabm.org/model/2518; use it and improve it!

The first phase of the VEP, reported in this book, was supported by National Science Foundation Grant BCS–0119981. The National Conservation Landscape Conservation System Research and Science Program provided funding for the color plates in the book. Washington State University and Crow Canyon Archaeological Center made the rest possible.

CONTRIBUTORS

FUMIYASU (FUMI) ARAKAWA
Department of Anthropology
New Mexico State University
Las Cruces, NM
farakawa@nmsu.edu

R. KYLE BOCINSKY
Department of Anthropology
Washington State University
Pullman, WA
bocinsky@gmail.com

SARAH M. COLE
Red River Archaeology LLC
Dallas, TX
s_cole498@yahoo.com

KEVIN COOPER
Department of Mathematics
Washington State University
Pullman, WA
kcooper@wsu.edu

JASON A. COWAN
Cultural Resource Consultants, Inc.
Bainbridge Island, WA
jcowan42@gmail.com

STEFANI CRABTREE
Department of Anthropology
Washington State University
Pullman, WA
stefanicrabtree@gmail.com

BEN FORD
Department of Anthropology
Washington State University
Pullman, WA
bdford@wsu.edu

DONNA M. GLOWACKI
Department of Anthropology
University of Notre Dame
Notre Dame, IN
Donna.M.Glowacki.3@nd.edu

C. DAVID (DAVE) JOHNSON
Bureau of Land Management
Arcata, CA
cdjsmail@gmail.com

ZIAD KOBTI
School of Computer Science
University of Windsor, Ontario
kobti@uwindsor.ca

TIMOTHY (TIM) A. KOHLER
Department of Anthropology
Washington State University
Pullman, WA
Santa Fe Institute and Crow Canyon Archaeological Center
tako@wsu.edu

KENNETH (KEN) E. KOLM
Integral Consultants, Inc.
Louisville, CO
kkolm@integral-corp.com

SCOTT G. ORTMAN
Santa Fe Institute and Crow Canyon Archaeological Center
Cortez, CO
ortman@santafe.edu

SCHAUN M. SMITH
Chevron Energy Technology Company
Houston, TX
SchaunMSmith@chevron.com

MARK D. VARIEN
Crow Canyon Archaeological Center
Cortez, CO
MVarien@crowcanyon.org

AARON M. WRIGHT
Center for Desert Archaeology
Tucson, AZ
aaronwright1@hotmail.com

ONE

Emergence and Collapse of Early Villages in the Central Mesa Verde

AN INTRODUCTION

Timothy A. Kohler and Mark D. Varien

TWO HUNDRED AND FORTY YEARS after the last Pueblo people left Colorado, Hernán Cortés de Monroy y Pizarro—or maybe it was Quetzalcoatl—stepped onto the shores of Veracruz. Arriving at the Aztec capital of Tenochtitlan less than seven months later, the Spanish had no trouble recognizing kings and slaves, temples and markets, gods and warriors, and all manners of artifacts and other institutions, none of which had existed when the ancestors of these two societies last lived together in Eurasia. What accounts for the surprising mutual intelligibility of social forms between two societies that had shared their last common biological and cultural ancestors tens of thousands of years ago?

Undoubtedly the shared biology (proximate by the standards of biological evolution) and perhaps the shared culture (which, however, was very distant by the more rapidly changing standard of culture itself) had some effect on channeling development in certain directions rather than others. But another possibility that needs to be considered is that winnowing of less-efficient forms by selection through competition among groups may have shaped these otherwise independent historical trajectories. Some solutions to the puzzle of how to induce people to cooperate with large numbers of other unrelated people probably work better than others, and, if invented independently, might be maintained. This logic helps us understand, for example, why markets appeared alongside older reciprocal exchange systems as Pueblo peoples began to concentrate in the northern Rio Grande region of New Mexico in the A.D. 1300s (Kohler, Van Pelt, and Yap 2000). Over long enough periods, solutions that are markedly less workable may tend to be supplanted, whether through lethal intergroup competition, nonlethal cultural group selection, selective migration, or related processes (see review in Salomonsson 2010).

But such arguments are plausible only if we can show that adaptation, a concept we are borrowing from biology, has some relevance for understanding cultural behavior. This might seem a modest claim: we are not, for example, proposing that natural selection is a "universal acid" explaining all aspects of culture and its contents, as Daniel Dennett (1995) sometimes

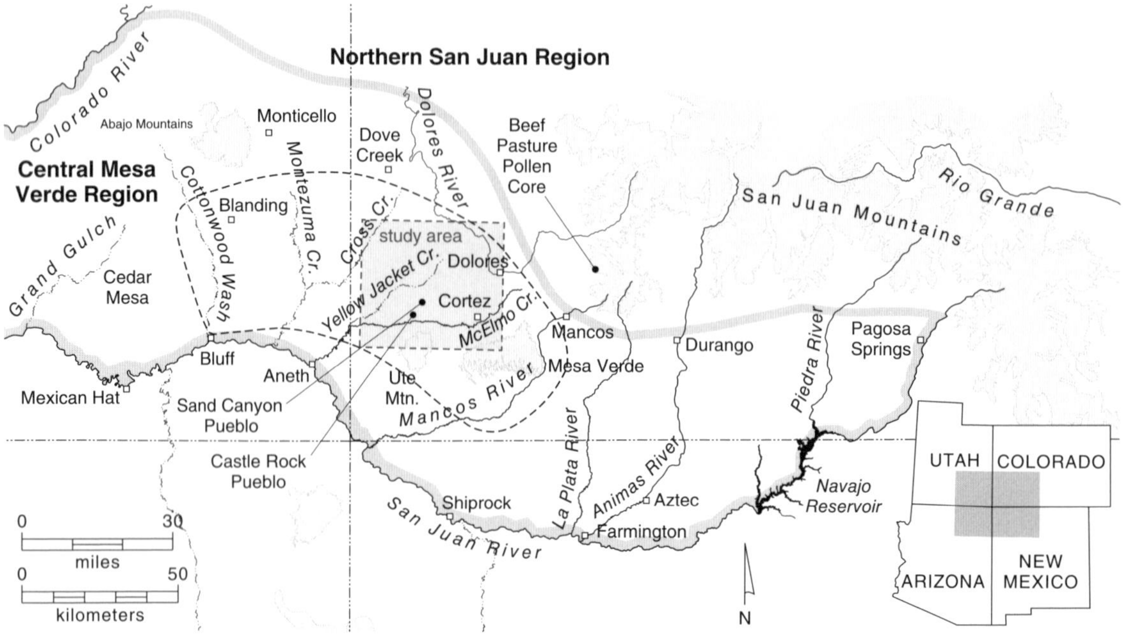

FIGURE 1.1 The VEP I study area lies within the central Mesa Verde portions of the northern San Juan region.

seems to do. But as we shall see, even this modest claim would be dismissed by some.

Societies are likewise in constant interaction with the environments they inhabit. They modify those environments by accident and on purpose, affecting their short- and long-term productivity. How could it not be the case that this also affects the success of the societies? One common characteristic of Neolithic societies is the rapid population growth they experience. If per capita use of resources remains constant as population grows, resources that regenerate slowly will be drawn down. This may soon require societies to move to previously uninhabited areas, if they are available—but eventually, this option will be impossible and new patterns of resource usage must be developed. Societies unable to innovate these patterns will be displaced by those that can.

The general growth of Neolithic populations also favors those groups that can effectively coordinate the largest numbers of people, since large cohesive groups can displace smaller or less cohesive groups, or resist displacement by others. These two pressures—building larger sedentary groups and not depleting the environment—are at odds with each other. This contributes to the dynamic character of the Neolithic record in most areas; in finding an effective compromise between these two opposing forces, societies are driven to a position on a metaphorical fitness landscape where they become extremely vulnerable to external perturbations such as climate change.

Eventually we would like to examine how these processes play out in Neolithic societies all over the world. (We suspect that they are completely general.) Unfortunately, there are few places where the Neolithic sequences are known well enough to make this possible in any convincing detail. In this book we analyze these social and "socionatural" processes through an extended case study set in southwestern Colorado between A.D. 600 and A.D. 1300. This is a fascinating 700 years that takes us, in its first two centuries, from the arrival of small groups of maize farmers in an almost uninhabited mesa and canyon country to the emergence of some of the Southwest's earliest and largest villages. But around 100 years later, by A.D. 900, most of the villages in our study area had disbanded and their inhabitants departed for the south or west. The remaining households once again lived mostly in small hamlets until about A.D. 1080, when a new wave of colonists arrived from the south, bringing a differently organized village lifeway backed by a novel, powerful political and religious system that united most of the eastern Southwest. Within 200 years this wave, too, receded, again to the south, leaving behind the famous ruins of Mesa Verde National Park and the less-well-known but more populous district that includes Canyons of the Ancients National Monument, where our work centers (Figure 1.1).

Even before the project on which we report here began in 2001, archaeologists knew more about the development of societies in this area than was known about almost any other comparable Neolithic society anywhere in the world. For that we thank generations of researchers, recently including Hayes (1964), Kane (1986), Lipe et al. (1999), Rohn (1977), Varien (1999a), and Varien and Wilshusen (2002), among many others.

Why, then, another book? The answer lies in an exciting opportunity that the National Science Foundation (NSF) presented in 2001 for a Biocomplexity in the Environment Special Competition in "coupled natural and human systems." NSF recognized that understanding how humans interacted with ecosystems over long periods required both deep interdisciplinary collaboration and modeling. They also realized that such research was not being generated by their existing programs. When we saw this call for proposals, we recognized it as providing precisely the sort of research program we had been trying to put together. We also understood that we needed help to carry out the kind and scope of research that NSF was requesting, and we assembled a team to create a research proposal that came to be known as the Village Ecodynamics Project (VEP).[1]

Fortunately we were not starting from a blank slate. Timothy Kohler, an external professor at the Santa Fe Institute (SFI), had recently hosted a conference there on agent-based modeling (Kohler and Gumerman, eds. 2000), at which Bob Reynolds presented a paper on his research with Joyce Marcus and Kent Flannery on the role of conflict in the prehispanic emergence of chiefdoms and states in the Valley of Oaxaca (Reynolds 2000). Reynolds is nearly unique among computer scientists in having a long history of collaboration with archaeologists: complexity pioneer John Holland and archaeologist Flannery had co-chaired Reynolds's dissertation committee at the University of Michigan. By then a computer scientist at Wayne State University, Reynolds agreed to join our team as a principal investigator on the proposal, and among the many assets he brought to the collaboration was Ziad Kobti, then Reynolds's Ph.D. student. In Chapter 13 of this book, Kobti reports on VEP efforts to model local exchange among households in our simulations.

Given NSF's charge to understand how societies were shaped by their changing environments, and how those societies in turn altered their environments, we also needed a partner who could bring a broad understanding of humans in their landscapes. A second acquaintance from an SFI workshop, hydrologist Ken Kolm, agreed to take a leading role in this research to help us understand the interactions through time between people and water, which we thought would be especially important to model on this semiarid landscape. Kolm, then of the Colorado School of Mines, brought along his Ph.D. student Schaun Smith. In late 2001 this team drafted a successful proposal to fund what we came to call VEP I (NSF BCS-0119981). Kohler also brought a group of graduate students at Washington State University (WSU) to the team, and Varien involved the staff at the Crow Canyon Archaeological Center,

1. Throughout, we will use the term *VEP* to refer to this research project broadly defined. Occasionally we will refer more specifically to the work reported in this book as *VEP I*, since beginning in 2008 we have begun an extension of this research, with additional funding from NSF.

especially Scott Ortman, who assembled the database of archaeological sites discussed in Chapters 2 and 14.

We have been at it ever since. As we write this in 2011, our research is supported by the NSF grant program Dynamics of Coupled Natural and Human Systems, which grew out of the special competition that funded the first phase of the VEP. VEP II is organically connected to VEP I because it has many of the same researchers and most of the same research interests. Many students from the first team, however, have now entered the professional workforce: Fumi Arakawa is embarking on an assistant professorship at New Mexico State University; Sarah Cole is president of Red River Archaeology in Dallas; Jason Cowan works as an archaeologist for Cultural Resource Consultants Inc. (CRC) in Washington state; Donna Glowacki is an assistant professor of anthropology at the University of Notre Dame; Dave Johnson is an archaeologist and a tribal liaison for the Bureau of Land Management in Arcata, California; Scott Ortman holds a dual position as the Lightfoot Fellow at Crow Canyon Archaeological Center and the Omidyar Fellow at SFI; Schaun Smith is a staff scientist in the Environmental Unit of Chevron Energy Technology Company in Houston; and Ziad Kobti is an associate professor of computer science at the University of Windsor.

Over the years we have seen the stream of research undertaken in 2001 lead to some fundamentally new ways of thinking about and seeing the past. Part of this is due to the interdisciplinary nature of our research, part is due to modeling the past using computer simulation, and part is due to amassing synthetic databases on all known archaeological sites in our study area. Together these allow us to develop "model-based approaches" (Kohler and van der Leeuw 2007) to generate and test expectations about what the past might have looked like if certain processes were dominant. For the VEP, it was a case of being in the right place at the right time. There are few areas in the world where the past environment can be reconstructed in such detail and linked so precisely to ancient subsistence practices; this reconstruction of ancient environments and subsistence practices was a fundamental aspect of the computer simulation. Further, there are few places where we could assemble so much previous archaeological research and analyze it in productive ways, though we had to develop new approaches to make that possible (Ortman et al. 2007). As a result, the VEP can use the computer simulation to generate models and evaluate them through quantitative comparisons to the archaeological data.

A SOCIONATURAL RESEARCH AGENDA

The VEP conducts socionatural research (van der Leeuw and Redman 2002); we examine the long-term interactions between humans and their environments. The network of scientists making this possible includes archaeologists, computer scientists, ecologists, an economist, geologists, and hydrologists, among others. We focused this research on an 1817 km^2 study area in southwestern Colorado, in the heart of the central Mesa Verde region, one of the most densely occupied portions of the prehispanic Pueblo world. For analytical purposes we divided this area into 45,400 cells that are 200 m on each side, or 4 ha in size, and selected the period between A.D. 600 and A.D. 1300 as the time frame for our research. An important goal of the VEP is historical research that seeks to explain the development of Pueblo Indian society in this area during these seven centuries. Equally important, the VEP seeks to use what we learn about these societies to better understand the general processes that underlie human social evolution, especially during the Neolithic period, that pivotal epoch when humans first adopted domesticated food production.

Two major efforts define the VEP research. The first is an agent-based computer simulation that reconstructs past climate and environment, models how annual variation in precipitation and temperature affected productive resources, and examines how humans use these domesticated and wild resources. The agents in this computer simulation are households, although the simulation tracks the fortunes of each individual in these households as well. These virtual households are let loose on our reconstruction of the prehispanic landscape of southwestern Colorado in A.D. 600. For the next 700 years, the households farm maize, obtain drinking water, collect wood for fuel, hunt mule deer, jackrabbits, and cottontails, and—in some versions of the simulation—exchange maize and meat with each other.

The second major VEP effort has been archaeological research to summarize the ancestral Pueblo occupation of the study area. This was accomplished by constructing and analyzing a database of all known archaeological sites in our study area—about 9,000 in all—supplemented by a database of all tree-ring dates from the region. VEP archaeological research focused on the 3,176 sites determined to be habitations, including 92 sites we termed "community centers" based on their larger size, longer occupations, and distinctive architecture, which included public or civic structures. Additional research, including new fieldwork, was conducted at some of these community centers.

The broad questions about coupled natural and human systems emphasized by the VEP led us to focus on three aspects of the ancestral Pueblo occupation of the study area. First, we sought to understand how farmers located themselves and used resources in this landscape. Second, we wanted to examine the exchange of subsistence goods among households and whether this exchange caused households to aggregate into villages in certain times and places and disperse into smaller settlements during other times. Finally, we wanted to determine why our study area and the surrounding Mesa Verde region was depopulated in the late A.D. 1200s.

These general questions required us to answer other, more specific, questions. What were the population dynamics during the 700 years when the study area was occupied? Could farmers continue to grow maize in sufficient quantities during the driest and coldest years? Did domestic water supplies disappear during prolonged drought? Was the landscape depleted of wood resources during the long-term occupation of the study area? Could people sustainably meet their needs for protein by hunting deer, jackrabbits, and cottontails, or would they tend to seriously depress the number of these animals over time? How was exchange affected by the distribution of resources? What was the history of conflict in the region?

These and other questions have shaped the VEP research program, and we determined that they could be answered only by integrating simulation and archaeological analyses. Some problems were more directly addressed through simulation, others through archaeological studies. Ultimately, the strength of the VEP was our ability to play the simulation and the archaeological analyses off of each other. The organization of this book reflects this integrated effort. The construction and analysis of the simulation are presented in Chapters 4 through 11, while results of archaeological analyses are addressed in Chapters 2 and 12 through 14.

THE VEP: ANTHROPOLOGICAL AND EVOLUTIONARY ISSUES

One motivation for this book is that VEP research enables us to contribute a new perspective on one of the oldest and what is still one of the most troublesome issues in the social and biological sciences: the question of adaptation. Differing beliefs about the importance of adaptation have riven communities of anthropologists since at least the 1970s. Moreover, this conflict is connected to a much older contest between "historical" and "structural" approaches in the social sciences, evident in Western thought by the nineteenth century but present even long before (McAllister 2002). In the remainder of this chapter we will briefly review where anthropology stands on the concept of adaptationism, explain our focus on settlement systems as the main database in this book, describe some local antecedents to this project, and situate the local prehistory in regional and worldwide contexts.

Adaptation and Optimality Debated

In the simulations discussed in detail in later chapters, the rules we impose on our agents lead them to live in places that are approximately optimal, in the sense and within the constraints defined more carefully later in this chapter. Optimality principles have guided theory formation in physics from at least the seventeenth century. In 1662 Pierre de Fermat formulated the principle of least time, which states that "a ray of light, moving through an arbitrary medium . . . will follow, out of all possible paths, that path for which the transit time of the ray is a minimum" (Rosen 1967:2–3). Two hundred years later, Darwin articulated the principle of natural selection for living systems, which predicted that in conditions of competition, entities that in some sense compete more effectively than others will better survive that competition and leave more offspring. Furthermore, if the characteristics that led to that comparative advantage are heritable, then those characteristics will spread in the population.

Of course, even within biology there has been lively debate between those who accord adaptation a more, or less, central role in evolution—compare, for example, John Maynard Smith (1978) and Ernst Mayr (1983) with S. J. Gould and R. C. Lewontin (1979). In the social sciences, principles of optimality have enjoyed their greatest authority in economics, generating a huge literature on topics such as individual optimizing behavior, and, in macroeconomics, optima within the consumer and producer sectors.

In anthropology and archaeology, on the other hand, we encounter profoundly conflicting positions on the usefulness of optimality either as a guide to research or as a principle whose past or current potency can be counted on. Within archaeology specifically, most processualists probably agreed with Patrick Kirch (1980:102), who characterized adaptation as "a robust concept capable of integrating disparate methodological orientations and of relating them to a central theme of culture as man's unique method of meeting environmental challenges." Human behavioral ecologists also subscribe to the concept of adaptation by asserting that humans have evolved to make approximately optimal choices in changing environments with respect to behaviors that plausibly affect fitness, including prey choice, patch choice, resource defense, mating strategies, and signaling strategies (e.g., Bliege Bird and Smith 2005; Kelly 1995; Smith 1992). Other strands of evolutionary thinking in anthropology, such as evolutionary psychology and dual-transmission theory, handle optimization in ways that typically offer quite variable predictions while retaining a foundation in optimality thinking.

Marshall Sahlins (1976), on the other hand, is one of many anthropologists who have argued that human cultures do *not* arise from practical, utilitarian considerations that lead to adaptive advantages. Sahlins argues that the "decisive" quality of culture "is not that [people] must live in a material world, circumstance [shared] with all organisms, but that [people] do so according to a meaningful scheme of [their] own devising, in which capacity mankind is unique" (1976:viii). The course of archaeological theory since the 1970s has shown that Sahlins could not have been more prescient in finding that "the contest between the practical and the meaningful is the fateful issue of modern social thought" (1976:ix).[2]

Some postprocessualist approaches to archaeology, for example, make no room for either optimality or competition among their foundational principles. Michael Shanks and Christopher Tilley (1987:51), for example, consider "rationality" (in which there is an implied economy or efficiency of behavior) to have valid

2. But we agree with Lee Cronk (1999:x) that cultural anthropology erred by following Sahlins in completely sealing off the concept of culture from any evolutionary logic.

referents only in contemporary capitalist societies. They argue that most human action has no direct survival value and that our task as archaeologists is to understand the free choices that people made and the stylistic record that those created through time (1987:56).

In the face of such positions, we make three main arguments for emphasizing optimality in our models. The first is that people clearly live in a material world, which makes certain demands. People meet these demands in part by attaching meanings to their world and to the utterances of others; this gives them leverage in the material and social worlds. But systems of meaning are difficult for archaeologists to decipher, and we ought to take advantage of every opportunity to study them. Settlement systems almost certainly contain information relevant to this task (Thomas 1999), but to be able to interpret them appropriately we propose (pace Thomas) that one ought to assess the degree to which they might be explicable by practical considerations.[3]

Second, it is, or ought to be, an empirical question as to "how much" of a settlement system, or any other aspect of ancient practice, is attributable to practical considerations such as efficiency of resource access. This is an extraordinarily difficult thing to assess, though, and it is not surprising that most archaeologists have taken an a priori stance on the theoretical issue at stake rather than attempt an empirical assessment of the adaptation model for particular settlement systems. Such assessment, however, can bear several fruits. If we can evaluate the extent to which adaptation shapes any particular settlement system, then we may be able to identify times when these pressures are strongest, and propose reasons for those differences through time, as we do in Chapter 10. This requires us to consider the empirical archaeological record with the same care that we devote to constructing models of it (see, especially, Chapters 2 and 14).

Finally, we recognize that even if simple optimality models are relevant, they will fail to explain important aspects of settlement systems—as they have for other human and non-human behaviors to which they have been applied (Maynard Smith 1978:52–53; Smith 1992:54). As Robert Foley concluded in his review of optimality theory in anthropology, "The main strength of an optimality model lies in the fact that its methodology provides a template for comparing behaviours. The point to stress about optimality theory is not that behaviours should conform to the template, but that it provides a standard of measurement and comparison against which deviations can be assessed" (Foley 1985:222–223). The ways in which such models fail will be of interest because they will contain information on the many social, political, and symbolic considerations affecting settlement systems that were not incorporated into the optimality model. In short, a model does not have to be complete, or completely correct, for it to provide a useful foot through the explanatory door.

We think that the common, though partial, success of human behavioral ecologists and other social scientists who make predictions based on assumptions of adaptation warrants an approach that is less interested in determining the *existence* of adaptive behavior than in trying to understand *how central* adaptation has been in various domains of human action. Here we find useful a framework proposed by philosopher of science Peter Godfrey-Smith. Godfrey-Smith (2001) recognizes three kinds of adaptationism in current biological thought, but his definitions are general enough to be applicable to anthropology as well. The first is empirical adaptationism, the strongest form, which holds that "natural selection is a powerful and ubiquitous force .... [T]o a large degree, it is possible to predict and explain the outcome of evolutionary processes by attending only to the role of natural selection" (2001:336).

The second is explanatory adaptationism, which is a bit trickier. Godfrey-Smith defines it as follows: "The apparent design of organisms

3. We do not pursue this program here, but we think that this is how such studies ought to be carried out.

[we might also say societies or cultures] and the relations of adaptedness between organisms and their environments, are the big questions, the amazing facts in biology. Explaining those phenomena is the core intellectual mission of evolutionary theory. Natural selection is the key to solving these problems: it is the big answer. Because it answers the biggest questions, selection has a unique explanatory importance among evolutionary factors . . . even if it is rare" (2001:336).

The third possible position is methodological adaptationism. This one is easy: "The best way for scientists to approach biological systems is to look for features of adaptation and good design. Adaptation is a good 'organizing concept' for evolutionary research" (2001:337). Unlike the first two varieties of adaptationism, this is simply a policy recommendation, not an ontological claim.

On the basis of evidence to be presented in Chapter 10, we will argue in Chapter 15 that both explanatory and methodological adaptationism make sense in the study of small-scale human societies in which households retain considerable autonomy of decision making. But we will also argue that much interesting variation in the archaeological record remains unexplained as household-level adaptive behavior, and that our next logical task is to develop methods to study how much social, political, and religious behavior can be explained as adaptive behavior of competing groups—and how much cannot.

A Focus on Placement and Landscape

In this book, our chief database consists of the location and sizes of habitation sites. These farmsteads, hamlets, and villages were formed of varying numbers of households that constructed particular facilities and placed their settlements on the landscape in particular ways that changed through time (Chapters 2, 10, and 14). This requires us to develop a method—agent-based simulation—that can cope with the fact that the locations of resources, too, are constantly changing because of both variability in climate and human use. Two chapters, though—Chapter 12 by Fumi Arakawa and Chapter 13 by Sarah Cole—develop other databases to help us understand the changes we see in settlement practices in our area. These chapters are particularly important precisely because the social behaviors they allow us to infer do not (for the moment, at least) emerge from or figure in our household-level agent-based models; we need to infer these behaviors directly from the archaeological record.

We focus much of our analytical attention on settlement patterns because there is good evidence that these spatial arrangements of households—and how they change by choice or necessity—are packed with information about ecological and social relations, and it is up to us to learn how to decode it. Archaeologists (and geographers) have taken the locations and sizes of settlements as something to be explained since at least the 1950s. Gordon Willey is among those most responsible for focusing archaeologists on settlement patterns, which he described as the "way in which man disposed himself over the landscape in which he lived. [The term] refers to dwellings, to their arrangement, and to the nature and disposition of other buildings pertaining to community life. These settlements reflect the natural environment, the level of technology on which the builders operated, and various institutions of social interaction and control which the culture maintained" (Willey 1953:1). In this research stream an interest that emerged quite early was the way in which settlement locations are tied to subsistence activities, which therefore inform us, indirectly, about how people provisioned themselves (see historical reviews in Kohler 1988a; Parsons 1972).

Over the years, however, archaeologists have explored other parts of Willey's charge, becoming aware that site sizes and types, the distribution of sizes, the longevity of households and sites, and their placement relative to landscape features and to each other contain additional information about a wealth of things we seek to know, including economy beyond subsistence (e.g., Kohler, Van Pelt, and Yap 2000), political processes (e.g., Lipe 2002; Spielmann 1994; Upham and Reed 1989), social organization (e.g., Adler 1994; Varien 1999a), level and type of resource control (Adler 1996; Kohler 1992a), cultural norms and preferences (Thomas 1999), and even underlying cognitive metaphors (Ortman and Bradley 2002). In fact, settlement systems are potentially affected by—and therefore may contain information about—virtually all the domains of culture and behavior that archaeologists aim to reconstruct. Moreover, this information is embedded in a trajectory of change that provides additional opportunities for inferring causes for the changes in the patterns we see. None of these claims would have surprised Willey, though he would have been interested in the new and sometimes ingenious methods for establishing them.

Unfortunately, the information richness, or multidimensionality, of settlement patterns poses a severe problem for archaeological interpretation. When settlement practices can in theory be affected by many factors, how do we know which factors in what mix are in fact effective in any given case? Traditionally, and quite rightly, archaeologists have turned to excavation to gain independent data about subsistence, economy, and social organization that can help us select among alternative understandings. Excavation, however, is slow, expensive, destructive, increasingly difficult politically in many contexts, and severely limited in its spatial scale. It seems inarguable that we need to understand settlement at larger scales, and to do that we have to learn to extract more information from survey data that in many areas are quite extensive.

The problem of how to understand the processes responsible for a settlement pattern, and why that pattern changes, is common in the historical sciences. Like all scientists, we must always work in two directions: inference of processes from patterns, and deduction of patterns from processes. For roughly a century, archaeologists have been exploring how far we can go along the first path, and we may now be bumping up against its limits. Consequently, in this book we will concentrate on the second strategy: we create an agent-based simulation, which is an engine for deducing patterns through time from the processes we program, and then compare the outcomes of those simulations to what we know about the archaeological record.

In theory, at least, we can specify whatever processes we wish in the simulation, and then view and analyze the patterns that result from them. This method will allow us to make well-informed claims about the processes that generated particular patterns. These are claims of "generative sufficiency," in the jargon of some modelers (Epstein and Axtell 1996:19–20), which means that if the specification of the model allows us to generate in silico a phenomenon of interest, we have determined one way in which that phenomenon might have been produced in vivo. The VEP is devoted to understanding the processes that produced the changing settlement patterns, through time, of ancient Pueblo farmers.

DEEP HISTORY OF THE VEP

At the beginning of this chapter we described how we assembled the current team in response to NSF's call for proposals, and how the VEP brings together two main raw materials: (1) spatially and temporally organized environmental and archaeological data, and (2) agents (computer processes) that act on those data using rules supplied by the programmers. Since archaeologists have always been in the business of collecting data and organizing it temporally and spatially, this project can draw on a century of accumulated research in or near our study area in southwestern Colorado (Figure 1.1).

Most likely the VEP would not have come to be without the Dolores Archaeological Project (DAP). Between 1978 and 1985, archaeologists from the University of Colorado and WSU investigated the northeastern corner of the VEP area along the soon-to-be-dammed Dolores

River. The DAP became famous for the numerous Basketmaker III (BM III, A.D. 600–750) and Pueblo I (A.D. 750–900) habitations in the Dolores area, and provided both editors of this book with their first experiences in the Southwest.

This takes us back to a time when the first geographic information systems were just becoming available, and their use was still extremely limited; spatial data analyses were still very difficult and time consuming. Richard Darsie's WSU master's thesis (1983), which came out of the DAP, shows this by example. For 538 points distributed across the Dolores area, Richard collected, by hand from various maps, data on eight soil variables, generating four classes of agricultural suitability whose distribution varied during three periods distinguished by their climatic regimes. Using SYMAP, a computer-mapping tool common in the 1970s and 1980s, and a mainframe mapping program called VICAR, he generated three continuous surfaces (estimates) of agricultural suitability from these point data, one for each of these periods. He then compared the proportions of these classes in the catchments of field houses against habitations through time. He found strong patterning: for example, there was a steep decline in the availability of the best class of land in his last period, from A.D. 880 through A.D. 950, and a shift from arable land as a determinant of habitation-site location before A.D. 880 to a determinant only of field-house placement after that. Darsie's research helped us understand the functional differences between these site categories and characterize the problems faced by the local inhabitants as conditions for agriculture deteriorated in the late A.D. 800s.

As Darsie was completing his thesis in Pullman, Barney Burns was completing a monumental dissertation at the University of Arizona, Tucson, on a related problem. Burns used a combination of historical weather-station data, historical production data, and tree-ring data to demonstrate that ratio-level estimates of potential maize productivity could be made on an annual basis for portions of southwestern Colorado (Burns 1983). For the time frame from A.D. 652 through the thirteenth-century depopulation, he demonstrated that many periods of surplus production were also periods of intense construction activity, particularly of public or civic structures like great kivas. He also demonstrated that the period between A.D. 1276 and A.D. 1299 "was the worst period of storage shortfalls to have plagued southwestern Colorado's Anasazi" (Burns 1983:295) and was exceeded in this distinction only once, from A.D. 1528 through A.D. 1551, long after farmers had left southwestern Colorado. In Chapter 6 we continue to build on Burns's methods (see also Berry and Benson 2010).

Shortly thereafter, Janet Orcutt (1987) published an examination of DAP-area agricultural ecology that considerably refined and expanded Darsie's efforts. She used a small FORTRAN program to collect catchment data for sites and random points from a series of mapped data organized by square cells 200 m on a side. She demonstrated that good agricultural land was a more important and consistent determinant of habitation site placement than was cold-air drainage risk or vegetative diversity. Field houses, on the other hand, were frequently quite responsive to avoiding cold-air drainage risk in their placement.

At the same time, Kohler et al. (1986) were examining whether population loss or gain at DAP sites, from period to period, could be explained according to the principles of least cost, where the only cost considered was the travel distance between habitation and fields. This entailed simulating, for each habitation site in each period, how far the occupants would have had to walk to their fields. In addition to requiring estimates of the number of households by period for each site, we created a grid-based reconstruction of agricultural productivity for each period and devised a spatial simulation for "claiming" fields in grid cells at increasing distances from the habitations until the caloric needs of the estimated number of occupants were satisfied. Then the costs for each site could be regressed against the number of households at each site in that period. This analysis allowed us to identify places where agricultural production was either relatively expensive or relatively cheap based on this measure. If minimizing costs for agriculture was a primary motivator of movement, then we would expect relatively "cheap" places to gain population across period boundaries and relatively expensive places to lose population.

To our surprise, in most transitions from one period to the next the opposite was the case: people aggregated in places even though, in so doing, they increased their distances to fields. It seemed clear from our results that the process of aggregation in the DAP area was not in any simple sense an attempt to minimize costs, at least for agriculture. In fact, aggregation in general had the opposite effect. Subsequent research with these simulations (Orcutt et al. 1990) showed that the main local episode of village formation, in the ninth-century Pueblo I sites, had the effect of dramatically lowering conflicts over agricultural catchments, at the expense of slightly raising per-household agricultural costs.

For Kohler, this research proved formative, and it nicely illustrates two principles underlying the present work. In many ways, in fact, the VEP is an expansion of the research program in these two papers, taking into account a number of additional resources and operating on a much larger area and through a longer span of years. First, the DAP simulations strongly suggested that competitive social pressures were sometimes more important than simple minimization of agricultural costs for causing households to bunch together to form large and well-bounded communities. Second, it showed that controlling for resource-acquisition cost effects via simulation allowed us to isolate these social effects much more convincingly than we could through any verbal or statistical argument alone.

The same research, though, also illustrates the slipperiness of the concept of "adaptation." In treacherous times, living in large sites or in defensible locations far from fields may provide a better fitness outcome than living near one's farming plots. Fitness responds to the net effects of all one's practices. Adaptation ("behavior," if you prefer) typically cannot optimize all considerations at once; more commonly, some are optimized at the expense of others.

Later in this book, therefore, we develop a measure of how well habitation site sizes and locations simultaneously minimize costs of acquiring several different resources. At present we are unable to measure how well they minimize social risks, such as the possibility of being dislodged by another group. At best we can hope to identify these risks in the residuals from our expectations, in combination with various kinds of empirical evidence.

The work of Carla Van West provided another key precursor to the research we report here. In her 1990 WSU dissertation (published 1994), Van West was able to spatialize the sorts of annual maize production estimates that Burns had produced for southwestern Colorado as a whole. This was a critical step toward constructing a dynamic landscape that agents could explore and settle.

Thus Van West and Kohler (1996; Kohler and Van West 1996) could begin to employ a body of theory that made strong, nonintuitive predictions about how exchange between households should respond to variable amounts of production. As explained by Eric Smith (1988) and Smith and Robert Boyd (1990), this theory has roots in microeconomics, evolutionary ecology, and game theory. Under certain not very restrictive assumptions about the shape of households' utility functions, we could predict that households seeking to maximize their marginal utility should exchange maize with other households in times when local production was generally high but also spatially and temporally highly variable. On the other hand, when production was generally poor but still highly variable across space and through time, households should hoard their maize, not exchange it. If we further assumed that early Pueblo villages were built on interhousehold exchange, we should find that villages form in periods of high but variable production and fall apart when production is low but variable. These predictions were quite accurate (Kohler and Van West 1996). We discovered, however, that population size was a confounding factor that needed to be taken into account: aggregation (which we took to be a proxy for exchange) was more common than expected when populations were high, and it was less common than anticipated by the production model when populations were low.

But even this work did not fully exploit the possibilities of the data set that Van West had constructed, since the analyses were nonspatial: they just worked with maize production means and variances across the entire landscape. A more interesting examination of these predictions would be to simultaneously look at the situations of many households located in different portions of the landscape. Would such households—equipped, of course, with the appropriate sharing rules—spontaneously form and abandon villages in the times and places where we actually see these behaviors in the archaeological record? The VEP was formulated partly as an attempt to rigorously examine that model (see Chapter 15).

Our initial VEP simulation results were reported in a 1997 symposium held at SFI and funded by the Wenner Gren Foundation (grant 4799; Kohler and Gumerman, eds. 2000). In that work, we construed agents as nuclear families, as we still do, and loosed them onto Van West's paleoproductivity landscapes, to which we had added the known water resources. The history of the simulation program itself, which we often just call "Village," is reviewed in Chapter 4.

As a result of that first round of research, we concluded that rules governing site locations in the Pueblo II (A.D. 900–1150) and III (A.D. 1150–1280) periods in our study area were rather different, though in both eras the best-fitting rule sets required households to take into account the distribution of both water resources and the most productive lands for dry farming. We also thought we had evidence that the actual location of Pueblo III households was more inefficient than the Pueblo II household location because Pueblo III sites were located further from the "ideal free distribution" simulated by our agents. Finally, we noted that simulations in which agents degraded lands slightly through farming provided a better fit between the simulated and the actual household-location patterns than did no-degradation scenarios.

This brings us back up to 2001, when we found these results sufficiently encouraging to propose the VEP to NSF. The VEP extended the approach taken by Kohler et al. (2000) back to A.D. 600, expanded the suite of resources that our agents tracked, and enabled simple economic interactions among the households. Just as important, it funded a large empirical data collection effort reviewed in the next chapter and in Chapter 14.

LARGER CONTEXTS: THE VEP IN THE SOUTHWEST AND THE SOUTHWEST IN THE WORLD

Farmers first moved into the central Mesa Verde region (Figure 1.1) in significant numbers around A.D. 600. This itself is something of a puzzle, since maize farming had arrived in some other areas of the Southwest slightly before 2000 B.C., aided by irrigation, floodplain aggradation, increased effective moisture, and a preexisting pattern of repetitive land use, focus on annual seed plants, and grindingstone technology that facilitated maize farming adoption (Roth and Freeman 2008). Maize was present in the northern Rio Grande region by 1000 B.C. (Vierra and Ford 2007) and first appeared in the Four Corners area remarkably early—around 2000 B.C.—though a commitment to farming as the primary food production strategy was not made until about 400 B.C. At that time, early farmers occupied the areas to the west and east of the VEP area. Mona Charles and Sally Cole (2006) identify eight clusters of BM II (prepottery farming) sites in the Four Corners region, seven of which were present by 400 B.C. The settlement cluster closest to the VEP study area is near Durango, Colorado, and it is among the youngest, with clusters of tree-ring dates in the second, third, and the fourth centuries A.D. (Charles and Cole 2006:173). Aaron Wright points out in Chapter 3 (see also Wright 2006:Figure 21) that his new pollen-based low-frequency temperature reconstruction for southwestern Colorado suggests that temperatures were quite cold from 100 B.C. until almost A.D. 600. Though the general trend is toward warmer temperatures through time, there is also a century of warmer conditions centered around A.D. 325 that might have been critical to the success of maize farming in the more northern BM II locations, like Durango.

What prevented the spread of these early farmers into the relatively high, cool, and deep-soiled country of the central Mesa Verde? At least four explanations are possible, and they may be complementary. Perhaps it was the cool conditions prevailing for most of the early first millennium A.D. Perhaps the available landraces of maize before the A.D. 600s were more suited to the water table and the overbank flood farming that seem to have been staples of BM II cultivation than to the dry farming that became so productive in the central Mesa Verde region; or perhaps the practice of dry farming itself was simply not yet a proven part of the cultural repertoire. We also know that the BM II farmers located west and east of the VEP area were of different historical origin and (almost certainly) from different linguistic groups (Matson 2002); the central VEP region may have served as an unoccupied buffer zone between these groups. It may have even been a hostile buffer zone: we have evidence of warfare in the western region (Hurst and Turner 1993), and there is imagery in Basketmaker rock art evoking warfare (Farmer 1997).

In any case, around A.D. 600 some sort of threshold was reached in the Southwest in general—including the Pueblo area—that facilitated rapid population growth for some 600 years (Kohler et al. 2008). Similar population expansions have been identified in Europe and the Near East, where Jean-Pierre Bocquet-Appel (2002) coined the term "Neolithic Demographic Transition" (NDT) to refer to both this phase of high growth and the subsequent period of lower growth, when increases in mortality catch up with the earlier increases in fertility. Bocquet-Appel attributes the earlier transition—toward higher fertility—to changes that improved mothers' energy balances. Women began to consume less low-calorie food from hunting (and fishing, in some areas) and more high-calorie food from farming, even as they were able to devote less energy to mobility and child transportation (Bocquet-Appel 2008).

It is interesting and somewhat puzzling that in the Southwest this fertility transition takes place so long after the introduction of maize. In Europe, by contrast, the NDT begins with the first local appearance of new domesticates. But in the Southwest the high-growth phase of the NDT appears only when maize farming was accompanied by several other innovations, including well-fired ceramic containers, beans, new maize landraces, dry farming, and the bow and arrow. All these come together in the middle of the first millennium A.D., when the VEP study area was colonized by substantial populations of farmers. The development of villages— which on one hand reflect a more competitive social environment, favoring high fertility, and on the other hand facilitate exchange among unrelated households, contributing to high fertility—was also important to the NDT, although the first villages appear at about A.D. 780, slightly after the earliest apparent increases in growth rates (Kohler and Varien 2010).

Working from a global sample of early Neolithic societies, Matt Bandy (e.g., 2008) sees an interesting relationship between the formation of large villages and the growth phase of the NDT. For Bandy, "large villages" have at least 300 inhabitants, so our Pueblo I villages, some of which have at least 500 inhabitants, qualify. Bandy divides the first villages around the world into two main types. "Type 1" villages appear as part of a system of settlements that are more or less equivalent in size and are relatively autonomous, as in our area. "Type 2" villages first appear as chiefly centers. Bandy demonstrates that Type 1 villages almost always develop during the first, high-growth phase of the NDT, whereas Type 2 villages almost always develop during the second, slow-growth phase of the NDT. His explanation for this is worth quoting at length:

> Rapid population growth during the NDT presented a challenge to early village social organization. Growth in community size produced rapidly increasing levels of internal conflict . . . at a rate approximately proportional to the square of the village population . . . . [A] critical rate of social stress was quickly reached. . . . [V]illage communities were [then] presented with two options: (1) they could fission into two or more daughter communities, each smaller than the critical threshold size, or (2) they could develop some social mechanism that regulated and managed internal conflict in such a way as to make fissioning unnecessary. These conflict management mechanisms were frequently of a religious or ritual character . . . but we must imagine that the variety of possible solutions to the problem is as large as the variety of early village cultural diversity and historical experience. However, only the development of novel institutions of social integration at a suprahousehold level could make possible the emergence of villages larger than the critical population threshold, here provisionally defined as approximately 300 persons [Bandy 2008:341].

To summarize, Type 1 villages could arise and then maintain themselves because their inhabitants had successfully innovated social mechanisms that allowed them to exist at population sizes larger than could have been maintained with only the social mechanisms available to earlier smaller settlements in the same region. And something about the rapid growth rates in the first phase of the NDT, when such villages appear, made crossing that threshold much more likely than it would have been in earlier or later phases of slower growth. Large villages never appear in some sequences, of course, and occasionally they may appear simultaneously with the arrival of agriculture, when immigrants bring both farming and a habit of living in large settlements.

Any model that purports to be applicable worldwide is necessarily very general, and this model is not a bad skeleton on which we can begin to drape the richer set of historical circumstances involved in the first local round of village formation, the apparent reasons for the demise of those first villages, and the later appearance of more persistent villages, followed ultimately by their demise as well. By the final chapter, these efforts will allow us to present a more comprehensive historical narrative for village formation and demise in our area for others to compare in other Neolithic settings.

TWO

The Study Area and the Ancestral Pueblo Occupation

Scott G. Ortman, Donna M. Glowacki, Mark D. Varien, and C. David Johnson

MOST OF THE CHAPTERS in this book present models of the ancient environment and society of the central Mesa Verde region. To a greater or lesser extent, all these models seek to account for aspects of the actual history of ancestral Pueblo settlement in this region. The archaeology of the region has been studied intensively for more than a century, but in order to compare model results directly to archaeological patterns on the ground, it is necessary to systematize the diverse archaeological data resulting from this long-term research. Thus, a major effort of the Village Ecodynamics Project (VEP) involved translating the archaeological record of the study area into quantitative summaries of the actual ancestral Pueblo settlement history using explicit and repeatable criteria. This chapter introduces the study area, explains how we translated the archaeological record into quantitative data, and presents the basic outlines of the resulting settlement history derived from these data.

The VEP study area is rectangular and encompasses 1,817 sq km in Montezuma and Dolores counties, Colorado (Figure 1.1). The southwest corner of the study area lies approximately 25 miles north of the Four Corners, where the southwest corner of Colorado meets the southeast corner of Utah, the northeast corner of Arizona, and the northwest corner of New Mexico. The largest contemporary settlement within the study area is the town of Cortez, with a 2010 census population of about 8,500. The 2010 census population of Montezuma County was approximately 25,000.

The study-area boundaries were originally defined by Carla Van West for her dissertation research on the effects of climate variation for prehispanic agricultural productivity (Lipe and Van West 1992; Van West 1994). In her work, Van West focused on a rectangular area encompassing 12 contiguous U.S. Geological Survey (USGS) 7.5-minute quadrangle maps from southwestern Colorado (Ruin Canyon, Pleasant View, Yellow Jacket, Trimble Point, Negro Canyon, Woods Canyon, Arriola, Dolores West, Bowdish Canyon, Battle Rock, Mud Creek, and Cortez). These 12 quads were chosen for three reasons. First, because of modern agricultural land use in this area, the necessary data for reconstruction of agricultural productivity were available in the mid-1980s when Van West initiated her research. Second, this rectangle encompassed that portion of the Mesa Verde archaeological region that was most densely populated during prehispanic times, and Van West reasoned that effects of climate variation on agricultural production, and thus human settlement, would be most pronounced in this most densely settled area. Third, this area encompassed the study areas of several significant archaeological research projects—including the Dolores Archaeological Program (DAP) (Breternitz et al. 1986), the Mockingbird Mesa Survey (Fetterman and Honeycutt 1987), and the Sand Canyon Project (Lipe 1992)—that were ongoing when Van West initiated her research. This large volume of new research provided an excellent understanding of prehispanic occupation within this area. Intensive archaeological research has continued within Van West's study area ever since (e.g., Kendrick and Judge 2000; Kohler et al. 2010; Lightfoot 1994; Lipe et al. 1999; Varien 1999a, 2000; Varien and Wilshusen, eds. 2002). Because VEP researchers wanted to build on Van West's pioneering and influential work, we carried her study area forward for the VEP (also see Kohler et al. 2000).

The VEP study area lies just north of latitude 37° north, on the western slope of the San Juan Mountains and the eastern edge of the Colorado Plateau. It also occupies a portion of the northern reaches of the San Juan drainage. Most drainages in the study area flow to the south, southwest, and west, eventually reaching the San Juan River after crossing into Utah. The San Juan, in turn, merges with the Colorado River in the area of present-day Lake Powell, approximately 240 km to the west, and eventually empties into the Gulf of California. Major landforms that ring the study area include the Sleeping Ute laccolith to the south, the north escarpment of the Mesa Verde cuesta containing Mesa Verde National Park to the southeast, the La Plata Mountains rising to nearly 4,000 m to the east, and the Dolores river valley to the northeast. The San Miguel, La Sal, and Abajo mountains are visible on the northern horizon, and Elk Ridge and Cedar Mesa are visible on the western horizon. Elevations within the study-area range from more than 2,200 m along the eastern edge to fewer than 1,500 m in the southwest corner. E. B. Ekren and F. N. Houser (1965) provide an introduction to the geology of the area.

The environment of the study area is best characterized as a semiarid plateau, and the dominant natural vegetative community within it is Great Basin conifer woodland (Adams and Petersen 1999; Brown 1982:52–57). During a typical year the middle elevations receive approximately 450 mm of precipitation, with about half falling as snow during the winter months, and the other half as monsoonal summer storms. The relatively flat uplands across much of the study area are covered by several meters of aeolian loess deposited by annual winds that transport material from the Monument Valley area to the La Plata piedmont. Most of this material accumulated during the Pleistocene, but accumulation continues today. These deep red soils possess outstanding moisture-retention properties and are sufficiently productive to support modern, mechanized, direct precipitation or "dry-land" farming of beans, alfalfa, winter wheat, and other crops. They also support communities of sagebrush, shrubs and grasses, cactus, pinyon pine and juniper trees, mule deer, and a variety of lagomorphs, rodents, birds, and reptiles. Chapters 6 and 7 provide more detail about the soils and the plant and animal communities they support.

The maize grown as a dietary staple by Pueblo farmers on the high and dry Colorado Plateau does not tolerate temperatures below −2°C (28°F), and such temperatures can occur on cold evenings early and late in the growing season. So in addition to precipitation and soils, a third important variable that affected ancient agriculture in the study area was the length of the growing season. At lower elevations temperatures are generally warmer, leading to longer growing seasons, and vice versa as elevations increase. However, the relationship between elevation and temperature is complex. Because of cold-air drainage, growing seasons can in fact be shorter in lower-elevation canyon bottoms than in higher-elevation areas away from major drainages. Also, because of passive solar gain, south-facing slopes tend to be warmer and drier and north-facing slopes cooler and wetter at any given elevation. Many, but not all, of these factors are taken into account in the agricultural productivity model developed in Chapter 6 of this book.

Although precipitation is sufficient to obtain a crop in most years, the distribution of surface water for domestic use is much spottier than the distribution of arable land. The only river that passes through the study area is the Dolores, which flows west out of the San Juan Mountains to the modern town of Dolores before making a sharp turn northward, toward its confluence with the Colorado River north of Moab, Utah. Perennial streams also flow through several of the larger and longer canyons, including McElmo and Yellow Jacket, but most are somewhat alkaline and have been enhanced by modern irrigation of the adjacent uplands. Thus, these streams may have been used to irrigate crops but were probably of limited use for direct human consumption. The highest-quality drinking water is found in springs and seeps in the canyons that drain the study area toward the southwest. Most of the largest and most reliable springs occur at the point of contact between the Dakota sandstone and underlying shale layers in the upper reaches of these canyons and side canyons. Ancestral Pueblo people also constructed reservoirs and catch basins to capture runoff from exposed bedrock expanses for domestic use (Wilshusen et al. 1997). Kenneth Kolm and Schaun Smith develop a paleohydrological model of surface water availability in Chapter 5 of this book.

Overall, then, prehispanic agricultural settlement in the study area was responsive to a number of factors, including soil fertility, precipitation, length of frost-free growing season, availability of domestic water, and the abundance of useful plants and animals. The simplest way to characterize the geography of these factors is to draw a diagonal line across the study area from the southeast corner to the northwest corner (see Figure 2.1). This diagonal roughly parallels the course of U.S. Highway 491, which skirts the heads of numerous canyons as it travels northwest from Cortez to Dove Creek, Colorado. The area north and east of the diagonal is of generally higher elevation, receives more annual precipitation, is relatively undissected by canyons, and possesses more continuous expanses of deep loess soils. The Dolores River also flows through the northeast corner, but because of cold-air drainage out of the high mountains to the east, the river valley itself has rarely been suitable for agriculture (Peterson and Clay 1987). Also, there are relatively few reliable springs in this half of the study area because appropriate geologic settings are less common. To the south and west of the diagonal, in contrast, elevations are generally lower, annual precipitation is less, and lands covered by deep loess are increasingly dissected by canyons that expose underlying geologic strata. Springs and seeps are numerous in these canyons, and the canyons expose a variety of stone and clay resources that were regularly used by Pueblo people. As a result of this geography, the central portion of the study area extending south of the diagonal, and the terraces above the Dolores river floodplain, provided the best balance of precipitation, soils, growing season, domestic water availability, and canyon resources for prehispanic agricultural settlement, and indeed the highest densities of archaeological sites are found in these areas.

THE ARCHAEOLOGICAL RECORD

Most archaeological remains in the VEP study area belong to the archaeological tradition known as the Northern San Juan or Mesa Verde Anasazi. The modern Pueblo Indians of Arizona and New Mexico are among the descendants of the inhabitants of these sites, and thus archaeologists have begun using the term

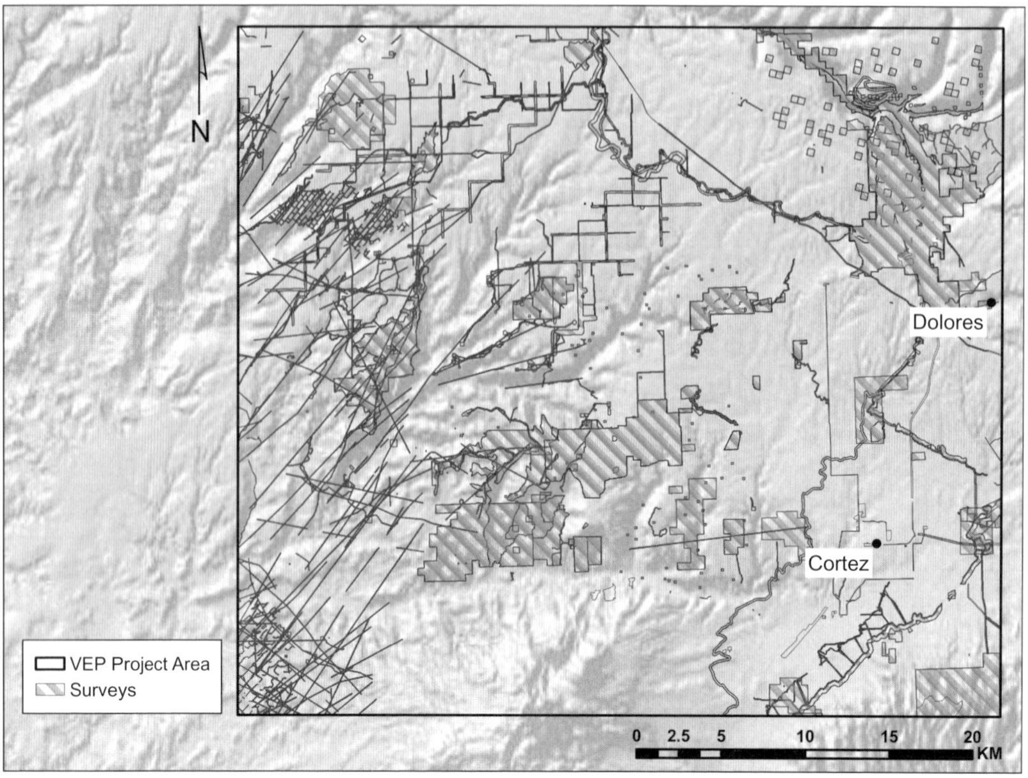

FIGURE 2.1 Surveys within the VEP study area.

"ancestral Pueblo" or just "Pueblo" to describe these sites in recent years. Remains of the Mesa Verde tradition of ancestral Pueblo Indians are found primarily around the northern tributaries of the San Juan River. The date of building construction on these sites can be established with remarkable precision using tree-ring dating. By matching growth patterns of preserved timbers to a master sequence maintained at the University of Arizona's Laboratory of Tree-Ring Research, it is often possible to establish the exact calendar year during which a specimen was harvested for use in construction. Because of the considerable amount of excavation in this region, the common preservation of charred construction timbers in open-air sites, and the preservation of both charred and uncharred timbers in dry alcove sites, the ancestral Pueblo society of the Mesa Verde region is quite likely the most precisely dated nonliterate village agricultural society in the world.

The precise dating of archaeological sites within the VEP study area in particular has allowed us to develop a comprehensive settlement history that specifies the resident populations of known archaeological sites using chronological periods of comparable duration to a single human life span. It also allows us to date the arrival and departure of ancestral Pueblo Indians with comparable precision. Using a database of tree-ring dates originally compiled as part of Mark Varien's dissertation research (Varien 1999a) and updated through 2003 ($n = 5{,}446$; also see Varien et al. 2007), the oldest timber that can be dated to the year of harvest from an ancestral Pueblo archaeological site within the study area dates to A.D. 445 (Figure 2.2). The distribution of cutting dates is spotty before A.D. 600, but from this point multiple cutting dates are available for nearly every decade through the A.D. 1270s. The most recently harvested timber in this database

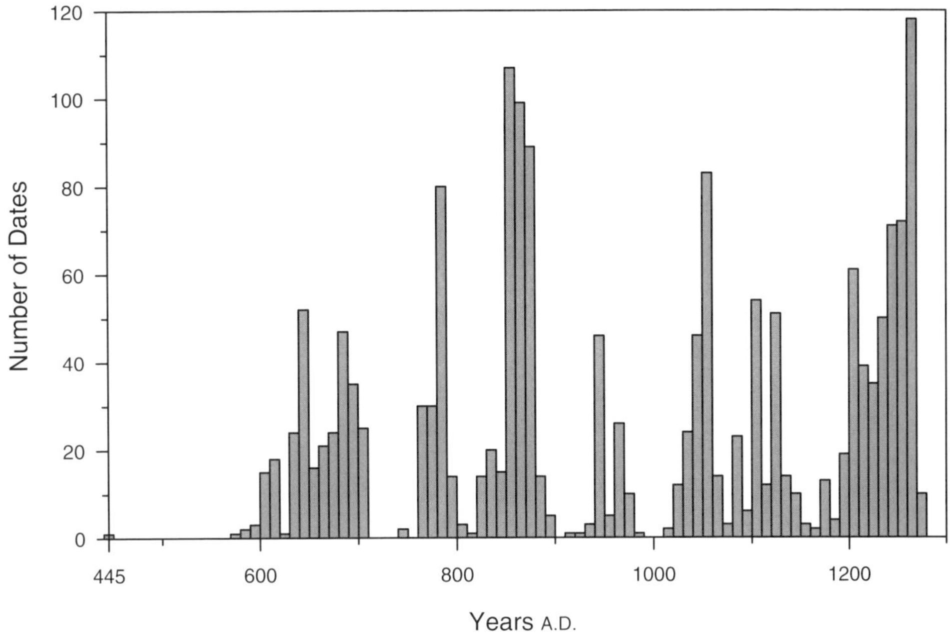

FIGURE 2.2 Bar chart of tree-ring cutting dates (and the latest date of any type from each site), by decade, for all ancestral Pueblo archaeological sites within the study area ($n = 1{,}794$).

dates from some time after A.D. 1277, and the latest cutting date from the Mesa Verde region overall is A.D. 1281. Although it is reasonable to assume that inhabitants of certain sites could have persisted for many years without harvesting additional construction timbers, the fact that no timbers from this very large sample postdate A.D. 1281 leads us to conclude that the entire Pueblo Indian population had moved out of the study area within a few years of this date. Based on these data, we have defined the ancestral Pueblo occupation of the VEP study area as beginning in A.D. 600 and ending at A.D. 1280.

In addition to being well dated, the Pueblo Indian archaeology of southwestern Colorado is remarkably well preserved, in a double sense. First, the relatively arid climate, alkaline soils, dry alcoves, and relatively shallow time depth of ancestral Pueblo sites all encourage preservation of bone, plant remains, fiber artifacts, and other organic materials, especially in the famous cliff dwellings (see Fewkes 1999[1909, 1911]; Nordenskiold 1990[1893]; Osborne 2004). Second, and more important for the purposes of the VEP, relatively few sites have been destroyed because this landscape has witnessed limited use since the Pueblo exodus. For several centuries following the Pueblo departure, southwestern Colorado was only lightly occupied by ancestors of the Utes and Navajos. These peoples pursued a relatively mobile lifestyle, trod lightly on the landscape, and in general did not disturb Pueblo sites. After A.D. 1600, southwestern Colorado became part of the northwestern periphery of New Spain and was visited by Spanish explorers (Warner 1995:16–17), but the area remained largely beyond the limits of Spanish colonial settlement. As a result, the first Americans to survey the newly acquired Colorado territory encountered a landscape where there was nothing but the passage of time between themselves and once-thriving Pueblo villages (Jackson and Holmes 1981[1876, 1878]).

Recent land use has also contributed to the preservation of ancestral Pueblo settlements. Approximately half the land within the study area was not offered for homesteading, and it is administered today as Canyons of the Ancients National Monument, Hovenweep National Monument, and the San Juan National Forest.

In fact, several parcels were withdrawn from homesteading specifically to protect the archaeological sites they contained. The sections that were homesteaded tended to be those with the greatest potential for agriculture, and as a result many sections close to canyons, and the seeps and springs around which Pueblo settlements clustered, remained in federal hands. In addition, most late Pueblo villages were built of stone masonry, and until recently Anglo farmers have tended to leave such sites in place and plow around them. Even more important, the central architectural space of a house throughout the ancestral Pueblo sequence was a subterranean pit structure that was excavated below the modern plow zone. So even when recent farmers bulldozed surface architecture and plowed over entire settlements, the pit structures and artifact scatters have remained for study. Finally, Anglo land use has been primarily agrarian and remains largely so to this day. So even though the surfaces of many sites have been affected by modern farming, very little of the study area has been paved or covered by modern buildings. The early twenty-first-century population, in fact, is comparable to the ancestral Pueblo population of the thirteenth century but is more dispersed today.

A final benefit of the study area is that so much archaeological research has taken place within it. Lipe (1999) provides a thorough account of the history of archaeological research in southwestern Colorado, starting in the mid-nineteenth century. In the following pages we provide a condensed version of this history that emphasizes research on the largest sites because these sites are the focus of subsequent analyses (see Chapter 14), and one of the primary goals of the VEP is to explain their histories. We also highlight some of the contributions various projects have made to the accumulation of conceptual tools and site-level data that form the foundation of VEP empirical work. Because of the history of research outlined in the paragraphs that follow, as well as the other factors already put forth here, we find it difficult to imagine a more productive setting for archaeological research on Neolithic village societies than the VEP study area.

PREVIOUS ARCHAEOLOGICAL RESEARCH IN THE VEP STUDY AREA

The earliest accounts of VEP study-area archaeology derive from explorations of the Four Corners region in the late 1800s and early 1900s. William H. Jackson and William H. Holmes visited the area during the Hayden Survey (1875–1877) and described a number of large sites, including Mud Springs, Yucca House, and Yellow Jacket Pueblo (Holmes 1878; Jackson 1876a, 1876b). Subsequent explorations included T. Mitchell Prudden's (1903) inventory of major sites in the northern San Juan drainage. In his article "Prehistoric Ruins of the San Juan Watershed in Utah, Arizona, Colorado, and New Mexico," Prudden synthesized information on site type, size, degree of preservation, relationship to other features (e.g., pictographs), and relationship to arable water and land, referring to the largest sites as "large communal pueblos" (Prudden 1903:241–242). Among other sites, he specifically mentions several of the largest sites in the VEP study area, including Yucca House (Aztec Spring Ruin), Goodman Point Ruin, Yellow Jacket Pueblo, Cannonball Pueblo, and the Hovenweep sites.

In 1908, Sylvanus Morley and Alfred V. Kidder surveyed an area surrounding the confluence of McElmo and Yellow Jacket canyons in the southwest corner of the VEP study area. In their publication of this survey, Morley and Kidder (1917) provided information on large sites such as Fortress Spur, a large Pueblo I village, and the Pueblo III sites of Cannonball and Jackson's Hovenweep Castle. Cannonball Ruins captured their particular interest, and as a result Morley (1908) excavated the southern room block at that site with the Colorado Branch of the Archaeological Institute of America. In addition to recording architectural traits and artifacts, Morley also noted that Cannonball Pueblo grew by accretion over time. This observation represents one of the earliest acknowledgments that the large pueblos of the Mesa Verde region had histories (Morley 1908:599–600).

During this same period, Jesse Walter Fewkes, a researcher for the Bureau of American Ethnology at the Smithsonian Institution, excavated Cliff Palace and Spruce Tree House in the newly created Mesa Verde National Park to prepare them for public visitation (Fewkes 1909, 1911). Although he is best known for these investigations, Fewkes also documented large sites in the VEP study area. In 1917, he explored the McElmo drainage because he believed the cliff dwellers of Mesa Verde originally came from that area. His work in the McElmo drainage proved fruitful, and as a result Fewkes expanded his survey to include the Yellow Jacket drainage in 1918. These explorations produced detailed descriptions of such important sites as Yucca House and Yellow Jacket Pueblo and are reported in his book *Prehistoric Villages, Castles, and Towers of Southwestern Colorado* (1919).

One of the earliest well-documented and systematic excavations in the VEP study area took place in the early 1930s. Under the auspices of a Field Museum Archaeological Expedition, Paul S. Martin excavated Lowry Pueblo to learn more about its occupation and its community (Martin 1936, 1938). Martin was drawn to southwestern Colorado because relatively little research had been done despite the dense concentration of sites. He selected Lowry because he thought the site would contain deep, stratified middens that would allow him to establish a pottery sequence. Martin was also interested in understanding the site's Chacoan architectural traits, which he believed were locally adopted and modified attributes. Martin's work at Lowry demonstrated that the site occupation was dynamic, with multiple building episodes and periods of short-term abandonment, reoccupation, and widespread remodeling. He also suggested Lowry Pueblo may have been an important religious center because the great kiva at Lowry had a long use-life and was the only one in the immediate vicinity (Martin 1936:207–209). Other excavation projects at large sites in the VEP study area took place in the 1930s (e.g., Hurst and Lotrich 1932), but Martin's work was by far the most systematic, rigorous, and completely reported.

From the 1930s to the 1970s, few projects in the VEP study area focused specifically on large sites. Instead, fieldwork emphasized small sites within the study area (e.g., Joe Ben Wheat's excavations of 5MT1, 5MT2, and 5MT3 beginning in 1954) or localities outside the study area, including Chapin and Wetherill mesas on Mesa Verde (Cattanach 1980; Hayes 1964; Lancaster et al. 1954; Lister 1964, 1965, 1966; Rohn 1971; Swannack 1969), Alkali Ridge and Glen Canyon in southeastern Utah (Brew 1946; Sharrock et al. 1961), and the Navajo Reservoir area in northwestern New Mexico (Dittert et al. 1961; Eddy 1966).

The next large-scale project to occur within the VEP study area was the DAP, undertaken between 1978 and 1983 by researchers from the University of Colorado and Washington State University on behalf of the Bureau of Reclamation. This project was, and remains, the largest survey and excavation project completed to date in the VEP study area (Breternitz et al. 1986). Sites surveyed and excavated during the DAP dated from A.D. 600 to historical times, with the majority dating to the first cycle of ancestral Pueblo occupation and aggregation, A.D. 600 through A.D. 980 (Breternitz et al. 1986:Table 1.12). Among the DAP's many contributions was the refining of sampling methods and the development of a site typology based on function and size. This typology differed from earlier "discovery and description" schemes, which tended to classify sites on the basis of their settings. Habitation sites documented during the DAP were defined as "residential centers of the community [that] contain the dwellings (households and interhousehold clusters), integrative structures, and other facilities and spaces integral to the community cluster" (Kane 1986:358). Within this category, large villages were defined as any site with more than 20 dwelling units in five or more room blocks (Kane 1986:Table 5.1). This definition worked well for the early sites that were most common in the DAP study area because these sites typically consist of multiple linear room blocks. The work provided detailed excavation data on some of the largest Basketmaker III and Pueblo I sites in the VEP study area and the northern U.S. Southwest overall (e.g., Grass Mesa and McPhee Pueblos), and the results have shaped research in the central Mesa Verde region ever since (e.g., Wilshusen and Ortman 1999).

Concurrent with the end of DAP fieldwork, Robert Powers and his colleagues (1983) conducted a comprehensive inventory of Chacoan outliers across the Colorado Plateau. Their goal was to assess the distribution and layout of such sites in order to evaluate the relationship between these outliers and the great houses in Chaco Canyon itself. In the VEP study area, Powers and colleagues collected detailed architectural and artifact data at Wallace, Ida Jean, Escalante, Lowry, and Yucca House. Their efforts represent the first time the Chaco-period (Pueblo II) occupation was synthesized for the VEP study area and for the Colorado Plateau as a whole.

Fieldwork in the middle to late 1980s and early 1990s continued to advance knowledge of the VEP study area and shifted the focus back to the large Pueblo III period villages. Building on the methods developed during the DAP, researchers at the newly created Crow Canyon Archaeological Center began excavation of the Duckfoot site, a small Pueblo I hamlet (Lightfoot 1994), and of the large Pueblo III period village known as Sand Canyon Pueblo (Bradley 1992, 1993; Kuckelman 2007; Lipe 1992; Ortman and Bradley 2002). The Duckfoot site was completely excavated and provided an invaluable data set for fine-grained behavioral and temporal analyses, such as pottery discard rates (Varien and Mills 1997; Varien and Ortman 2005; Varien and Potter 1997). These studies proved critical for the VEP data analysis and modeling described in this chapter and elsewhere (Ortman et al. 2007).

The excavations at Sand Canyon Pueblo systematically sampled kivas, rooms, and civic-ceremonial architecture to learn more about village construction and community organization. This work demonstrated that the pueblo was constructed over a relatively brief period according to a plan that included specific architectural zones within the pueblo (e.g., residential versus civic-ceremonial precincts). The excavations at Sand Canyon Pueblo eventually became the centerpiece of a larger project that included settlement pattern surveys (Adler 1992), test excavations at a number of small settlements (Varien 1999b), and intensive excavations at Castle Rock Pueblo (Kuckelman 2000). One emphasis of the larger Sand Canyon Project was to define residential communities as being distinct from villages because of the distribution of integrative (civic-ceremonial) architecture and the density of the surrounding settlement (Adler 1990, 1994, 2002; Adler and Varien 1994). Additional goals were to learn more about social and political relationships between the investigated sites (Driver 1996; Glowacki et al. 1995, 1998; Muir and Driver 2002, 2004; Ortman 2008; Pierce et al. 2002) and to understand the role of warfare in Pueblo III society (Kuckelman 2002; Kuckelman et al. 2000).

An outgrowth of the Sand Canyon Project, which culminated in a working conference on the Pueblo III period across the Colorado Plateau (Adler, ed. 1996), was the initial creation of a large site database for the central Mesa Verde region (see Adler and Johnson 1996:262–264; Varien et al. 1996:86–90). This synthesis marked the first attempt to map the distribution of communities across the Mesa Verde region and to understand how this distribution structured regional interaction during the Pueblo III period. In the process, Varien and colleagues also established the idea of community core areas that included aggregated villages with 50 or more structures, including rooms, kivas, and towers.

Varien's involvement in this work became the foundation for his doctoral dissertation in which he synthesized data on all known large, Pueblo II and III sites in the central Mesa Verde region (Varien 1997). In this work, Varien introduced the term "community center" to describe "densely settled area[s] usually associated with public architecture" (Varien 1999a:4). He also identified more than 60 sites within the VEP study area that contain 50 or more structures as community centers (Varien 1999a:Tables 7.1–7.3). This work thus marks the emergence of the community-center concept that became the organizing principle of the VEP Community Center Survey (see Chapter 14 in this book). Varien's work also reinforced earlier suspicions that Mesa Verde region communities had long histories of growth, occupation, and relocation.

In a contemporaneous synthesis of the settlement history of southwestern Colorado, Lipe and Varien (1999:345) again used 50 or more structures as the defining criterion for community centers, but added that sites with six to eight kivas probably also represent centers because four to 10 surface rooms typically were associated with each kiva in excavated sites. They further emphasized that these sites typically had civic-ceremonial architecture and were surrounded by smaller sites, which is in large part the rationale for inferring that these sites functioned as community centers.

This synthetic work established the research trajectory emphasizing large sites that has been followed by researchers at the Crow Canyon Archaeological Center for the last 15 years. Community-center data in the VEP study area has been greatly improved by the Village Mapping Project (Lipe and Ortman 2000) and by survey, excavation, or both at Woods Canyon Pueblo (Churchill 2002; Ortman et al. 2000), Yellow Jacket Pueblo (Kuckelman 2003; Ortman et al. 2000), Shields Pueblo (Duff et al. 2010), Yucca House (Glowacki 2001), Albert Porter Pueblo (Ryan 2008, 2010), and Goodman Point Pueblo (Kuckelman et al. 2009). These projects have refined our knowledge of community centers by providing tree-ring dates from key architectural contexts and large, well-provenienced pottery assemblages. They have also shown that, despite broad similarities across sites, each village had its own unique configuration and history (Kuckelman 2010a, 2010b; Ortman et al. 2000).

Over the past several decades, consulting and federal archaeologists have also contributed substantially to the documented archaeological record of the VEP study area. Much of this work has emphasized survey (e.g., Fetterman and Honeycutt 1987, 1994; Hovezak et al. 2004), and the establishment of Canyons of the Ancients National Monument in particular has resulted in a surge of new work to inventory sites for management and preservation (e.g., Hovezak et al. 2003). The resulting site data have contributed substantially to the database developed for the VEP, as described in the next section. Another, perhaps even more important contribution of cultural-resource management work has been the excavation and reporting of numerous small sites, many of which are precisely dated to relatively short periods (see Ortman et al. 2007:Table 2 for references). These data proved critical for the methods developed by the VEP to estimate the occupational histories of all ancestral Pueblo sites in the study area, from the largest village to the smallest hamlet.

With this background on the study area in place, the remainder of this chapter focuses on the archaeological site database we developed for the VEP, the methods we used to analyze these data, and summary statements on the ancestral Pueblo settlement history of the area.

CONSTRUCTING THE ARCHAEOLOGICAL DATABASE

The basic goal of the VEP is to examine the historical ecology of ancestral Pueblo farmers within the study area, especially the social and environmental factors that influenced settlement patterns during the seven centuries of intensive occupation by Pueblo Indians. To reach this goal we needed to synthesize the archaeological record of the study area and specify the distribution of population across the study area over time. Although there is a rich history of excavation in southwestern Colorado (Lipe et al. 1999), the vast majority of ancestral Pueblo sites are known only from information recorded on survey forms. Thus, our primary task was to locate the sites and model their occupational histories using information recorded by surveyors.

Our first step toward this goal was to assemble a database of information on all recorded sites within the VEP study area. We began by compiling existing digital data sources, including digital site records from the Colorado Office of Archaeological and Historic Preservation, catalog records from the Anasazi Heritage Center, the DAP archaeological database (Wilshusen 1999), the Crow Canyon Archaeological Center Research Database, and the personal files of a number of local researchers. Michael Spitzer then reviewed the associated paper records for these sites, determined how greater consistency could be achieved, and edited the existing digital data as needed. The resulting database contains information on 8,948 sites, most of which were recorded during one of 336 surveys conducted in the project area between 1950 and 2003 in the context of problem-oriented research or cultural-resource management work. These surveys cover approximately 15 percent of the study area and include large block surveys, transect surveys, and a variety of smaller surveys (Figure 2.1). Although resurvey would undoubtedly discover additional sites within certain of these blocks, we assume that all habitation sites within the surveyed areas have been discovered and that only a portion of such sites have been recorded outside these surveyed areas.

The first step in organizing these data was to assign each site to one or more functional categories (isolated public architecture, single habitations, multiple habitations, field houses, and a variety of limited-activity site types) and to distinguish ancestral Pueblo sites from sites of other general periods. In Table 2.1 we present the general periods of use and size classes for all habitation sites in the VEP site database. Our data show that the study area has been used by humans from the Paleoindian Period to historical times, but that 97 percent of sites for which a period of use can be determined date from ancestral Pueblo times. We used the presence of a trash midden and one or more pit structure depressions as evidence that a given settlement was a year-round residence for at least one household; the analysis of features and assemblages at excavated sites with these characteristics supports this interpretation (Varien 1999a; Varien, ed. 1999). In total, 3,176 ancestral Pueblo sites in the database possess these features and are accordingly interpreted as single habitations, multiple habitations, or community centers.

Based on recent research in our study area (Cater and Chenault 1988; Lightfoot 1994; Lipe 1989; Ortman 1998; Varien, ed. 1999), we infer that each pit structure was the central architectural space used by a single household consisting of individuals related by descent or marriage, two to three generations deep. About two-thirds of the habitation sites in the database contain only one pit structure and thus represent farmsteads occupied by a single household (Figure 2.3). Approximately 850 sites contain evidence of between two and eight pit structures and are termed "multiple habitations."

Finally, the largest sites are called "community centers." Building from the history of research already outlined in this chapter, the VEP defines community centers as settlements with nine or more pit structures, 50 or more total structures, or some form of public architecture. We interpret sites meeting these criteria as community centers because their inhabitants could not have all been lineally related, they often contain public architecture such as a great kiva or plaza, and they are typically the largest site within a cluster of sites (Adler and Varien 1994; Lipe and Varien 1999b:345; Ortman and Varien 2007; Varien et al. 1996). These centers also generally have longer use-lives than the smaller habitations in the region (Ortman et al. 2000; Varien 1999a:202–207; Varien and Ortman 2005). Finally, excavations at these centers indicate that social, economic, and political activities involving multihousehold social groups took place in these sites (Adler 1994; Bradley 1988, 1993, 1996; Driver

TABLE 2.1

Average Momentary Household Estimates for Habitation Sites in the VEP Archaeological Site Database

Modeling Period			Occupation Span		Span / Period Length		Total Households		Momentary Households		
Beginning	*End*	*Length*	*Small Sites*[a]	*Community Centers*[b]	*Small Sites*	*Community Centers*	*Small Sites*	*Community Centers*	*Small Sites*	*Community Centers*	*Total*
600	725	125	8	28	.064	.224	1567	10	100.3	2.2	103
725	800	75	13	28	.173	.373	545	105	94.5	39.2	134
800	840	40	18	28	.450	.700	436	97	196.2	67.9	264
840	880	40	18	28	.450	.700	483	149	217.4	104.3	322
880	920	40	18	28	.450	.700	158	154	71.1	107.8	179
920	980	60	18	28	.300	.467	240	61	72.0	28.5	100
980	1020	40	18	28	.450	.700	362	38	162.9	26.6	190
1020	1060	40	21	40	.525	1.000	326	35	171.2	35.0	206
1060	1100	40	21	40	.525	1.000	721	269	378.5	269.0	648
1100	1140	40	40	40	1.000	1.000	591	309	591.0	309.0	900
1140	1180	40	40	40	1.000	1.000	579	394	579.0	394.0	973
1180	1225	45	45	45	1.000	1.000	487	598	487.0	598.0	1085
1225	1260	35	35	35	1.000	1.000	763	967	763.0	967.0	1730
1260	1280	20	20	20	1.000	1.000	343	850	343.0	850.0	1193

a. Small-site occupation spans after Varien and Ortman (2005:138–140).

b. Community-center occupation spans for modeling period 6–12 based on the ratio of total household years from the Bayesian analysis and the accumulation of cooking pottery at Grass Mesa (Kohler 1988b) and Rio Vista (Wilshusen 1986a) villages.

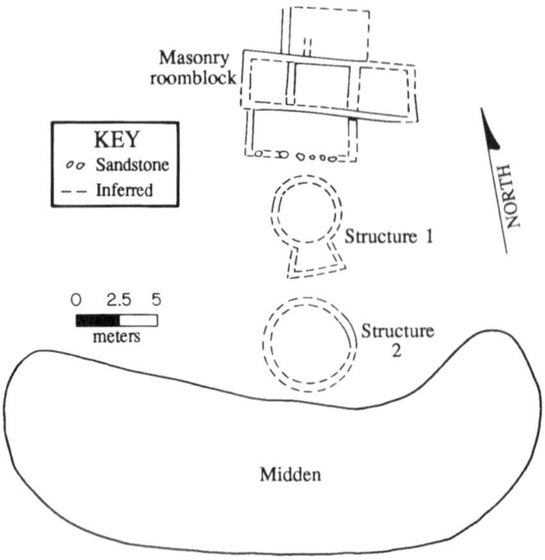

FIGURE 2.3 Single habitation at Roy's Ruin, illustrating the building blocks of ancestral Pueblo settlement in the VEP study area. Structure 1 is a subterranean pit structure (or kiva) and Structure 2 is an above-ground masonry tower. The midden lies in a plowed field and thus covers a larger area than in undisturbed sites. Reproduced courtesy of Crow Canyon Archaeological Center.

1996; Lipe 2002; Muir 1999; Muir and Driver 2002; Ortman and Bradley 2002; Potter 1997, 2000; Potter and Ortman 2004).

As a result of 15 years' effort to document large settlements in the greater Mesa Verde region (see Varien 1999a; Varien et al. 1996), we have identified 106 community centers in the VEP study area. As we compiled the database, we evaluated existing records for these centers, and a team led by Donna Glowacki conducted new fieldwork at 58 poorly documented centers. As a result, we believe that we have a fairly complete data set of all community-center sites in the VEP study area, regardless of whether the center is within a block-survey area. Glowacki and Scott Ortman discuss the archaeological record of community centers in greater detail in Chapter 14.

Our task in analyzing the data for ancestral Pueblo habitation sites was to use explicit and repeatable criteria to determine the period or periods during which each site was occupied, as well as the resident population during each period of occupation. Assignment of sites to time periods and population categories has long been standard in archaeological survey, and surveyors had recorded such data previously for most of the sites in our database. However, the bases of these summary interpretations were too variable across surveyors and have changed too much over time for us to rely only on these data to infer the occupational histories of sites in our research. We needed to reinterpret the observations recorded for individual sites in light of our current knowledge of the local archaeological record.

For a number of reasons, we pursued a formal (algorithmic) approach to this problem. First, the number of recorded sites is so large that it would have been impractical to revisit sites or evaluate each site form individually. Second, the available data for many sites in our study area are so uneven, and in some cases so meager, that it would be difficult to determine how best to interpret them using ad hoc, judgmental methods. Third, it was our goal to develop repeatable and explicit criteria for our interpretations, and this could be accomplished only by developing formal methods to analyze data of such variable quality and quantity. Finally, the formal methods we developed allow us to update our interpretations of sites as knowledge of the archaeological record improves in the future.

The method we developed for analyzing archaeological survey data draws heavily on Bayesian statistical concepts. It uses the rich excavation records of the study area to calibrate change through time and to correlate surface evidence with likely excavation results. Ortman and colleagues (2007) present the details and mathematics of this method. Here we merely summarize this method on a conceptual level. In a nutshell, our approach formalizes and makes explicit the ad hoc reasoning process followed by fieldworkers when they record new sites. We use the known populations, periods of occupation, pottery assemblages, and architectural characteristics of excavated sites in our study area to quantify the relationship between surface evidence and occupational histories. We then use these relationships as prior knowledge in a Bayesian statistical framework (Buck et al. 1996; Iversen 1984; Robertson 1999) to model the occupational histories of unexcavated sites known only from surface evidence.

We started with a calibration data set of 80 excavated and well-dated site components; these were used to estimate the frequencies of 18 architectural attributes and 24 pottery types in sites dating to each of 14 modeling periods between A.D. 600 and A.D. 1280. We defined these periods to be as short as possible given our current understanding of observable changes in the surface expressions of archaeological sites in the study area. These periods vary from 20 to 125 years, but the modal span is 40 years. The calibration data were used to calculate probability density functions for each pottery type and architectural attribute; these functions specify the probability that a site exhibiting one of these types or attributes was occupied during each of the 14 modeling periods.

These probability density functions were combined with the sample data from each site to calculate composite probability density distributions for three categories of observations: architectural characteristics, undecorated pottery, and decorated pottery. A probability density function was also calculated to reflect the surveyor's assessment of when each site was occupied. For example, if the surveyor assigned an occupation to the Pueblo III period, the probability density function for this assessment has null values for the first 10 modeling periods and a value of .25 for the final four modeling periods (those within the Pueblo III period). Null rather than zero was entered as the probability of occupation for modeling periods outside the surveyor's date range because we did not want the original surveyor's assessment to outweigh evidence supporting a different conclusion with the benefit of current knowledge. We also constructed a probability density function to characterize the distribution of tree-ring dates for the 101 sites that have those data. Finally, we weighted and combined probability density functions based on samples of 11 or more decorated pottery sherds for all sites within 7 km of each habitation site, and used this information to specify the occupational history of the "neighborhood" around each site. These neighborhood distributions are critical for our analysis because the available data for about half the sites in our database were insufficient for us to determine their most probable period of occupation, and previous settlement pattern surveys have shown that contemporaneous ancestral Pueblo habitations occur in spatial clusters that represent residential communities (Adler 1990; Adler and Varien 1994; Fetterman and Honeycutt 1987; Lipe 1970; Mahoney et al. 2000; Ortman and Varien 2007; Rohn 1977). Once we calculated these six probability density functions for each site—architecture, plain pottery, decorated pottery, the surveyor's estimate, absolute dates, and the neighborhood—we averaged the available lines of evidence (the neighborhood and surveyor's assessment were included only for sites with fewer than 11 decorated sherds) to produce a mean probability density function for each site.

VEP archaeologists use households to track population, and as mentioned earlier, several lines of evidence suggest that a pit structure was the central architectural space used by a household throughout our sequence (Cater and Chenault 1988; Lightfoot 1994; Ortman 1998; Varien 1999a). We therefore estimated the peak populations of habitation sites using a two-step process. First, we estimated the total number of pit structures present at each site, using as a basis the number of depressions on the modern ground surface, the size of the surface room block, and the total site area (Adler 1990; Kuckelman 2000; Schlanger 1985). Then, we applied multiple regression analysis to data from excavated sites, where the peak populations are known, to predict the proportion of total pit structures that were occupied during the period of peak population, using their total pit structure estimates and the characteristics of their mean probability density functions.

In the last step, we integrated the peak population estimate with the mean probability density function for each site. Excavations demonstrate that sites with one pit structure typically were occupied only during one period (Varien, ed. 1999). For sites of this size, we simply assigned one household to the period for which the probability of occupation was highest and assigned an additional household to periods corresponding to secondary modes in the mean probability density function. Excavations also show that larger sites typically were occupied for multiple periods (Kohler and Blinman 1987; Ortman et al. 2000; Varien and Ortman 2005). To determine the periods of occupation at these larger sites, we conducted a second multiple regression analysis using survey and excavation data for 35 multiple habitations and community centers. This allowed us to use the size of the site and characteristics of its mean probability density function as a basis to predict the minimum probability value that signifies occupation. At these sites we assigned households to each of the periods that, according to the ratio of the nonpeak probability values to the peak value, exceeded this threshold.

A MODEL OF ANCESTRAL PUEBLO SETTLEMENT HISTORY

Our analysis of the site database resulted in a spreadsheet (Crow Canyon Archaeological Center 2004) in which each row represents a habitation site and there are fourteen columns corresponding to each of our modeling periods. The value in each cell indicates the number of households that resided at a given site during a given modeling period. Figures 2.4 through 2.10 present a series of 14 maps illustrating the distribution of population in recorded sites within the study area during each of our 14 modeling periods.

With occupational histories for more than 3,000 sites in hand, we proceeded to reconstruct the ancestral Pueblo population history of the study area. The first step in this analysis is to calculate the momentary population, in households, of sites in the database for each of the 14 modeling periods. Momentary household estimates take into consideration the use-lives of houses in small sites and community centers vis-à-vis modeling-period lengths. During the final few centuries of our sequence, the average use-life of a house in both small sites and community centers was equal to or greater than the length of our modeling periods, so for these periods the momentary populations of sites are equal to the total populations. For earlier periods, however, the average use-life of a house was shorter than the length of the modeling period, so for these periods we need to take the variable use-lives of houses in small sites and centers into account.

We estimated house use-life for early small sites by measuring the accumulation of cooking pottery in excavated sites, following the methods developed by Varien and others (Varien 1999a; Varien and Mills 1997; Varien and Ortman 2005). Excavations and the analysis of pottery assemblages indicate that houses in community centers were used for longer periods than houses in small sites (Kohler and Blinman 1987; Ortman et al. 2000), so we developed a new method for estimating the use-life of houses in early community centers. This method employs the probability density analysis and point estimates for the total accumulation of cooking pottery at two early community centers excavated using stratified random samples: Grass Mesa (Kohler 1988b) and Rio Vista villages (Wilshusen 1986a). To calculate the average use-life of houses in these two centers, we divided the total number of households inferred for all periods from the probability

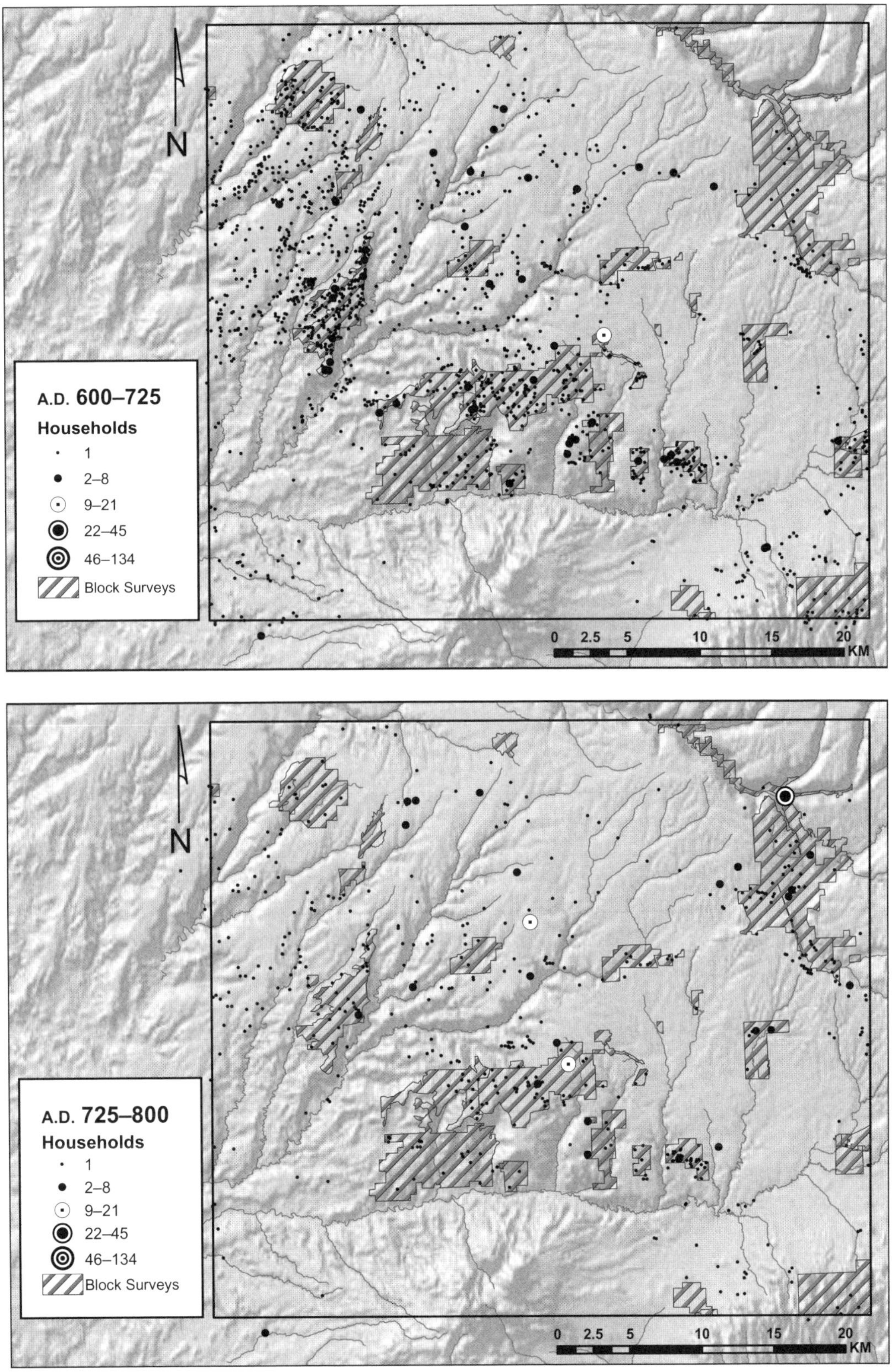

FIGURE 2.4 Site distributions in the VEP study area, modeling periods 6 and 7.

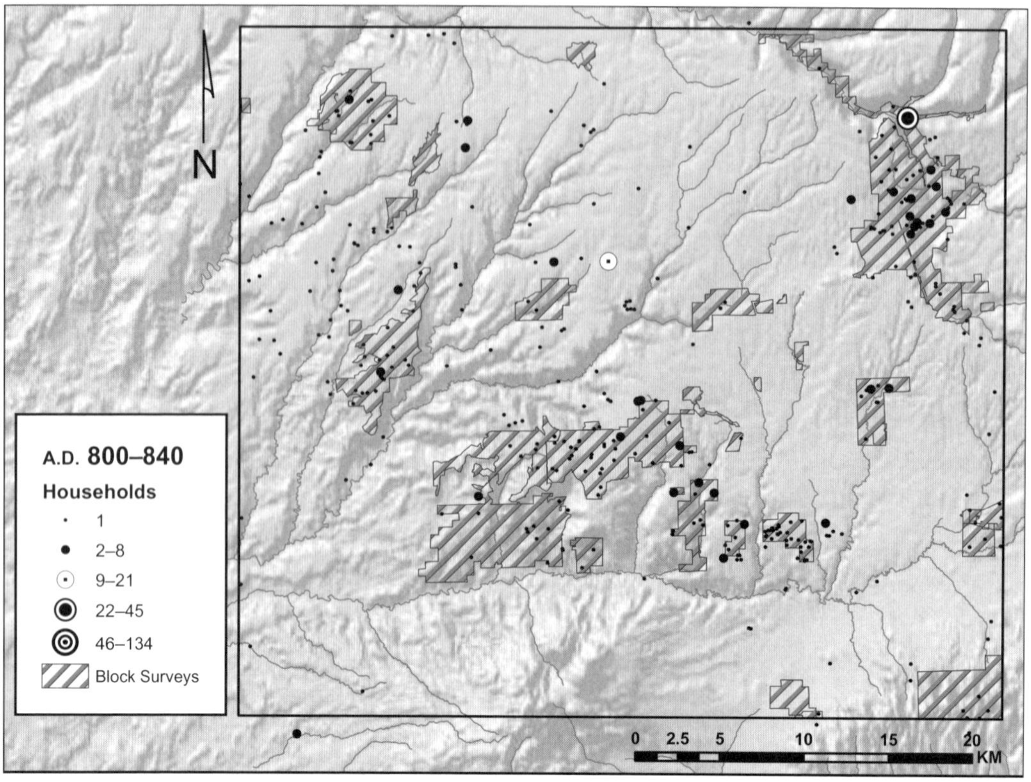

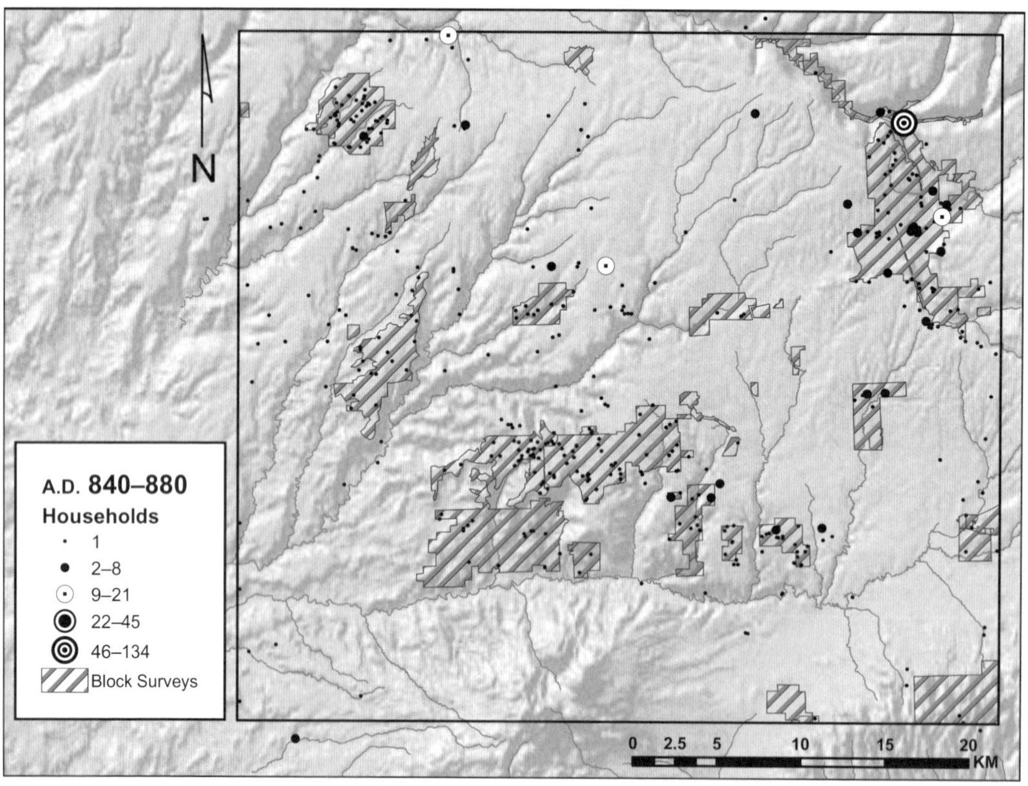

FIGURE 2.5 Site distributions in the VEP study area, modeling periods 8 and 9.

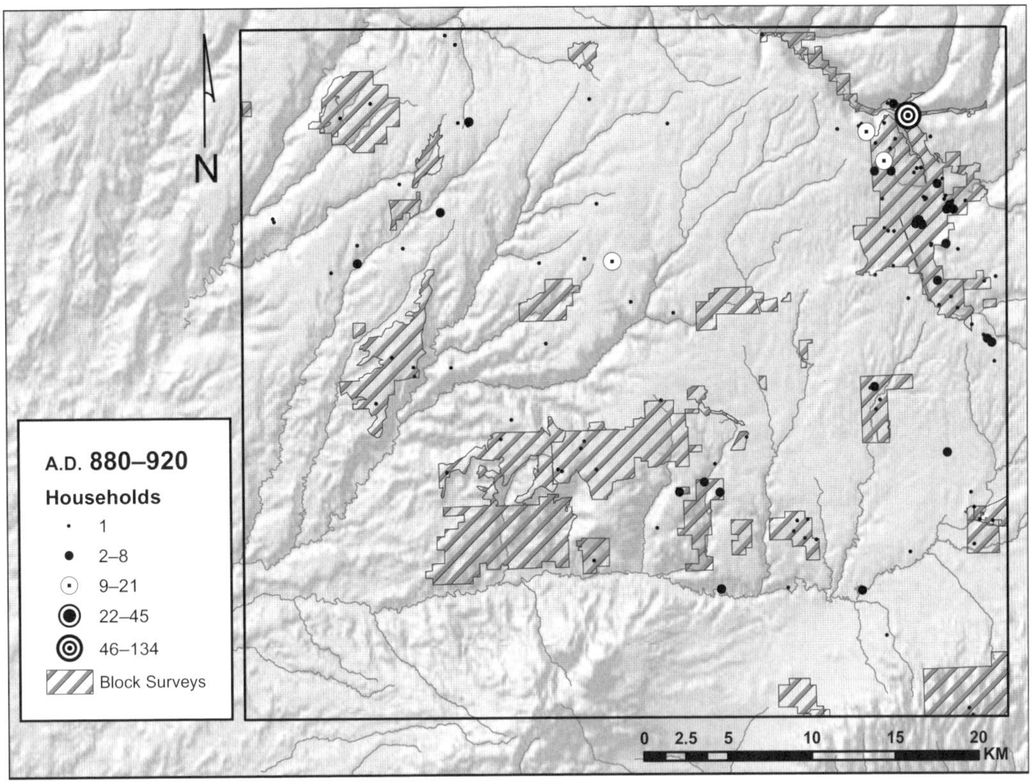

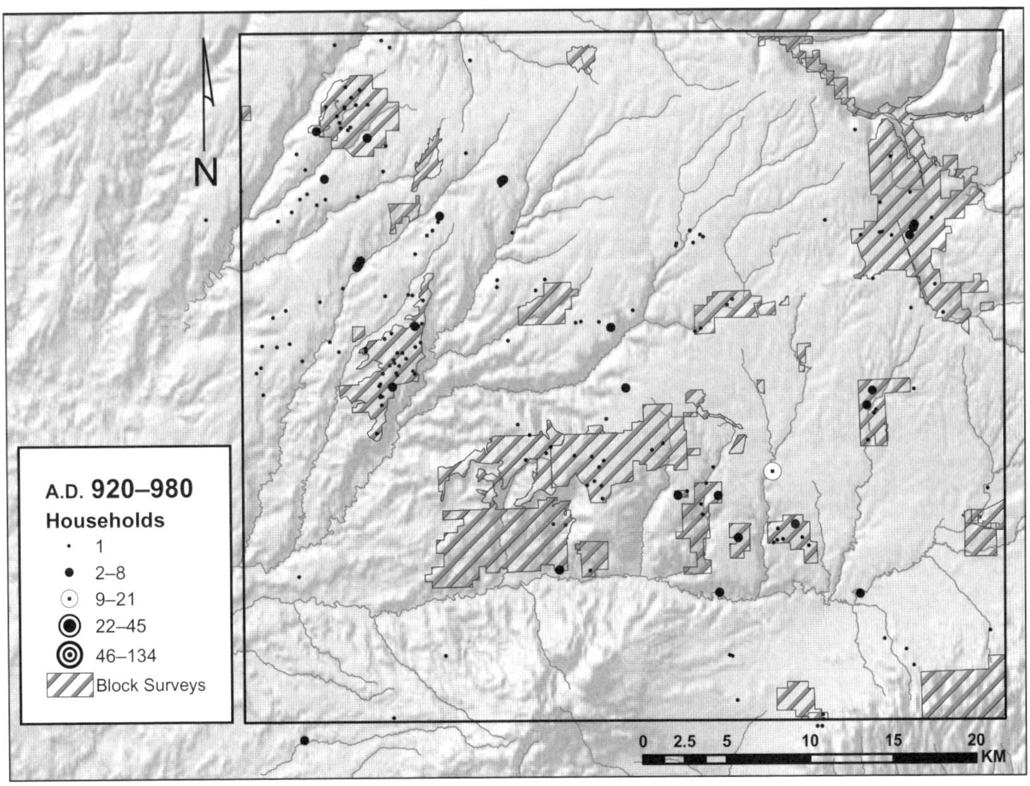

FIGURE 2.6 Site distributions in the VEP study area, modeling periods 10 and 11.

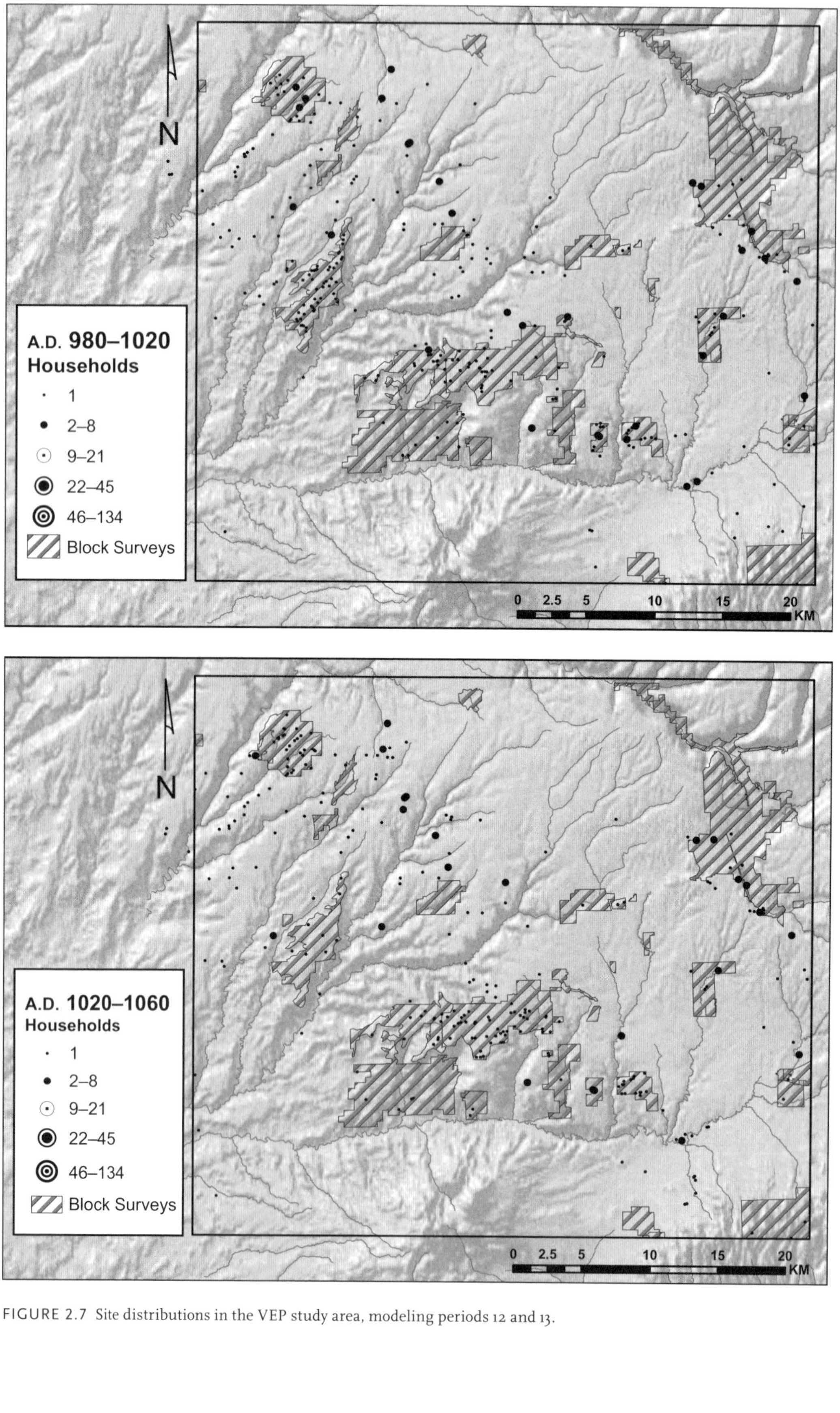

FIGURE 2.7 Site distributions in the VEP study area, modeling periods 12 and 13.

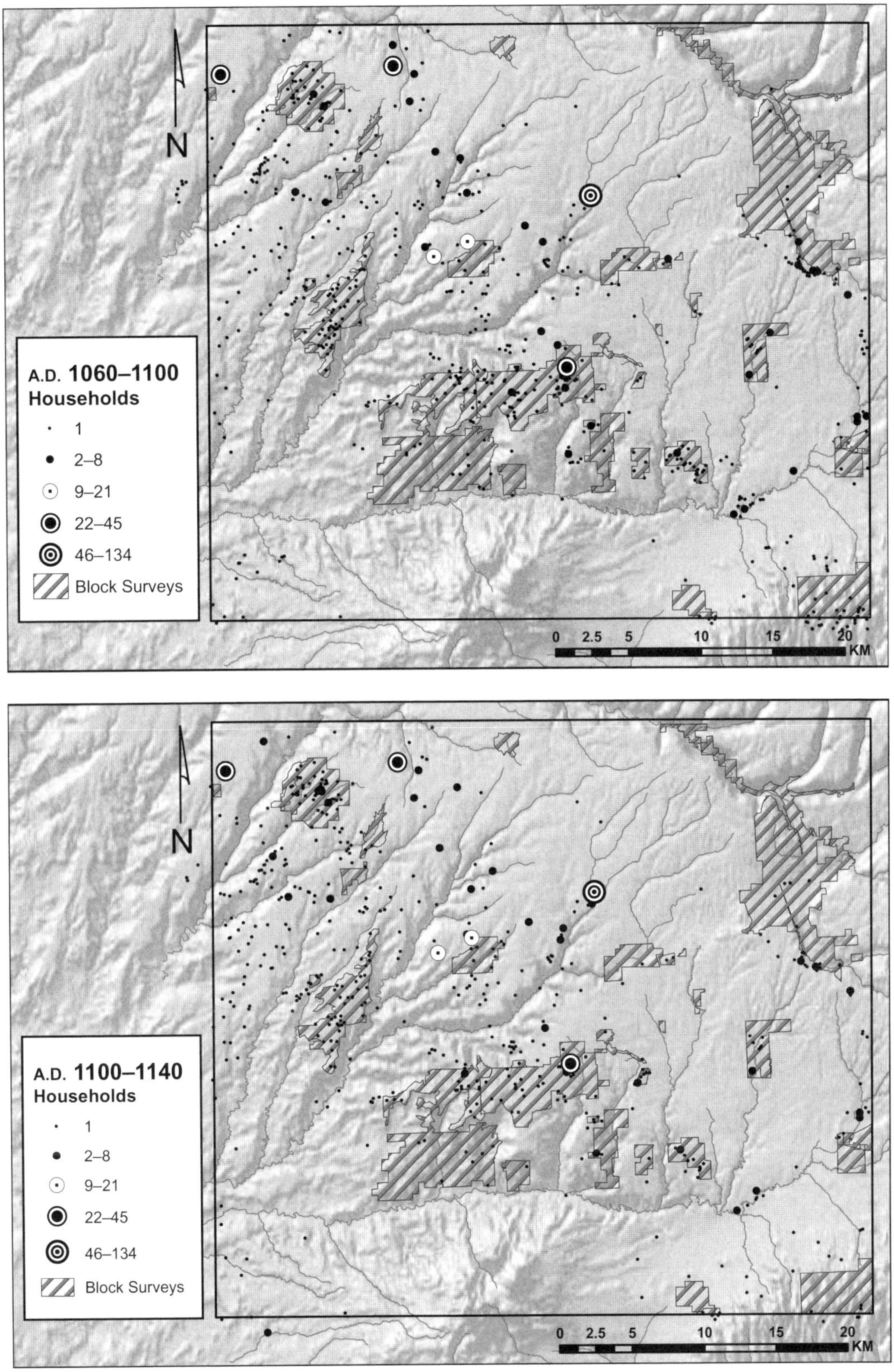

FIGURE 2.8 Site distributions in the VEP study area, modeling periods 14 and 15.

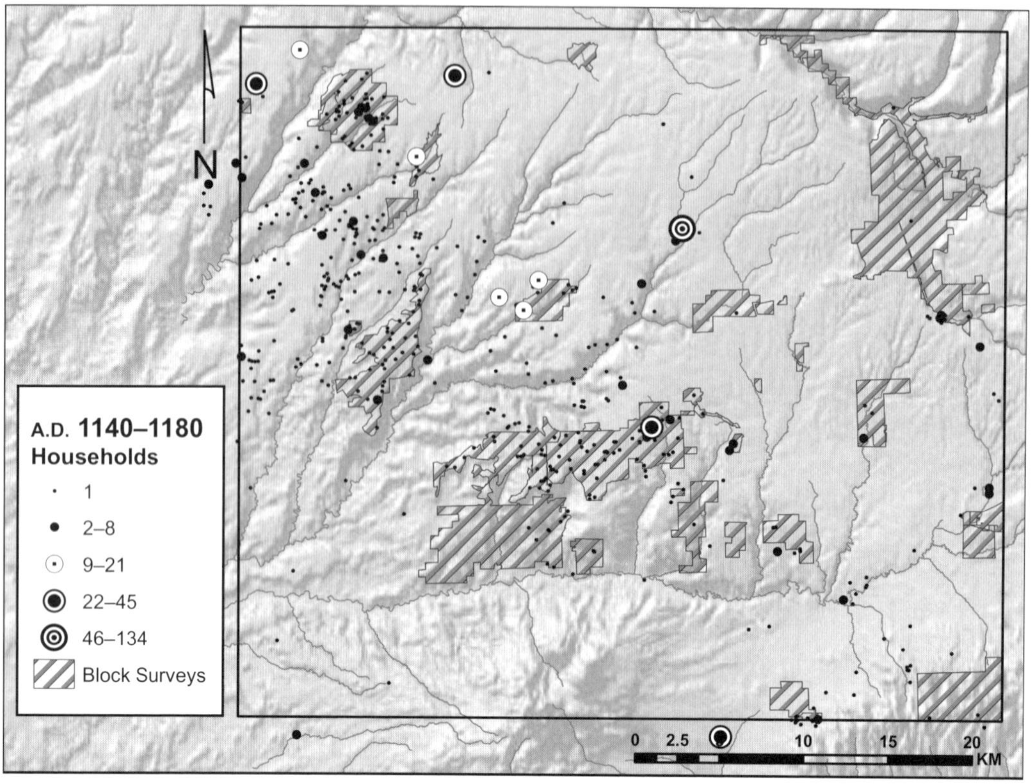

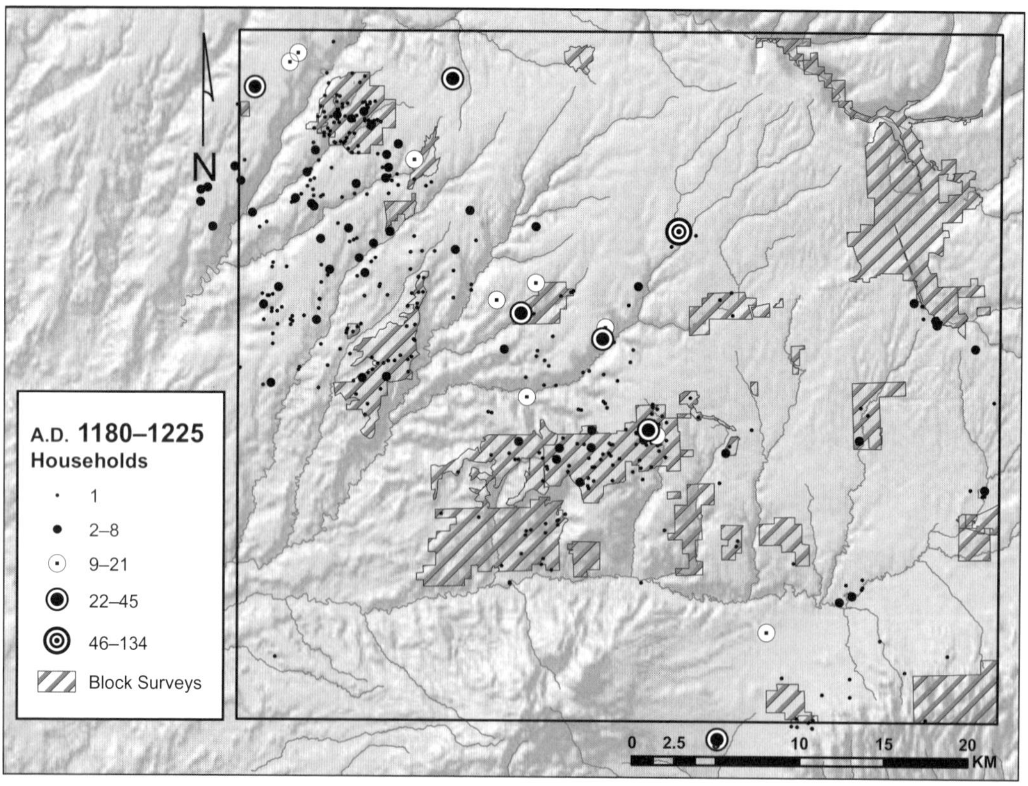

FIGURE 2.9 Site distributions in the VEP study area, modeling periods 16 and 17.

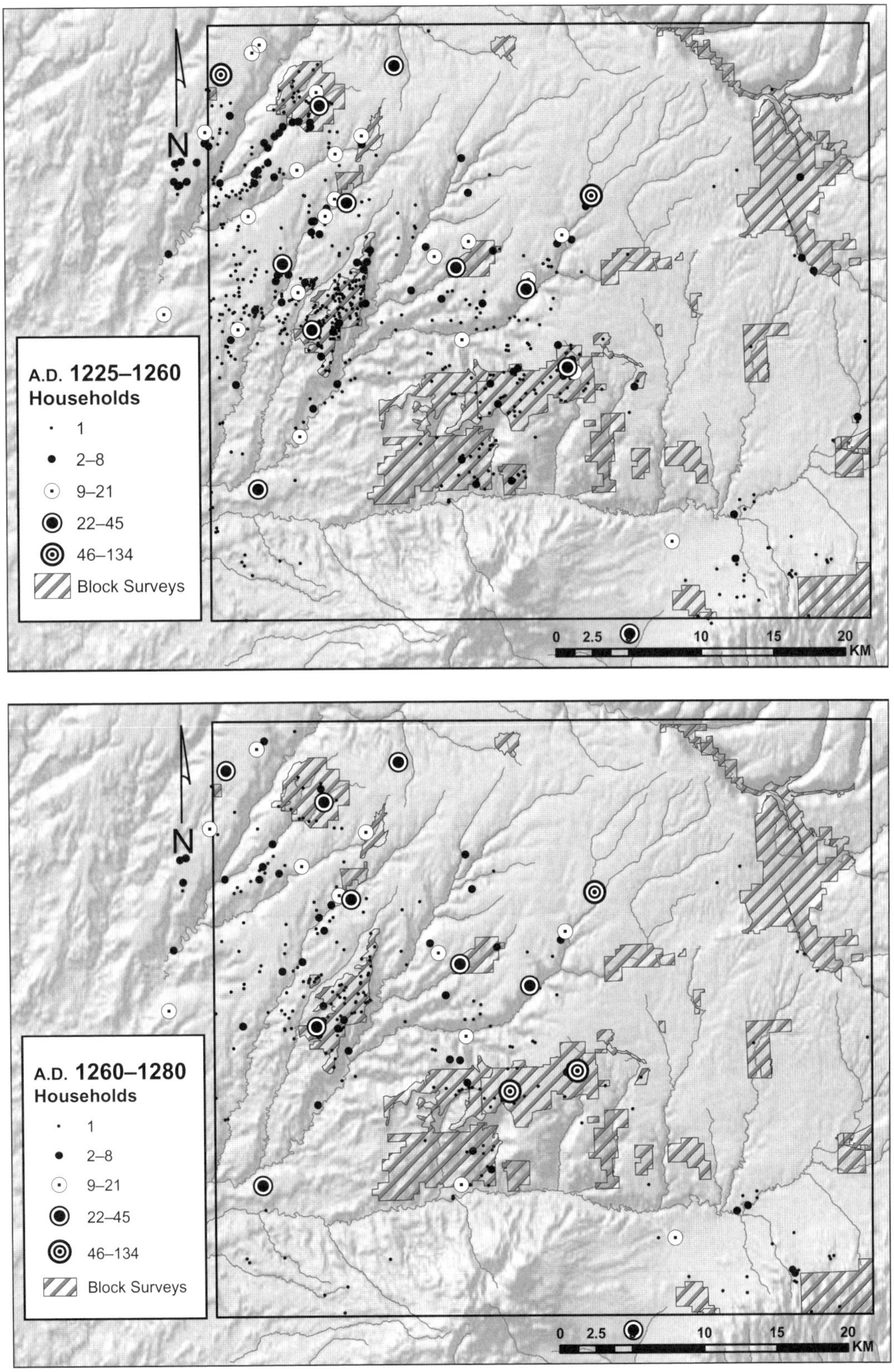

FIGURE 2.10 Site distributions in the VEP study area, modeling periods 18 and 19.

density analysis by the total household-years of occupation, based on point estimates of the total cooking pottery accumulation and Varien's (1999a:107) accumulation rate per household-year. Table 2.1 uses these average use-life estimates and modeling period lengths to calculate the average momentary population of sites in the database for each modeling period.

The next step in demographic reconstruction is to estimate the total population of the study area for each modeling period. Because population estimates are at the mercy of the assumptions used in calculating them (see Powell 1988), we made total population estimates using three different methods and decided on the middle-range estimates as the most reasonable (see Varien et al. 2007:280–281). Our preferred method incorporates two assumptions: (1) that the small-site momentary household density for the block-surveyed area is representative of the entire study area; and (2) that we have a nearly 100 percent sample of community centers. First, we determine the proportion of the study area covered by block surveys, as well as the momentary population of small sites in these areas. Second, we multiply the small-site momentary population in surveyed areas by the inverse of the sampling fraction to get total momentary household estimates for small sites. Third, we add the total number of momentary households in community centers to the small-site total to estimate the total number of momentary households in the study area. Fourth, we convert the number of total momentary households into estimates of the total number of people who resided in the study area during each period by multiplying the total momentary household figure for each period by six, based on Ricky Lightfoot's (1994) ethnographic review of household sizes in historical Pueblo groups. Finally, we use these total population figures to calculate the population density of humans per square kilometer of potentially arable land in the study area. These calculations and estimates are presented in Table 2.2.

The results of our demographic reconstruction indicate that the Pueblo occupation of the VEP study area occurred in two cycles, each approximately three centuries long (Figure 2.11A). The first began after A.D. 600 and ended around A.D. 920, and the second began after A.D. 980 and ended around A.D. 1280. Population levels during the late cycle were much higher than at any point in the early cycle and peaked during the period between A.D. 1225 and A.D. 1260, when approximately 19,500 persons, or about 11 persons per square kilometer of land below 2400 m elevation, lived in the study area.

Population Dynamics

Figure 2.11B translates our total momentary population estimates into population growth rates for each period. The dashed lines on this chart reference positive or negative annual growth rates of .7 percent (an intrinsic rate of natural increase, *r*, of .007). Although we recognize this is a conservative limit (see Kirch 2010:138; Richerson et al. 2001:396–397), we follow Cowgill (1975) in interpreting values above or below this threshold as indicating movement of people into or out of the study area. By this criterion, settlement in the region included several periods of in-migration and out-migration. This dynamism offers local support for the contention that population movements played an important role in sociocultural change throughout Pueblo history (e.g., Berry 1982; Cameron 1995; Clark 2001; Lyons 2003; Naranjo 1995; Ortman 2009, 2010; Ortman and Cameron 2011; Stone 2003; Wilshusen and Ortman 1999).

The first episode of migration into our study area occurred when it was initially settled by Pueblo farmers around A.D. 600. Although there are more single-household habitation sites dating to this period than there are in any other period, the momentary population of this initial period was quite low, a finding attributable to the relatively coarse chronological resolution of the archaeological record (the initial modeling period is 125 years long) and the relatively short use-life of a house for this period. Nevertheless,

TABLE 2.2

Total Momentary Population Estimates for the VEP Study Area

Period		Database Momentary Households		Study Area Momentary Households			
Begin	*End*	*Small Sites in Block Surveys*	*All Community Centers*	*Small Sites*[a]	*Total*	*Total Persons*[b]	*Population Density*[c]
600	725	40.4	2.2	302.2	304.4	1826	1.03
725	800	38.3	39.2	286.6	325.8	1955	1.10
800	840	102.6	67.9	767.7	835.6	5013	2.82
840	880	123.8	104.3	925.9	1030.2	6181	3.48
880	920	35.1	107.8	262.6	370.4	2223	1.25
920	980	34.8	28.4	260.4	288.8	1733	.98
980	1020	83.7	26.6	626.2	652.8	3917	2.21
1020	1060	85.1	35.0	636.3	671.3	4028	2.27
1060	1100	149.1	269.0	1115.6	1384.6	8307	4.68
1100	1140	218.0	309.0	1631.1	1940.1	11641	6.55
1140	1180	225.0	394.0	1683.5	2077.5	12465	7.02
1180	1225	231.0	598.0	1728.4	2326.4	13958	7.86
1225	1260	303.0	967.0	2267.1	3234.1	19404	10.93
1260	1280	123.0	850.0	920.3	1770.3	10622	5.98

a. Block survey small sites per proportion of study area surveyed.

b. Total number of households multiplied by 6, after Lightfoot (1994).

c. Total number of persons per 1776 km^2 of land below 2400 m (7900 ft.) in study area.

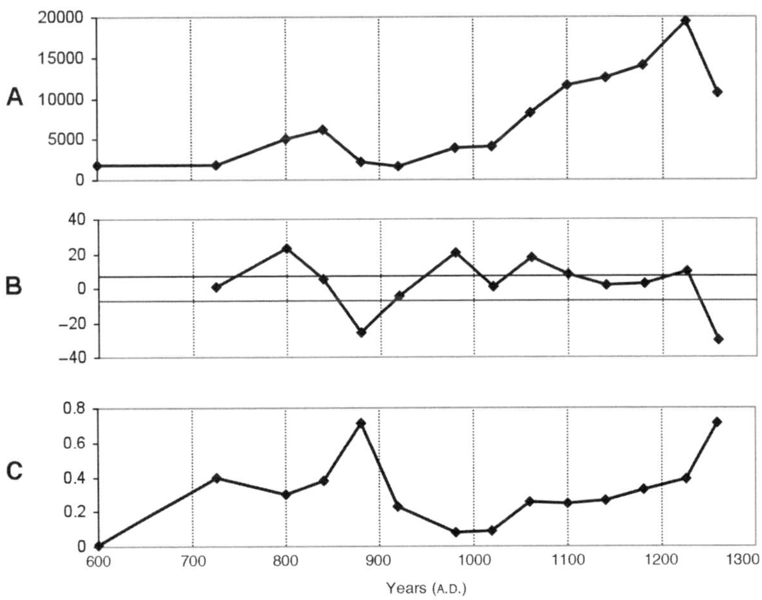

FIGURE 2.11 Ancestral Pueblo population history in the VEP study area. Points in all graphs represent the initial date of a modeling period. The *x*-axis represents calendar years (A.D.), and the *y*-axis represents: A) Total population (individuals); B) Rate of population change (per 1000 per year, dashed lines mark limits of in situ growth/decline); C) Proportion of block-survey households in centers.

our model suggests that approximately 300 households settled in the VEP study area at that time. We know very little about the reasons for this migration or the source of the migrants, and these are important problems for future research.

The second episode occurred as the well-documented villages of the Dolores River valley formed in the northeast corner of our study area (Kane 1986, 1989; Kohler 1992a; Orcutt et al. 1990; Potter 1997; Schlanger 1988). Sarah Schlanger (1988:787) was the first to argue for large-scale migration into the Dolores River valley during the mid-ninth century. More recently, Richard Wilshusen and Ortman (1999) argued that these villages were created by people with distinct cultural and historical backgrounds, with one source in the upper reaches of the San Juan drainage to the east and another in southeastern Utah to the west. Our reconstruction suggests that these groups moved into the study area primarily during the period between A.D. 800 and A.D. 840 and then coalesced into the large villages of the Dolores River valley in the subsequent period.

One interesting detail of our reconstruction is that the Dolores River Valley was a focus of early village development during the first settlement cycle but was largely unoccupied during the second. The Dolores River is a good source of domestic water but the valley floodplain is plagued by cold-air drainage problems and thus was a risky location for agriculture (Peterson and Clay 1987). Inhabitants of the Dolores River valley therefore likely farmed the terraces and slopes above the cold-air pool (Kohler 1992a). Why these locations went largely unused during the second settlement cycle is unclear, although one possibility is that early Pueblo people settled the Dolores River valley to be close to upland resources such as large game (Kohler and Reed 2011). James Potter and Ortman (2004:177–179) compared total artifact assemblages from McPhee Village, an early center in the Dolores River Valley, and Sand Canyon Pueblo, a late center on the uplands in the western half of the study area, and found that items related to hunting and hide processing—especially projectile points, bone awls, flaked stone tools, and artiodactyl bones—are much more common in the early Dolores-area village (see also Kohler and Reed 2011; Potter 2004). Varien and colleagues (2007) and Timothy Kohler and Meredith Matthews (1988) also present evidence of deforestation over the course of the early settlement cycle in the Dolores River Valley. Perhaps anthropogenic impacts to this landscape, including overhunting of large game, swidden-type clearing of hill slopes for agriculture, and over-harvesting of trees for construction timbers, made this landscape less attractive for settlement during the second cycle. We will return to this question in the final chapter.

The early cycle of village formation was followed by the first large-scale emigration from the study area between A.D. 880 and A.D. 920. Possible destinations for these migrants include Mesa Verde proper (Glowacki 2007), northwestern New Mexico (Wilshusen and Wilson 1995), and the San Juan geologic basin (Judge 1989:216; Windes 2007). Several recent studies have argued that this movement from north to south contributed to the initial development of the Chacoan regional system (Varien 2001:53; Varien et al. 2007; Wilshusen and Ortman 1999; Wilshusen and Van Dyke 2006; Windes 2005). The next major episode of population movement, this time back into the study area, occurred between A.D. 1060 and A.D. 1100 (Figure 2.8). This is a surprising and significant finding because most evidence of influence from the emerging regional center at Chaco Canyon occurs in sites established during this same period (Cameron 2005; Lipe 2006; Varien et al. 2007).

A fourth period of population growth and possible in-migration occurred during the period between A.D. 1225 and A.D. 1260, when the ancestral Pueblo population peaked in the study area. This increase in population density has been largely unrecognized in previous studies, but it is associated with several well-recognized changes in settlement patterns. First, settlement aggregation intensified: both the number of community centers and the proportion of people who lived in centers increased. Second, most of these new centers were built in canyon settings that had less-productive catchments than established centers in the uplands. Third, the new centers in canyon settings contained domestic water sources, especially springs, and had ready access to stone and timber resources for construction and fuel (see Fetterman and Honeycutt 1987; Glowacki 2006; Lipe and Ortman 2000; Ortman et al. 2000; Varien 1999a, 1999b; Varien et al. 1996). These trends are discussed in further detail in Chapter 14.

The packing of population into the western half of the study area in the middle A.D. 1200s was a precursor to the largest change of all: the collapse of the settlement system and emigration of the remaining population during the late thirteenth century. Andrew Duff and Wilshusen (2000) argued that emigration from the larger region, of which the VEP study area is a part, was a long-term process that began in the late A.D. 1100s or the early A.D. 1200s, and they note that robust intrinsic growth could increase the overall population of an area even in the context of limited emigration. In a separate study, Wilshusen (2002:118–119) argued that population levels in southwestern Colorado remained high until the middle to late 1200s. Although it may not apply to the entirety of the larger areas studied by Duff and Wilshusen, our reconstruction supports elements of both models. It suggests that the final emigration from the VEP study area was well under way by the middle A.D. 1200s, but it also suggests that more than 10,000 individuals remained in the VEP study area after A.D. 1260 (see Varien 2010). These people either emigrated or died over the following two decades.

Community Centers, Aggregation, and Migration

Figure 2.11C presents an index of settlement aggregation over time. It uses data from block surveys, where we have a 100 percent sample of both small habitation sites and community centers, to estimate the proportion of households living in centers during each period. This chart shows that the formation of centers also occurred in two cycles, with peak aggregation at the end of each cycle, during the periods from A.D. 880 through A.D. 920 and A.D. 1260 through A.D. 1280. These data emphasize several points. First, in contrast to previous views (e.g., Adams 1989; Lekson and Cameron 1995; Schlanger and Wilshusen 1993; Wilshusen and Blinman 1992), our analysis suggests that many community centers had long histories and persisted for several generations after their initial founding. Second, it appears that the passage of time promoted population aggregation on this landscape to a greater extent than did high-frequency climatic cycles. We suspect social factors drove this pattern.[1] Third, the proportion of population living in centers peaked during the initial phase of out-migration at the end of each cycle. This suggests that the initial migrants at the end of each cycle came disproportionately from small sites.[2]

There are several possible explanations for this third pattern. Perhaps the larger settlements exerted a stronger hold on their inhabitants than did isolated farmsteads because of differing "sunk costs" in these two settlement

1. For example, the agent-based model has the ability (not exploited in Chapter 11) to allow the growth of exchange networks with "hub-nodes" at community centers that would have promoted long-term occupation of these centers. Also, as discussed by Michael Adler (1990) and Varien (1999a), the development of land-tenure systems would have allowed people living in aggregated villages to claim use rights to land without living directly on that parcel. These and other factors behind this historical trend in aggregation are discussed further in Chapter 14.

2. This does not necessarily mean that community-center populations were stable during these periods. Test excavations at Yellow Jacket Pueblo, for example, have recovered evidence of a reduced village population during the period of initial out-migration at the end of the second cycle (Varien and Ortman 2005; Varien et al. 2008). The areas of this site that were vacated first were also constructed relatively late in its history, around the edges of the core settlement. Ortman (2003:168) argues, on the basis of Pueblo ethnography and artifact assemblage composition, that inhabitants of these peripheral areas were of lower social standing than the occupants of the core settlement area.

types (Janssen et al. 2003), or perhaps inhabitants of centers—especially those in the "core" kin groups—had more secure access to the best lands for farming (Varien et al. 2000). It is also possible that inhabitants of centers were more secure during times of conflict (Haas and Creamer 1996; Kuckelman 2002; Kuckelman et al. 2002; LeBlanc 1999; Lightfoot and Kuckelman 2001; Wilcox and Haas 1994), although the analysis by Sarah Cole in Chapter 13 suggests that aggregation also increased during many periods in which conflict was not pronounced. In reality, it is likely that many factors were involved as Pueblo people made the individual decisions that resulted in these peaks in aggregation during periods of overall population decline.

SUMMARY

In this chapter we have introduced the study area and attempted to characterize it in broad terms suitable for cross-cultural comparison and for comparison with agent-based simulation results. We have also highlighted the environmental factors that exerted the strongest influence on Pueblo settlement, including soils, growing-season length, precipitation, and domestic water sources. Each of these factors and their influence on settlement patterns is examined in greater detail in the chapters that follow.

We have also developed an outline of the ancestral Pueblo occupation between A.D. 600 and A.D. 1280 through an analysis of all recorded sites in the VEP study area. We used the large number of excavated and well-dated sites in the study area to calibrate change through time and to correlate surface evidence with excavation results. As a result, we can estimate the resident populations of unexcavated habitations at a chronological resolution comparable to a human life span. This in turn allows us to reconstruct the population history of the study area with great precision. We have identified two cycles of population growth and decline over the course of the Pueblo occupation, and several periods of in-migration and out-migration that correlate with significant changes in settlement size, layout, and location. We have also reconstructed two cycles of village or community-center formation that correlate to some degree with population, and found that the initial emigrants at the end of each settlement cycle came disproportionately from small sites.

These analyses of the settlement data provide an introduction to the basic issues surrounding ancestral Pueblo historical ecology that the remaining chapters of this book address in various ways. They also provide the "pattern of resistance" against which models of climate change, agricultural potential, water availability, wood, stone and game resources, exchange, warfare, and settlement decisions are evaluated throughout this book. Given the precision with which the "real" settlement history of the study area can be specified, these models are held to a much higher standard of validation than those created for the archaeological study of ancient societies just about anywhere else. The fact that these models do account for some of the patterns identified through the archaeological work of the VEP is thus a testament to the usefulness of the models and of a model-based approach to archaeology in general.

THREE

Low-Frequency Climate in the Mesa Verde Region

BEEF PASTURE REVISITED

Aaron M. Wright

AROUND A.D. 600, early farmers set deep roots in the high desert country of the Mesa Verde region, dry-farming the loamy canyon floors and mesa tops and building hamlets and villages that, over time, culminated in an intricate array of settlement clusters inhabited by tens of thousands of people. Although these settlements were centered, in most periods, on the region's most productive agricultural lands (Varien 1999b; Varien et al. 2000), their Pueblo residents ultimately left during the thirteenth century in favor of places peripheral to the San Juan region. One long held explanation for this migration has been an adaptive response to a "Great Drought" during the latter half of the thirteenth century (e.g., Douglass 1929; cf. Hewett 1908; Kidder 1924). Yet, more recent research (e.g., Varien et al. 2000), drawn heavily from results of Carla Van West's (1994) agricultural paleoproductivity model for the central Mesa Verde region, suggests this drought would not have limited regional agricultural potential to such an extent that all established communities could no longer be supported by means of a farming lifeway.

These recent studies, however, have not entirely closed the debate. The latest paleoproductivity model developed by the VEP (Kohler 2010, Chapter 6 in this book; Kohler et al. 2007) reconstructs maize production as being considerably lower than Van West's estimates. While even these revised estimates suggest farming productivity would have been sufficient to support smaller populations in some areas, they underscore the relevance of poor climate and depressed farming yields for understanding what transpired during the "turbulent 1200s" (Lipe 1995).

One of the VEP's objectives is to reevaluate and, if necessary, revise existing paleoproductivity reconstructions in light of advances in paleoclimate modeling and theory. This objective has expanded research beyond simple drought-dependent scenarios to consider a range of climatic factors related to maize productivity. For example, there is the possibility that the local onset of a widespread and long-lasting cooling phenomenon, often referred to as the Little Ice Age and variably dated somewhere between A.D. 1250 and A.D. 1850 (Grove 1888;

Matthews and Briffa 2005), may have also reduced or eliminated maize production in portions of the northern Southwest at certain times (Petersen 1988, 1994; see also LeBlanc 1999). Although directly applying the notion of a Little Ice Age to the paleoclimatic regime of the Greater Southwest is controversial and beset by a series of problematic assumptions (Dean 2010; Salzer and Kipfmueller 2005; Van West and Dean 2000), temperature conditions—however they are classified and labeled—need to be accounted for in paleoproductivity models. Ken Petersen (1988) realized that variability in maize productivity in the Mesa Verde area—probably more so than in lower, more southern areas—is dictated by temperature as well as precipitation. Some earlier studies (e.g., Berlin et al. 1977; Martin and Byers 1965; Smiley 1961; Woodbury 1961) had also hypothesized, albeit briefly and in passing, that cooling may have influenced the Mesa Verde regional depopulation during the thirteenth century, and temperature has been given increased weight as an explanatory factor in more recent investigations (e.g., Salzer 2000a).

Earlier paleoproductivity models (e.g., Burns 1983; Van West 1994), however, did not explicitly account for temperature variability in relation to maize productivity. The VEP developed a unique paleotemperature record (Kohler et al. 2007:65) to model the potential impact cold temperatures may have had on farming success, and this is a contributing factor to why the VEP's productivity estimates are lower than those of previous studies. Since they are based on tree-ring data, however, the paleoproductivity models of Burns, Van West, and the VEP contain relatively little low-frequency climatic variability (see Dean 1988, 2010; Dean et al. 1985; and Euler et al. 1979 for more detail on the scales of climatic change). Climatic variability at this temporal scale also influences productivity potentials, yet it is poorly expressed in tree-ring records (Cook et al. 1995; Dean 1988, 1996; Fritts 1991). Therefore, these tree-ring-based models may either overestimate or underestimate agricultural productivity when low-frequency conditions depart markedly from their long-term mean. Low-frequency climatic fluctuations, the focus of the present study, are better expressed by other ecological and geologic proxy measures, such as well-dated stratigraphic pollen sequences from montane environments.

Given the limitations of tree-ring data to inform on low-frequency climatic processes (those of a greater than 25-year duration), the VEP wanted to develop alternative proxy measures to model long-term and broad-scale climate patterns over the past 2,000 years in southwestern Colorado. This chapter details the methodology and results of a new pollen-based reconstruction of low-frequency temperature and precipitation developed specifically for the Mesa Verde region. This new reconstruction refines previous pollen-based reconstructions (discussed in greater detail in the paragraphs that follow) through increased temporal resolution. However, since this paleoclimate reconstruction was developed near the end of VEP I, it has yet to be quantitatively integrated into the VEP paleoproductivity estimates. This chapter provides the low-frequency paleoclimatic context for the other chapters in this book.

I begin by briefly reviewing previous low-frequency climate research in southwestern Colorado and then summarizing the methods used to develop pollen proxies to reconstruct fluctuations in regional temperature and winter precipitation. These reconstructions are then presented and compared to previous high- and low-frequency reconstructions. This comparison reveals periods of prolonged favorable farming conditions from approximately A.D. 600 to A.D. 800 and A.D. 1000 to A.D. 1200. Each of these periods was followed by long intervals of below-average temperature and precipitation. I then briefly discuss how these cycles correspond to cycles of population growth and decline in the Mesa Verde region. The correlation between climate and demography suggests that the two cycles are related, evidently through a filter of maize productivity, but perhaps in an unexpected way. It appears that the central

Mesa Verde region—with its better soils and more reliable water sources—may have served as a refugium for farmers displaced by deleterious farming conditions in the A.D. 800s, 900s, and 1200s.

A LEGACY OF PALYNOLOGY IN THE LA PLATA MOUNTAINS

The coupling of palynological research to archaeological questions has a considerable history in the La Plata Mountains, beginning with the Salmon Ruins Archaeological Project (Petersen and Mehringer 1976) and continuing through the Dolores Archaeological Project (DAP) (Petersen 1988). Both relied on continuous stratigraphic pollen sequences from subalpine wetland environments. Cores from Twin Lakes were used to reconstruct regional temperature patterns; Beef Pasture cores were used to reconstruct precipitation (Figure 1.1). Petersen (1988) used these sequences along with tree-ring and historic vegetation data to develop a model of prehistoric fluctuations in the regional dry-farming belt (the minimum and maximum elevations where farming was possible). Petersen's research was thorough but perhaps less influential than it should have been, since the period between A.D. 600 and A.D. 1300 was dated by just two traditional ^{14}C dates in each of sequences analyzed (Petersen 1988:Table 3; Petersen and Mehringer 1976:280–283). This led Petersen to resort to the more chronologically secure tree-ring record from Almagre Mountain to reconstruct temperature during Pueblo occupation of the Mesa Verde region.

Petersen (1988:123–127) reconstructed significant climatic variability during this occupation and suggested the following demographic responses: (1) before farmers moved into the region around A.D. 600, growing seasons were too short for successful maize agriculture; (2) during the A.D. 700s and A.D. 800s, a period of continuous population increase, climates were relatively favorable for farming; (3) a period of drought and cold temperatures coincided with a significant population decrease between A.D. 875 and A.D. 950; (4) climatic conditions improved in the late A.D. 900s, and people subsequently returned in the A.D. 1000s; and (5) an interval of cooling around A.D. 1200 and a subsequent drought beginning in A.D. 1275 effectively "pinched out" the region's dry-farming belt, leaving the entire farming population of the northern San Juan region to find suitable land elsewhere.

Our chronology and knowledge of culture change and settlement dynamics in the Mesa Verde region have become significantly more precise since these paleoclimatic reconstructions were generated. The VEP, for example, recognizes 14 periods between A.D. 600 and A.D. 1280, with an average length of about 50 years (standard deviation, 25 years) (Chapter 2). This in turn makes achieving a higher temporal resolution for the regional paleoclimatic reconstruction attractive. This higher resolution is possible, given analysis of more closely spaced pollen samples and additional ^{14}C samples dated by accelerator mass spectrometry. To identify climatic variability on a time scale more compatible with that recognized in the local archaeological record, the new reconstruction from Beef Pasture (Wright 2006) yields a temporal precision of about 30 years—actually higher than that recognized during most of the archaeological record to which it will be compared.

A NEW POLLEN SEQUENCE FROM BEEF PASTURE

Beef Pasture is a subalpine fen located at an elevation of 3,060 m asl along the western flank of the La Plata Mountains, approximately 25 km northwest of Durango, Colorado, and about 30 km east of the eastern border of the VEP study area. Since this location flanks the northeastern edge of the San Juan basin, climatically induced changes in local vegetational communities should correlate with significant regional climatic fluctuations across the northern San Juan region. Beef Pasture is currently surrounded by a subalpine forest with a

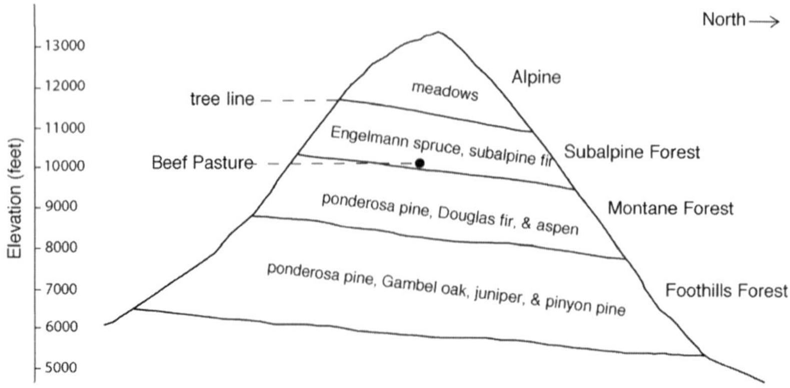

FIGURE 3.1 Dominant overstory species of La Plata Mountains forest zones (adapted from Maher 1961:Figure 2; Petersen 1988:Figure 4).

characteristic overstory dominated by Engelmann spruce (*Picea engelmanni*) and subalpine fir (*Abies lasiocarpa*) (Figure 3.1). The understory includes numerous plant species; however, those belonging to the Cyperaceae, Asteraceae, Amaranthaceae, Chenopodiaceae, and Poaceae families dominate the local microenvironment. Slightly below the elevation of Beef Pasture, at approximately 3,000 m asl, the subalpine forest begins to grade into a mixed-conifer montane forest, dominated by ponderosa pine (*Pinus ponderosa*) and Douglas fir (*Pseudotsuga menziesii*). Since climatic factors dictate the elevation of this boundary zone between the subalpine and montane forests, the fen's proximity to this transition warrants using pollen of climatically sensitive and locally abundant plants deposited in Beef Pasture as a proxy for regional paleoclimatic changes.

As Petersen and Peter Mehringer recognized, the uppermost levels of Beef Pasture's sediment, which consists of 4 m of woody sedge peat deposited over the past 10,000 years (Petersen 1988:Table 3), offers an ideal context for continuous pollen deposition and preservation. Petersen's (1988) paleoclimatic reconstruction spanned nearly the entire Holocene; the VEP's resampling of Beef Pasture focused on the last 2,000 years; and this chapter emphasizes the period between A.D. 500 and A.D. 1400.

A single core containing the uppermost 1.45 m of peat was extracted to obtain sediment dating to the period of interest (Figure 3.2). Starting at 4 cm below surface, 71 pollen samples, each measuring 1 by 2 by 3 cm and spaced every other centimeter, were extracted and processed from the core. This sampling strategy was selected for two primary reasons. First, such high-resolution pollen sampling permits a more accurate assessment of low-frequency paleoenvironmental change over short periods. Sedimentation rates calculated by Petersen (1988:Figures 15 and 18, Table 6) indicated that approximately 7 to 10 cm of sediment accumulated at Beef Pasture every 100 years. Accordingly, each centimeter of sediment was predicted to contain less than 15 years of pollen deposition, with sample intervals likewise representing temporal durations of less than 15 years. Second, sampling every other centimeter is an effective means of maintaining a high temporal resolution while minimizing the potential for misrepresentation due to any postdepositional translocation of pollen.

The chemical extraction of pollen from these samples followed closely the methodology of other palynologists (e.g., Faegri and Iversen 1989; see Wright 2006 for more details). At least 500 pollen grains were counted in each sample to ensure reliable representation of past vegetation (Barkley 1934) and to provide sample sizes amenable to statistical analyses (Crabtree 1968; Maher 1972a). Excluding mosses and ferns, at least 46 plant genera are represented

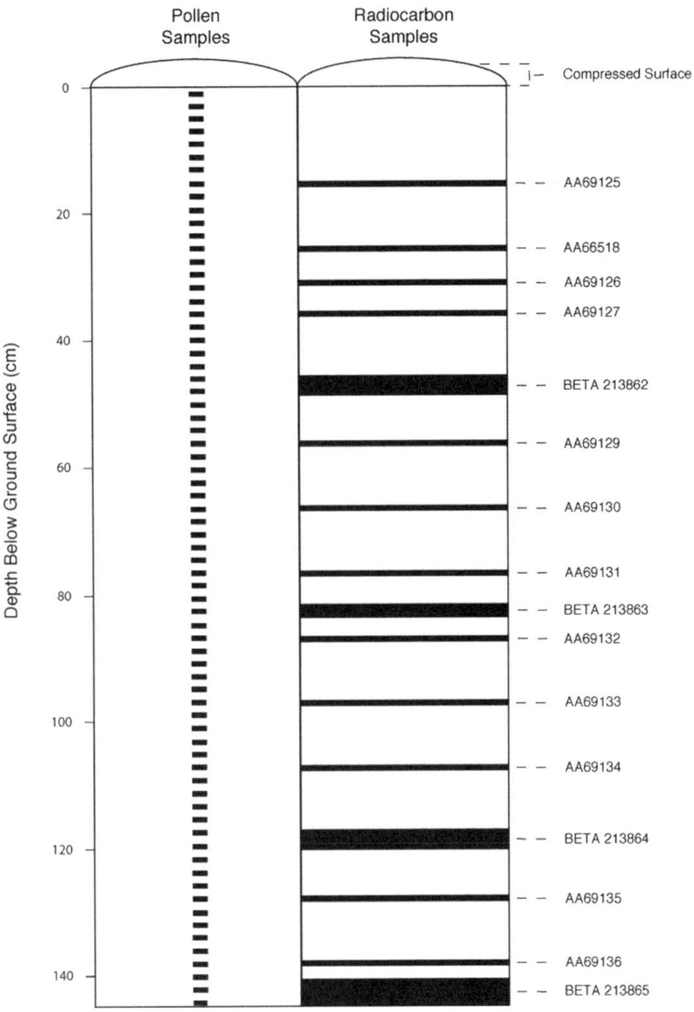

FIGURE 3.2 Beef Pasture sediment core showing pollen ($n=71$) and ^{14}C ($n=16$) samples by depth.

in the Beef Pasture pollen profile (Table 3.1), and Figure 3.3 shows changes in the percentages of the most abundant pollen types throughout the core profile.

A deposition rate for Beef Pasture sediment, required for calculating pollen influx rates and assigning each pollen sample an approximate date, was calculated from 16 ^{14}C samples (Figure 3.2, Table 3.2) and a surface date. Uncorrected dates were calibrated using the revised CALIB radiocarbon calibration program (Stuiver and Reimer 1993; Stuiver et al. 2005). A two-step linear regression procedure, with depth below surface (cm) as the independent variable and calibrated date as the dependent variable, provided a means to identify the most likely calibrated date range for those samples with multiple date ranges at the 1-σ level. In the first pass, only those samples with one calibrated date range at the 1-σ level were used along with the date for the modern ground surface ($n=7$) to predict the most likely calibrated date ranges of the other samples ($n=11$) (in bold in Table 3.2). These results were then incorporated in a second pass, where Maher's (1972b:540–544) method was applied for calculating a deposition rate from all 16 samples. This method involves regressing the midpoints of the calibrated date ranges onto the samples' depth below surface. Although deposition rates may

TABLE 3.1

Plants Identified in Beef Pasture Pollen Record, Excluding Algae, Moss, and Ferns

Pollen Type	Scientific Name	Common Family Name	Common Name	Habitat Class
Abies	*Abies lasiocarpa*	Pine	Subalpine fir	Arboreal
Alnus	Betulaceae *Alnus icana*	Birch	Alder	Arboreal
Apiaceae	Apiaceae	Parsley	Various	Nonarboreal
Artemisia	*Artemisia* sp.	Sunflower	Sage varieties	Nonarboreal
Asteraceae, high-spine	Asteraceae	Sunflower	Various	Nonarboreal
Asteraceae, low-spine	Asteraceae	Ragweed / goldenrod	Various	Nonarboreal
Betula	*Betula fontinalis*	Birch	Water birch	Arboreal
Carex	*Carex* sp.	Sedge	Various	Aquatic
Celtis	*Celtis reticulata*	Elm	Hackberry	Arboreal
Cirsium	*Cirsium* sp.	Sunflower	Thistle	Nonarboreal
Cheno-am	Chenopodiaceae/ *Amaranthus*	Goosefoot / pigweed	Various	Nonarboreal
Ephedra torreyana	*Ephedra torreyana*	Ephedra	Mexican tea	Nonarboreal
Ephedra viridis	*Ephedra viridis*	Ephedra	Mormon tea	Nonarboreal
Fabaceae	Fabaceae	Pea	Various	Nonarboreal
Fragaria	*Fragaria* sp.	Rose	Wild strawberry	Nonarboreal
Geraniaceae	*Geranium* sp.	Geranium	Wild geranium	Nonarboreal
Juncus	*Juncus* sp.	Rush	Various	Aquatic
Juniperus	*Juniperus* sp.	Juniper	Various	Arboreal
Lamiaceae	Lamiaceae	Mint	Various	Nonarboreal
Liguliflorae	*Liguliflorae* sp.	Lettuce	Various	Nonarboreal
Lythraceae	*Lythrum* sp.	Loosestrife	Various	Nonarboreal
Onagraceae	Onagraceae	Evening primrose	Various	Nonarboreal
Opuntia	*Opuntia* sp.	Cactus	Prickly pear cactus	Nonarboreal
Parthenocissus	*Parthenocissus* sp.	Grape	Various	Nonarboreal
Pinus edulis	*Pinus edulis*	Pine	Pinyon pine	Arboreal
Pinus ponderosa	*Pinus ponderosa*	Pine	Ponderosa pine	Arboreal
Picea engelmannii	*Picea engelmannii*	Spruce	Engelmann spruce	Arboreal
Poaceae	Poaceae	Grass	Various	Nonarboreal
Polemoniaceae	Polemoniaceae	Phlox	Various	Nonarboreal
Polygala	*Polygala* sp.	Milkwort	Various	Nonarboreal

(continued)

TABLE 3.1 *(continued)*

Pollen Type	Scientific Name	Common Family Name	Common Name	Habitat Class
Populus	*Populus tremuloides*	Willow	Quaking aspen	Arboreal
Pseudotsuga	*Pseudotsuga menziesii*	Pine	Douglas fir	Arboreal
Quercus	*Quercus gambelii*	Beech	Gambel oak	Arboreal
Ranunculaceae	Ranunculaceae	Buttercup	Various	Nonarboreal
Rosaceae	Rosaceae	Rose	Various	Nonarboreal
Rubus	*Rubus idaeus*	Rose	Wild raspberry	Nonarboreal
Salix	*Salix brachycarpa*	Willow	Barren ground willow	Arboreal
Sarcobatus	*Sarcobatus* sp.	Goosefoot	Greasewood	Nonarboreal
Thalictrum	*Thalictrum fendleri*	Meadowrue	Fendler's meadowrue	Nonarboreal
Typha	*Typha latifolia*	Cattail	Common cattail	Aquatic
Verbenaceae	Verbenaceae	Verbane	Various	Nonarboreal
Unknown Types ($n=7$)	Unknown	Unknown	Unknown	Unknown

vary through time, in this case we found that a curvilinear relationship between depth and age did not provide a better enough fit to justify the use of additional, higher-order, independent variables. We used weighted least-squares regression (Figure 3.4) to give more weight to those samples with higher chronological precision (smaller standard deviations).

As can be seen in Figure 3.4, a few of the ^{14}C dates do not correspond exactly with those predicted by the regression, even at the 2-σ level. Those falling within the date range of interest here, however, do match the predicted values very well, suggesting that the linear regression is reliable for the A.D. 500 through A.D. 1400 segment of the pollen sequence. This regression relationship was used to estimate a date for each pollen sample, and a deposition rate of .071 cm / year was calculated by dividing the difference in depth (144.5 cm) by the temporal span between the modern surface and bottommost pollen sample (2,047 years). This is nearly identical to the .072 cm / year deposition rate Petersen calculated for the past 2,830 years (1988:Table 6). Each sample used for the present reconstruction, therefore, represents approximately 14 years of pollen deposition, and the intervals between samples represent a similar span of time.

POLLEN AS PROXY FOR LOW-FREQUENCY TEMPERATURE FLUCTUATION

Aaron Wright (2006:34–44) provides a comprehensive description—summarized here—as to how variability through time in subalpine pollen types may serve as proxies for climatic variability. Studies using subalpine pollen sequences to infer fluctuations in regional climate usually employ pollen frequencies, influx rates, and ratios as measures of the tree species that demarcate forest boundaries. For temperature, the underlying premise is that forest boundaries and tree lines move in tandem with

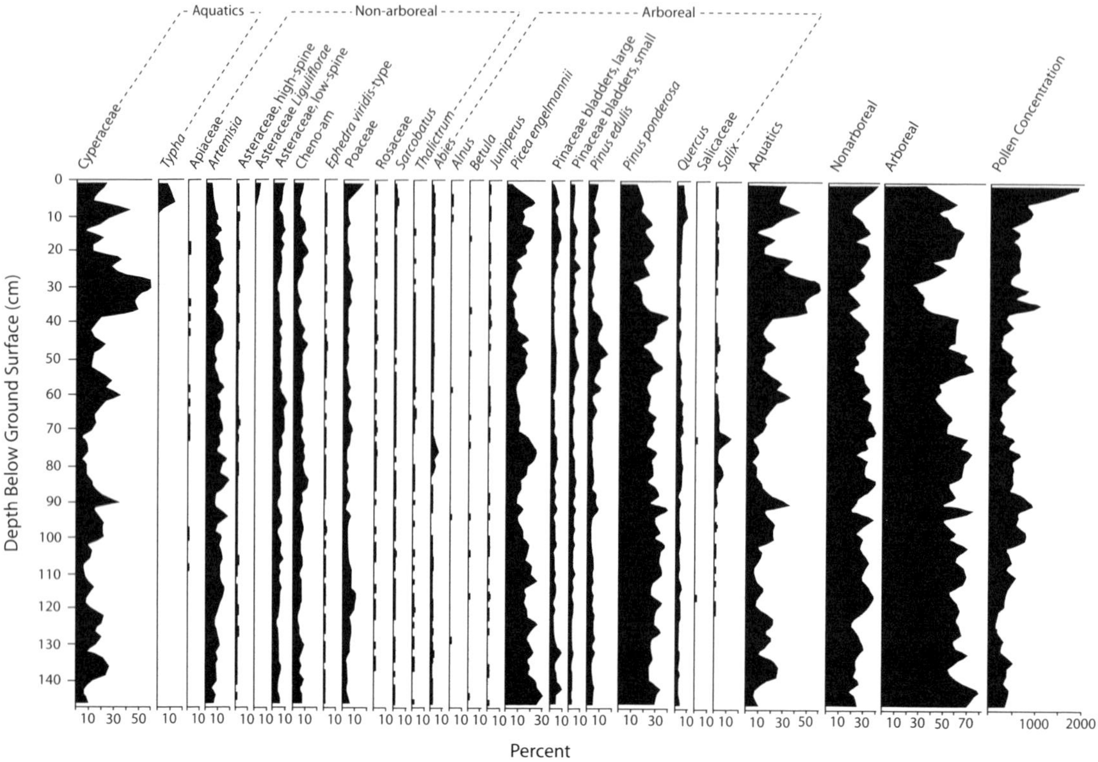

FIGURE 3.3 Pollen percentage diagram showing fluctuations of most abundant pollen types by depth at Beef Pasture.

the elevation of a particular summer isotherm (Arno 1984; Daubenmire 1954; LaMarche 1973; Scuderi 1987). Therefore, changes in the elevation of the tree line indicate prolonged changes in the elevation of this isotherm, and the climatic oscillation causing such ecological shifts is then extrapolated into a regional context. For example, a lowering of the elevation of the isotherm implies cooler summers and annual temperatures; increases in the elevation of the isotherm are coupled with higher regional annual temperatures.

In the La Plata Mountains, Engelmann spruce dominates the subalpine forest overstory (Figure 3.1). The elevational distribution of this species is dictated largely by the duration of winter snowpack along mountain slopes (Daubenmire 1954:128–129; Dix and Richards 1976; LaMarche 1973:637; Lindsay 1971:Table 1; Marr and Marr 1973; Pearson 1931:Figures 14 and 15; Wardle 1968:Figure 3). Temperature is the major factor controlling snowpack duration at the upper boundary of Engelmann spruce distribution, and both temperature and winter precipitation rates determine snowpack duration at the lower boundary. As a result, past elevational changes in Engelmann spruce distribution serve as proxies for past fluctuations in regional trends in temperature and winter precipitation. Such elevational changes can be discerned from pollen sequences from wetland environments located near the upper and lower subalpine forest boundaries (e.g., Andrews et al. 1975; Fall 1985, 1988, 1997; Maher 1961, 1972a; Markgraf and Scott 1981; Petersen 1988; Petersen and Mehringer 1976; Short 1985; Vierling 1998), especially since spruce pollen is rather large and poorly equipped for long-distance aerial transport. The distribution of spruce pollen is considered to be highly localized (Janssen

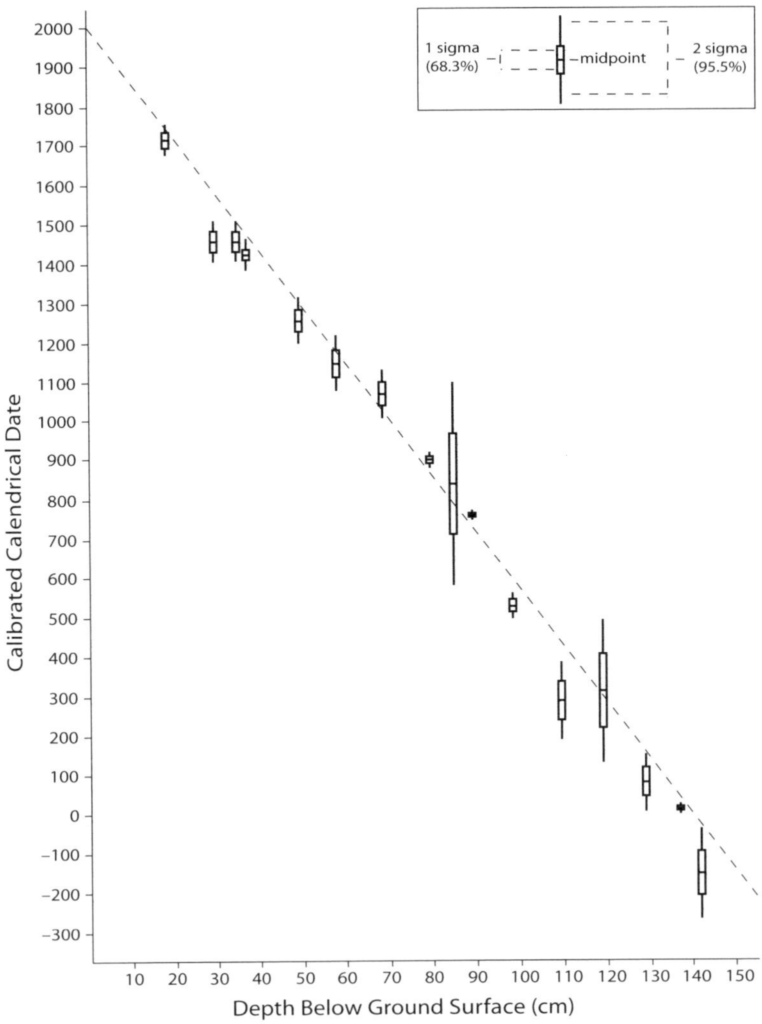

FIGURE 3.4 Weighted least-squares regression of calibrated ^{14}C dates on depth used to derive age estimates for each pollen sample and to calculate a deposition rate (year A.D. = 2000.18 − 14.32*cm below surface, $r^2 = .994$, $p < .0001$).

1966:816; Maher 1961), with the vast majority deposited within 30 m of the source tree (Dyakowska 1937; Jonassen 1950; Wright 1952, 1953).

Although the upper boundary of Engelmann spruce is controlled largely by temperature, the lower boundary—such as the area just below Beef Pasture—is controlled by both temperature and the amount of winter precipitation. This led Petersen (1988) to compare pollen sequences from both the upper (Twin Lakes) and lower (Beef Pasture) subalpine forest boundaries to distinguish between the effects of changes in temperature and winter precipitation. We will distinguish between these climatic variables using just the Beef Pasture sequence, near the lower subalpine forest boundary, by comparing changes in Engelmann spruce pollen deposition against those of other taxa whose abundance and distribution are dependent solely on rates of winter precipitation.

This study uses changes in the deposition of sedge (*Carex* sp.) and Cheno-am pollen types as independent proxies for winter precipitation. Here, Cheno-am denotes any plants belonging to either the Chenopodiaceae family or the *Amaranthus* genus whose pollen grains are indistinguishable with binocular microscopes. Sedges thrive in wet and waterlogged sediment

TABLE 3.2

Beef Pasture ^{14}C Samples

Boldface indicates calibrated dates selected for this study on the basis of a linear regression of samples with only one option, at 1σ.

THE MIDPOINTS OF THE DATES IN BOLD FELL CLOSEST THE POINT ESTIMATED BY THE REGRESSION EQUATION.

Depth (cm)	Laboratory[a]	Technique[b]	Lab No.	Uncorrected Date (B.P.) at 1σ	Calibrated Dates (A.D. / B.C.) at 1σ[c]	Calibrated Dates (A.D. / B.C.) at 2σ[c]
18– 19	UA	AMS	AA69125	94 ± 37	1893 ± 26	1870 ± 68
					1860 ± 00	1758 ± 05
					1846 ± 06	1710 ± 29
					1826 ± 13	
					1711 ± 16	
28–29	UA	AMS	AA66518	422 ± 35	1460 ± 25	1605 ± 15
						1579 ± 02
						1470 ± 50
33–34	UA	AMS	AA69126	426 ± 37	1459 ± 26	1606 ± 15
						1579 ± 03
						1469 ± 52
38–39	UA	AMS	AA69127	496 ± 36	1426 ± 14	1423 ± 29
						1335 ± 08
48–51	BETA	TR	213862	780 ± 50	1252 ± 34	1226 ± 67
58–59	UA	AMS	AA68129	846 ± 38	1250 ± 02	1209 ± 59
					1237 ± 04	1131 ± 08
					1194 ± 34	1067 ± 19
68–69	UA	AMS	AA69130	914 ± 34	1154 ± 08	1202 ± 05
					1131 ± 13	1111 ± 80
					1073 ± 30	
78–79	UA	AMS	AA69131	1079 ± 37	1009 ± 05	956 ± 63
					974 ± 25	
					909 ± 10	
83–85	BETA	TR	213863	1190 ± 90	940 ± 22	1009 ± 03
					833 ± 66	831 ± 165
					729 ± 15	
88–89	UA	AMS	AA69132	1342 ± 37	**757 ± 05**	754 ± 18
					669 ± 21	683 ± 46

(continued)

TABLE 3.2 *(continued)*

Depth (cm)	Laboratory[a]	Technique[b]	Lab No.	Uncorrected Date (B.P.) at 1σ	Calibrated Dates (A.D. / B.C.) at 1σ[c]	Calibrated Dates (A.D. / B.C.) at 2σ[c]
98–99	UA	AMS	AA69133	1567 ± 37	**522 ± 17** 464 ± 30	493 ± 79
108–109	UA	AMS	AA69134	1743 ± 37	292 ± 49	309 ± 96 183 ± 04
118–121	BETA	TR	213864	1740 ± 80	311 ± 96	523 ± 07 501 ± 12 261 ± 177
128–129	UA	AMS	AA69135	1914 ± 40	**91 ± 39** 34 ± 04	109 ± 106
138–139	UA	AMS	AA69136	1929 ± 39	111 ± 14 **71 ± 22** 34 ± 07	202 ± 07 162 ± 07 67 ± 74 –23 ± 16
141–144.5	BETA	TR	213865	2150 ± 60	**–157 ± 57** –223 ± 07 –322 ± 31	–208 ± 161

a. UA = University of Arizona AMS Laboratory, Tucson; BETA = Beta Analytic, Inc., Miami, FL.

b. AMS = radiometric via accelerator mass spectrometry; TR = traditional radiometric with extended counting.

c. Calibrated with CALIB® Rev. 5.0.1 (Stuiver and Reimer 1993; Stuiver et al. 2005).

and are characteristic of Rocky Mountain alpine fens (Herman 1970). Patricia Fall (1983, 1988, 1997) and Lee Vierling (1998) also view sedge pollen deposition as indicative of moisture conditions at a wetland. Ecological research has shown that sedges are more dependent on winter than on summer precipitation because they favor locations with high rates of spring moisture derived from winter snowmelt (Colorado Natural Areas Program 1996; Cooper and MacDonald 2000; DeBenedetti and Parsons 1984). Vierling (1998) compared sedge and Cheno-am pollen percentages to differentiate between perennially wet and seasonally dry meadows because Cheno-am varieties favor disturbed environments, such as fens experiencing unusually low rates of ground moisture. Thus, comparing fluctuations in sedge and Cheno-am pollen deposition to that of Engelmann spruce provides a basis to distinguish between the effects of winter precipitation and temperature on the distribution of the subalpine forest. Specifically, we will expect the following at Beef Pasture: (1) an increase in spruce pollen indicates a downward expansion of the subalpine forest because of an increase in winter precipitation, a decrease in temperature, or both, whereas a decrease in spruce pollen indicates the opposite conditions; (2) an increase in sedge pollen indicates an increase in winter precipitation, and vice versa; and (3) an increase in Cheno-am pollen indicates a reduction in winter precipitation, and vice versa. Therefore, for example, we expect spruce and Cheno-am to increase together under increasingly cold and xeric conditions.

THE REVISED LOW-FREQUENCY PALEOCLIMATIC RECORD

In my original study (2006), I divided the entire 2,000-year pollen sequence into 11 biostratigraphic zones using a stratigraphically constrained cluster analysis. The intent was to identify periods of relative homogeneity in the pollen record, and samples falling into each period would serve as populations on which statistical measures of difference could be applied. The cluster results were strongly influenced by the primary taxa of concern for the present study: Engelmann spruce, sedge, and Cheno-am. Statistical procedures were applied to three measures from the pollen record: pollen percentages, pollen influx rates, and pollen ratios of these indicator taxa. Each of these measures provides slightly different inferences for changes in the forest structure, but when combined, they provided a more robust picture of past vegetational changes. Pollen percentages are useful for analyzing a taxon's distribution in relation to the entirety of the vegetational assemblage. Changes in the percentage of a pollen type, however, may not necessarily be the result of changes in that taxon's distribution. Pollen influx rates provide a more appropriate measure of changes in a taxon's distribution because these measures are not influenced by the abundance of pollen of other taxa. When comparing pollen types, such as those indicative of the two forest zones around Beef Pasture, pollen ratios have proved useful for elucidating elevational changes in tree lines and forest boundaries (e.g., Maher 1961, 1972b; Petersen 1988). While Wright (2006) analyzed all three of these measures and found a general conformity among them, here I will emphasize the pollen ratios.

To infer temperature changes from the Beef Pasture pollen record, a winter precipitation curve must be compared to the depositional history of Engelmann spruce. Figure 3.5 contrasts the sedge-to-Cheno-am pollen ratio with the ponderosa pine-to-spruce pollen ratio. Ponderosa pine was selected as the comparative taxon to Engelmann spruce because it is the dominant overstory plant of the montane forest just below Beef Pasture. Increases in this ratio indicate an upward expansion of the montane forest because of reduced winter precipitation, increased temperatures, or both; decreases in the ratio indicate the reverse. Increases in the sedge-to-Cheno-am pollen ratio indicate greater amounts of winter precipitation, and vice versa. Simultaneous increases in both ratios thus suggest warmer conditions; simultaneous decreases in both ratios indicate colder temperatures.

As Figure 3.5 shows, there is general agreement between the fluctuations of the two pollen-ratio curves. In fact, Wright (2006:87–89) observed only one period (A.D. 290–575) in which these two curves diverged to such an extent that a distinction between a decrease in winter precipitation or an increase in temperature could not be positively identified. Temporal variation in winter precipitation and temperature evidenced in Figure 3.5 can be subsumed within the following sequence of five periods of relative low-frequency climatic homogeneity that were differentiated by reversals in major climatic trends; four of these correspond to statistically significant differences between the pollen zones originally identified (Wright 2006).

1. Before A.D. 600, the Mesa Verde region was characterized by relatively low temperatures and low winter precipitation.
2. After A.D. 600, as sedentary farmers became common in the northern Southwest, temperatures and winter precipitation increased significantly. These favorable conditions persisted until the late A.D. 700s, when the region experienced major low-frequency reductions in temperature and winter precipitation.

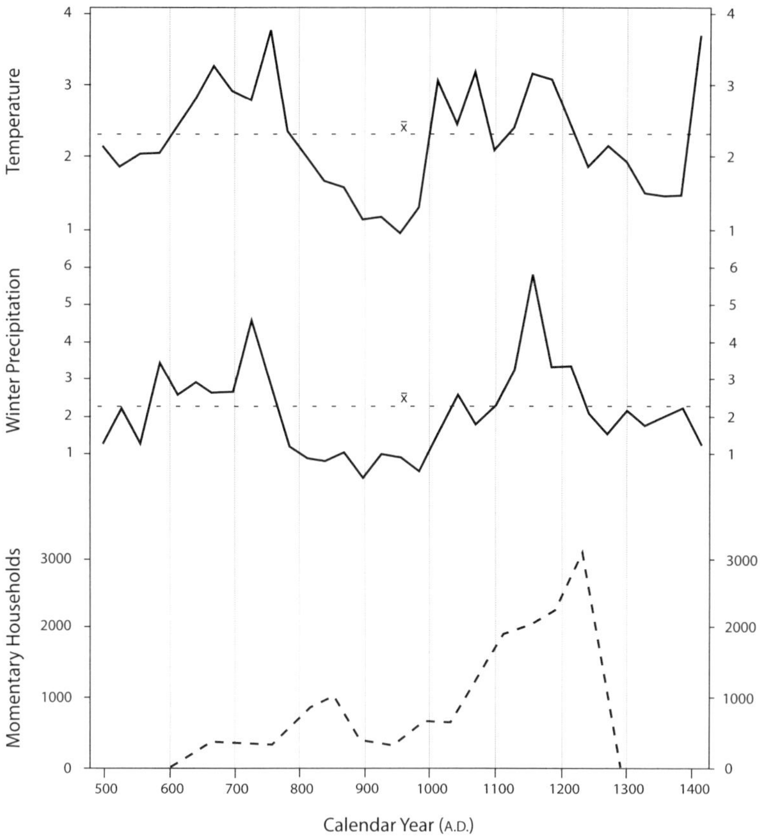

FIGURE 3.5 Comparison of low-frequency temperature and winter precipitation curves to the number of momentary households in the central Mesa Verde region. Temperature: ponderosa pine / spruce pollen ratio (mean = 2.26); Winter Precipitation: sedge/Cheno-am pollen ratio (mean = 2.13). Momentary household estimates are from the VEP paleodemographic reconstruction (Chapter 2).

3. These poor farming conditions continued until the late A.D. 900s, when both variables increased significantly once again.
4. Relatively high temperatures and rates of winter precipitation characterized the Mesa Verde region until the early A.D. 1200s; this period corresponds with the well-known Medieval Warm period (Hughes and Diaz 1994; Lamb 1977). Short periods of low temperatures and reduced winter precipitation in the late A.D. 1000s and early A.D. 1100s, however, appear in the Beef Pasture record. It is interesting that the "megadrought" of the middle A.D. 1100s (Benson et al. 2007; Berry and Benson 2010; Herweijer et al. 2007; see also Cook et al. 2004, 2007; Stahle et al. 2000; Wright 2010) does not register in the Beef Pasture pollen profile, which may suggest that the local expression of this drought in the Mesa Verde region involved reduced summer (monsoonal) precipitation rather than shortfalls in winter or annual rates.
5. After the middle A.D. 1200s, both winter precipitation and temperature dropped considerably and remained relatively low throughout the fourteenth century. The earlier end of this precipitation shortfall correlates with the Great Drought of the late A.D. 1200s, and the reduced temperatures may correspond to a local onset of a prolonged period of hemispheric or global cooling (Little Ice Age; Bradley 2000:1353; Grove 1988; see also Petersen 1994), although if so its timing and expression in the Southwest are clearly very different from what they are in other areas (Salzer and Kipfmueller 2005; Van West and Dean 2000:27–28).

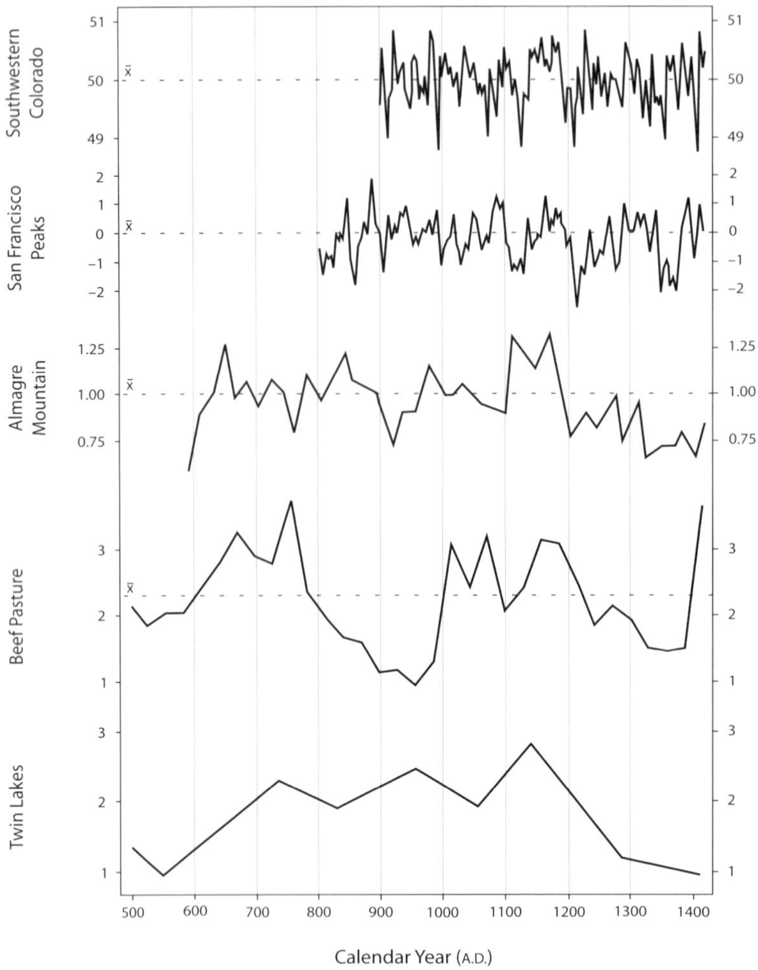

FIGURE 3.6 Comparison of regional temperature reconstructions. Southwestern Colorado: eight-year smooth of Mesa Verde Douglas fir ring-width index series expressed in degrees F, adapted from Dean and Van West (2002:Figure 4.1); San Francisco Peaks: ten-year smooth of bristlecone pine ring-width index expressed in Z-scores, adapted from Salzer (2000b:Figure 3); Almagre Mountain: twenty-year smooth of bristlecone pine ring-width index, adapted from Petersen (1988:Figure 39); Beef Pasture: ponderosa pine / spruce pollen ratio (Wright 2006); Twin Lakes: conifer / non-arboreal pollen ratio, adapted from Petersen (1988:Figure 39).

To cross-check the new paleotemperature reconstruction from Beef Pasture, Figure 3.6 compares the new temperature curve to other regional reconstructions based on both tree-ring and pollen records. Although high-frequency climate change is not emphasized in this chapter, Figure 3.6 presents two such reconstructions—one from southwestern Colorado and another from the San Francisco Peaks in northern Arizona—to offer a visual contrast between climate change measures derived from different proxies and of variable chronological resolution. With regard to low-frequency change, the three measures in Figure 3.6 show notable similarities in the timing and direction of temperature oscillations, especially with below-average conditions at the beginning (A.D. 500s) and end (A.D. 1300s) of the records. The most obvious differences concern the midsection of the Twin Lakes record, where the trajectory of temperature change appears independent of the other records. Although we should expect the Beef Pasture and Twin Lakes records to show the greatest correspondence, since they are only several kilometers apart, the discrepancies between these two records are not as real as the figure suggests. The Twin Lakes record was coarse-grained enough—both with regard to sampling interval and temporal control—that Petersen (1988:100) was reluctant to rely on it and instead referred to a smoothed version of the Almagre Mountain tree-ring record (also presented in Figure 3.6) to infer low-frequency temperature changes between A.D. 500 and A.D. 1970.

The Beef Pasture temperature curve correlates well with the smoothed version of the Almagre Mountain tree-ring record; many of the peaks and dips are matched in both records. Three main discrepancies can be noted among these records. First, the Beef Pasture record expresses a temporal lag in temperature fluctuations when compared to the Almagre record. As should be expected, the forest structures around Beef Pasture did not respond as fast to temperature changes as did annual growth in the bristlecone pine tree-rings in the Almagre Mountain record. Several years, or possibly even decades, of persistent climatic changes apparently were required for the forest structures to reshuffle around Beef Pasture and for this ecological reconfiguration to be evidenced in the pollen record. If the fluctuations in the major temperature trends recorded in the Beef Pasture record actually began 10 to 30 years earlier, this needs to be considered as we infer these effects of this variability.

Another discrepancy is that the Almagre Mountain record exhibits a temperature peak in the A.D. 800s that contrasts with a temperature trough in the Beef Pasture record. This temperature low, however, is visible in the San Francisco Peaks tree-ring record. The poor match between the two tree-ring records may be the result of scalar differences between the smoothing splines applied to each index (10 years for the San Francisco Peaks and 20 years for Almagre Mountain), but it is more likely that these two regions were experiencing different low-frequency climatic patterns at this time. In fact, the climate of the Mesa Verde region should correspond with different records at different times because of complicated jet stream dynamics (Petersen 1988:14–19); this is why the VEP extracted the first principal component of these two records to develop a temperature record unique to the central Mesa Verde region (Kohler et al. 2007; Varien et al. 2007). The persistent temperature low between the late A.D. 700s and the late A.D. 900s expressed in the Beef Pasture record does correlate with the two tree-ring records exactly, but does so with each at various times. The San Francisco Peaks record demonstrates lows in the early and middle A.D. 800s and early A.D. 900s, whereas the Almagre Mountain record demonstrates another major low persisting throughout the first half of the A.D. 900s. While of much higher frequency, the regional southwestern Colorado Douglas fir record also shows several intervals of relatively low temperatures throughout the A.D. 900s. Other than these early inconsistencies, the new low-frequency temperature reconstruction does correlate rather well with the longer-term trends evidenced in the existing tree-ring records.

The third discrepancy between the Beef Pasture and the Almagre Mountain records, as well as the other two tree-ring records, is that the magnitudes of fluctuations are different. This is not surprising, because only one of these is calibrated to temperature; moreover, it can hardly be expected that variability in pollen ratios and tree-ring-width indices will match in magnitude. There are parameters within which tree growth can serve as an adequate proxy for climatic variables, but the more extreme the climatic condition, the less reliable tree-ring width becomes as a proxy for such conditions. Likewise, although tree rings express high-frequency year-to-year fluctuations very accurately, low-frequency trends (on the order of several decades or more) tend to be reflected poorly because of the tree's adaptive responses to climatic change and because of the "segment-length curse" (Cook et al. 1995). Finally, at a population level, trees may select for differing microenvironments within localities over long time scales. Together these factors may obscure, and probably often minimize, the magnitude and duration of climatic variability within tree-ring records (Esper et al. 2002). Changes in the structure of vegetational communities as witnessed in pollen records provide an independent measure of long-term temperature changes that—though not problem-free—is at least beset by a different suite of limitations and assumptions. Obviously it is desirable to find a way to combine the strengths of both proxies to fully understand the interaction of the high- and low-frequency components in climatic variability (Swetnam et al. 1999). This should be an objective of future research into the paleoclimate of southwestern Colorado and elsewhere.

DISCUSSION AND CONCLUSION

Since adequate amounts of winter and summer precipitation are necessary for successful agricultural production in the Greater Southwest, and since lengthy growing seasons are critical to maize farming in the Mesa Verde region (Bellorado 2007; Petersen 1988; Wright 2010), fluctuations in these variables likely affected the ability of prehistoric (and historic) farmers in this region to dry farm successfully. Jeffrey Dean and colleagues (1985) suggest that most archaeologically visible demographic processes would respond more closely to low- rather than high-frequency environmental changes. A comparison of momentary population to the low-frequency paleoclimatic variables in the Mesa Verde region supports this suggestion. Long-term trends in settlement in the central Mesa Verde region—as measured by the peaks and dips in the momentary number of occupied households in Figure 3.5—correspond to periods of expected high and low agricultural potential as measured by the low-frequency climatic variables presented here.

Considering the demographic and paleoclimatic trajectories of the central Mesa Verde region in Figure 3.5, it is interesting to note that population levels increased dramatically shortly after two major downturns in favorable farming climate: the A.D. 800s and A.D. 1200s. Since the central Mesa Verde region appears to contain the most productive agricultural land in the northern San Juan region, these population spikes may represent farmers from other portions of the northern San Juan region moving into the project area to access more reliable and productive farmland. This is supported by evidence that some of the social transformations accompanying population aggregation in both periods were aimed at coping with climate-induced resource scarcity and diminished agricultural productivity. For example, central Mesa Verde farmers reconfigured land-tenure systems in the A.D. 800s (Kohler 1992a), perhaps in an effort to feed more people with a limited supply of farmland or under increasingly cold and xeric conditions. A large proportion of this population, however, migrated out of the region after this climatic deterioration proved to be prolonged and severe. In the late A.D. 1100s and A.D. 1200s, local farmers developed new technologies for water management, such as innovative water storage features and runoff irrigation (Haase 1985; Rohn 1963; Schlanger 1988; Wilshusen et al. 1997), and during the few decades before complete depopulation in the late A.D. 1200s Mesa Verdeans repositioned many of their villages around perennial water sources (Lipe and Ortman 2000; see also Kolm and Smith, Chapter 5 in this book) and in settings that maximized the absorption of solar energy (Salzer 2000a).

The processes of depopulation in the A.D. 800s and A.D. 1200s clearly followed different historical trajectories, yet significant downturns in aspects of climate relevant to farming success preceded both. This suggests that a concern with agricultural productivity and resource abundance—at least to some degree—was intrinsic to both depopulations, and that climatic downturns were persistent problems with which farmers in the Mesa Verde region had to contend. Although each depopulation was unique, both witnessed social, technological, and settlement shifts that attempted to buffer against subsistence and resource shortfalls that were due to both poor climatic conditions and increased demand induced by population aggregation. In the end, such attempts to mitigate the effects of climatic change on agricultural practices, systems of exchange and cooperation, and the overall social fabric of a farming lifeway in the increasingly aggregated villages of the central Mesa Verde region proved ineffective.

The general correspondence of the new low-frequency climatic reconstruction with the local demographic trajectory underscores the interrelatedness of these variables, but this relationship is rather complex, nonlinear, and multifaceted (Kohler et al. 2008). It must have been difficult to support increased population loads reconstructed by the VEP under the climatic deterioration of the thirteenth century. The pressures of aggregation within the refugium of the central Mesa Verde region may also have bolstered the attraction of apparently more socially and economically viable communities at lower elevations south of the central Mesa Verde region. And yet, if the depopulations of the A.D. 800s and A.D. 1200s were both responses to climatic deterioration, then why were the details of these demographic shifts and their longer-term results so different? Although climatic stress was the common denominator in both cases, aspects of each cultural response were unique and involved different social and demographic transitions. The intricacies of the historic relationships between climatic change, resource stress, social upheaval, and migration at various periods in the Mesa Verde region warrants continued investigation.

FOUR

Simulation Model Overview

Timothy A. Kohler

THERE IS NO SUBSTITUTE for a map if you are going someplace you have never been before. Most archaeologists have little or no experience with simulation; in this chapter we provide "maps" to the collection of Swarm code that constitutes the Village simulation model in the form of flowcharts summarizing the main features of the simulation. We also explain how the demographic and household location routines work and how they interact with other parts of the simulation.

Some of the data used by the simulation have long stories behind them, and in Chapters 5 through 7 we discuss how we generated those inputs. In Chapters 9 through 11 we discuss the results we obtain by running the model and make some comparisons between those results and the archaeological record.

But here we are much more interested in the code—that collection of over 17,000 lines (about 450 Kb) of Objective-C that defines what happens in the simulation, when it happens, and how the stream of circumstances and events generated by the simulation is displayed and stored. This may be more information than most readers of this book want to know, but it will be important for anyone who plans to reuse or improve this code. Its story begins during the year between 1992 and 1993. I had the good luck to spend that year on sabbatical at the Santa Fe Institute just as Chris Langton, Dave Hiebeler, and others were beginning to develop a series of libraries (originally called PGas for Process Gas, but later retitled Swarm) that were eventually written in an object-oriented computer language called Objective-C. This language was developed in the early 1980s as a superset of C, was used by NeXT, and today provides the foundation for the Macintosh OS X and iPhone OS. Object-oriented languages provide a useful framework for agent-based modeling because they conceptualize a program as consisting of objects, composed of both behaviors and states, that communicate with each other through messaging. They also offer the possibility of reusable components because of their tidy modularity. To this framework Swarm adds flexible abilities to build collections of objects, or agents, schedule their activities, collect data on them, and display data about them. Over that year I had many opportunities to talk with Langton and his codevelopers about what features would make such software useful to archaeologists, and I developed a basic understanding of how Swarm worked.

The next important step was taken in 1995 when Eric Carr—then a gifted undergraduate at Stanford who was spending the summer at the Santa Fe Institute, and now vice president of Location Technologies at loopt—wrote a quick first draft of Village based on an early prepublic version of Swarm. At that time our agents were concerned only with agricultural land. To represent the farmable lands, we used the paleoproductivity landscapes that Carla Van West had created for the Pueblo II and III periods in her dissertation (published in 1994). For several years this proof-of-principle version of Village was packaged with Swarm as a demonstration application and served as the basis for some preliminary publications (Kohler and Carr 1997; Kohler et al. 1996).

Since then, many others have contributed to the Village code that we are now releasing as v2.8 on www.OpenABM.org. Assisted by a grant from the National Center for Preservation Technology and Training (R-47), Jim Kresl—then a Ph.D. student in archaeology at Washington State University (WSU) and now director of the University of Washington Office of Research Information Services—contributed code and data planes that located water sources of various types on the landscape, and that also allowed agents to use water and consider its distribution in their decisions about where to live. We reported on this version in Kohler, Kresl, Van West, Carr, and Wilshusen (2000).

More recently, Jason Cowan, then an M.A. student in archaeology at WSU and now an archaeologist working in the private sector, made substantial additions to the program. Working closely with Dave Johnson, Cowan wrote the code enabling vegetation other than maize to be "grown" on the landscape and consumed by animals (or, in the case of woody vegetation, as fuel by people). He also wrote the current version of our hunting algorithm, explained in detail in Chapter 8, and crafted a difficult algorithm enabling agents to pick the least costly location when moving, where costs are assessed simultaneously for several variables. I discuss this algorithm at the end of this chapter.

As we were working on this part of the code at WSU, at Wayne State University a Ph.D. student in computer science, Ziad Kobti, was completing his dissertation on exchange in the Village world under the guidance of Robert Reynolds, a computer scientist introduced in Chapter 1. Their code is explained and demonstrated in Chapter 11 of this book. Most recently, Kyle Bocinsky, currently a Ph.D. student in archaeology at WSU and a trainee in IPEM, our Integrative and Graduate Education and Research Traineeship (IGERT) Program in Evolutionary Modeling, has become the chief code troubleshooter, getting it ready for the sweeps reported in this book, and cleaning and documenting earlier code. As this book is being completed, Bocinsky and Denton Cockburn, a Ph.D. student of Kobti, who is now a professor at the University of Windsor, are making important additions to the code base to allow agents to raise turkeys and develop economic specializations. Cockburn has also completed a port of the code to a newer simulation platform, RePast, which is itself a development from the Swarm platform, but written in Java. So a second generation of researchers is now continuing to develop Village, and the results of their contributions will be reported elsewhere.

A YEAR IN THE LIFE

The easiest way to get a basic understanding of what Village does is through a narrative description of the action in any year from the point of view of the agents (households). We will assume a typical year, and not the first year of the simulation. Then we will discuss some specific issues in more detail. Those wanting more information will probably want to inspect or run the code itself.

Overview of Scheduling

The master schedule for the model is built and executed in AgentModelSwarm. As the model executes a particular year, it begins by updating each of the 45,400 cells (200 E-W by 227 N-S) that define the world the agents occupy. For example, the rates governing water flow for cells that have springs, derived as described in Chapter 5, are read in from an external file, as are the rates for potential productivity for maize (see Chapter 6). Before agents begin any actions in a new year, their order in the list in which they are maintained (AgentList) is randomized to minimize possible effects arising simply from order of execution. Then, in random order, each agent executes -step_procure as defined in the next paragraph, and possibly -step_move, also discussed there. Then deer are made to diffuse across cells according to the algorithm described in Chapter 8, dead households (those in which all members have died) are removed, various statistics are written to external files, and AgentModelSwarm advances the year counter by 1 and writes some summary information to the console. Finally, cells update how many agents they have in them. This general schedule repeats 700 times from year A.D. 600 to year A.D. 1299, when the simulation ends.

From the perspective of the household, a year begins in the spring, when -step_procure executes (Figure 4.1). -Step_procure is one of many actions (or "methods" whose names will have a "-" in front of them, as they do when they are defined in the code) defined for each agent in the file Agent.m. This is by far the largest of the many files that are compiled to form the run-time program and contains almost half of the total code, reflecting the fact that the behavior of these households is complicated. Agent.m is our main focus. -Step_procure begins by setting many internal variables to either zero or the correct current value. For example, the agents' year counter is advanced by 1, the number of calories produced that year is reinitialized to zero, and so forth. Figure 4.1 shows the main initial actions in -step_procure and then, after initialization, the main agent actions that take place in the spring. Agents discover whether they may exchange with other agents using -get_economy: this returns the value for the parameter COOP which, if greater than zero, enables one or more of several types of exchange (see Chapter 11). Table 4.1 lists key parameters kept constant, as well as the parameters changed in the sweep of 512 runs reported herein. In this book, any parameter name—e.g., COOP—will be in small capital letters. For balanced reciprocal exchange, there may be a wide range of potential exchange partners, and an agent's trading partners can be influenced by the cultural algorithm (see Chapter 11). Then spring begins.

Spring

Each season begins by having the household consume enough maize from their storage for three months. The exact amount needed depends on the number of family members; whether they are adults or children; and, if they are adults, whether they are men or women. How much they work over the course of the season also affects how much maize they consume.

Planting takes place in the spring. Households can choose to plant more or less maize by increments of "plots" (1-acre fields). Each 4-ha cell is allowed to contain up to nine of these, leaving some space in each cell for houses or other uses. We require households to have their plots either in the cell they are living in or in one of the eight neighboring cells. That means that we do not allow a field-house strategy, where households could have fields far from their homes. Households do not literally take up space in a cell, so that having many households does not reduce the number of potential plots. However, we cap the number of households that may occupy a cell to a value set in Village.h (our parameter file) for HOUSE_LIMIT. Here we have used 200, but in fact cells never begin to approach that number of households because of the requirement that they have their fields in the home cell or the eight cells surrounding it (its Moore neighborhood).

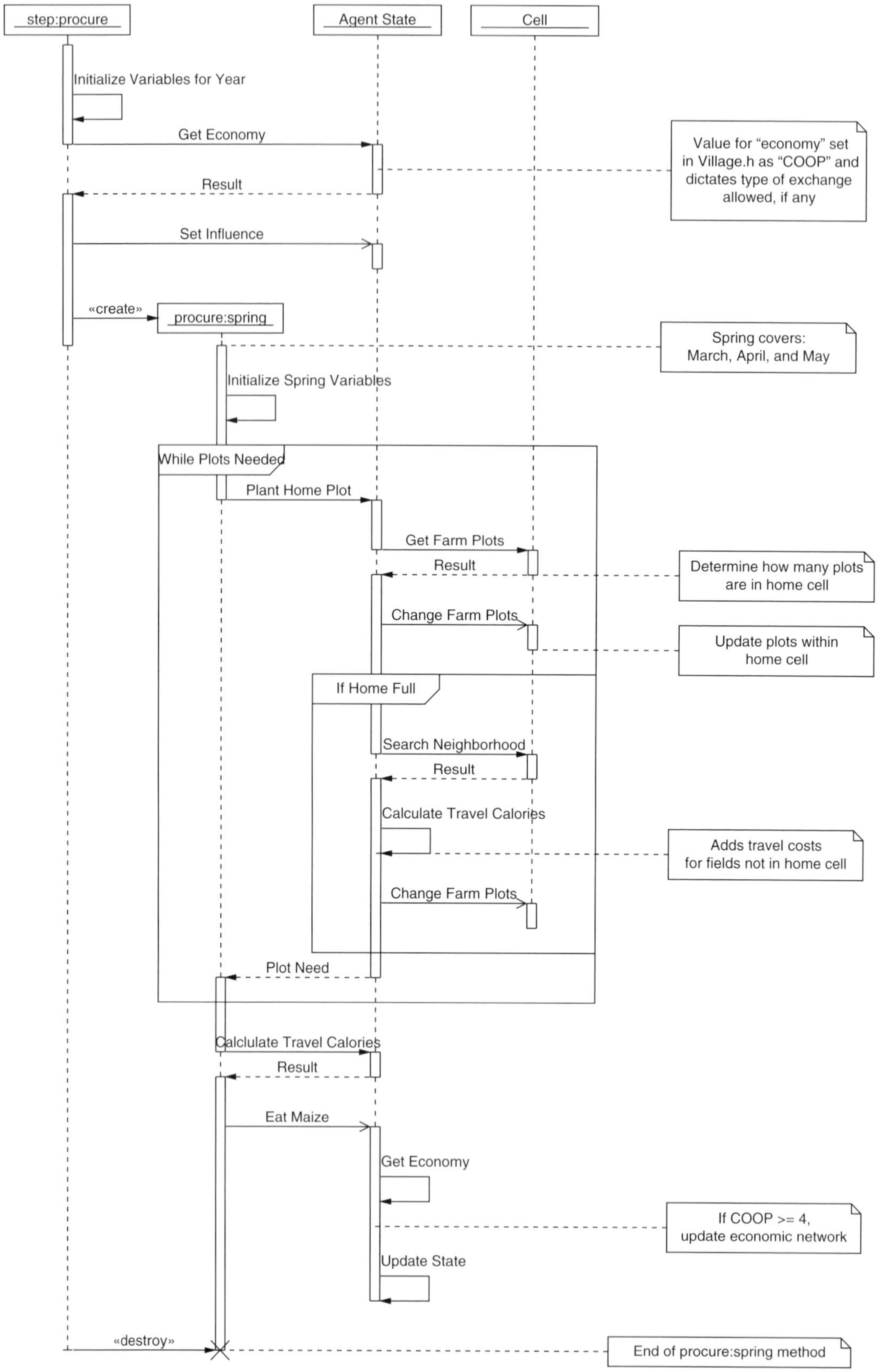

FIGURE 4.1 Sequence diagram for agent activities at start of year and in spring.

TABLE 4.1

Key Parameters in Simulation

Values in bold are the alternative settings explored in the sweep in this book.

Parameter Name	Value(s)	Comment
SOIL_DEGRADE	**1, 2**	1 = mild degradation, to 70% of potential; 2 = more severe degradation, to 40% of potential. See also Chapter 6.
COOP	**0, 4**	Governs exchange behaviors of households: 0 = disable all exchange between households; 1 = requests; 2 = gifting; 3 = both 1 and 2; 4 = balanced reciprocity (and 3); 5 = hubnet (no exchange); 6 = hubnet with regional exchange. See also Chapter 11.
PROTEIN_NEED	**10, 25**	Target meat consumption in g / person / day. See also Chapters 8, 9.
PROTEIN_PENALTY	**0, 1,** 2	0 = no penalty for protein deficiencies; 1 = removal of STATE_GOOD bonus if protein needs not met (reversion to rates in life table); 2 = kills household after it has been protein deficient for DEATH_YEARS. See also Chapters 8, 9.
HUNT_SRADIUS	**20, 40**	Radius for hunting (in cells; 20 cells = 4 km). See also Chapters 8, 9.
HARVEST_ADJUST_FACTOR	**0.75, 0.95**	Acts as a divisor to maize harvest in Agent.m; hence, the .75 setting boosts production by 33% over the baseline; the .95 setting boosts production by about 5% over the baseline. See also Chapter 6.
NEED_MEAT	**0, 1**	0 = agent will go to cell that minimizes costs for obtaining all resources, even if hunting is unproductive within HUNT_SRADIUS; 1 = agent is constrained to move to cells from which hunting is possible within HUNT_SRADIUS, even if that is not the least costly alternative. See also Chapters 7, 8, 9.
STATE_GOOD	**0.1, 0.2**	When agent state is good (2), increments birthrate by 10% and 20%, respectively, and decrements death rates by 10% and 20%, respectively, relative to those provided in life table.
H_2O_TYPE	**1, 3**	Allows agents to get water from any source of that type or higher. 1 = intermittent stream; 2 = perennial stream; 3 = rivers; 4 = springs.
DEATH_YEAR	n	How many years an agent is allowed to be protein deficient before household perishes, when PROTEIN_PENALTY = 2.
TRAVEL_SPEED	15	Speed at which individuals walk, in cells / hour; 15 = 3 km / hour.
MOVE_RAD	20	Agent search radius in cells when considering relocation.
BUD_OFF	MOVE_RAD/2	Maximum distance in cells from relevant parent when new household is formed by marriage.

(continued)

TABLE 4.1 *(continued)*

Parameter Name	Value(s)	Comment
BASE_CAL_MAN	170,820	Base caloric requirements for men, 1,872 cal / day * 91.25 days.
BASE_CAL_WOM	142,350	Base caloric requirements for women, 1,560 cal / day * 91.25 days.
BASE_CAL_KID	91,250	Base caloric requirements for child, 1,000 cal / day * 91.25 days.
WORK_CAL_MAN	240	Caloric increment per hour work, men.
WORK_CAL_WOM	200	Caloric increment per hour work, women.
WORK_CAL_KID	92	Caloric increment per hour work, child.
MAIZE_PER	0.7	Proportion of calories assumed to come from maize.
WOOD_NEED	1,130	Kg / person / year.
WATER_NEED	3,650	Liters / person / year (10 liters / day).
HH_SIZE	10	Maximum individuals in household
MAX_REMARRIAGE_COUNT	2	Maximum number of remarriages within a household.
HOUSE_LIMIT	200	Maximum households / cell.
PLOTS	9	Maximum plots (corn fields) of .4 ha (about 1 ac) / cell.
CARRY_CAPACITY	20	Amount able to be carried in one trip, in kg.
INITIAL_AGENT_COUNT	200	Number of households seeded randomly onto landscape at initialization (A.D. 600).
MAX_COOP_RADIUS_GRN	30	Distance (in cells) allowed for generalized reciprocal exchange.
MAX_COOP_RADIUS_BRN	40	Distance (in cells) allowed for balanced reciprocal exchange.

How households calculate the number of plots to plant is somewhat complicated. The number is initially set to 2 + (number of children / 2), but later the calculation takes into account both the number of workers in a household (all occupants seven years or older are workers) and the "state" of the household. (Basically, households in a good state, or condition, may be able to shed a plot, whereas households in poor condition due to insufficient current calories, or with a projected shortfall, will try to add plots.) In the simulations reported here, we allowed households to plant one more plot than their number of workers; this is controlled through the parameter AD_PLOTS. Households needing plots first try to add them in their home cell, and then, if that is not possible, in a cell in the first tier of neighbors. If that too is impossible because all the available plots in the accessible cells are in use, or because the available plots are not projected to supply enough maize, an agent may try to move. Households accrue costs (in calories) for planting, weeding, and harvesting each plot, and those costs are slightly higher, because of travel time, if plots are outside the home cell. We follow C. Daryll Forde (1931:390ff) in charging 17 eight-hour days per plot for preparing and planting fields. Adding a new plot costs additional calories for clearing. Households that still have living

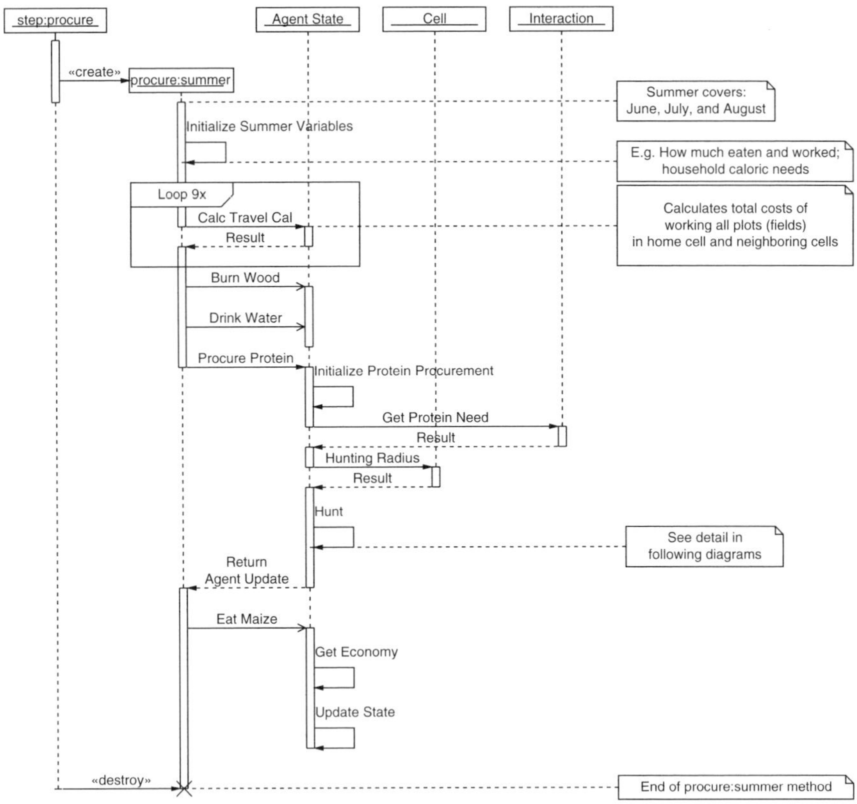

FIGURE 4.2. Sequence diagram for agent activities in summer.

members at the end of the season evaluate and update their states and proceed to summer. We will discuss how households update their states after presenting the round of seasonal activities.

Summer

In addition to continuing to work in their fields, as a computational convenience we imagine that in June, July, and August agents accrue all the costs for collecting firewood, carrying water from the nearest available source, and hunting for the entire year (Figure 4.2). The most complicated of these is hunting, described in Chapter 8.

For fuels we make a distinction between standing crop and deadwood. It takes longer to collect fuel from a living forest than to pick up deadwood; we assume that it takes one person an hour to collect one load—20 kg—of deadwood, but four hours to cut and gather the same amount of wood from growing shrubs or trees. Agents will satisfy their fuel needs with deadwood if they can collect enough within a radius of FW_SEARCH_DISTANCE cells (10 cells, or 2 km, in these simulations); otherwise, they begin to chop down the living forest, starting in their home cell and proceeding outward. In these simulations we assume that households need 1,130 kg of wood / person / year (see rationale in Johnson et al. 2005). Households keep track of the caloric cost of gathering wood and bringing it home. Wood use by each agent depletes wood in each cell, as tracked by the cells. However, cells also regrow woody fuels at a rate determined by the current-year precipitation proxy, and by characteristics of the soil, as described in Chapter 7. We assume that households clearing plots for fields make no use of any woody fuels they may remove, and that fields in use do not produce any woody

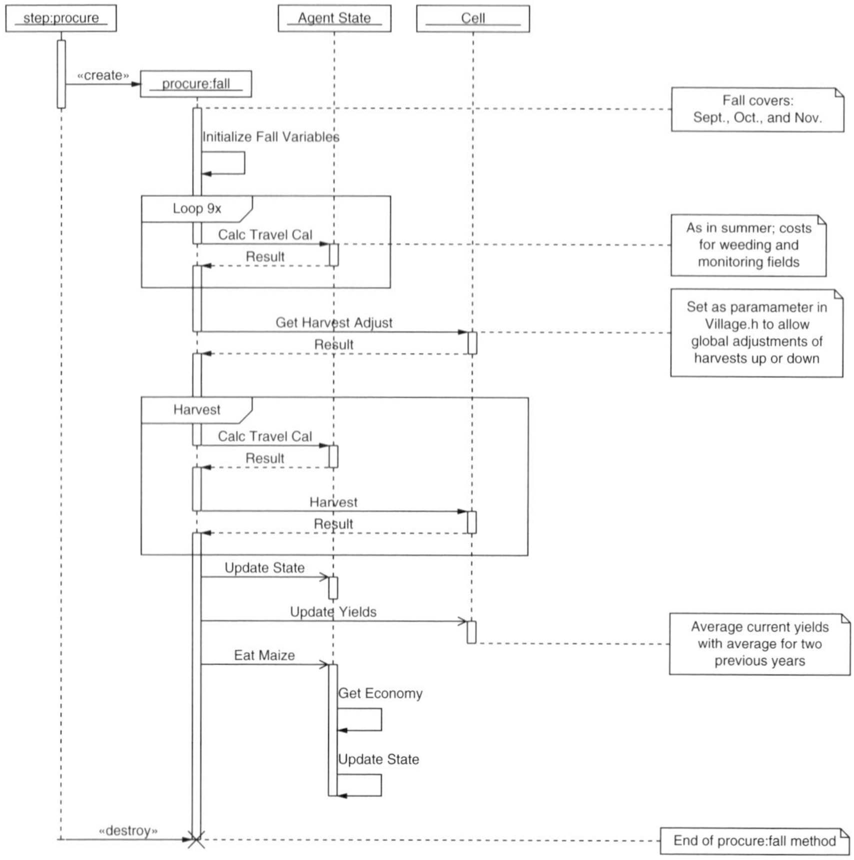

FIGURE 4.3 Sequence diagram for agent activities in fall.

fuels. On the other hand, we allow households to use all woody biomass in other areas for fuels, which presumably included large trunks that would have been impractical to process with stone axes.

In the present simulation we assume that each household needs 3,650 liters / person / year, or 10 liters / day, of potable water. In these simulations we explored the differences in settlement resulting from allowing agents to use water from a fairly restricted set of supplies (when H_2O_TYPE = 3, this includes our only permanent stream, the Dolores River and 267 springs within the project area for which data were located by Ken Kolm and Schaun Smith) versus a more ubiquitous set of sources (when H_2O_TYPE = 1, this also includes intermittent streams such as the McElmo and the stream in Yellow Jacket Canyon). We model all streams and some springs (those tapping into the lowest aquifer) as inexhaustible, but for springs developing from higher aquifers we allow households to draw only as much total water in any year as the paleohydrological model estimates to be available (see Chapter 5).

Fall

Routine farming activities continue for the first 15 days of September, and then fields are harvested. We assume that each 25 kg of maize harvested costs two hours of work, plus travel and cartage costs. The harvest size is used to update the average calories produced over the last three years, and this in turn is used by the agents in updating their states (Figure 4.3).

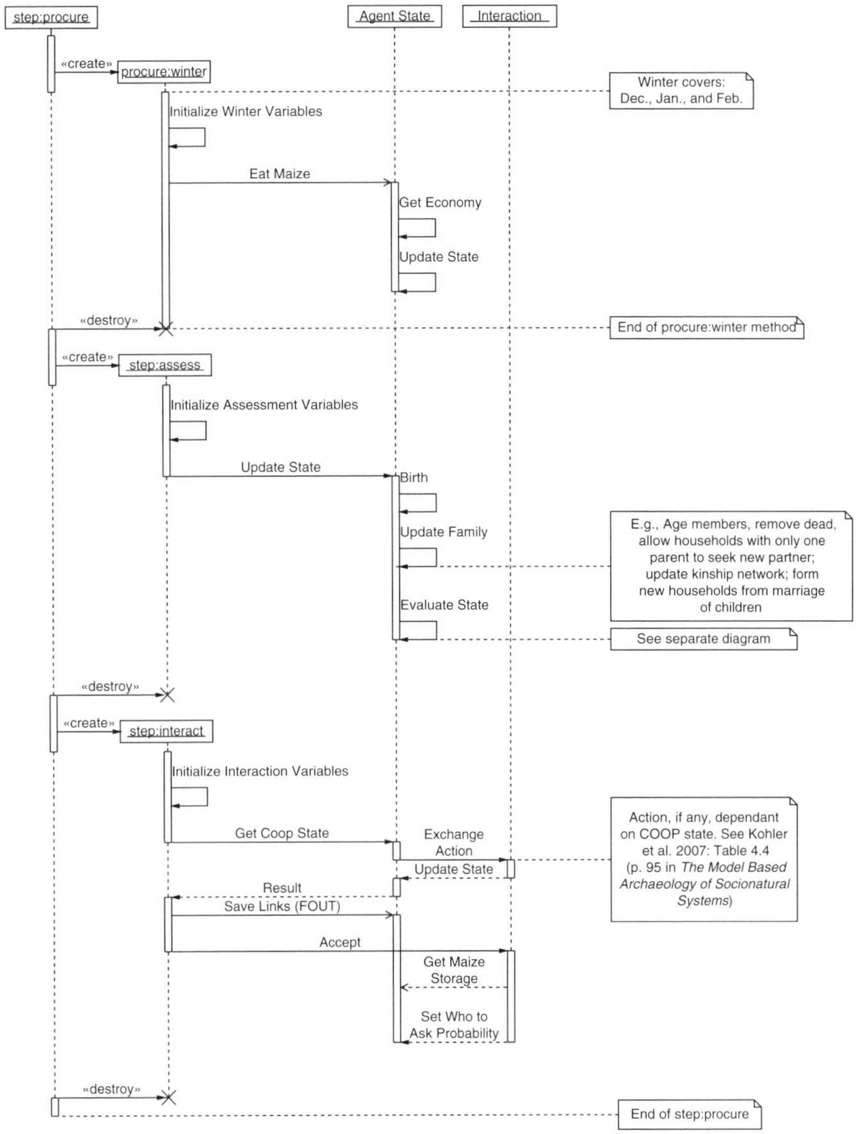

FIGURE 4.4 Sequence diagram for agent activities in winter and at end of year.

Winter

Winter is a slow season, but because it is also the end of the agent's year, in Figure 4.4 we also present many activities done once a year in -step_assess (births, deaths, updating family size) and -step_interact, in which all exchange activities (if COOP > 0) are done for the entire year. This is a computational convenience, similar to putting all the hunting, firewood, and water collection into the summer. We defer a discussion of exchange until Chapter 11.

DEMOGRAPHY

Although households are the main agents in the model, unlike another agent-based model (ABM) examining settlement practices in the U.S. Southwest (Axtell et al. 2002; Dean et al. 2000) Village households contain individuals who age, marry, give birth, and die according to a realistic timetable. Following advice from demographer Alan Swedlund (personal communication), we adopted the age-specific mortality and fertility schedule given by Kenneth Weiss's (1973:156) 27.5–55.0 model table (reproduced in Kohler, Kresl, Van West, Carr, and Wilshusen 2000:Table 4). Births are handled by –birth, which first checks to see that both a mother and a father are present in the household. If so, and if the mother is between 14 and 50 years old and has had fewer than eight children, her age is checked against the model table to generate her probability of giving birth. This probability may be affected by the household's state (if STATE_GOOD ≠ 0) and also by protein deficits (if PROTEIN_PENALTY > 0). Any newborns are added to the household. Deaths are handled by –mortality. As with births, individuals are evaluated against the life table by their ages to determine their probability of dying. Should an individual slip past the probabilities given in the life table, he or she dies at age 75. Probability of death in any interval may be affected by the household's state (if STATE_GOOD ≠ 0) and also by protein deficits (if PROTEIN_PENALTY > 0). Both –birth and –mortality are called for each household annually by –step_assess, but individuals are evaluated for births (in the case of women) or death only once in the intervals defined by the life table. –Step_assess itself is called by –step_procure_conclude at the end of each year. Should both parents in a household with children die, the children die too unless at least one of them is older than seven, in which case the oldest is promoted (-promoteChild) to be a parent. Households in which all members have died are removed from the simulation. Parents in households where the other parent has died may be able to remarry up to a limit set by MAX_REMARRIAGE_COUNT (here, two).

Marriage

Beginning at age 16, regardless of sex, with a probability of .5 in each year a child will search the list of people eligible to be married. If he or she finds someone of the opposite sex, they get married and form a new household, and they are removed from the eligibility list if they were on there. Individuals seeking a mate for the first time who are unsuccessful are added to the eligibility list. This list is general to the entire project area, which we assume is potentially a single mating pool. New households have kinship links to the households of both the woman and the man. If the person who successfully searched the eligibility list was female, then the new household is formed either in the cell where her parents live, or if that is full, in a cell within radius BUD_OFF (set to half the normal move radius) from there. If, on the other hand, the person who successfully searched the eligible list was male, the new household would form near his parents. Thus we model a bilateral, ambilocal scheme. It would be easy to change this if specific information seemed to warrant it, and Kohler and Reed (2011) have recently speculated that at least some VEP Pueblo I populations may have been patrilineal and virilocal.

How a Household's "State" Is Calculated

At the end of each year, households evaluate how well they are doing in -evalState. State may take on a value of zero, 1, or 2 (higher numbers denote better conditions). State is also reset (to 1) when a household moves. State as set here should not be confused with a variable called coop_state that is set in the method –updateState; coop_state determines exchange actions explained in Chapter 11.

The default value for state is 2, but this gets lowered to 1 if the current maize in storage is less than that needed for the current year plus that expected to be needed for next year, or if the maize just harvested is less than next year's anticipated needs (taking into account the ages and sexes of the family members). If state is 1, households will attempt to add another plot if they have enough labor to do so. If this is not possible—either because plots are unavailable locally or because of labor limitations—

households try to move (-moveHouse). Other than in –evalState and -moveHouse, a household's state may be changed in –updateFamily. Here, a household's state will get set to zero when its last person dies, or if both parents are dead and there are no children older than seven years of age who can be promoted to parent; either case results in death (which affects a household, whereas –mortality affects individuals). When a household dies, all of its plots are returned to the commons, any links to other households are removed, any members who may have been on the eligible list are removed, and the world updates its population accordingly.

How State May Affect Demography

When STATE_GOOD = 0, the probability of births and deaths is calculated from the life table, as already explained. This results in an approximately stable population. If STATE_GOOD > 0, a household's state will affect the probability of births and deaths in that household. In the simulations reported here, we explored the results of two different values (.1 and .2) for STATE_GOOD. These have no effect when a household is in state 1, but when a household is in state 2, these values get added to 1 and the result acts as a multiplier on the probability of births in that household. In other words, when STATE_GOOD = .1, households in state 2 will experience 10 percent higher natality, on average. Individuals in that run and household also experience a similarly decreased probability of death (here, 10 percent) in each interval. When STATE_GOOD = .2 these effects operate at 20 percent instead of 10 percent.

PROTEIN_PENALTY likewise has potential effects on population when it has a nonzero value. In the runs reported here, we explored the results of setting PROTEIN_PENALTY to zero and to 1. (Even when PROTEIN_PENALTY = 0, households pay some price for not being able to meet their PROTEIN_NEED, since they may expend fruitless energy hunting within their HUNT_SRADIUS.) When set to 1, the bonus a household receives for being in state 2 is erased if that household fails to meet its PROTEIN_NEED: probabilities of natality and mortality revert to those in the life table. When set to 2 (not explored here), the potential penalty is more extreme: a household that has not met its PROTEIN_NEED for DEATH_YEARS in a row will perish along with all its members.

In Plate 4.1 (see color insert in this book) we examine how these two parameters affect the populations generated in our model by coloring differently each of the four possible combinations of their two values across all our runs, allowing the other parameters to vary as set forth in Appendix A. As one would expect, runs with STATE_GOOD = .2 and PROTEIN_PENALTY = 0 (black) develop the largest populations, whereas runs with STATE_GOOD = .1 and PROTEIN_PENALTY = 1 (green) develop the smallest populations. In general, the penalty to population of STATE_GOOD = .1 is more severe than that of PROTEIN_PENALTY = 1.

Are the population growth rates we generate realistic? Run 132, which grew more households than any other run (more than 2,700 households in the second cycle—Plate 4.1) attained a specific growth rate of .0075 from A.D. 608 (460 people in 115 households) to A.D. 991 (8,013 people in 2,442 households); from A.D. 608 to A.D. 1133 (when there are 9,059 people in 2,782 households), the specific growth rate is .0057.[1] The higher of these two rates corresponds to a doubling time of 92 years. Andrew Chamberlain (2009:281) cites instantaneous growth rates of .005 to .02 recently observed in small groups of Ache, Agta, Asmat, Hadza, and Yanomama peoples over short periods. Demographer Jean-Pierre Bocquet-Appel (2008:52) considers .1 to .2 percent (presumably a specific growth rate of about .001 to .002, which is considerably slower than our highest rates) to be "a growth rate typical of pre-industrial populations."

---

1. Computed using individuals rather than households, as $r = (\ln Nt_2 - \ln Nt_1) / (t_2 - t_1)$ (Odum 1971:181). Note that the household sizes generated by Village are rather small. The overall household size for all 512 runs is 3.3 people. The average household has had .7 remarriages. The average age of the mother across all simulated households in all runs is 27.6 years; for fathers, it is 29.2 years.

However, the rate of increase in Run 132 is very close to that actually achieved during the growth phase of the first population cycle, although some of that growth possibly came from immigration. Overall, we do not think that even the highest growth rates we model are out of the question for Neolithic pioneers spreading into fertile new territory.

We will return to Plate 4.1 in the final chapter to discuss why the modeled populations and those estimated from survey are different, especially after the first population cycle.

HOW HOUSEHOLDS DECIDE WHERE TO LIVE

Households move automatically after they are placed in a random location at A.D. 600. They also move when their rate of maize production, or amount of maize storage, will not meet their anticipated needs and the easier fix of simply adding another local plot is impossible. Finally, households formed from marriage must "decide" where to live.

Village contains several different algorithms for household movement. At the top level is a switch set in the parameter file for SOCIAL: when this is ≠ 0, agents execute –searchSocialNeighborhood, which locates exchange partners (assuming COOP ≠ 0) and searches in their neighborhoods for a promising place to move. In the runs described here, however, SOCIAL = 0, so relocation was not biased in this way. A second switch is set by GREEDY. When GREEDY = 1, households use a method called –searchNeighborhoodGreedy, which in turn asks –evalCell about the resources available in the cells within the permissible radius. Then –evalCell returns information about the presence and type of water, as well as the amount of potential maize production, in each cell that is available (cells that are already fully planted are unavailable). Households move to the best available cell within the search radius on the basis of these two criteria.

In the runs reported here, however, we set GREEDY = 2, which requires households to move to that cell within their permissible radius which allows them to meet their caloric needs *and* minimizes the costs of obtaining their required amounts of water, fuelwood, and protein from game. This employs –searchNeighborhoodAll, which first orders the cells within the search radius according to their current potential maize productivity, making a list of the top S_ARRAY (here set to 100) of these. Then the method computes the costs for obtaining water (-searchdrinkWater), wood (-searchburnWood), and protein (-searchHunt) from each of those cells. (Note that this entails a search around each of these cells, not just a query to the specific cell.) Finally, by subtracting the costs of farming and obtaining water, fuel, and protein from the expected returns to farming, the method computes the net energy return for living in each of those cells. Ordinarily, then, the household will move to the top-ranked cell.

If, however, NEED_MEAT = 1, requiring households to move to a cell from which hunting is possible within HUNT_SRADIUS, then –search NeighborhoodAll will step through the ranked list of length S_ARRAY until it finds the top-ranked cell from which hunting is possible. If there is one, the household moves to that. If there is not, which may happen if the landscape is completely locally depleted of game, then the household simply moves to the highest-ranked cell, as it would have if NEED_MEAT had been zero.

In Chapter 10 we present analyses showing how the parameter settings we explored affected household locations in the simulation. We also describe the degree of similarity between the real and simulated household placements in each period.

EFFECT OF VARIABILITY IN RANDOM-NUMBER SEEDS ON SIMULATED NUMBER OF HOUSEHOLDS

Before we conclude, one more issue needs to be briefly discussed. It is common practice for simulators to run thousands of replications using various random-number seeds to explore

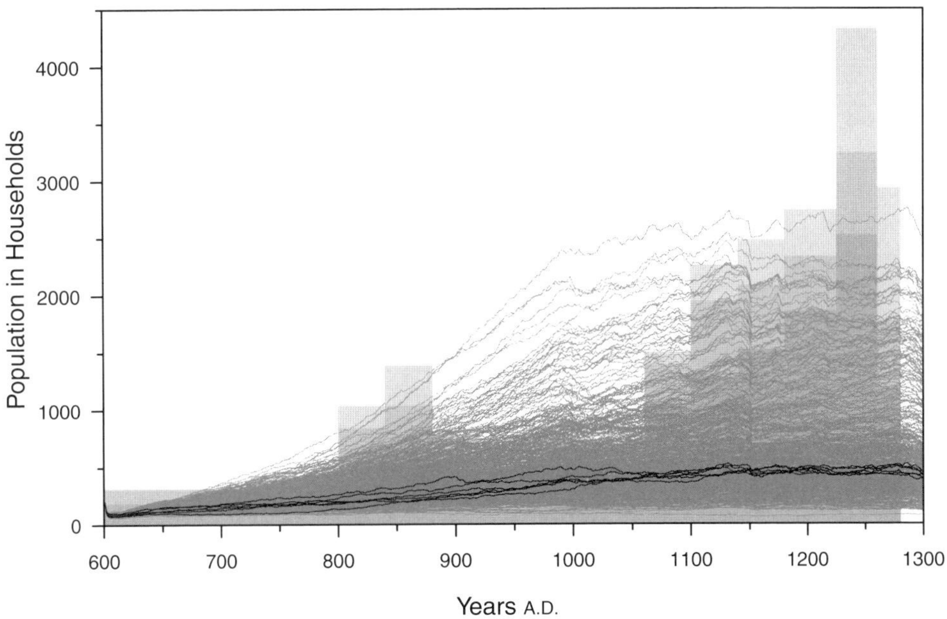

FIGURE 4.5 Like Plate 4.1, except that the population histories for all the 512 runs except for Run 142 are shown in light gray; the population histories from Run 142 and five additional runs with the same parameters but different random seeds are shown in black.

the effects of this variability on their results. We did not do that here, primarily because these simulations consume lots of computer time; depending on how many agents the runs generate, and also on whether exchange is implemented, runs can take from an hour or two to a week or two on fast processors. All 512 runs in the sweep reported here used the same random-number seed, so the differences among the runs are due solely to the differences in the parameters in each run.

Nevertheless, we did some exploration to see how much variability different random-number seeds introduce into the simulation, since we use random-number generators in many places. Within appropriate limits, they control such things as the numbers and age distributions of individuals in households at the beginning of the simulation and where households are placed at the beginning of each run; with whom households exchange; who gets married, and to whom, in a specific year; whether widows (or widowers) get an opportunity to remarry; when children are born; when individuals die; and how many animals a hunter encounters in a cell with game, as well as how many if any of those get acquired.

Figure 4.5 shows the variability through time in the number of agents for six runs with identical parameters (those we used for Run 142 in the sweep; see Appendix A) but different random-number seeds. We chose a run with exchange (COOP = 4), since many exchange behaviors have a random component; and we chose Run 142 because it produced the median number of agents at the peak of the average simulated populations, in year A.D. 1131. The variability in population size among these six runs is not inconsiderable (s and cv at A.D. 980 = 64 and 17.4; at A.D. 1200, 22 and 4.7) but is relatively small compared to the variability stemming from different parameter values. There is little or no evidence of path dependence in population size within these runs; runs that begin with relatively few households may become top ranked, only to be overtaken in population size by yet another. This is what we would expect of a random process.

One final observation can be made on these runs: when productive conditions are generous, as they were, for example, in the late A.D. 900s, the random processes produce relatively high variability in population histories. This variability gets pruned, however, when conditions are less propitious, as they were around A.D. 1200. We likewise expect a great deal of random (or neutral) variability in site location when productive conditions accommodate many placements, followed by pruning when conditions deteriorate sufficiently for households in inferior locations either to become less viable or to be forced to move lest they perish. We will report evidence for this effect in Chapter 10.

CONCLUSIONS

In this chapter I have briefly reviewed how Village works. This body of code has been built by several people over many years. For the most part the coders have not been computer scientists. Indeed, we have lived the experience in support of which Chris Langton started the Swarm project: to provide scientists who are not computer scientists with tools to build dynamic, agent-based simulations. As this book was being written we moved this project to RePast, a process that has revealed many minor inconsistencies that we have left in place in the Swarm code, since it documents the state of the simulation as reported herein. We hope that scrutiny by many eyes will allow this code to be improved and adapted for use in other settings. Ultimately models such as these should become tools in a comparative archaeology.

We now turn to a discussion of how we built a dynamic model for the distribution and flow rates of various water sources in the VEP area, and how that model was incorporated into Village.

FIVE

Modeling Paleohydrological System Structure and Function

Kenneth E. Kolm and Schaun M. Smith

OVER THE SEVEN CENTURIES of prehispanic occupation, the settlement and subsistence strategies of the Pueblo inhabitants of the central Mesa Verde region (Figure 1.1) underwent several well-documented transformations, including two cycles of population aggregation and disaggregation (Huckleberry and Billman 1998; Ortman et al. 2000; Varien et al. 2007); changes in farming and irrigation strategies (Schlanger 1988; Van West 1994; Wilshusen et al. 1997); movement of large population centers to canyon heads, which offered protection and springs as a source of water (Varien et al. 1996); and finally, the complete depopulation of the region, accompanied by the intensification of warfare (Chapter 13, this book; Kuckelman et al. 2000; Lipe 1995). However, significant questions remain about the interactions of this ancient farming society with the region's natural resources—specifically, its supply of drinking water from springs.

This book reports how our interdisciplinary team worked to link the social and environmental processes affecting this prehispanic civilization into a unique, unified framework, and to integrate ecological and archaeological data with computer simulation (Kohler et al. 2007; Ortman et al. 2007; Varien et al. 2007). This research allows us to address the impact of environmental factors, such as temperature and precipitation variability, and their combined effect on agricultural productivity; wood use and deforestation; and hunting and the depletion of key animal species on human population dynamics.

A necessary part of this effort was a quantitative assessment of the paleohydrological fabric of the Pueblo landscape to determine whether this society reached the limits of its drinking-water supply, ultimately resulting in depopulation of the region. We adopted a Hydrologic Systems Analysis (HSA) approach (ASTM International 2002; Kolm 1996; Langer and Kolm 2001) to develop an answer. HSA incorporates analytical and quantitative modeling tools as well as scientific and engineering principles to conceptualize and mathematically assess both modern and prehistoric springs. We used these methods to compare the spatial distribution of water resources to the distribution of Pueblo settlements and their population estimates, thereby providing the first quantitative assessment of the availability of domestic water to the Pueblo inhabitants of the region. To the extent that we had preconceptions, we began this study expecting to find that drought depleted spring discharge and contributed, at least, to the emigration at the end of the thirteenth century.

This analysis and modeling required six innovative and challenging steps to examine the drought hypothesis. We had to (1) determine how the modern-day hydrological system works in the region; (2) build a three-dimensional representation of this hydrogeologic system; (3) quantify the hydrological processes; (4) simulate drinking-water supplies from A.D. 600 to A.D. 1300; (5) estimate the number of ancient inhabitants living in the region; and, finally, (6) compare the estimated population of Pueblo peoples with the drinking-water supplies. Each step required a distinct set of approaches, methods, modeling strategies, and hypotheses that are detailed elsewhere (Kohler et al. 2007; Ortman et al. 2007; Smith 2008; Varien et al. 2007). Summaries of the initial five steps and their solutions, including methods and critical figures, are presented as a foundation for discussion of the final step later in this chapter.

CONTEMPORARY HYDROLOGICAL SYSTEM

The first challenge was to determine how the modern-day hydrological system works by specifying how the surface and groundwater systems supply drinking water.

This arid region lacks significant surface water bodies, and groundwater information is meager with virtually non-existent aquifer engineering data records associated with water wells. Therefore, HSA (ASTM International 2002; Kolm 1996; Langer and Kolm 2001) was used (1) to conceptualize and characterize the surface and subsurface of the natural landscape by integrating climate, topography, vegetation type and distribution, surface water, groundwater, geology, and geomorphology into a holistic, three-dimensional, conceptual site model; and (2) to determine the functionality of the hydrogeologic systems at multiple scales. Ultimately, the goals were to develop an accurate, estimated water budget and to establish appropriate surrogate parameter values for paleohydrological modeling.

Results of the analysis quickly established that surface sources of drinking water were not significant in this region because of the arid climate and probable poor water quality. Groundwater resources did appear to be the naturally sustainable source for drinking water (Kohler et al. 2007; Smith 2008).

THREE-DIMENSIONAL REPRESENTATION OF THE MODERN AND PALEOHYDROGEOLOGIC SYSTEM STRUCTURE

The next challenge was to visualize how the groundwater system works by constructing a three-dimensional block model (Plate 5.1; see color insert in this book) of the region's hydrogeologic framework, incorporating data obtained from the HSA. Hydrogeologic framework and hydrological-system characterization data were combined to evaluate formation dimensionality and geologic continuity, as well as to interpret the confined and unconfined aquifer conditions and hydrological response parameters of groundwater recharge and discharge. Because direct engineering measurements such as aquifer tests and groundwater well elevations (heads) were not available, surrogate values for parameters such as hydraulic conductivity, specific yield, saturated thickness, and water-table levels (heads and spring levels) were estimated on the basis of the properties of the hydrogeologic materials and on properties estimated or determined in studies conducted by other investigators on similar hydrogeologic units in nearby areas (Smith 2008).

The block model illustrates the spatial dimensionality and continuity of the aeolian and Dakota-Burro Canyon (D) aquifers, the main hydrogeologic units in the Village Ecodynamics Project (VEP) area (Plate 5.1). The model clearly demonstrates that the D aquifer is topographically and hydrogeologically incised by deep canyons. Because of this predominant dissection, the D aquifer is discontinuous; it is a local, community-scale aquifer in the southern and western parts of the study area. However, to the north and east, the aquifer is subregional in scale, remaining quite continuous. Most of the dynamic groundwater discharge zones observed in the field (springs and seeps) are associated with the D aquifer (Plate 5.2). It is likely that ancestral Pueblo people used these springs as their primary sources of drinking water. This inference is supported strongly by archaeological evidence that shows the association of Pueblo settlements and the locations of springs mapped by the Bureau of Land Management (Lipe and Ortman 2000; Ortman et al. 2000; Varien 1999a).

The subregional and local- or community-scale hydrological systems are conceptualized for the combined aeolian and D aquifer systems in Plate 5.3. Extensive archaeological evidence indicates that Pueblo settlements were located on mesa tops or Dakota sandstone canyon rims in topographic and hydrological settings where groundwater recharge and discharge occurs in both the aeolian and D aquifers. Groundwater recharge by infiltration of precipitation occurs in these mesa top environments, where most Pueblo farming occurred. Groundwater flow paths from the mesa tops to the springs can be observed along the canyon rims.

QUANTIFICATION OF MODERN-DAY HYDROLOGICAL SYSTEM FUNCTION

The next step was to quantify the hydrological processes observed in the current landscape. To accomplish this we built a mathematical, deterministic computer model based on the conceptual hydrogeologic block model; estimated hydrological model inputs using surrogate parameters determined through use of HSA; created steady-state spring flow simulations confirming water-supply distribution and amounts; and employed robust statistical methods for constraining the uncertainty of selected surrogate values and model results (Kohler et al. 2007; Smith 2008).

Conceptual or estimated parameters (or both) of the three-dimensional hydrological-system model include the distribution of groundwater recharge and discharge, flow path vectors, type and distribution of boundary conditions, potentiometric surfaces, hydraulic conductivity, formation saturation thickness, and specific yield values. These parameters were discretized as matrices in the mathematical model. A uniform grid cell size (200 × 200 m) was chosen to match the spatial resolution of the Village simulation. The block-centered, finite-difference MODFLOW code (McDonald and Harbaugh 1988) was used for simulating steady-state and transient conditions.

The mathematical models simulate discharge from 61 known springs (as head-dependent flux drains) within the specified nodes (zones) where the springs occur. Each of the drain outputs simulates the availability of drinking water for each spring, seep, and groundwater discharge zone, and this water availability is calculated by the model as a unique mass balance quantity. A robust sensitivity and statistical error analysis was performed on each of the iterative model simulations for both the steady-state and transient conditions.

The potentiometric surface (or pressure head; Plate 5.4a) calculated by the steady-state model simulation illustrates the three hydrological systems that operated at different scales in the combined aeolian and D aquifer. In the southeastern portion of the VEP study area, there is a regional hydrological system, with groundwater flow from the House Creek fault to McElmo Creek near Cortez, resulting in a predominant southeastern discharge at McElmo Creek and underflow that continues as a regional system into the deep San Juan basin of New Mexico. In the northwestern portion of the study area there is a subregional hydrological system that stretches from the Dolores River divide to the deeply incised canyons in the southwest portions of the area. The modeled groundwater flow here is from northeast to southwest, with discharge at springs, many of which are adjacent to concentrations of prehistoric settlement sites, such as Lowry Pueblo and Yellow Jacket Pueblo (Plate 5.4a). Last, there are smaller, community-scale hydrological groundwater flow systems, as illustrated in the Sand Canyon Pueblo area (Plate 5.4b). Here, modeled groundwater flow is in an inward, radial pattern from the nearby mesa top to the model drain nodes, which are the foci for discharge that represents the drinking-water supply.

SIMULATION OF DRINKING-WATER SUPPLIES FROM A.D. 600 TO A.D. 1300

The fourth step was to simulate past drinking-water supplies. Prehistoric climate patterns and precipitation rates were translated into available groundwater recharge amounts, and computer simulations of long-term, transient spring discharge were generated to reveal the distribution and volumes of the drinking-water supply. Parameter surrogates for recharge in the simulations included the use of precipitation data developed from tree-ring proxies as reported by Jeffrey Dean and Carla Van West (2002) and Van West (1994). Van West's precipitation data employed the Mesa Verde Douglas fir tree-ring width index series (also used in Chapter 6 of this book), calibrated for retrodiction against Palmer Drought Severity Indices (Palmer 1965) generated using historical climate data.

These annual precipitation estimates were plotted from A.D. 600 to A.D. 1300 (Plate 5.5). To aid in the visualization of trends, various moving average curves are shown in this figure instead of the high-frequency annual data. The unsmoothed precipitation data suggested that during the 700-year study period, the combined aeolian and D aquifer system might not exhibit many long-term droughts, as most dry periods (below-average annual precipitation) were of short duration (one to three years) followed by wetter years of above-average precipitation. The wet years might be able to recharge the aquifer continuously over time, mitigating significant effects of short-term droughts on the potable water supply.

A series of transient MODFLOW computer models simulated temporally variable paleohydrological spring discharge at five-year intervals, for every 100 years, to examine this possibility and to give us estimates of spring flows for use in the agent-based model. Resultant data sets of spring discharge and mass balance output were extracted from the model output files and generated for each of the 140 five-year intervals. A total of 8,540 discrete paleohydrological spring discharge values were produced, and the calculated head surface for each 100-year interval from A.D. 600 to A.D. 1300 was generated. Discharge rates simulating the available drinking-water supply from the seventh to the thirteenth centuries were then ready for comparison with the known settlement dynamics, and also were input into the VEP agent-based computer model.

ESTIMATING REGIONAL PREHISPANIC POPULATIONS

Before this comparison could be made, however, the fifth research step had to be completed: estimating the population of inhabitants who lived in the VEP area. Estimates of the numbers of ancestral Pueblo people living in the study area were made by Scott Ortman et al. (2007) and Mark Varien et al. (2007), beginning with a database of 8,948 recorded archaeological sites in the VEP study area (Chapter 2, this book). Functional classifications of archaeological sites allowed Ortman and Varien to identify Pueblo-era habitation sites and use these to estimate population size between the seventh and thirteenth centuries. Specific features—including accumulation of cooking pottery, number of pit structure depressions, trash middens, estimated size and number of rooms, and presence of public architecture such as a great kiva or plaza—were used to determine the function and size of habitation sites and evaluate when these sites were occupied (Varien 1999a; Varien et al. 2007). The average momentary population of the settlements was initially calculated on the basis of a population probability density analysis for reconstruction of occupational histories (Ortman et al. 2007), which ultimately defines the total number of households at each location during an estimated period. From the tally of total households, the average momentary population was calculated by considering dwelling and pueblo use and life at small sites and larger community centers (Kohler et al. 2007; Smith 2008; Varien 1999; Varien and Mills 1997; Varien and Ortman 2005; Varien et al. 2007).

COMPARISON OF PAST DRINKING-WATER SUPPLIES AGAINST NUMBER OF INHABITANTS

The final research step to determine the role of paleohydrology in the VEP-area depopulation puzzle was to compare the drinking-water supplies in the past with the number of inhabitants estimated in the two archaeological population cycles. This was accomplished by graphically, spatially, and statistically comparing prehistoric population frequencies, based on archaeological data, with predicted mathematical amounts of spring discharge from the combined aeolian and D aquifer system that was available as drinking water throughout the 700-year period. To facilitate this comparison, we examined three interdependent village-spring couples. These consisted of communities comprising villages and farmsteads that relied on closely spaced groups of springs for their drinking water (Lipe and Ortman 2000; Ortman et al. 2000). These communities were selected for their varying positions on the natural landscape and for the differing scales of the corresponding hydrological systems that sustained the drinking-water supply. The communities we selected were centered in three large villages, each of which was associated with a spring: Lowry, Yellow Jacket, and Sand Canyon Pueblos.

A 2-km community catchment around these villages and springs was used to connect populations to these three spring locations (Adler 1994; Adler and Varien 1994; Bradfield 1971; Kohler et al. 1986; Stone 1992). The size of these communities was calculated by identifying the total number of households in each catchment area (Ortman, personal communication 2006) and assuming that a Pueblo household consisted of six individuals (Lightfoot 1994). We assumed that people required 10 liters / person / day (Gleick 2000; Kohler et al. 2007). Finally, the dynamics of the paleohydrological system and prehistoric settlement populations were linked by graphically and statistically comparing modeled spring discharge rates, estimated momentary populations, and water required to sustain the collective population within the catchment (Plate 5.6a and 5.6b; Table 5.1). For ease of description, the populations calculated for the 2-km radii are referred to by their community-center names: Lowry, Yellow Jacket, and Sand Canyon Pueblos. We will show the results graphically only for Yellow Jacket and Sand Canyon Pueblos, where we plot the community populations against a Fourier expansion of the regional paleoprecipitation estimates rather than their local spring flow rates.

The groundwater system of Lowry Pueblo fluctuated widely in magnitude, and spring discharge declined from an initial maximum in the period between A.D. 600 and A.D. 725 to a low between A.D. 1020 and A.D. 1060, when spring flow was virtually nonexistent (Table 5.1). However, the discharge volume rebounded tremendously to a second maximum between A.D. 1100 and A.D. 1140 before showing a steady decline in the waning years of settlement (A.D. 1260–1280). By comparison, the population, which is characterized by a bimodal cycle, actually increased during the second, drought-induced decline of the drinking-water supply (Table 5.1): the Lowry population increased dramatically during the second population

TABLE 5.1

Collective Maximum Populations, Consumptive Water Use, and Spring Discharge Data of 2-km Community Catchments for Lowry, Sand Canyon, and Yellow Jacket Settlement Areas

For ease of description, each 2-km catchment is referred to as a pueblo in this chapter.

Archaeological Phase	Years A.D.	Average Spring Discharge (m^3 / d)	Discharge Total Liters over Period	Required Total Liters over Period Based on Population (10L / d / person)	Max Population Based on Number of Households (1HH = 6 persons)
			Sand Canyon Pueblo		
PH6	600–725	11.99	547,171,500	186,150,000	408
PH7	725–800	12.31	336,881,883	29,565,000	108
PH8	800–840	12.52	182,730,356	16,644,000	114
PH9	840–880	12.43	181,443,933	21,900,000	150
PH10	880–920	12.30	179,520,210	4,380,000	30
PH11	920–980	12.20	267,257,492	9,198,000	42
PH12	980–1020	11.42	166,777,098	18,396,000	126
PH13	1020–1060	11.64	169,931,022	17,520,000	120
PH14	1060–1100	12.61	184,075,178	26,280,000	180
PH15	1100–1140	13.22	192,961,711	13,140,000	90
PH16	1140–1180	11.56	168,769,511	21,900,000	150
PH17	1180–1225	12.29	201,832,043	26,608,500	162
PH18	1225–1260	12.33	157,539,703	23,761,500	186
PH19	1260–1280	12.58	91,831,080	42,048,000	576
PH20	1280–1300	12.06	88,032,160	–	–
			Yellow Jacket Pueblo		
PH6	600–725	31.75	1,448,396,650	35,587,500	78
PH7	725–800	33.30	911,643,961	3,285,000	12
PH8	800–840	40.24	587,520,222	876,000	6
PH9	840–880	40.74	594,852,667	876,000	6
PH10	880–920	37.68	550,126,378	–	–
PH11	920–980	34.37	752,758,592	7,884,000	36
PH12	980–1020	27.39	399,900,489	876,000	6
PH13	1020–1060	22.29	325,498,889	876,000	6
PH14	1060–1100	28.99	423,231,289	81,468,000	558
PH15	1100–1140	44.33	647,226,111	95,484,000	654
PH16	1140–1180	37.74	550,961,822	106,872,000	732
PH17	1180–1225	39.56	649,750,005	132,057,000	804
PH18	1225–1260	36.78	469,920,391	104,244,000	816
PH19	1260–1280	37.76	275,683,040	50,370,000	690
PH20	1280–1300	35.35	258,084,200	–	–

(continued)

TABLE 5.1 *(continued)*

Archaeological Phase	Years A.D.	Average Spring Discharge (m^3 / d)	Discharge Total Liters over Period	Required Total Liters over Period Based on Population (10L / d / person)	Max Population Based on Number of Households (1HH = 6 persons)
			Lowry Pueblo		
PH6	600–725	32.49	1,482,426,513	71,175,000	156
PH7	725–800	40.77	1,116,075,328	6,570,000	24
PH8	800–840	67.38	983,764,222	9,636,000	66
PH9	840–880	63.99	934,301,044	28,032,000	192
PH10	880–920	47.51	693,621,504	876,000	6
PH11	920–980	39.87	873,201,854	15,768,000	72
PH12	980–1020	38.50	562,172,446	13,140,000	90
PH13	1020–1060	2.73	39,882,171	22,776,000	156
PH14	1060–1100	36.28	529,754,511	21,900,000	150
PH15	1100–1140	90.71	1,324,348,156	42,048,000	288
PH16	1140–1180	42.67	622,912,244	35,916,000	246
PH17	1180–1225	52.41	860,873,670	39,420,000	240
PH18	1225–1260	48.63	621,308,931	50,589,000	396
PH19	1260–1280	58.28	425,436,700	20,586,000	282
PH20	1280–1300	44.18	322,499,400	–	–

cycle, with people residing in the 2-km community radius around the water source while spring-based drinking water was already declining. This suggests that protection of water supplies for these community-based populations may have been the driving force for the development of villages immediately adjacent to the springs during the thirteenth century. Ultimately, the area was depopulated *before* a critical decline of spring volume. The water supplies remained adequate to the catchment population during the depopulation between A.D. 1260 and A.D. 1280, so the decline in spring flow is not a satisfying explanation for the depopulation and migration. Except for the A.D. 1020 and A.D. 1060 period, the spring discharge volume was adequate to supply the populations in and near this pueblo (Table 5.1).

The groundwater system serving Yellow Jacket Pueblo did not fluctuate widely in magnitude. Spring discharge volume declined from an initial maximum between A.D. 600 and A.D. 725 to a minimum volume from A.D. 1020 through A.D. 1060, although it was an adequate source of drinking water for the entire time (Plate 5.6a). The spring volume increased after A.D. 1060 before showing a decline in the waning years of settlement (A.D. 1260–1280). Comparing these flows to the population estimates for Yellow Jacket Pueblo, we see that population actually increased during the second, drought-caused decline of the drinking-water supply between A.D. 1180 and A.D. 1260 (Plate 5.6a and Table 5.1). Ultimately the community was depopulated *before* a critical decline of spring volume. Like the 2-km community centered in Lowry Pueblo, the Yellow Jacket Pueblo community increased dramatically in size, and the population increasingly concentrated closer to the water source during the second population cycle, when spring-based drinking water was already in decline. Again, protection of water supplies may have been the driving force for this population aggregation, but not for the later depopulation. By our analysis, during all periods the spring discharge volumes were adequate to supply the populations at and immediately around this pueblo (Table 5.1).

The groundwater system of Sand Canyon Pueblo, like that of Yellow Jacket, was not widely fluctuating in magnitude, and spring volume declined from an initial maximum to become a consistent source of drinking water (Plate 5.6b and Table 5.1). However, the spring shows a steady decline in the waning years of settlement (A.D. 1260–1280). By comparison, the population, which is characterized by a bimodal distribution, actually increased during the second, drought-caused decline of the drinking-water supply (Plate 5.6b and Table 5.1), and ultimately the area was abandoned *before* the critical decline of the spring volume. The first population cycle appears to be sensitive to the amount of drinking water available from the Sand Canyon spring, and the people of Sand Canyon most likely functioned as a community in the early stages of settlement, since they were constrained by their unique landscape (a small, topographically constrained mesa top). As with the Lowry and Yellow Jacket communities, the Sand Canyon population increased dramatically up to approximately A.D. 1260. The population concentrated in the area around the canyon rim near the water source during the final periods of occupation in the second population cycle, while spring-based drinking water was already in decline. This also suggests that protection of water supplies may have been an important force for aggregation at that location, but since water supplies remained adequate for the size of the community it was not a reason for its ultimate depopulation. In fact, throughout all times, the spring discharge volume was adequate to supply all populations at this community (Table 5.1).

DISTANCES TRAVELED FOR POTABLE WATER THROUGH TIME

When the data on spring location and discharge, quantified using the modeling results described in this chapter, and surface water type and distribution depicted on the U.S. Geological Survey maps for this area are combined with the Village settlement simulation, we are in a position to make a contribution unique in the archaeological literature by answering questions like "How far, on average, did people have to travel for their drinking water in this semiarid area? Does this change through time as springs and populations waxed and waned?"

One of the parameters that was varied in the sweep reported here was H_2O_TYPE (Table 4.1), which specifies the minimum source type that could be used by households. Intermittent streams were coded as 1, perennial streams such as most of Yellow Jacket Canyon as 2, permanent streams (only the Dolores River) as 3, and all of our mapped springs as 4 (see also Chapter 4, this book). We explored the consequences of letting households use any of these sources (H_2O_TYPE = 1) as well as the consequences of imposing a much more restricted view of what could have been used (H_2O_TYPE = 3, which allows households to use only the Dolores River and the springs). In Figure 5.1 we graph how far households needed to travel, on average, in the 256 runs where they were required to use a type 3 source (*light gray*) versus the distances agents traveled for water if we allowed them to use a type 1 source (*dark gray*). Note that there is no overlap between these two distributions, even at the beginning of the simulation. For all households accessing type 1 sources, the mean distance from water was only .22 km, which does not change perceptibly through time. However, the mean distance to water for households constrained to type 3 sources was 1.79 km, and this distance tends to increase as the average simulated population size (calcu-

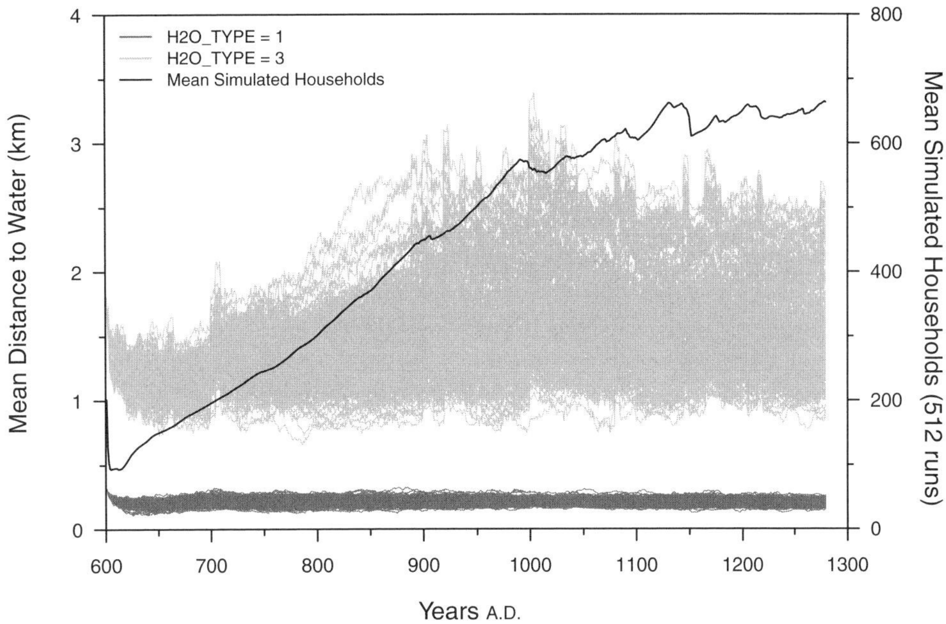

FIGURE 5.1 An average household's distance to water in the 256 runs where they were required to use a type 3 or higher source (light gray), compared to the distance if they were able to use any water source (dark gray).

lated across all 512 runs—see the black line in Figure 5.1) increases.

Unfortunately, our mean simulated populations tend to greatly underestimate the populations in the second cycle, and even those at the peak of the first cycle, for reasons to be discussed later in this book. (On the other hand, they slightly overestimate populations during the trough between A.D. 920 and A.D. 1060.) We can correct for this, however, using information from the simulation on the relationship between population size and distance to water. That is, we can estimate the maximum distance that people might have had to go for water at the population peak between A.D. 1225 and A.D. 1260 (for which we use our middle estimate of 3,234 households, which we consider to be the most reliable; see Chapter 2) if the relationships on population size and distance to water *among the simulations* holds at higher population levels.

In Figure 5.2 we regress mean distance to water (from the 256 runs where H_2O_TYPE = 3) on the simulated populations from those runs at A.D. 1004 (when we see the maximum range in that distance). If we extrapolate from the strong linear relationship ($r=.82$; $p<.0001$) between these two variables to the distance to water expected when the population peaked from A.D. 1225 to 1260, then assuming those populations were using only type 3 sources they may have been going as far as 5.25 km for their water. This is probably too high—both because the simple linear regression does not account for a hint of downward concavity in the relationship in Figure 5.1, and also because some or all type 2 sources were probably usable—but it can serve as an upper bound on the one-way distances that households at the population peak might have had to travel to get water. Although this seems extremely far to us, ethnographies recount numerous examples of people in arid areas traveling considerably farther for water on a regular basis. Davis and Yawson (1945), for example, describe the Seri of Sonora as traveling 6 to 24 km (4 to 15 miles) for fresh water.

However, it is worth mentioning that in the simulation, runs with households limited to type 3 water sources develop considerably lower

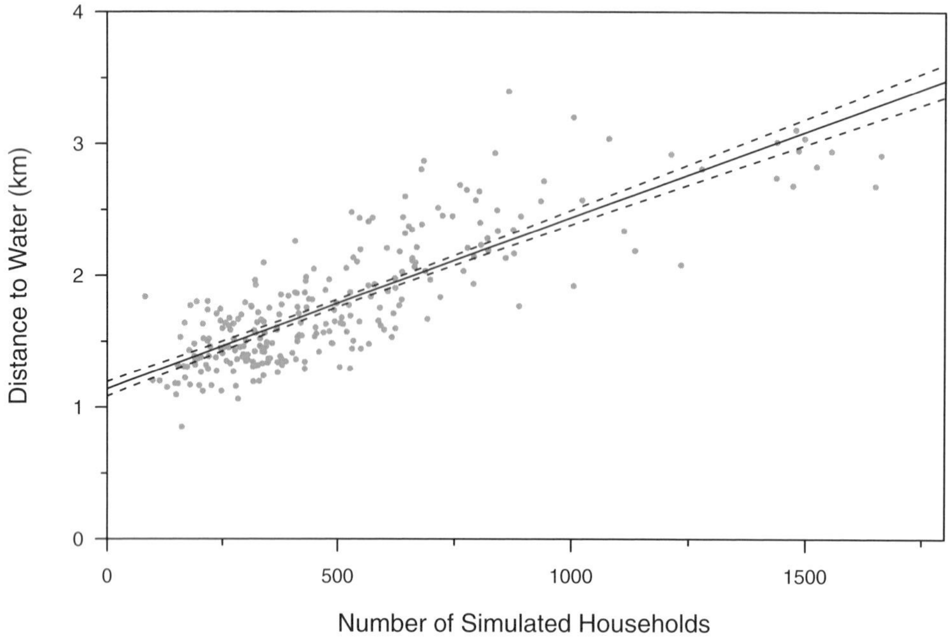

FIGURE 5.2 Regression of mean distance to water (from the 256 runs where H2O_TYPE = 3) on the simulated populations at A.D. 1004 from those runs ($r^2 = .67$; $p < .0001$).

populations than runs in which households are allowed to use type 1 sources: the extra energy spent takes its toll. At A.D. 1131, when the mean number of simulated households peaks, the mean number of households in the 256 runs constrained to type 3 sources is 596, whereas the mean when households can use any mapped water source is 728 households.

CONCLUSIONS IN THE CONTEXT OF SUSTAINABILITY

Our research in the VEP study area is the first time, to our knowledge, that quantitative modeling and analysis of the temporally dynamic populations and resources seen in the archaeological record have incorporated the structure and function of the paleohydrological system. Our analyses determined the amount and distribution of drinking-water supplies from spring discharge during the rise and fall of Pueblo populations in the study area.

Several basic assumptions were made in this study: (1) the inhabitants relied predominantly on the combined aeolian and D aquifer groundwater system as their primary source of drinking water; (2) annual precipitation is correctly retrodicted using tree-ring data; (3) annual precipitation data have been correctly used for estimating groundwater recharge rates throughout the period between A.D. 600 and A.D. 1300; (4) the conceptual and mathematical groundwater flow system models accurately characterize the modern and paleohydrological systems in the VEP study area; (5) archaeologists have accurately determined the levels of population that existed on the landscape over the seven centuries studied; and (6) Pueblo people utilized an average of 10 liters of water / day / person.

Given these assumptions, this study showed that the ancestral Pueblo populations stayed well within the sustainable levels, or "carrying capacity," of the drinking-water supply. Therefore, the possibility—which we among others took seriously—that drinking-water resources (spring discharge) were reduced by drought to the point that the reduction significantly contributed to the region's depopulation is not supported by our results. However, it appears to be possible, and even likely, that reduction in drinking water from spring discharge may have been a factor contributing to the aggregation of population around water sources in the thirteenth century. Episodic trends of declining groundwater supplies are noted, but there was drinking water to spare from springs in the thirteenth century. Other social or environmental factors, and not shortages of potable water, presumably caused the region to be completely depopulated by the late thirteenth century.

When our data on spring location and discharge—quantified using the modeling results described in this chapter, as well as surface water type and distribution depicted on the USGS maps for this area—are combined with the Village settlement simulation described in Chapter 4, we can predict that the inhabitants traveled a distance of less than 5.25 km for water supplies at the thirteenth-century population peak. The settlement simulations using the hydrological data developed here also suggest that if our "real" households were limited to using the water available in the permanent river and the springs modeled here, the accumulated energy penalty would have resulted in smaller populations than if they were able to use any of the mapped water sources. One service of the simulation—which can represent counterfactual situations—is to demonstrate that the energy expended to obtain water plausibly affects the number of households supportable on a landscape. Another service is to directly estimate the size of that effect.

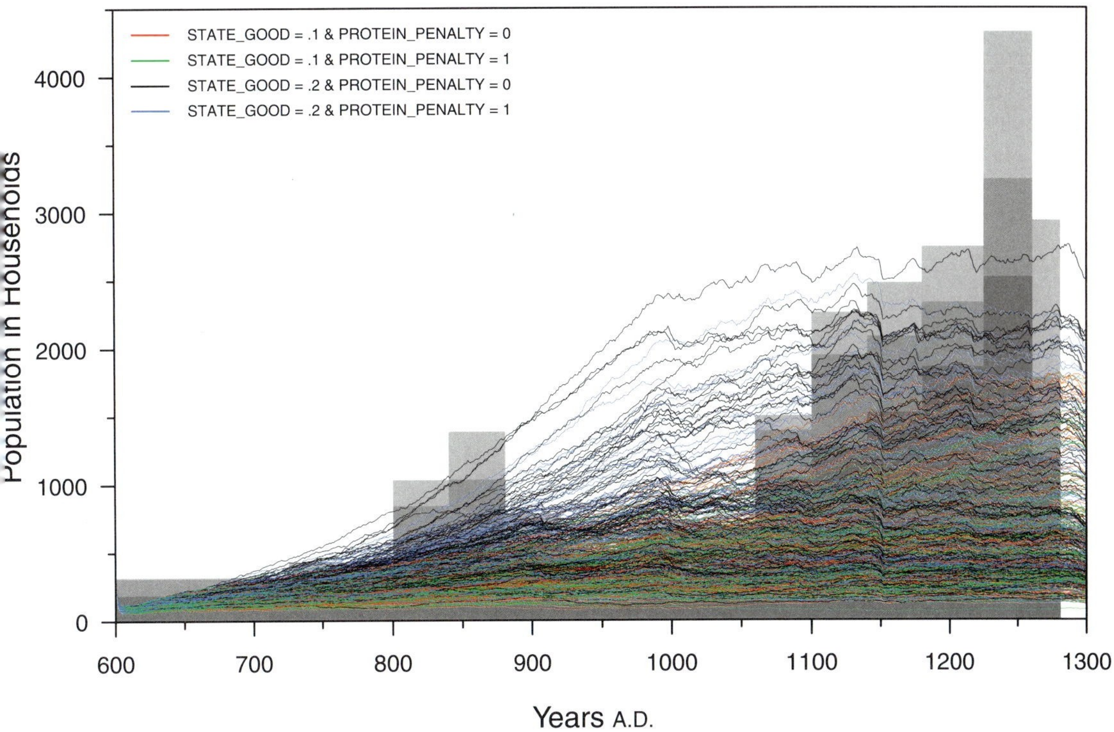

PLATE 4.1 Real and simulated populations in the VEP area. Simulated household counts are shown for every year from A.D. 600–1300 for each of the 512 runs whose parameters are given in Appendix A.

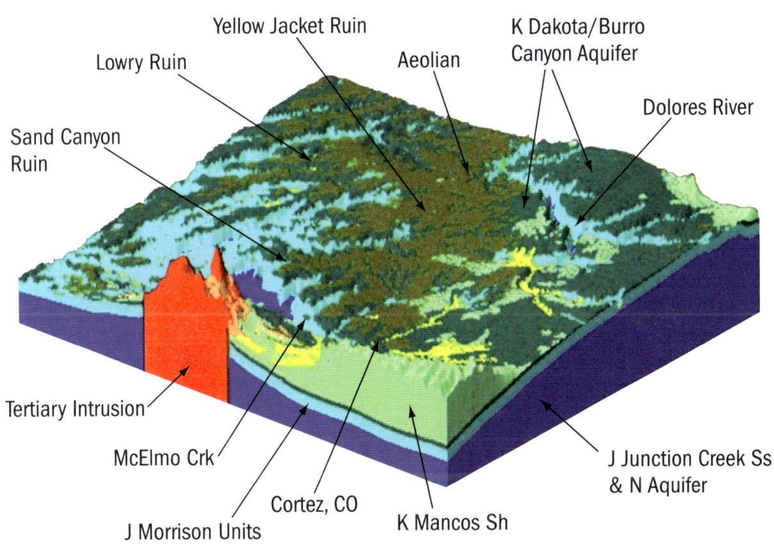

PLATE 5.1 Conceptual three-dimensional hydrologic block model of the study area at 200 m resolution. Note intense dissection of the Quaternary (Q) aeolian aquifer and the Cretaceous (K) D aquifer by the Dolores River canyon in the northeast and by numerous northeast-southwest–trending canyons in the southwest. The regional Jurassic (J) Junction Creek sandstone and N aquifer are only rarely exposed in the Dolores River, Yellow Jacket Creek, and McElmo Creek canyons.

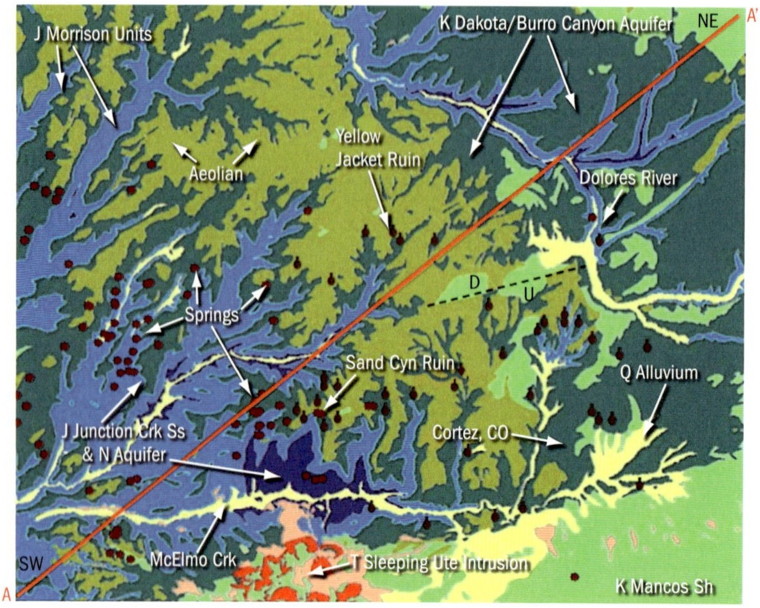

PLATE 5.2 Distribution of hydrogeologic units, selected springs (red dots), and Yellow Jacket and Sand Canyon pueblos in the study area. The aeolian (dark yellow) and D aquifer (gray-blue) hydrologic systems are both subregional and local in scale. These units are spatially discontinuous due to incising canyons that create natural hydrogeologic boundaries and divides. The J Junction Creek sandstone and N aquifer (dark blue) and J Morrison aquifer and confining unit (light blue) hydrologic systems are regional in scale, with exposures mainly in canyon bottoms. The House Creek fault is indicated in the east-central portion with downthrown (D) and upthrown (U) blocks.

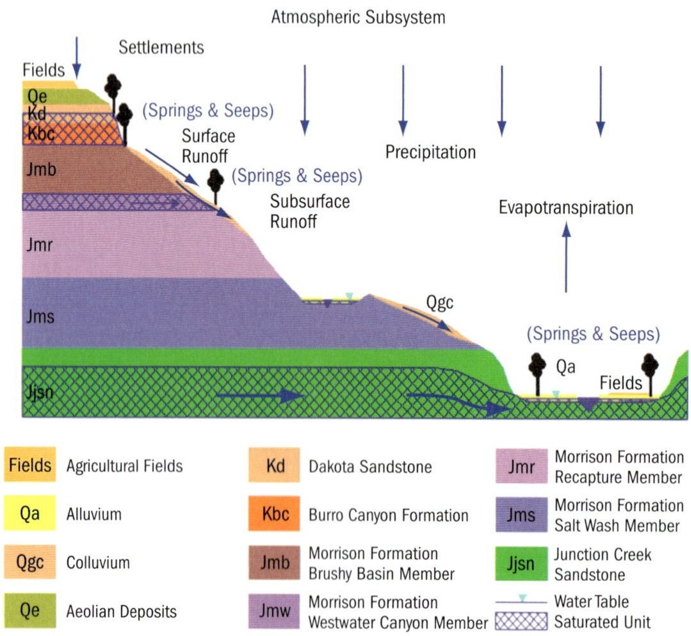

PLATE 5.3 General conceptual model of the regional, subregional, and local hydrologic and hydrogeologic systems in the McElmo Creek canyon environment. Location of spring discharge (phreatophytes), canyon rim, settlements, and fields are depicted. Subregional and local systems at mesa top and canyon rim, with recharge as infiltration of precipitation and discharge at the cliff face of the aeolian and D aquifers. Also shown are local alluvial systems recharged by surface runoff and subsurface colluvial interflow. Regional systems illustrated as J Morrison Formation (aquifers and confining units) spring discharge and Junction Creek sandstone and N aquifer groundwater underflow and discharge into gaining reach of McElmo Creek.

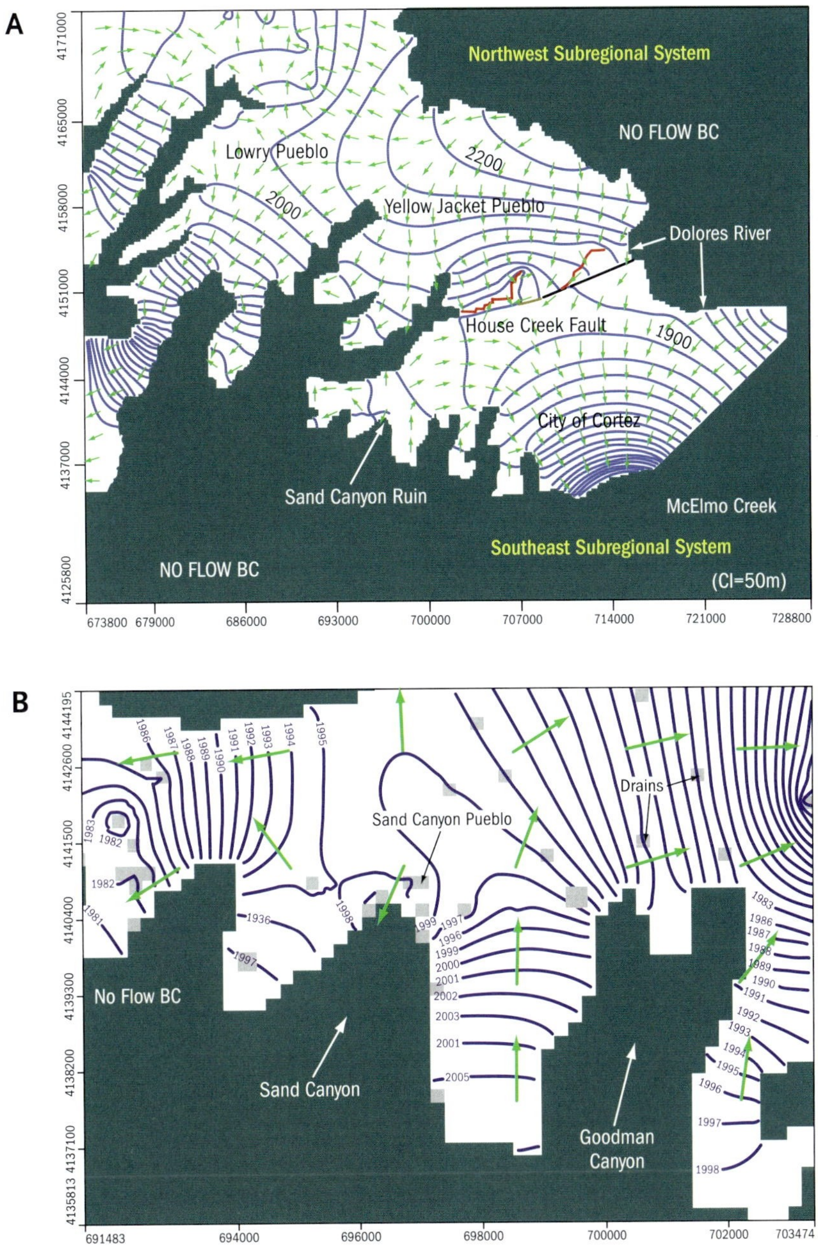

PLATE 5.4 Mathematical models of potentiometric surface of the combined aeolian and D aquifer groundwater flow system, generated using MODFLOW. Arrows show direction of groundwater flow. A) Regional-scale model of D aquifer; B) Local or community scale.

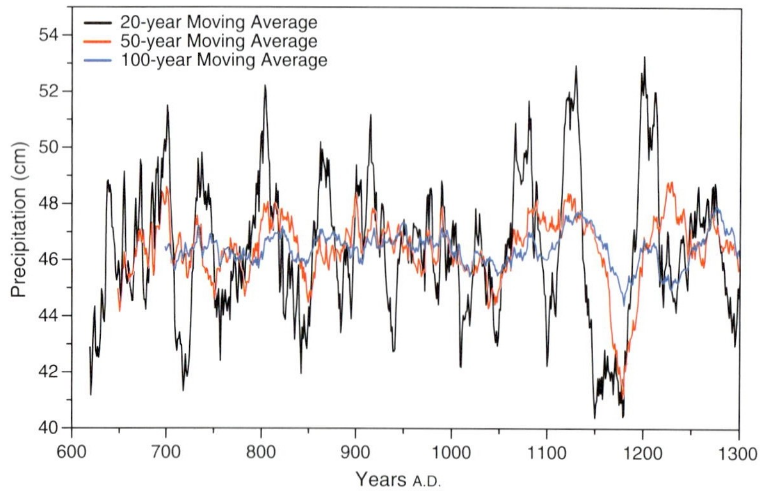

PLATE 5.5 Low-frequency filters applied to the annual paleo-precipitation data proxied by the Mesa Verde Douglas fir tree-ring-width index series from Dean and Van West (2002). See Plate 6.1 for unsmoothed values.

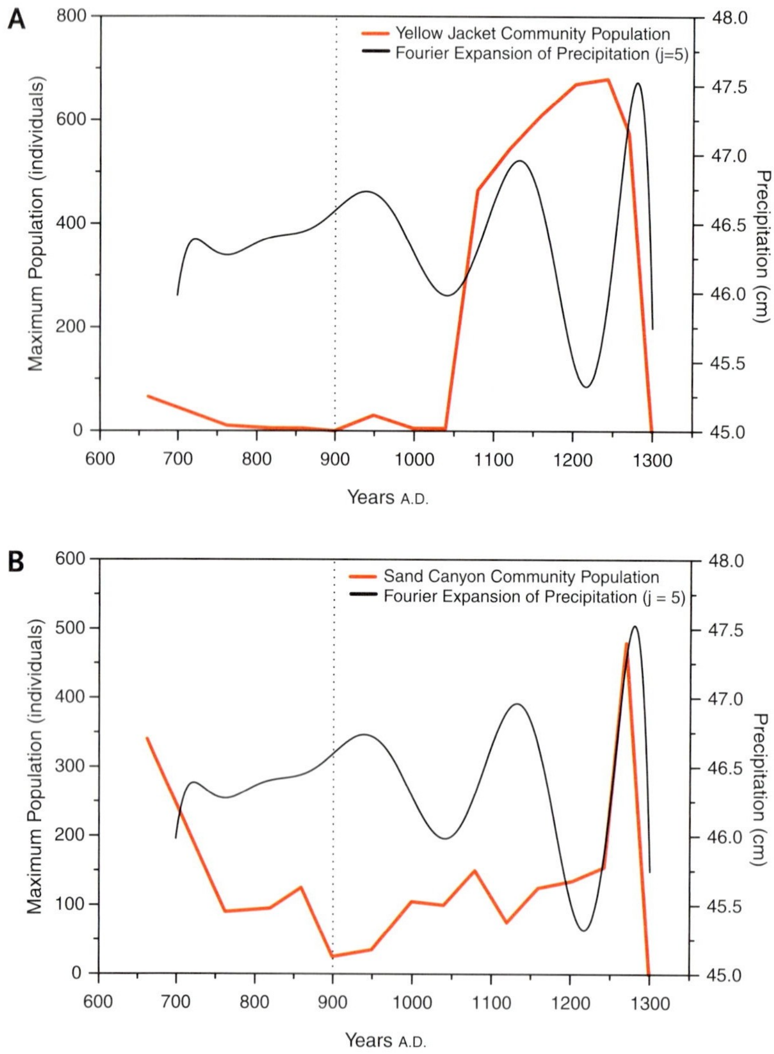

PLATE 5.6 Comparison of two community-level settlement population histories (red) with the regional 100-year moving average paleoprecipitation (as smoothed by a Fourier expansion j = 5 [black]) through time, with the two VEP population cycles delimited by the vertical dotted line. A) Yellow Jacket Community; B) Sand Canyon Community.

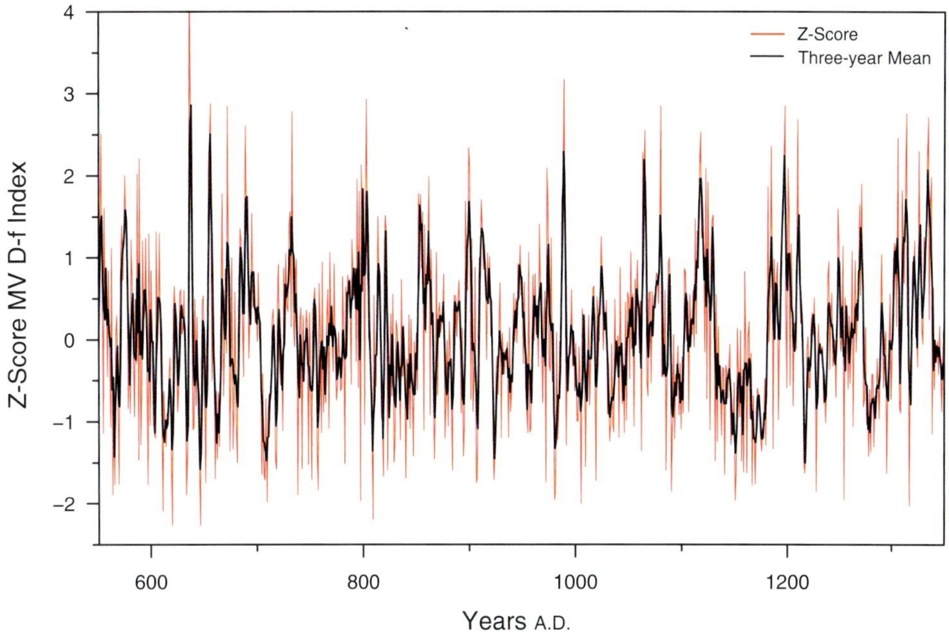

PLATE 6.1 Z-scores through time for the Mesa Verde Douglas fir ring-width indexed series. Red, annual values; black, three-year means (current and two previous years).

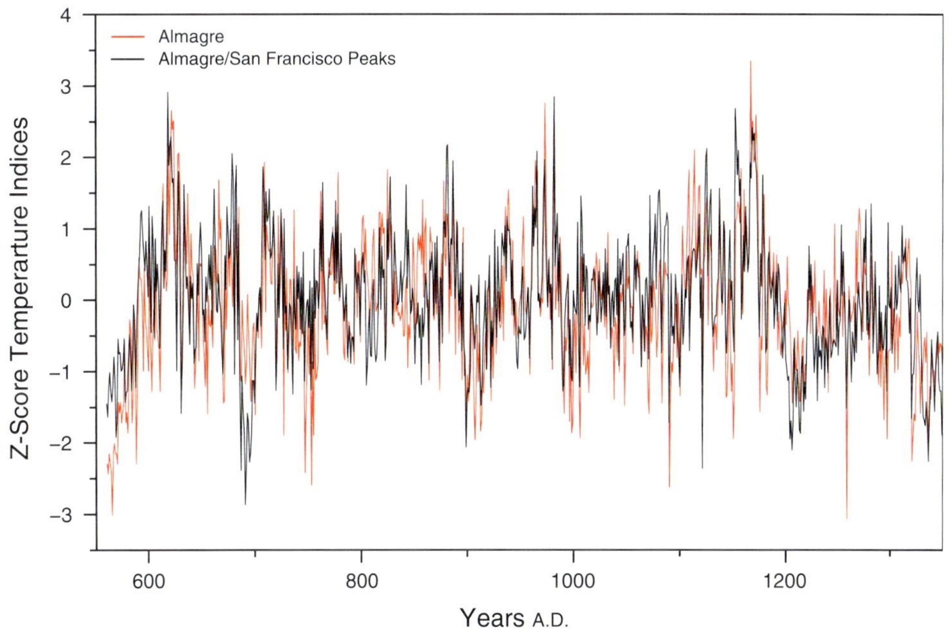

PLATE 6.2 Annual values for two bristlecone-pine-based temperature proxies. Red: Almagre; black: scores on first principal component of Almagre and San Francisco peaks series.

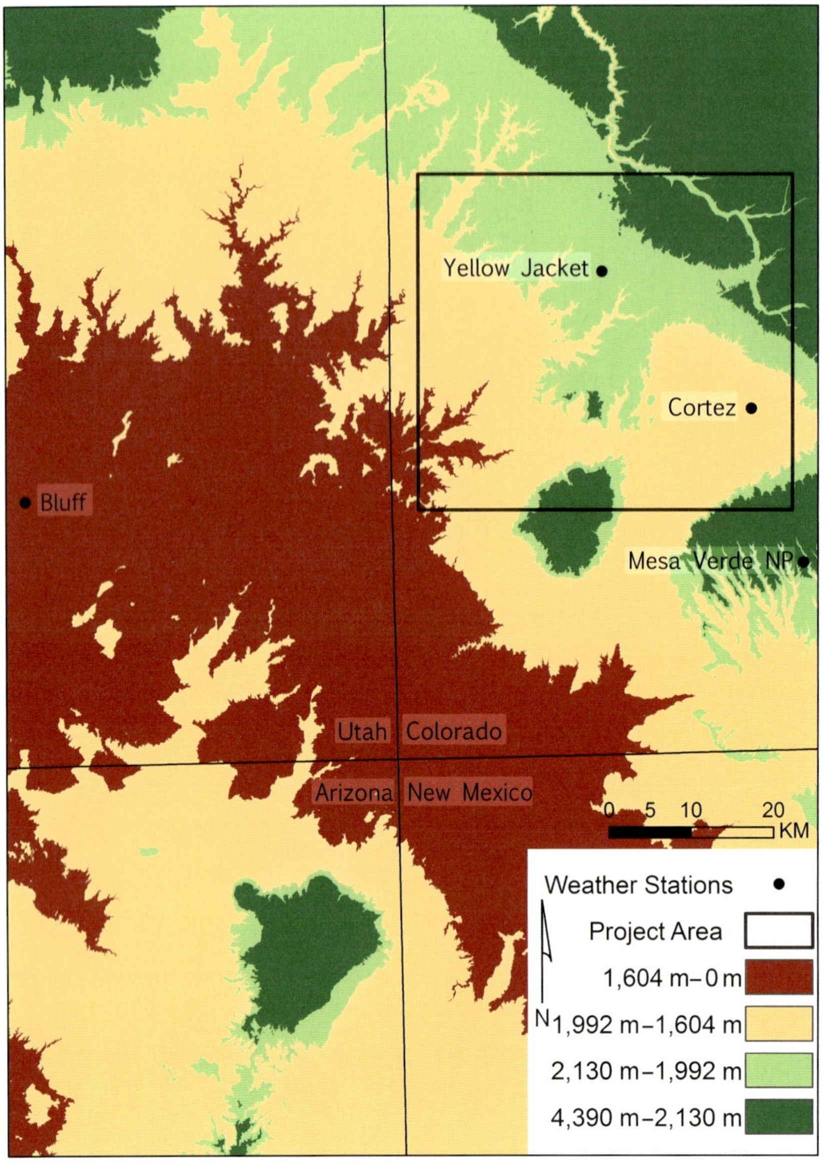

PLATE 6.3 Extent of each elevation band in and around the study area, and the location of the four weather stations used to calculate PDSIs during the 1931–1960 calibration period.

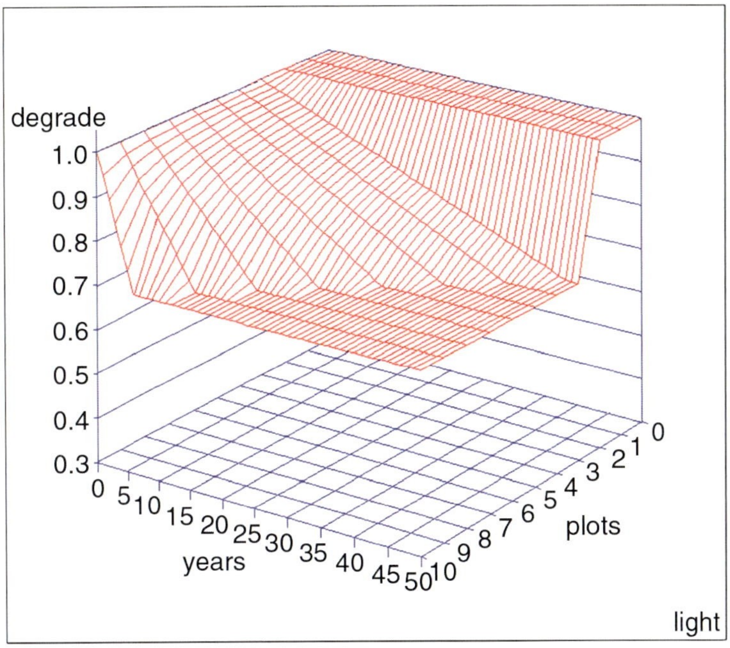

PLATE 6.4 The effect of the mild degradation (to 70 percent of full potential) as a function of time and number of 1-acre plots within a 4-ha cell.

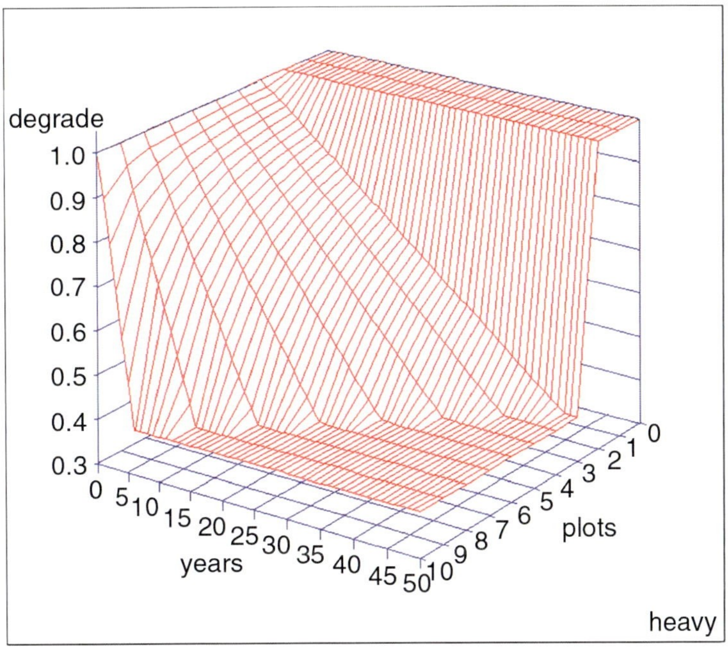

PLATE 6.5 The effect of the severe degradation (to 40 percent of full potential) as a function of time and number of 1-acre plots within a 4-ha cell.

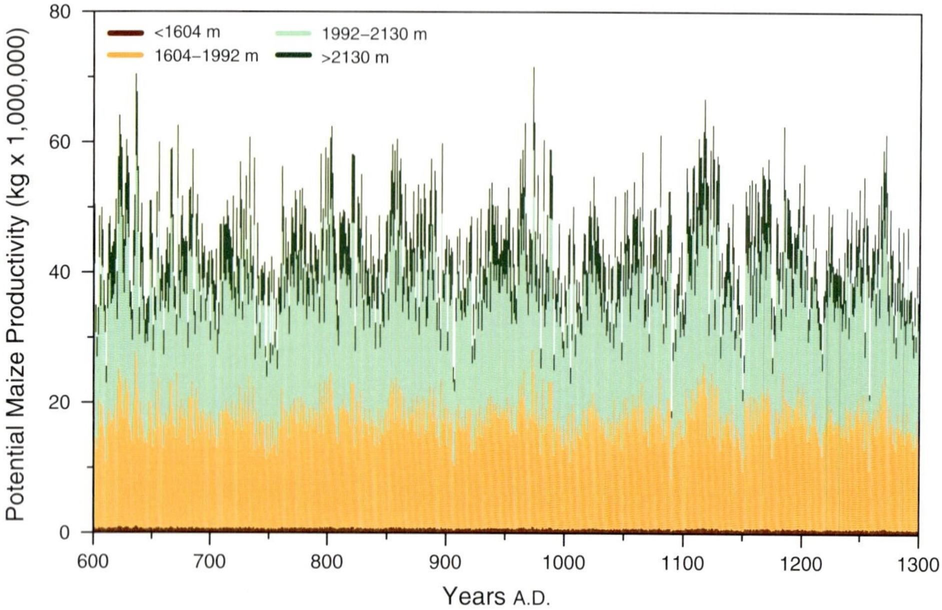

PLATE 6.6 Stacked plot showing potential annual maize productivity in each elevation band, cumulated across elevation. The top of the 1604–1992 band, for example, reflects the total production for all lands below 1992 m, and the top of the dark green (uppermost) band reflects total potential annual maize production in the study area.

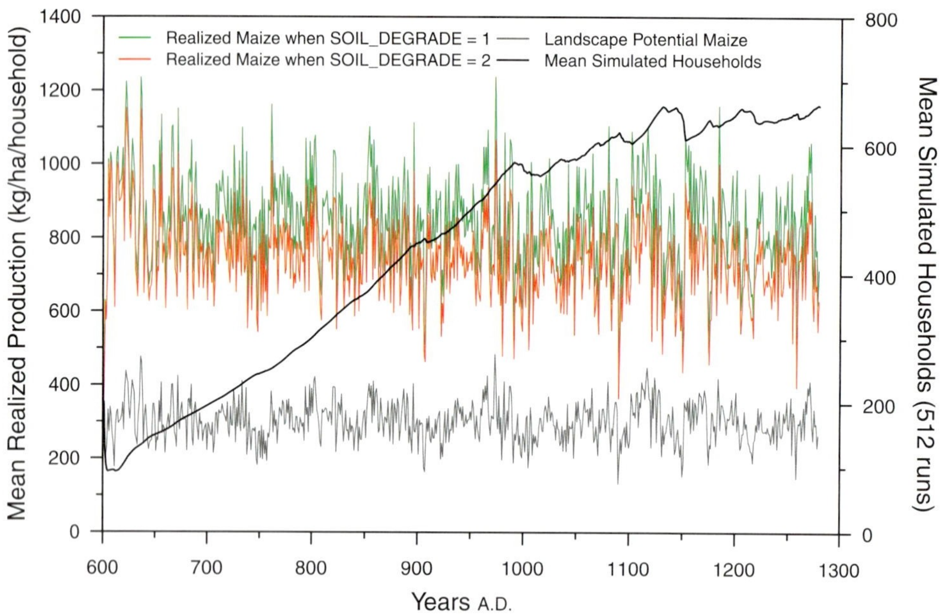

PLATE 6.7 Mean maize productivity realized by simulated households at different levels for SOIL_DEGRADE (left axis), vs. mean maize productivity of VEP landscape (left axis) and mean simulated number of households (right axis) through time.

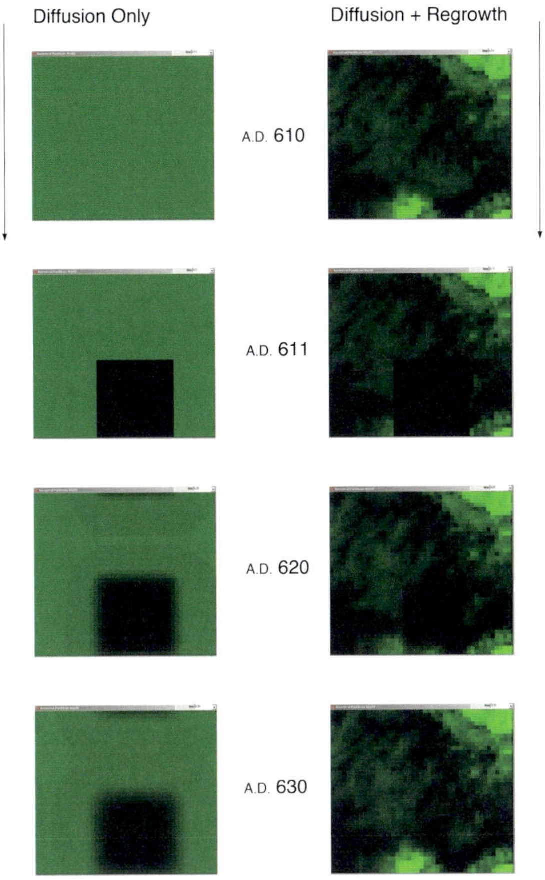

PLATE 8.1 Deer diffusion after an artificial localized catastrophe with only diffusion (no regrowth—left column) and with both regrowth and diffusion (right column). Brighter colors indicate higher deer densities.

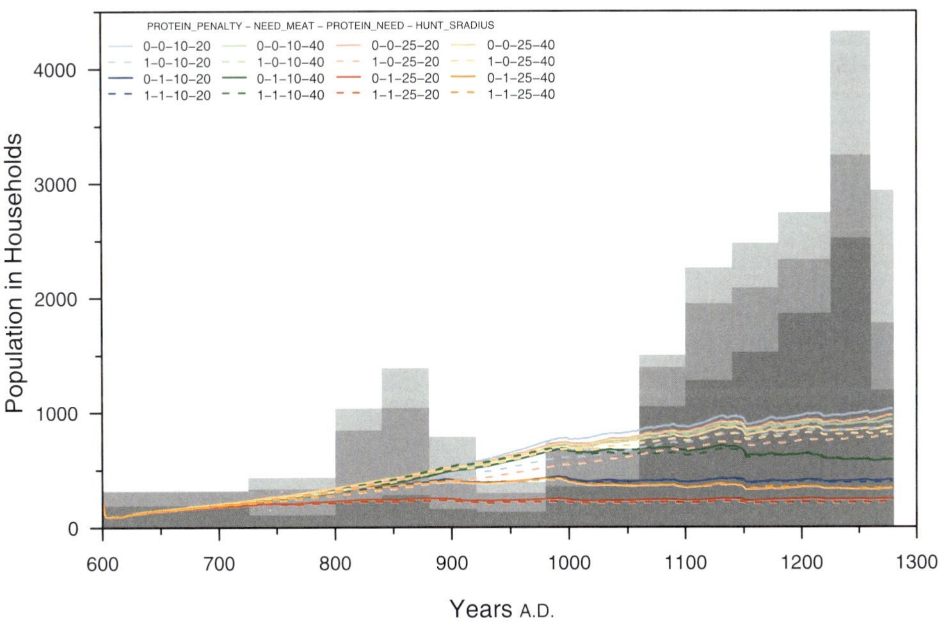

PLATE 9.1 Modeled human population size in households, by composite run, compared with estimates of local momentary human populations from Varien et al. (2007).

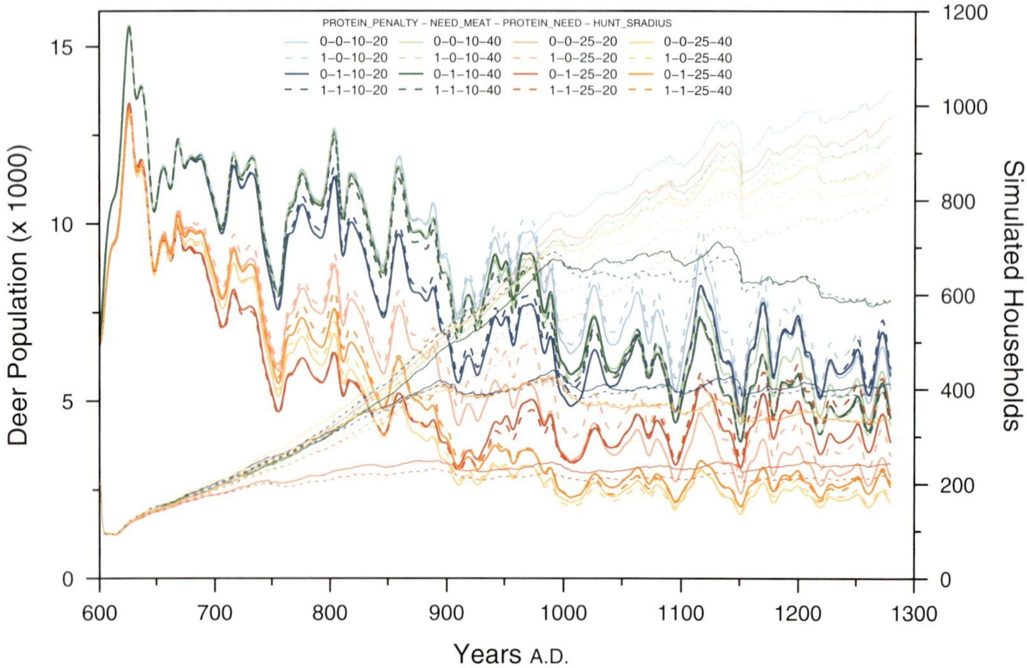

PLATE 9.2 Modeled deer (left axis, bolder lines) and agent (right axis, thinner lines) populations through time, by composite run. Deer populations are seeded at approximately 6900 deer at year 600; agent populations at 200 agents.

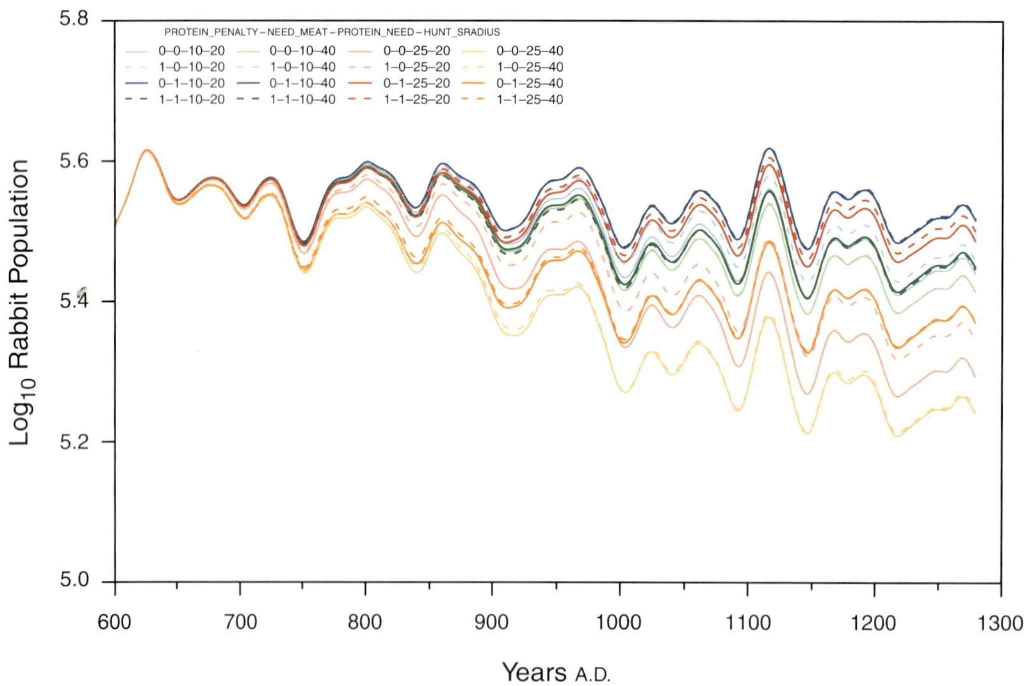

PLATE 9.3 Modeled rabbit populations, by composite run.

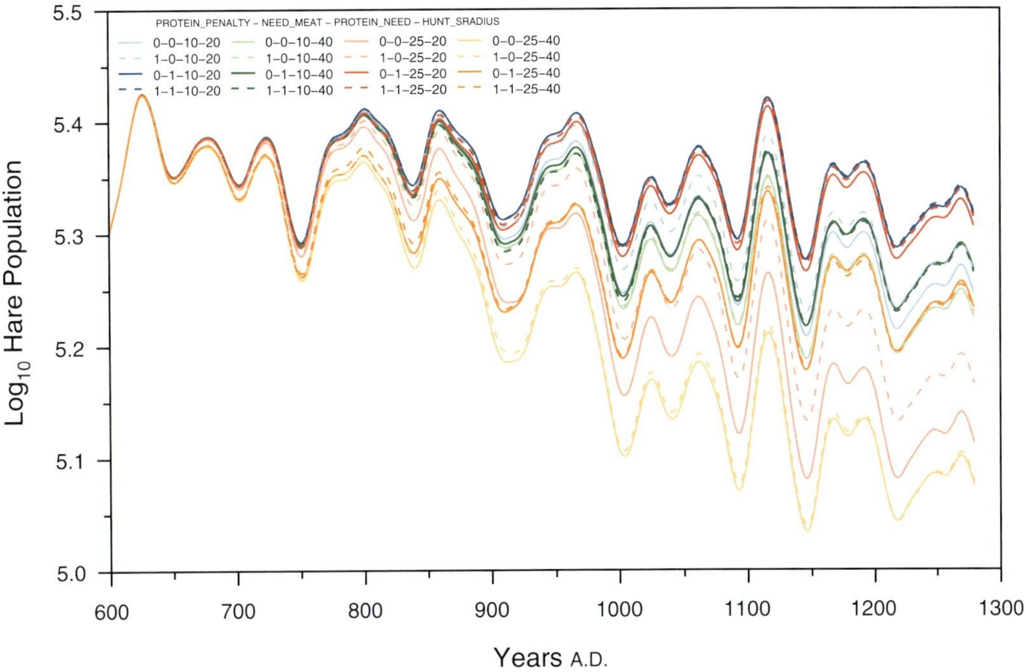

PLATE 9.4 Modeled hare populations, by composite run.

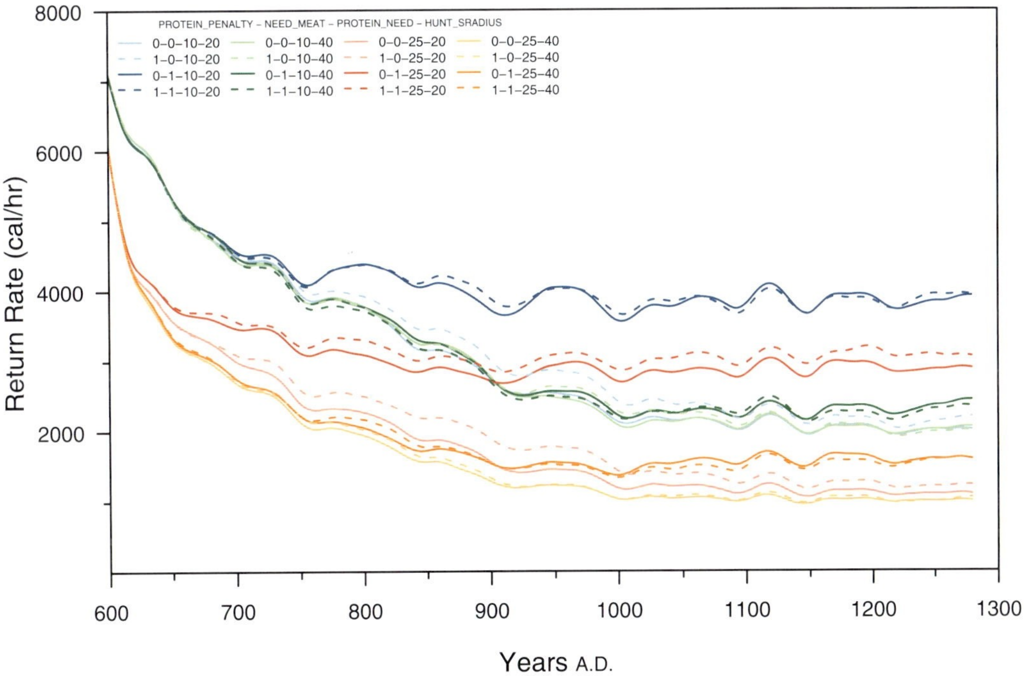

PLATE 9.5 Modeled return rates for deer, by composite run.

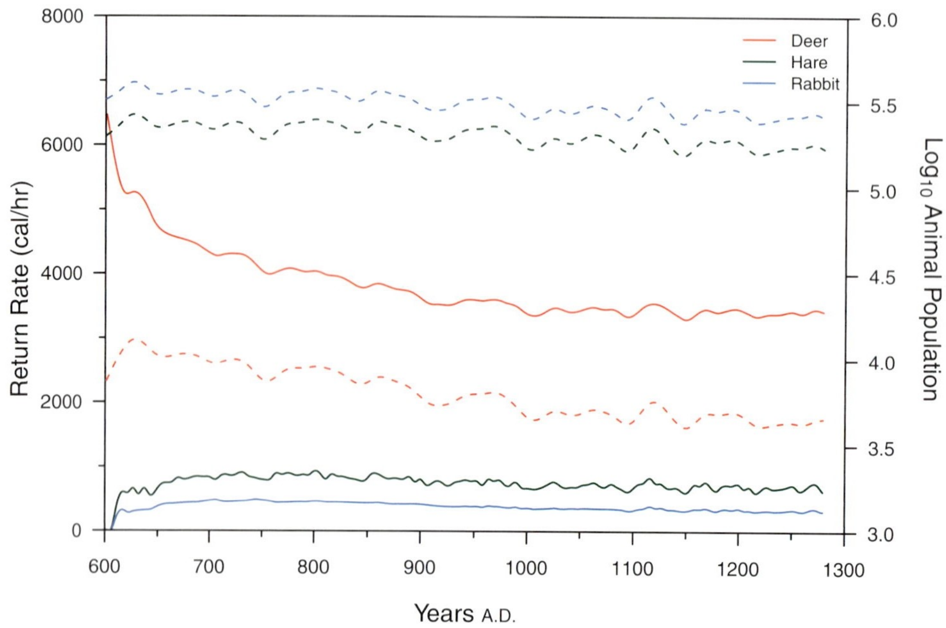

PLATE 9.6 Modeled average return rates for deer, hare, and rabbit for composite run 1. Dashed lines represent the $\log_{10}$ number of deer, rabbit, and hare on the landscape

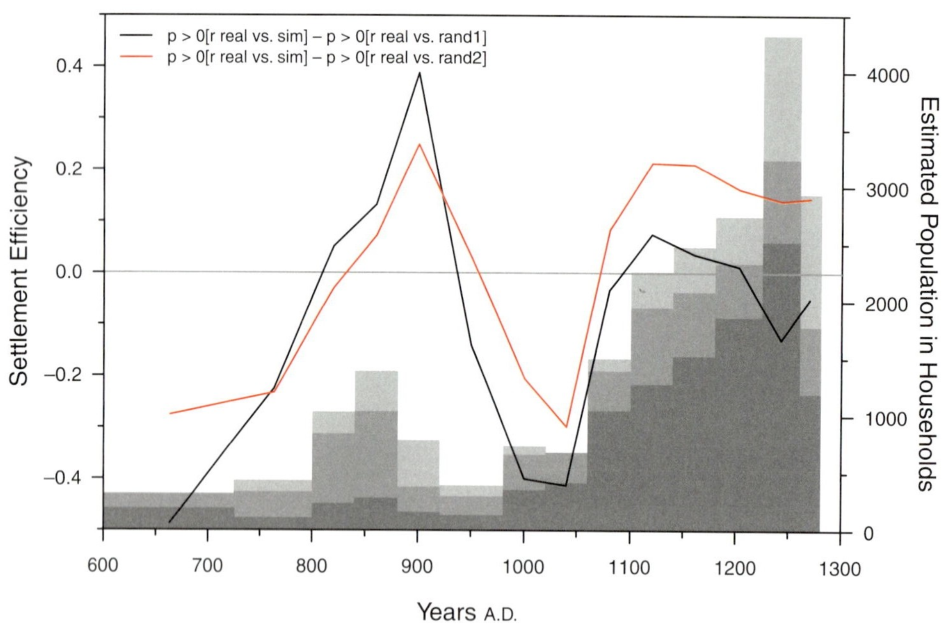

PLATE 10.1 Goodness-of-fit, relative to random expectations, between real and simulated site locations and sizes, using the correlation coefficient approach. The black line reports this comparison where the random expectation is formed by permuting households; the red line reports this comparison after permuting cells. The horizontal reference line demarcates efficiencies above (positive) and below the random expectations computed in these ways.

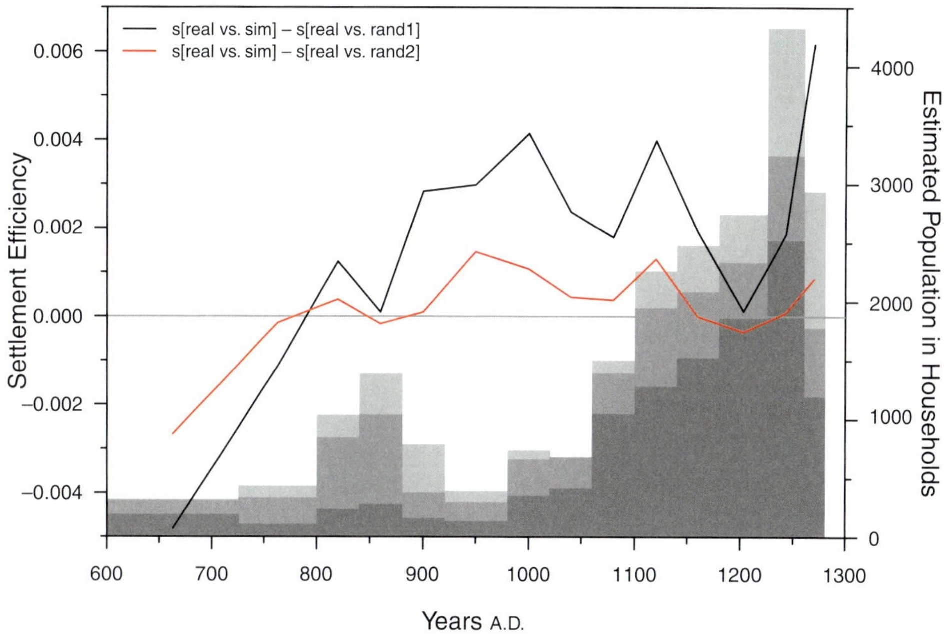

PLATE 10.2 Goodness-of-fit, relative to random expectations, between real and simulated site locations and sizes, as measured by the similarity coefficient *s*. The black line reports this comparison where the random expectation is formed by permuting households; the red line reports this comparison after permuting cells. The horizontal reference line demarcates efficiencies above (positive) and below the random expectations computed in these ways.

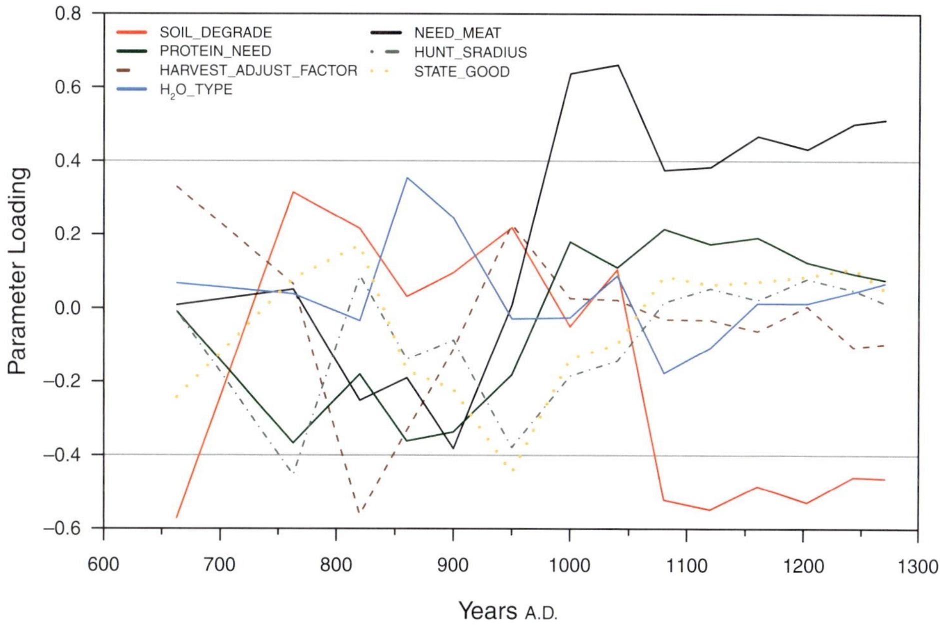

PLATE 10.3 Loadings through time for seven of the nine parameters varied in the sweep reported in this volume, plotted at the midpoint for each of the 14 periods.

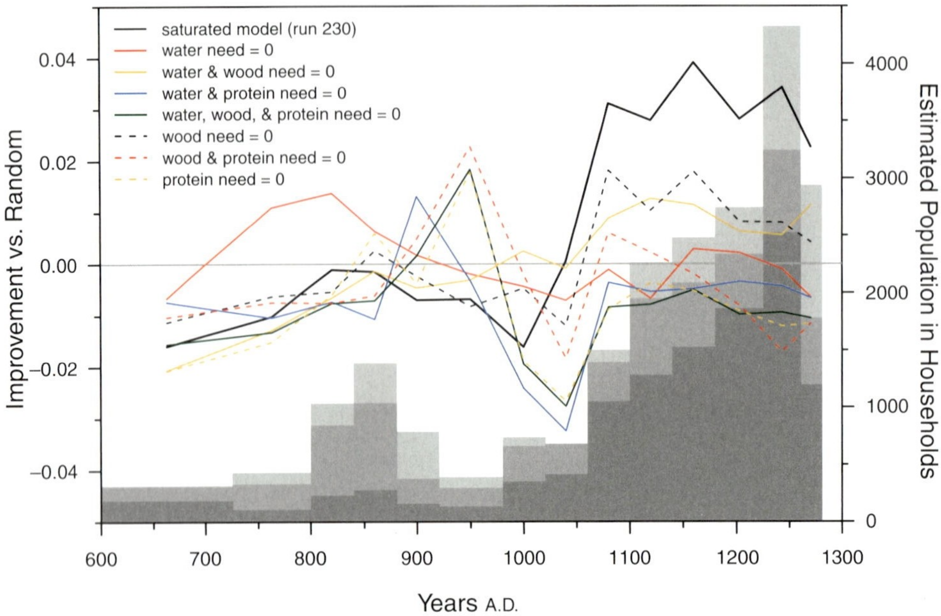

PLATE 10.4 Difference between the correlations obtained for each of the simulated simplified variants of Run 230, and Run 230 itself, and the randomizations that are specific to each of them.

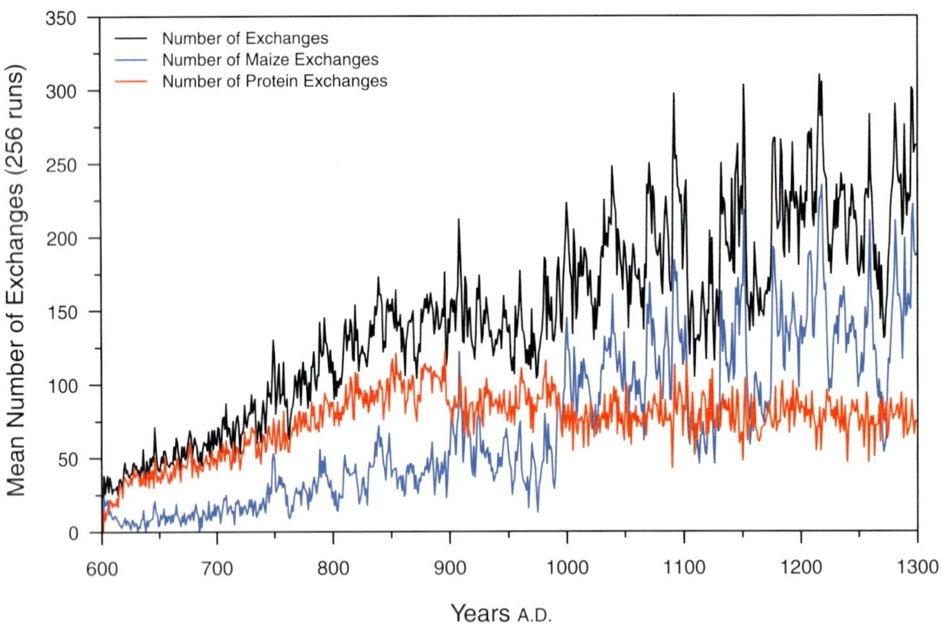

PLATE 11.1 The number of exchanges between households broken down in terms of protein (meat) or calories (maize), averaged across 256 runs of the simulation.

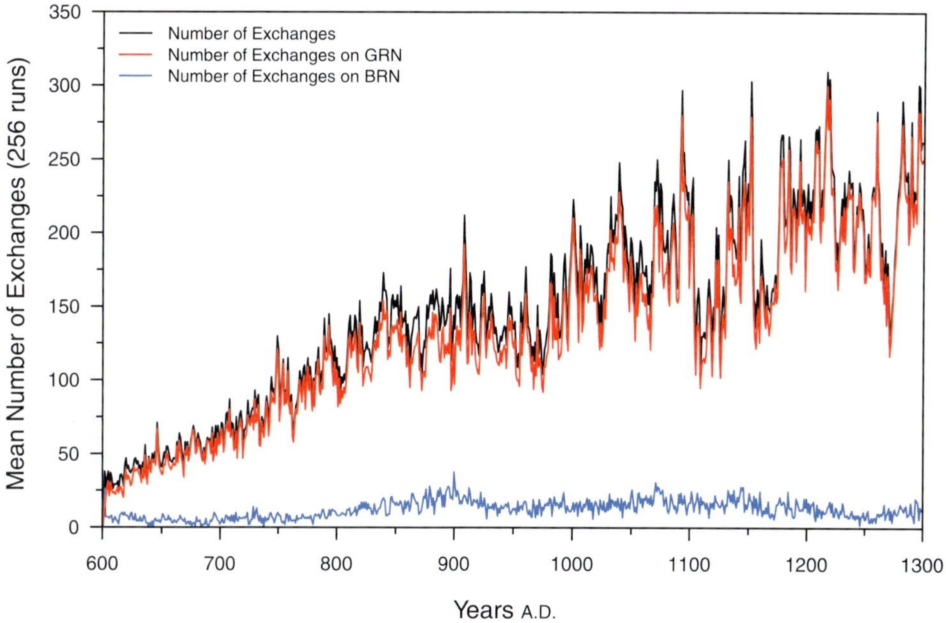

PLATE 11.2 The number of exchanges between households broken down by network type (BRN vs. GRN), averaged across 256 runs of the simulation.

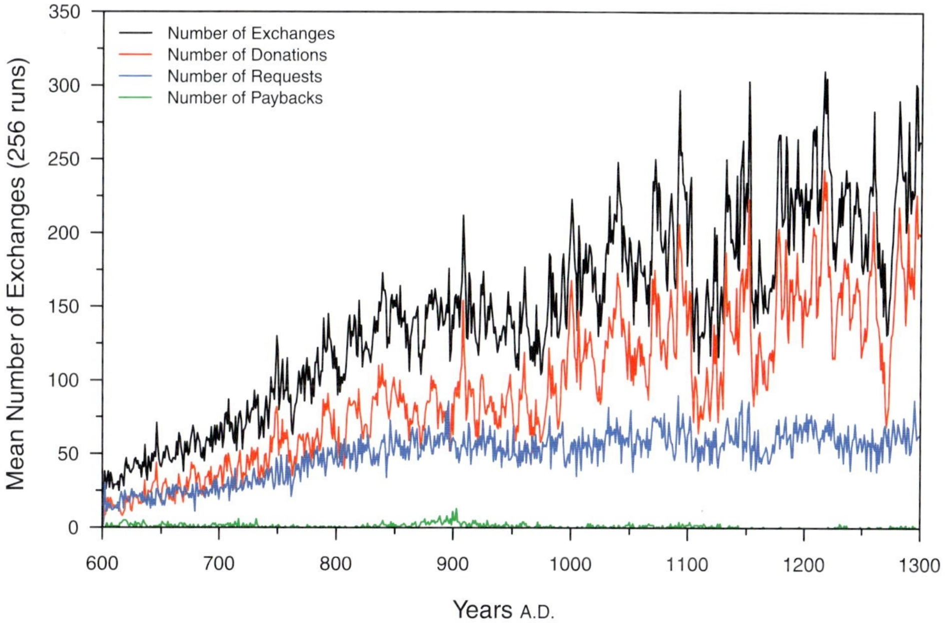

PLATE 11.3 The number of exchanges between households broken by exchange type (donations, paybacks, or requests), averaged across 256 runs of the simulation.

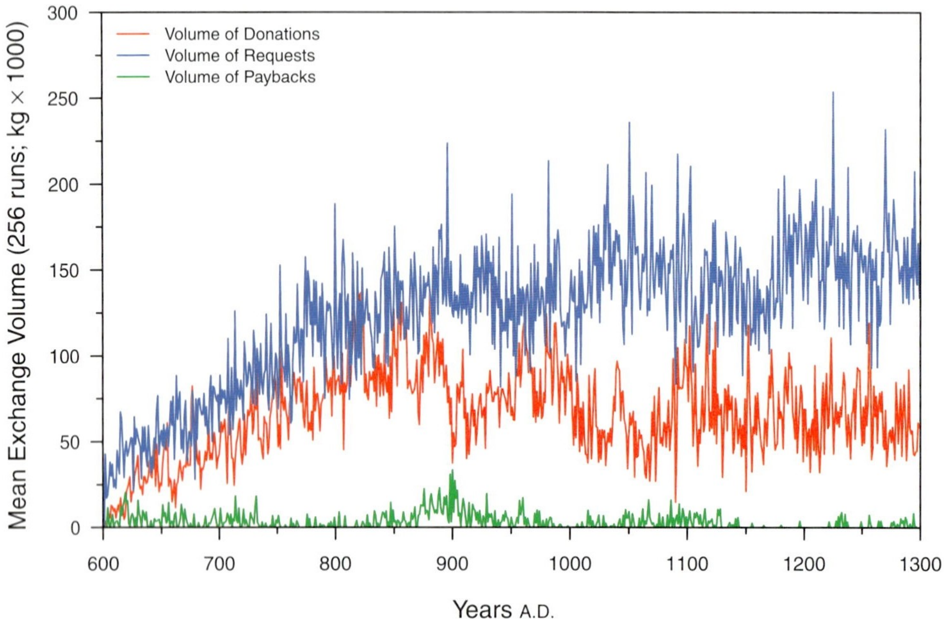

PLATE 11.4 The amount of exchange in any currency (protein or maize) traded over time broken down in terms of the exchange type—a donation, a payback or a request—averaged across 256 runs of the simulation.

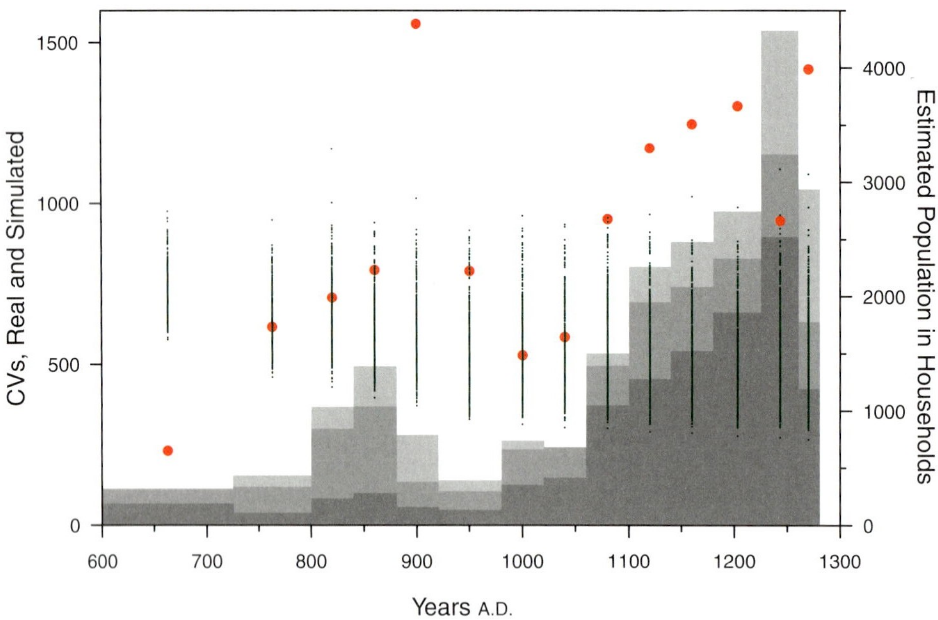

PLATE 15.1 Coefficients of variation for household years / cell in the real sites (red-filled circles) and each of the 512 runs (green dots), indexed on left vertical axis, on top of the block histogram for our three estimates of VEP population size, indexed on the right vertical axis.

SIX

Modeling Agricultural Productivity and Farming Effort

Timothy A. Kohler

TO MODEL SUBSISTENCE in the central Mesa Verde region from A.D. 600 through 1300 is—largely—to model maize production. Stable isotope studies indicate a heavy reliance on maize by both the eastern Basketmakers (of southwestern Colorado) and the western Basketmakers (of northeastern Arizona and southeastern Utah) by the first millennium B.C. Already in the first millennium B.C., up to 80 percent of the Basketmaker diet came from maize, with reliance on maize likely peaking at a level only slightly higher in the period between A.D. 500 and A.D. 1000 (Coltrain et al. 2006, 2007; Matson and Chisholm 1991).

Maize (*Zea mays* ssp. *mays*) arose from an annual grass, teosinte (*Zea mays* ssp. *parviglumis*), in the Río Balsas region of western Mexico, probably by the early seventh millennium B.C. (Piperno et al. 2009) and certainly by the middle of the fourth millennium B.C. (Blake 2006; Piperno and Flannery 2001). Recent evidence reinforces a view of maize as originally being at home in the seasonal tropical forest and not the cool, semiarid uplands as once thought. In the temperate, cool, and semiarid climate found in the central Mesa Verde region, maize production will increase, at least up to a point, as conditions become warmer and wetter. The challenge of modeling maize productivity is first to determine the variability in temperature and moisture in the central Mesa Verde region between A.D. 600 and A.D. 1300, and then to determine how this local variation in precipitation and temperature would have affected maize productivity.

Several fine discussions of Holocene paleoclimates in the U.S. Southwest are now available. Larry Benson and Michael Berry (2010), Jeffrey Dean (2007), Dean and Carla Van West (2002), David Gregory and Fred Nials (2007), and Aaron Wright (2010 and Chapter 3 in this book) all report recent research oriented toward understanding the effects of climate variability on humans in the prehispanic Southwest. The purpose of this chapter and the next is not to review this more general work, but to emplace the most relevant climatic phenomena onto the landscape of the central Mesa Verde region studied in this book. How specifically did variability in climate between A.D. 600 and A.D.

1300 affect the plants and animals on which human populations depended? We necessarily answer this question by building models that focus on some aspects of both production and climatic variability and not on others. We want both the strengths and weaknesses of our approach to be evident.

In addition to an unpublished dissertation that emphasizes how our agent-based model "grows" deer, hare, and rabbits and the plants they graze and browse (Johnson 2006), we have previously published three summaries of how we model agricultural productivity. Mark Varien et al. (2007:276–280) and Timothy Kohler et al. (2007:64–67) provide overviews of the entire process and graph the estimated potential production of maize through time, averaged across this landscape. In Kohler (2010), I discuss the history of attempts to model maize productivity in the northern Southwest and consider what our current estimates imply about the thirteenth-century depopulation of the northern Southwest in the context of other climatic and demographic data. This history is important, because near the beginning of this project we decided to refine and extend earlier approaches—specifically, that of Van West (1994)—rather than to attempt something radically different.[1] These previously published summaries will be adequate for most readers—and even too much for some. This chapter provides more details on how the reconstruction was made for those who wish to improve on it or apply similar techniques elsewhere. Generally following Kohler et al. (2007:65–67), I divide the process into a number of steps.

STEP 1: DETERMINE BEST TREE-RING PROXIES FOR LOCAL PRECIPITATION AND TEMPERATURE

In considering potential tree-ring indexed series, we needed to select from those (1) that extend back to at least A.D. 600 and forward at least through our calibration period from 1931 through 1960; (2) that are reasonably local to our area; (3) that are sensitive to aspects of either precipitation or temperature that are relevant to maize growth; and (4) that have as much sample depth (number of samples per year) as possible within the period of interest.

Considering precipitation first, a number of considerations pointed to the Mesa Verde Douglas fir indexed ring-width series (Dean and Robinson 1978:29–30) as the best candidate. Research by the Laboratory of Tree-Ring Research in partnership with the Wetherill Mesa Archaeological Project demonstrated that in this series, wide rings are strongly positively correlated with precipitation from the previous to the present June, and are also, though less strongly, inversely correlated with temperatures over the same period (Fritts et al. 1965). The variance in the tree-ring widths explained by combining these variables was very high (R^2 values from .80 to .88), using trees whose living members range in elevation from 2,000 to 2,500 m asl along the slopes of Navajo Canyon west of Wetherill Mesa, and weather as measured at an official station only 1.3 km away. Douglas fir chronologies were more consistent from tree to tree and from site to site than were either Utah juniper or pinyon pine chronolo-

1. One "radical" alternative to our present approach would have been to estimate the response of maize to changes in precipitation and temperature directly, rather than through intermediate regressions involving the Palmer Drought Severity Index and the historic record of maize production in this area. In that case we would have used the tree-ring records to estimate precipitation and temperature across the project area, taking topographic variability into account, and used published growth responses of maize to these variables in estimating productivity. (For example, L. A. Hunt et al. [2001] review published response curves of maize to temperature.) This is attractive, because the alternative that we used calibrates through maize-production estimates for Montezuma County as a whole. This undoubtedly masks a great deal of local variability in how specific locations within the county responded in specific years. The probable effect of this masking is that our regressions underestimate both highs and lows for specific locations in specific years. Nevertheless, the (alternative) approach of direct regression of maize growth on reconstructed climatic parameters has other problems—e.g., how to take local variability in soil fertility and water-holding capacity into account, and how to translate maize growth responses into quantities of maize per unit area per unit time.

gies, and were sensitive to both summer and winter precipitation, whereas the pinyon and juniper chronologies largely measured winter drought. As summarized by Harold Fritts et al. (1965:120), the climate conditions that produce narrow rings in Douglas fir are "a hot, dry previous June; a dry, late summer; a dry autumn; a dry winter; a dry, cool spring; and a dry [current] June." Each of these variables individually is significantly correlated with the tree-ring widths; the multiple regressions show that years when these conditions coincide would be particularly unfavorable. With the possible exception of dry autumns, these same conditions would also be detrimental to maize production. The number of samples on which this series is built is generally quite large (mean from A.D. 600 to A.D. 1300, 9.4, $s=5.9$), though there are unfortunately 33 years in the mid-A.D. 600s for which there is only one sample.

Dean and Van West (2002:85–90) chose this series as the independent variable from which to retrodict total regional precipitation across southwestern Colorado from the previous August through the current July, mean regional tree-year temperatures, and regional June Palmer Drought Severity Indices (PDSIs). They discussed only the period between A.D. 900 and A.D. 1500, though the series extends from A.D. 442 through A.D 1962. We also selected this series as our proxy for precipitation, and graph it in Plate 6.1 (see color insert in this book) from A.D. 550 through A.D. 1350 in two forms: annually, and smoothed by three-year running means (current and two previous years).

The choice of tree-ring proxies for temperature was less clear, though our goal was well defined: can we find a proxy that is linked to agronomically significant temperature variation in our region? We wanted to look beyond the Mesa Verde Douglas fir series since we were already exploiting that for its precipitation signal. Previous research indicates, quite logically, that the tree-ring records most likely to be sensitive to temperature variation will come from locations where temperature is likely to be limiting—e.g., at tree lines and at species' latitudinal limits (Lamb 1995:99). Given previous local research, especially that of Matthew Salzer and Kurt Kipfmueller (2005), we confined our search to records for high-elevation bristlecone pine (*P. aristata* or *longaeva*) in our general region whose series length met our requirements.

Our first task was to build a database of historical temperature measurements to compare against the tree-ring data. We compiled data from three local weather stations: Mesa Verde National Park (055531-2; 2167 m asl) just outside our area to the southeast, and Cortez (051886; 1884 m asl) and Yellow Jacket (059275-2; 2091 m asl), both within our area. Although the instrumented series for Yellow Jacket is more applicable to our highest-producing agricultural lands than is Cortez, the series is shorter (extending from 1962 to 2002) than that from Cortez (1929 to 2002) and Mesa Verde (1948 to 2002). The number of days in each month with minimum daily temperatures less than 32°F was obtained from the National Climate Data Center (see Bibliography). A few missing values for individual years in the temperature database were replaced using regression. Multiple consecutive missing values were not replaced. We also obtained or computed mean temperature in each month, though only correlations for growing-season months (May through September) are presented here. We view the Mesa Verde series of instrumented data as the best temperature record of the three, since it is unaffected by cold-air drainage problems that seem to affect the Cortez series, and it is almost three times as long as the Yellow Jacket series, whose correlations could be less reliable as a result.

In a recent, relatively local, calibrated and verified temperature reconstruction from the San Francisco Peaks (SFP) area of northern Arizona, Salzer and Kipfmueller (2005:470) found that, in general, temperature data (monthly mean-maximum temperatures) that were lagged one year correlated more strongly with tree-ring indices than did current-year data. Taking this as a possible lead, lag-1 variables (which

TABLE 6.1

Eigenvalues of the Correlation Matrix for the Principal Component Analysis of Almagre and San Francisco Peaks

Principal Component	Eigenvalue	Difference	Proportion	Cumulative
1	1.23	.46	.62	.62
2	.77	–	.38	1.00

TABLE 6.2

Eigenvectors from the Principal Component Analysis of Almagre and San Francisco Peaks Series

Series	Principal Component 1	Principal Component 2
Almagre	.71	.71
San Francisco Peaks	.71	–.71

moved copies of previous-year temperature data into the current year) were created for all the temperature variables.

The second data set we created consisted of five high-elevation bristlecone pine series known to be temperature sensitive, selected from among other such series in the western United States because of their relative proximity to the study area. Three of these were obtained from the World Data Center for Paleoclimatology (see Bibliography): Almagre Mountain B (Colorado; A.D. 560–1983; 38N, 104W; 3563 m; Graybill, D. A.), Sheep Mountain (E. California; A.D. 1–1990; 37N, 118W; 3475 m asl; Graybill, D. A.), and the White Mountains (E. California; 37N, 118W; 3113 m asl; Schulman, E.). An additional series, Cottonwood (E. California), was obtained from Chronology Series 1 of the Laboratory of Tree-Ring Research (Drew 1972). The SFP series (N. Arizona; 663 B.C.–A.D. 1997; 3536 m asl) was obtained from Jeffrey Dean, courtesy of Matthew Salzer.

We first looked for shared signals in these tree-ring indices that might correlate with study-area temperatures by extracting their five principal components. Only components 1 and 2 were of interest (they alone have eigenvalues greater than 1), and together they account for almost two-thirds of the variation in the data set. Neither component, however, was significantly correlated with agronomically meaningful local temperature variability.

We then tried to determine what temperature signal relevant to our area might be contained within only the two most local series, Almagre and SFP. (Almagre Mountain is located 350 km east-northeast of our area; the SFP are 335 km to the southwest.) We performed a second principal components analysis of just these two data sets. Table 6.1 presents the resulting eigenvalues and Table 6.2 the eigenvectors from this second principal components analysis. Component 1 is of most interest, since it is correlated with 62 percent of the variation in the data set.

Given the location of the study area halfway between these two series, and given the fact that they share a considerable amount of variance, scores on component 1 would be attractive for this research if they correlate with local temperature data relevant to maize or bean farming. Scores from both principal components were correlated with the same suite of local temperature data already described here. Significant correlations from this exercise ($p \leq .05$) are presented in Table 6.3.

TABLE 6.3

Significant (p ≤. 05) Correlations of Candidate Tree-Ring Proxies with Local Temperature Variables

The strongest correlation in each column is in bold.

ONLY CORRELATIONS WITH POTENTIALLY AGRONOMICALLY SIGNIFICANT VARIABLES ARE FULLY REPORTED.

Station / Variable	San Francisco Peaks	Almagre	Scores on 1st Principal Component	Scores on 2nd Principal Component
MVNP / Mean Temperature Previous September	.457 .000 60	.236 .042 60	.467 .000 60	–
MVNP / Mean Temperature Current September	.408 .001 61	.387 .002 61	.491 .000 61	–
Yellow Jacket / No. Days Previous September < 32°F	–	.457 .037 21	–	.505 .019 21
Yellow Jacket / Mean Temperature Previous June	.467 .033 21	–	+	–.504 .019 21
Yellow Jacket / Mean Temperature Previous July	.538 .012 21	**–.506** **.019** **21**	+	–.622 .003 21
Yellow Jacket / Mean Temperature Previous August	.446 .043 21	–	+	–.509 .018 21
Yellow Jacket / Mean Temperature Previous September	**.666** **.001** **21**	–	.495 .023 21	**–.684** **.001** **21**
Yellow Jacket / Mean Temperature Current June	.497 .019 22	–	+	–.477 .025 22
Yellow Jacket / Mean Temperature Current September	.517 .014 22	–	**.535** **.010** **22**	–

(continued)

TABLE 6.3 *(continued)*

Station / Variable	San Francisco Peaks	Almagre	Scores on 1st Principal Component	Scores on 2nd Principal Component
Cortez / Mean Temperature Previous July	–	–.288 .037 53	–.335 .014 53	–
Cortez / N Days Current May < 32°F	–	–.372 .030 34	–	–

NOTE: Cells contain *r*, *p*, and *n* of observations. The "+" and "–" indicate nonsignificant positive and negative correlations, respectively. MVNP = Mesa Verde National Park.

Bristlecone pines in both the Almagre and SFP series tend to put on wide rings when Mesa Verde experiences warm previous and current Septembers. Since local dry-farmed pinto beans and corn are still growing until mid-September, when, on average, they mature, warm Septembers tend to increase yields. In most other respects these two candidate proxies are somewhat at odds with each other. The SFP series is strongly positively correlated with several measures of previous and current summer warmth from the Yellow Jacket instrumented data set, whereas the Almagre series tends to be weakly negatively correlated with these same variables. Curiously, Almagre bristlecones respond negatively to previous warm Julys in Cortez, but more logically, they respond positively when Cortez experiences fewer days below freezing in the current May.

Given these explorations, as well as other analyses not reported here, we decided to move forward with two proxies for temperatures: the scores on the first principal component already reported here, and the Almagre series by itself (Plate 6.2). These seem to track slightly different aspects of growing-season warmth in our study area, though the scores on the first principal component seem to most reliably reflect late summer warmth on both Mesa Verde and in the Yellow Jacket area. We used each of these proxies to produce (different) sets of paleoproductivity data planes, and in the end we will be able to determine which produces data planes that provide the best fit between the simulated agent settlement behavior and the size and placement of settlements through time that we actually see in the archaeological record.

STEP 2: DEFINE GROUPS OF SOILS THAT ARE SIMILAR IN PRODUCTIVITY

Soils in our area can be divided into three large classes: aeolian, colluvial, and residual soils formed from soft shale bedrock (Ramsey 2003a). The study area contains 148 unique soil complexes, described in three separate Natural Resource Conservation Service (NRCS) soil surveys (Johnson 2006:124–147). The Cortez area soil survey (Ramsey 2003b) describes the vast majority of study-area soils, but the Animas-Dolores Area soil survey (Pannell 2003) adds some soils in the northeastern part of the study area, and the Ute Mountain Area soil survey (Ramsey 2008) adds some areas in the south-central portions of the study area. To make geographic information system coverages for the paleoproductivity models, some categories had to be changed as described by C. David Johnson (2006:126–129) to compensate for modern dams, the reservoirs they back up, and gravel pits.

In general, the most productive soils for agriculture in our area are the mesa top loesses

(Arrhenius and Bonatti 1965), whose advantages include greater depth than most other soils, especially toward the centers of mesas, and a gentle slope contributing to higher insolation, reduced erosion, enhanced buildup of organic matter, and enhanced water infiltration. These aeolian deposits may have been building up for the last 100,000 years and continue to arrive from the south and west during spring dust storms, though at very low rates—1,000-year-old ruins on mesa tops are not buried by recent deposits (Ramsey 2003a). Deposition appears to have been particularly rapid during the terminal Pleistocene and early Holocene (Reheis et al. 2005).

Our procedure for estimating productivity first requires estimating PDSIs (Palmer 1965) for each soil taxon. This in turn requires knowing the available water supply for the upper 15.2 cm (6 in) and lower 17.7 to 152.4 cm (7 to 60 in) for each taxon. These data are provided in the soil survey tables, but since most soil complexes contain more than one soil, each with distinct depth properties, we had to first derive average values for each complex, as explained by Johnson (2006:Appendix A). The available water supply of the lower horizon is especially significant for production in this semiarid area, and it tends to be more variable across soil complexes than that of the upper horizon.

In our procedure for estimating paleoproductivity, the only two soil variables of importance are current normal-year dry-weight productivity, and the available water supply of the lower horizon—both estimated by the NRCS and available in the soil surveys. Since many soil complexes have somewhat similar values for these two variables, we recognized that we could simplify our procedures by working at the level of groups of similar soils, rather than making separate calculations for each of the 148 complexes. I therefore performed an agglomerative hierarchical cluster analysis using average linkage, employing these two variables after standardizing them. For solutions with relatively few clusters, which would be more manageable for us in later work, I judged the best solution to be at 14 groups, which yielded local peaks in the cubic clustering criterion and the pseudo-F statistic. One "cluster" contained only one soil complex; the largest contained 27. In what follows, I use the median values of normal-year productivity and the available water supply of the lower horizon for each soil cluster, rather than the original values of each soil complex.

STEP 3: PRODUCE PDSI ESTIMATES FOR EACH SOIL CLUSTER BY ELEVATION BAND

The PDSI is a measure of accumulated relative soil moisture calculated from temperature and precipitation, given a specific latitude and available water supply. It ranges from approximately –5, indicating severe drought, to 5, indicating a very high positive water balance. Of course, temperature and precipitation vary greatly with elevation in our study area, since elevations range from about 1,500 m where McElmo Canyon exits to the west, to 3,040 m at the top of Ute Mountain on the southern study-area border. We generally followed the approach of Van West (1994:41–44) in using weather stations in or near the project area to represent the temperature and precipitation at various elevations. She used five stations, including Ft. Lewis and Ignacio, which we removed because both are rather far east of our study area. Instead we used four stations: Bluff (at 1,461 m, to represent soils below 1,604 m in elevation, constituting 2.6 percent of our area); Cortez (at 1,835 m, for elevations between 1,604 and 1,992 m, 49.7 percent of our area); Yellow Jacket (at 2,115 m, for elevations between 1,992 and 2,130 m; 32.5 percent of our area); and Mesa Verde (elevation, 2,169 m) for elevations above 2,130 m (15.2 percent of our area). (See Plate 6.3.) Experienced local dry-land gardener and archaeologist Linda Honeycutt (1995:373) identifies the best dry-farming lands in our study area as lying between 1920 and 2195 m, therefore largely contained within the band proxied by Yellow Jacket.

Using version 2.0 of the PDSI program developed at the National Agricultural Decision

Support System of the University of Nebraska-Lincoln (available through http://greenleaf.unl.edu/downloads/), we estimated the PDSI values for the first of every month for the duration of the instrumented record at each station, using the available water-supply parameters applicable to each of our 14 soil clusters and the latitude at the center of our study area. This program was thus run 4×14, or 56 times, resulting in 56 tables of monthly PDSI values, one for each combination of soil cluster and elevation band. Further use of these PDSI values was restricted to those expected on the first of June every year, following arguments by Van West (1994). Soil moisture in early June reflects the accumulated influences of temperature and precipitation over the last year; low values inhibit seed sprouting and critical early growth of maize.

STEP 4: FIND A TREE-RING PROXY FOR RETRODICTING PDSI ESTIMATES

To use these PDSI estimates as a basis for estimating productivity during the Pueblo period, we had to verify that they are significantly correlated with a tree-ring record that extended at least to A.D. 600. We chose the Mesa Verde Douglas fir indexed series (after standardization) for this, as already explained. Table 6.4 reports key statistics from the linear-regression model in which we attempt to predict the June PDSI values calculated from the weather station data for each of the four weather stations and each of the 14 soil clusters from the Mesa Verde Douglas fir indexed series value for that same year.

The high r^2 values in Table 6.4, and the low p values associated with them, indicate that the PDSI values calculated from historical weather station data, and the Mesa Verde Douglas fir ring widths, respond in a very similar way to changes in temperature and precipitation even as PDSIs change across the range of soil depths and elevations experienced in our study area. This makes sense, as both respond relatively slowly to changes in precipitation and temperature. (However, the extent to which Douglas fir responds more slowly than maize to such changes is a fundamental limitation of the high-frequency component of our paleoproductivity model.) The uniformly positive slopes indicate that higher PDSI values are tied to larger ring widths. Not surprisingly, the best fits tend to be found with the Mesa Verde-elevation soil complexes, since their PDSIs are calculated from weather data accumulated next door to the Douglas firs providing these rings. The lowest fits in general tend to be found with soil cluster 10. This is a group of the 19 shallowest soils in our area, with virtually no lower soil horizon and therefore little available water capacity. Since these soils are also very unproductive for agriculture, their relatively poor fit with the tree-ring proxy is of little concern. Overall, the r^2 values indicate that the tree-ring proxy explains about half the variability in the PDSI calculated from the historical data—and a little more in the higher-elevation, deeper soils most important for maize production. This gives us confidence that we can use this proxy to retrodict PDSI values into the Pueblo past. We therefore did that for each of these 56 combinations of elevation and soil group, using the Mesa Verde Douglas fir indexed series as the independent variable and the slopes and the intercepts reported in Table 6.4.

STEP 5: PRODUCE WEIGHTED, POOLED PDSI VALUES FOR "BEAN SOILS"

Our model for maize paleoproductivity depends on a data set laboriously collected from several sources by Barney Burns during his dissertation research at the University of Arizona (1983). Burns located data sets that were aggregated and published by the State of Colorado on nonirrigated maize and bean production from the 1920s through 1960 for five counties in southwestern Colorado. Van West (1994) used values from this data set from Montezuma County from 1931 to 1960, and we use these same values in forming the calibration between

TABLE 6.4

Goodness-of-Fit in the Linear-Regression Model Standardized Mesa Verde Douglas Fir Indexed Series Value = June PDSI Estimate, for Each of the 56 Combinations of Elevation and Soil Cluster

Soil Cluster	Intercept	Slope	r^2	$p>F$
		Bluff		
1	−.110	2.715	.424	.000
2	−.117	2.764	.439	.000
3	−.042	2.635	.396	.000
4	−.134	2.733	.426	.000
5	−.172	2.535	.432	.000
6	−.132	2.740	.429	.000
7	−.246	2.292	.459	.000
8	−.283	2.232	.444	.000
9	−.196	2.489	.436	.000
10	−.294	1.578	.317	.000
11	−.110	2.714	.424	.000
12	−.114	2.756	.434	.000
13	−.109	2.719	.425	.000
14	−.039	2.606	.319	.000
		Cortez		
1	.299	3.182	.582	.000
2	.238	3.038	.597	.000
3	.302	3.175	.579	.000
4	.297	3.170	.596	.000
5	.236	2.713	.553	.000
6	.217	3.083	.593	.000
7	.061	2.306	.477	.000
8	.047	2.283	.473	.000
9	.160	2.594	.538	.000
10	.049	1.719	.351	.000
11	.299	3.181	.582	.000
12	.229	3.060	.592	.000
13	.287	3.182	.582	.000
14	.164	2.994	.583	.000

(continued)

TABLE 6.4 *(continued)*

Soil Cluster	Intercept	Slope	r^2	$p > F$
		Yellow Jacket		
1	.552	2.855	.649	.000
2	.477	2.553	.627	.000
3	.557	2.862	.648	.000
4	.487	2.791	.650	.000
5	.352	2.246	.560	.000
6	.539	2.761	.657	.000
7	.000	1.710	.366	.001
8	–.045	1.646	.346	.001
9	.237	2.096	.507	.000
10	–.185	1.498	.354	.001
11	.551	2.857	.648	.000
12	.593	2.649	.641	.000
13	.547	2.852	.649	.000
14	.476	2.977	.662	.000
		Mesa Verde		
1	–.381	2.835	.669	.000
2	–.227	2.610	.639	.000
3	–.367	2.872	.671	.000
4	–.334	2.702	.644	.000
5	–.258	2.255	.554	.000
6	–.289	2.676	.647	.000
7	–.304	2.020	.506	.000
8	–.316	2.001	.498	.000
9	–.246	2.198	.543	.000
10	–.368	1.412	.323	.000
11	–.380	2.837	.669	.000
12	–.293	2.617	.627	.000
13	–.377	2.836	.669	.000
14	–.082	2.238	.576	.000

PDSI = Palmer Drought Severity Index.

proxies related to annual weather and crop production.[2]

Unfortunately, we do not know precisely where within Montezuma County the production appearing in these tables comes from and cannot specify which soils produced these yields. The best we can do is to presume that these yields generally apply to soils identified in the soil surveys as potentially high in yield and historically used for dry farming.

Therefore, the next step in our analysis was to group the soils likely to have been producing the dry-farmed crops in these tables. We refer to these as the "bean soils" because beans constitute their main contemporary crop, but we presume that they include the areas where the vast majority of both the dry-farmed maize and beans were produced during our calibration period between 1931 and 1960. We need to identify these soils within the county in order to correlate our measures of PDSI with yields in each year, since our measures of PDSIs need to be made on soils that are representative of the soils contributing to these yields.

For each of these "bean soils," therefore, we first calculated the proportion of that soil within the historical dry-farming niche in the county. We then multiplied that proportion against the PDSI estimate (as calculated from the historical weather station data) that pertained to that soil cluster, elevation band, and year. The net result of this process was a new annual estimate for PDSI that was weighted proportionally to the representation of soils that we believe were being used for dry farming.[3] For example, the soils in soil cluster 1, in the Yellow Jacket band, had a weighting of only .02 in producing the pooled estimate of bean-soil PDSI, since they were little used for agriculture. The soils in cluster 3 in the Yellow Jacket band, on the other hand, are the most important in contemporary dry farming and had a weight of .49, meaning that almost half of the annual estimates for PDSI for the bean soils were derived from the PDSI estimates for this combination of soil cluster and elevation. Overall, soils in the Cortez elevation band are little used for dry farming and contributed less than 5 percent to the pooled PDSI value for the bean soils. Soils in the Mesa Verde elevation band contributed 7.7 percent. The few soils in our area in the Bluff band are too low for dry farming and did not contribute at all to the pooled PDSI estimate for the bean soils. Soils in the Yellow Jacket band overwhelmingly dominate those used for dry farming today, and their total weight in calculating the pooled annual PDSI estimates was almost 88 percent. The pooled, weighted estimates for bean-soil PDSIs from 1931 to 1960 are shown in Table 6.5, along with local historical commentary collected by Van West (1994:Appendix B) on the weather and production in those years. These remarks show good qualitative agreement with the calculated values for the PDSI.

STEP 6: DETERMINE THE "BEST" MODEL FOR RETRODICTING MAIZE PALEOPRODUCTIVITY

One other problem with using the production series discovered by Burns is that U.S. farms increasingly relied on technological intensification throughout the twentieth century. Like their counterparts elsewhere, between 1931 and 1960 southwestern Colorado farmers used progressively more fertilizer, mechanized

2. Burns' figures (1983:310) for Montezuma County begin in 1919, but the yields he reports between 1919 and 1930 seem abnormally high (mean = 15.6 bushels / harvest acre) when compared against those for the period between 1931 and 1960 (13.7 bushels / harvest acre), most of which benefited from much greater technological inputs. Moreover, the yields in these early years do not fit the regression relationships between production and climatic variables that pertain to the period between 1931 and 1960. It is likely that some inconsistency was present in the way values were reported before, and after, about 1930. We follow Van West in dealing with this issue; we use only the 1931 to 1960 values in establishing the calibration between climate and production.

3. The weighted, pooled PDSI estimates that we produced for the bean soils, which we called Rec32, are highly correlated with a similar measure produced by Van West (1994), which she called Rec34 ($r = .952$ for the period between 1931 and 1960).

TABLE 6.5

Annual Pooled Estimates for "Bean-Soil PDSI" from 1931–1960 Along with Local Historical Comments Assembled by Van West (1994:Appendix B)

Year	Bean-Soil PDSI	Comments[a]
1931	.751	
1932	3.963	
1933	–.645	
1934	–3.020	Record drought struck Montezuma County, probably the driest year since its occupancy by the white man: very light winter snowfall, almost no spring storms, very scant summer rains, crop production at lowest levels ever known. U.S. government bought cattle to help ranchers. Grasshoppers, prairie dogs, and bean beetles added to the calamities. Rains fell during week of July 20; some beans and corn that had survived produced a good crop. 1934 is frequently mentioned by present and former Goodman Point residents as a drought year.
1935	1.884	Plenty of water and good crops reported throughout county. Drought was broken during winter of 1934–1935, and soil profile refilled with water.
1936	–1.081	
1937	1.294	Record snowfall in Mesa Verde National Park in winter 1936–1937 (367.6 cm, or 145.5 in.), and deep mud the following spring.
1938	1.998	
1939	.548	
1940	.041	
1941	4.497	
1942	2.499	
1943	.590	
1944	.661	
1945	.357	
1946	–2.357	Wheat harvest exceeded previous dry-year estimates; some nonirrigated wheat fields returned 15 bushels / acre, and even farmers getting only 3–5 bushel returns showed a profit because of high prices.
1947	1.020	Recalled by previous and current Goodman Point residents as an example of a good year.
1948	2.021	
1949	3.567	
1950	–.896	
1951	–3.765	Recalled by previous and current Goodman Point residents as a year of severe drought with almost total crop failure. County's yield of beans was the lightest per acre ever known. Yield, on average, was expected to be 20% of normal.
1952	1.890	
1953	–1.946	

(continued)

TABLE 6.5 *(continued)*

Year	Bean-Soil PDSI	Comments[a]
1954	–1.564	Winter of 1953–1954 received only 65% of the average snowfall, and a shortage of irrigation water ensued.
1955	–.174	
1956	–1.540	The Yellow Jacket locality produced one of the "shortest" crops ever. Summer was extremely dry. The past few years (1953–1956) had also been drier than average, but did not approach possibility of total crop failure.
1957	1.819	
1958	1.884	
1959	–4.057	Bean yield was the worst in years: one-half to 5 bags / acre, with an average of 2 bags / acre.
1960	1.318	

PDSI = Palmer Drought Severity Index.

a. These are highly condensed; see Van West (1994:239–240) for full quotations and sources.

equipment, pesticides, herbicides, hybrid varieties, and closer planting intervals. These practices tended to boost yields, or they would not have been continued.

Burns therefore removed what he called a "technology trend" by regressing historical yield data on pounds of fertilizer applied per harvested acre (drawn from data for the entire state of Colorado), which he considered to be correlated with other technological inputs. (His estimates for the amount of fertilizer applied per harvested acre are graphed in Kohler 2010:Figure 5.1). Then, for further analysis, he used the residuals from the regression of production on current and lagged fertilizer use. Van West (1994:103–104) developed an alternative procedure with a similar effect, which I also follow. She (and I) fitted an independent variable that can be understood as the technology trend that was simply the (standardized) year from which the yield figures came, as part of a multiple linear regression that also included proxies for PDSI and temperature. Table 6.6 reports the values for the independent and dependent variables in the regression model used here. This production series contains missing values from 1944 through 1947. Burns (1983:58–63) developed a multiple linear-regression procedure to fill those in, and we used the estimates for those missing years that he produced. I made one design decision in performing the original calibration that future workers might want to revisit (see also note 2). I followed Van West in using only the production figures for 1931 through 1960 in the calibration. Burns reports maize values back to 1919 for Montezuma County, though he states that the first two years are unusable. My own attempts to use the pre-1931 maize production figures resulted in poor goodness-of-fit in the regressions, so I used these data only from 1931 through 1960.

After some exploration of other regression models not reported here, for this calibration I settled on the three-independent-variable least-squares model already briefly reported in Kohler et al. (2007:65) and Varien et al. (2007:278). Because we have two candidate temperature proxies, I produced a calibration specific to each of them using the maize production data from 1931 through 1960 (Table 6.7).

Note that the Almagre temperature proxy (used in the first model shown in the preceding text) provides a slightly better fit to the historical maize production data, though both models

TABLE 6.6

Variables and Values Used in Determining Model for Maize Paleoproductivity

Year	Maize Montezuma County (Bushel / Harvested Acre)[a]	Year (Standardized)	Rec32[b]	Rec32 (Standardized)	Prin1[c]	Almagre Index[d]	Almagre Index (Standardized)
1931	9	−1.647	.751	.170	.683	1313	1.444
1932	12	−1.533	3.936	1.647	1.415	1280	1.299
1933	12	−1.420	−.645	−.477	1.790	1409	1.865
1934	4	−1.306	−3.020	−1.578	0.160	918	−.288
1935	10	−1.193	1.884	.696	1.453	1383	1.751
1936	9	−1.079	−1.081	−.679	2.183	1405	1.848
1937	12	−.966	1.294	.422	2.317	1427	1.944
1938	12	−.852	1.998	.748	1.512	1426	1.940
1939	9	−.738	.548	.076	1.317	1267	1.242
1940	12	−.625	.041	−.159	1.814	1390	1.782
1941	20	−.511	4.497	1.907	2.132	1554	2.501
1942	15	−.398	2.499	.980	2.777	1722	3.238
1943	16	−.284	.590	.095	3.187	1641	2.883
1944	12	−.170	.661	.128	1.878	1416	1.896
1945	16	−.057	.357	−.013	2.019	1403	1.839
1946	18	.057	−2.357	−1.271	1.982	1399	1.821
1947	20	.170	1.019	.294	3.044	1672	3.019
1948	13	.284	2.021	.759	2.448	1629	2.830
1949	19	.398	3.567	1.476	2.338	1559	2.523
1950	16	.511	−.896	−.594	1.880	1326	1.501
1951	15	.625	−3.765	−1.924	2.929	1667	2.997
1952	15	.738	1.890	.698	1.298	1069	.374
1953	14	.852	−1.946	−1.080	2.921	1278	1.291
1954	14	.966	−1.564	−.903	2.519	1327	1.506
1955	14	1.079	−.174	−.259	2.362	1319	1.470
1956	18	1.193	−1.540	−.892	3.290	1224	1.054
1957	19	1.306	1.819	.665	3.991	1631	2.839
1958	16	1.420	1.884	.696	4.728	1733	3.286
1959	15	1.533	−4.058	−2.060	3.804	1664	2.983
1960	16	1.647	1.318	.433	1.572	1224	1.054

NOTE: Year and Rec32 were standardized from 1931 through 1960 for this analysis. Almagre indices were standardized over the entire period of their availability.

a. Nonirrigated corn crop records from Montezuma County as originally reported by Colorado's Crop and Livestock Reporting Service, summarized by Burns (1983:43–65, Table 2-4).

b. "Bean-soil" PDSI (from Table 6.5).

c. Scores on first component from PCA of San Francisco Peaks and Almagre indexed series.

d. From Graybill's extension of the original Almagre series to 1983.

TABLE 6.7

Two Regression Models for Maize Productivity Employing Different Temperature Proxies

Variable	Label	Parameter Estimate	Standard Error	*t* Value	$p > t$
Model: Maize (bu / ac) = year Rec32 Almagre ($R^2 = .62$; Adj. $R^2 = .58$; $p > F = .000$)					
Intercept		11.38	1.15	9.91	.000
Year	Standardized date A.D.	2.28	.48	4.81	.000
Rec32	Standardized weighted bean-soil PDSI	1.10	.47	2.33	.028
Almagre	Standardized indexed series value	1.40	.55	2.53	.019
Model: Maize (bu / ac) = year Rec32 Prin1 ($R^2 = .59$; Adj. $R^2 = .54$; $p > F = .000$)					
Intercept		11.18	1.50	7.43	.000
Year	Standardized date A.D.	1.78	.63	2.82	.010
Rec32	Standardized weighted bean-soil PDSI	1.33	.48	2.78	.010
Prin1	Scores on first principal component	1.28	.64	2.01	.055

PDSI = Palmer Drought Severity Index.

provide acceptable fits. Collinearity is not an issue in either model. The signs of the parameter estimates for both models indicate that as technological inputs increase through time, maize production rises (this is the strongest effect in both models). Controlling for that effect, maize production also responds positively to higher PDSI values, as well as to higher values for the Almagre series by itself, or alternatively to higher scores on the component indexing the shared variance between the Almagre and SFP series.

Given that the calculation of PDSI depends both on precipitation and temperature—since both affect evapotranspiration and therefore accumulated water storage—it is interesting that the tree-ring proxies for temperature have a separate significant positive effect, holding PDSI constant. Higher summer temperatures, and longer growing seasons, depress PDSI, holding precipitation constant, but in this area where temperatures can often be limiting, longer growing seasons and warmer summers have an independent effect on increasing yields, holding PDSI constant. That is the fundamental justification for employing temperature (as proxied) as an additional independent variable in these reconstructions. In either model, as we might expect, a cold, dry growing season generates the worst possible yields, whereas a warm, wet summer provides the highest yields.

One obvious problem in using these regression equations to retrodict production is that the "year" was entered to hold the historical technology trend constant and of course cannot be used in the retrodiction. Since the year was standardized, taking on a mean value of zero between 1945 and 1946, the practical effect of not using this variable in the retrodiction is to retrodict maize production values that would be appropriate to the mid-1940s, when this variable takes on a value of zero. In step 9, we renorm our estimates to make them applicable to the technology we deem appropriate for the Pueblo period, rather than to the mid-1940s.

STEP 7: REWEIGHT PRODUCTION APPROPRIATELY FOR EACH SOIL CLUSTER

The estimates for maize productivity produced in step 6 allow us to estimate how much maize could have been grown on the "bean soils" as climates vary. But they do not apply to the other soils in our study area. Now we develop a ratio that will allow us to translate the production on the bean soils to the other soils in the study area (as lumped together in soil clusters). We do this by employing the only modern and fairly consistent estimator for production we have that applies to all the soils: the estimates in the soil surveys for the "total dry-weight production" of vegetation. Specifically, we used their estimates for the normal-year production in pounds per acre (e.g., Ramsey 2003b:Table 7).[4] According to the NRCS soil surveys, these soils as a group have a weighted mean normal-year dry-weight vegetation production of 1,093 lbs / acre. Soils in cluster 10, on the other hand, which are shallow, have a median dry-weight vegetation production of only 571 lbs / acre (Table 6.8). Maize is assumed to respond to these differences in vegetative productivity in the same proportion as other vegetation, so before any further steps are done, a weight was calculated for each soil appropriate to its soil cluster. For cluster 10, for example, maize production was reweighted by the proportion of its cluster's normal-year dry-weight production to that of the bean soils, or 571 / 1093 (=.523). The same procedure was completed for all 14 soil clusters.

STEP 8: ESTIMATE POTENTIAL MAIZE PRODUCTION FOR EACH YEAR, SOIL CLUSTER, AND ELEVATION BAND

We are now in a position to produce maize production estimates for each of the 45,400 200 × 200-m cells in the model, for every year from A.D. 600 through A.D. 1300. This is the last step done in our statistical software (SAS). The results of this step are read into the agent-based model, and further corrections are done on the fly as the model runs. Each cell, of course, is assigned to a soil cluster and an elevation band. For each cell, in each year, we compute the following: maize production = intercept (from step 7) + [partial slope (from step 7) * PDSI value appropriate to that year, soil cluster, and elevation band (from step 4)] + [partial slope (from step 7) * temperature proxy for that year (either Almagre or Prini)].

After the reweighting discussed in step 7, these estimates are written to 701 text files, one for each year. From there they are read into the appropriate cells as the model steps through the years. In the model, these are converted to kg / ha.

STEP 9: RENORM MAIZE PRODUCTION FOR PREHISPANIC VARIETIES AND CULTIVATION PRACTICES

The potential maize production values up to this point are what we might expect in each cell, in each year, given planting practices and the technologies available in the 1940s. Step 9 adjusts these values so the estimates of maize productivity are consistent with cultivation during the Pueblo period. (Of course, many of these soils could not be planted with tractors at all—or in some cases even by hand—and we account for such limitations in the next step). This and subsequent steps are performed in the model so that if changes need to be made, we can do so without having to produce new paleoproductivity data planes.

How can we correct these values to be appropriate to prehistory? To be frank, this process is imperfect, which is the main reason that we parameterize maize production in the simulation to make it easy to investigate the effects of different production levels. We approach this problem using guidelines from Hopi and Zuni ethnographies and ethnoagricultural experi-

4. We are pushing these data, which according to NRCS soils scientists are based on limited range clipping and fieldwork, beyond their intended use as a general guide for livestock stocking rates. We hope that soils work now being performed as part of Village Ecodynamics Project II will allow us to generate better estimates of potential soil variability for maize production.

TABLE 6.8

Reweighting of Maize Production by Proportion of Each Soil Cluster's Normal-Year Dry-Weight Production Relative to That of the Pooled Bean Soils (1093 lbs / ac)

Soil Cluster	*N* Soil Complexes in Cluster	Median AWS Lower Profile	Pooled Weighted Normal-Year Dry-Weight Production (lbs / ac)	Reweighting Factor
1	15	8.0	812	.73
2	26	5.3	845	.77
3	26	8.3	1,307	1.20
4	24	6.5	1,282	1.17
5	27	3.6	976	.89
6	4	6.2	2,831	2.59
7	19	2.4	378	.35
8	6	2.3	1,355	1.24
9	5	3.2	2,240	2.05
10	19	0.8	571	.52
11	9	8.0	2,109	1.93
12	5	5.8	2,247	2.06
13	3	7.9	3,360	3.07
14	1	9.0	336	.31

ments (e.g., Adams et al. 1999; Muenchrath et al. 2002). Recently, relevant demonstration gardens and experimental farming projects have been comprehensively reviewed and added to by Benjamin Bellorado (2007). We present some of the most applicable data in Table 6.9. Considering only those data from dry-farmed fields (some of which were located like akchin plots to receive subirrigated water—i.e., moisture due to runoff), yields are still highly variable, ranging from zero to over 1,300 kg / ha (about 21 bu / acre). Bellorado's yields from the Durango, Colorado area are among the highest, and he notes that his research "produced large yields because the conditions under which I planted and grew maize were prime for dry land farming techniques, with good soils, sufficient frost-free periods and heat units, and relatively high amounts of moisture available for developing maize plants" (2007:257)—though production in some of his plots was decreased by adverse cold-air drainage conditions.

Using these data, we chose a round figure of 500 kg / ha (about 8 bu / acre) for our "bean soils" in an average year, a figure that is in accord with—but not specifically determined by—the data. Keeping in mind that since the weighting factors in Table 6.8 will apply against this mean, the effect of this is to assign potential mean yields of about 155 kg / ha (about 2.5 bu / acre) to our least favorable soils (and this will be reduced further later in this chapter), and yields about three times that, or 1535 kg / ha (about 24 bu / acre), to our most favorable soils, in an average year. These yields of course are further modulated by retrodicted PDSI and temperature for any specific year. To achieve this renorming, we multiplied all yields produced in step 8 by .68. That is, we estimate that, on average, yields per unit area from A.D. 600 though A.D. 1300 were about a third less

TABLE 6.9

Data Guiding Our Renorming of Modern Maize Production Values to Be Appropriate to Prehistory

Project Name	Maize Variety	Mean (kg / ha)	Comments	Source
		Experimental Farming Projects		
MAÍS Southwest	Alpha Group 1	3,292	Irrigated & fertilized	Adams et al. 2006
	Alpha Group 2	1,840	Irrigated & fertilized	Adams et al. 2006
	Alpha Group 3	2,415	Irrigated & fertilized	Adams et al. 2006
	Alpha Group 4	2,169	Irrigated & fertilized	Adams et al. 2006
ALP Experimental Farming Project	Hopi Blue Flour	1,346	Not irrigated or fertilized but watered at planting	Bellorado 2007
	Hopi Red Flour	980	Not irrigated or fertilized but watered at planting	Bellorado 2007
	Navajo White Flour	1,066	Not irrigated or fertilized but watered at planting	Bellorado 2007
	Copper Canyon Blue Flour	432	Not irrigated or fertilized but watered at planting	Bellorado 2007
	Copper Canyon Red Flour	416	Not irrigated or fertilized but watered at planting	Bellorado 2007
	Copper Canyon Blond Flour	372	Not irrigated or fertilized but watered at planting	Bellorado 2007
Zuni	Zuni Blue Flour	572		Muenchrath et al. 2002
NMSU Agricultural Science Center, Los Lunas	Tohono O'odham "60-day" Flour	1,395[a]	Treatment 5, watered only at planting (average of two years)	Adams et al. 1999
		Ethnographic Reports		
Hopi		25–255	Widely spaced clumps, deeply planted seeds	Manolescu 1995; Dominguez and Kolm 2005
		630–756 on average (10–12 bu / ac)	Citing with approval earlier estimates by A. M. Stephens	Bradfield 1971:21
Zuni	Various "traditional, open-pollinated cultivars"	0–1,841 (mean from 5 fields, 572)	Fields located to profit from runoff or floodwater	Muenchrath et al. 2002

ALP = Animas-La Plata Project; MAÍS = Maize of American Indigenous Societies; NMSU = New Mexico State University.

a. Kg / ha not reported; this figure was extrapolated from total grain yields of 634 (in 1992) and 194 g (in 1993) from seed planted in 12 hills spaced 1 m equidistantly.

than yields would have been from those same areas in the mid-1940s.

STEP 10: MAKE HAND-PLANTING ADJUSTMENT

According to the soil surveys in our area, as well as our own familiarity with this landscape, many of the soil complexes in our area contain within them portions that are too steep, rocky, boggy, or alkaline to be suitable even for hand planting. The NRCS surveys report a hand-planting restriction code ranging from zero (no restrictions) to 1 (complete restriction—no planting possible). Most commonly, the reason for the restriction on hand-planting suitability is due to soil stoniness or lack of depth, both of which generally occur in soils developed on steeper slopes. Soils deposited on gentle slopes are usually deeper, more uniform, less stony, have higher available water-supply values, and are more productive for agriculture. The soil surveys also report degree of "crop suitability" somewhat independently of this hand-planting restriction. For soil complexes where the crop suitability is listed as "unsuitable," we multiply the yields from step 9 by a factor derived from the hand-planting restriction.[5] This step further reduces potential yields in the least favorable 53.2 percent of our study area.

STEP 11: MAKE COLD CORRECTION

The estimation procedure so far favors production in higher areas, and the higher the better because precipitation increases with elevation. Experienced local farmers and gardeners, however, recognize that the risk of experiencing a growing season that is too short or cool for crops to mature also generally increases with elevation. As mentioned, Honeycutt (1995:373) specifies 2,195 m as the local upper limit for obtaining a favorable combination of temperature and precipitation. The length of the growing season that maize needs to mature depends on the variety and on the temperature during the growing season. Deborah Muenchrath and colleagues (2002:20) report that varieties currently planted at Zuni require about 125 days to mature, and Maitland Bradfield (1971) reports that Hopi maize matures in 115 to 130 days.

Our attempts to calibrate this effect are limited by available data. We located some yield data from a facility in Hesperus, Colorado, originally called the Fort Lewis School of Agriculture, which at about 2,315 m (7,600 ft)—with farm plots at about 2,322 and 2,326 m—is near the upper elevation limit of the dry-farming niche, as the plots' spotty production indicates (Table 6.10). For the period between 1923 and 1959 covered by the table, the length of the growing season averaged 114 days (s=18). The way in which this number was computed was nonstandard, however, and by a more conventional measure the mean growing season is close to 100 days, which is often said to be the minimum possible length of the growing season for maize.[6] As a cross-check on the consistency

5. Actually, this further restriction is applied to all soils reported in the Cortez Area Survey (Ramsey 2003b) as "unsuited" for agriculture, and to *all* soils of both the Animas-Dolores Area (Pannell 2003) and Ute Mountain Area Surveys (Ramsey 2008). The specific multiplier used is obtained by subtracting the hand-planting restriction code from 1 and dividing the result by 2. Thus, a hand-planting restriction code of 1 results in a multiplier on the potential maize productivity of .5—in effect, halving the potential production. This is in recognition of the possibility that contemporary views of what is unsuitable even for hand planting may not be completely applicable to the prehispanic context.

6. From 1926 to 1949, short reports were issued from an organization located at Hesperus, variously called the Fort Lewis College of Agriculture, the Fort Lewis School of the Colorado Agricultural College, the Fort Lewis substation of the Colorado Agricultural College, the Fort Lewis substation of the Colorado State College of Agriculture and Mechanical Arts, the Fort Lewis Station, or the San Juan Basin Branch Station of Fort Lewis A & M College, all written by Dwight Koonce when the authorship was noted and all reporting yields from the same Hesperus location (and occasionally elsewhere). Reports continued through 1961, variously authored by Herbert Mann, William Paulson, or Vernon Cardwell, who noted after a failed maize crop in 1959 that "there was no reason to continue tests of maize: no adapted variety is available." In the early 1960s the Hesperus station began to be phased out in favor of the Yellow Jacket loca-

TABLE 6.10

Local Data Relevant to Establishing Cold Correction to Potential Yields, from the Hesperus–Ft. Lewis Experimental Station

Year	Crop	Yield	Frost-Free Period (Days)[a]	Almagre Index	Prin1 Index[b]	Comments
1923			91	–.442	–1.460	From 1929 report
1924			85	–.858	–1.061	From 1929 report
1925			94	–.613	–1.168	From 1929 report
1926			136	–.468	–1.124	Frost-free period exceptionally long
1927			117	–.157	–1.061	
1928			119	.418	–.935	
1929			103	1.181	–.239	Crop destroyed by hail
1930	Corn	62 bu / ac	114	.984	.204	Pinto yields "very good"
1931	Corn	50 bu / ac	113	1.444	.683	
1932	Corn	50 bu / ac	94	1.299	1.415	Light frost on September 10 injured corn only slightly
1933			130	1.865	1.790	
1934			111	–.288	.160	
1935			120	1.751	1.453	
1936	Pintos	130 lbs	120	1.848	2.183	
1937			111	1.944	2.317	
1938	Corn	24.3 bu / ac	139	1.940	1.512	Distribution of rainfall unfavorable to pintos
1939	Corn	30.8 bu / ac	103	1.242	1.317	Shortage of irrigation water; cold & dry
1940	Corn	25 bu / ac	114	1.782	1.814	
	Pintos	989 lbs / ac			2.132	
1941	Corn	0 bu / ac	91	2.501	2.132	Very wet & cold
	Pintos	0 lbs / ac				
1942	Corn	35.7 bu / ac	112	3.238	2.777	
	Pintos	9.8 bu / ac				
1943			129	2.883	3.187	High pinto yields
1944	Pintos	652 lbs / ac	108	1.896	1.878	
1945	Pintos	1,337 lbs / ac	92	1.839	2.019	Corn yields not reported; crop failure or none planted?
1946	Pintos	2,018 lbs / ac	111	1.821	1.982	Corn yields not reported; crop failure or none planted?
1947	Pintos	1,224 lbs / ac	91	3.019	3.044	Corn yields not reported; crop failure or none planted?
1948	Pintos	451 lbs / ac	139	2.830	2.448	Very long growing season

(continued)

TABLE 6.10 *(continued)*

Year	Crop	Yield	Frost-Free Period (Days)[a]	Almagre Index	Prin1 Index[b]	Comments
1949	Pintos	486 lbs / ac	108	2.523	2.338	
1952			112	.374	1.298	No reports found for 1950 & 1951
1953	Corn	44 bu / ac	123	1.291	2.921	Hybrid
1954	Corn	78 bu / ac	129	1.506	2.519	Hybrid, irrigated
1955	Corn	47.1 bu / ac		1.470	2.362	Hybrid, no growing season length reported
1956	Corn	53.8 bu / ac	151	1.054	3.290	Irrigated; really long season but very dry
1957	Corn	43.6 bu / ac	86	2.839	3.991	Cold & wet. "Not at all feasible to grow corn for . . . grain production in this area."
1958	Corn	73.3 bu / ac	159	3.286	4.728	Hybrid; irrigated & fertilized
1959	Corn	0 bu / ac	119	2.984	3.804	Early frost & cool season

NOTE: Elevation of farm plots about 2,325 m (7,630 ft).

a. As computed by noticeable frost damage to crops. This method results in a computed frost-free growing season that is, on average, about 12 days longer than seasons computed on temperature readings of 32° F or less, according to the 1947 report.

b. Scores on first principal component of principal component analysis of Almagre and SFP indexes.

of the growing-season estimates from this facility with our own tree-ring-based cold proxies, we note that their estimates are, as expected, positively—but not significantly—correlated with both the Almagre z-scores ($r^2=.04$, $p=.24$) and the scores on the first principal component of Almagre and SFP ($r^2=.10$, $p=.07$). Unfortunately, yields and growing season lengths are not always noted; fields were sometimes irrigated and sometimes not; and whether they were irrigated, and how often they were irrigated, varied both with crop need and water availability. Crop varieties changed through time. Either there were no reports in 1950 and 1951, or we have not been able to locate them.

Nevertheless, for the 14 years with both maize yields and growing-season estimates, yields do rise with longer growing seasons ($r^2=.17$), though not significantly ($p=.14$). (Introducing a standardized year as a second independent variable, to help control for technology trend, does not improve the fit: $r^2=.18$, $p=.34$.) Qualitatively, we can see that crop failures at this elevation are not uncommon and that the reason frequently given is short, cold growing seasons.

We find ourselves in a situation that is similar to that in step 9: we can build a function to decrease production at high elevations in cold years using available but imperfect data, but our correction function is necessarily more specified than warranted by the available data. Nevertheless, we think that it is better to have an imperfect correction for cold than no correction at all. Here is what we do:

- First, we disallowed any production on soils above 2,395 m (about 7,860 ft) in any year. This 2,395-m limit is about 70 m above the

tion, which at about 2,118 m (6,950 ft) is much less subject to adversely short growing seasons and therefore less interesting for our purposes here.

Hesperus location and 200 m above the elevation that Honeycutt (1995) considers optimum locally. This completely removes the highest 2.2 percent of our study area from maize production.

- Next, we decided that soils below 2,150 m (about 7,055 ft) did not have any special risk for short growing seasons. Of course, the regression equations in step 6 apply to these soils, which constitute 86.9 percent of the study area, so negative values on the temperature proxy will decrease production at these lower elevations, though rather modestly.
- Finally, we identified a band of soils between 2,150 and 2,395 m (10.9 percent of the study area) for which we compute a function, for each cell in each year, that decreases production more than that specified by the step 6 regression formula. In years when our temperature proxy is negative, we compute a "cold correction" for these soils as follows:

$$coldcorr = ([2395 - elev]/245) * ([2.986 + PC1]/2.986)$$

Coldcorr ranges from zero – 1.0 and acts as a multiplier on production. The first right-hand term ranges from zero (when elevation = 2,395) to 1 (when elevation = 2,150) and takes on progressively lower values as the elevation approaches 2,395. The second right-hand term ranges from zero (when PC1 = –2.986, its lowest value) to 1 (when PC1 = 0). Obviously, this term takes on lower values as PC1 gets more negative (indicating colder or shorter summers). PC1 is the score on the first principal component. A slightly different form of the equation, with the same effect, is used for the Almagre proxy.[7]

The net result of this procedure is to decrease production progressively as elevation increases within this band—beyond that already dictated by the regression equation—and as summers become shorter or cooler. No correction is made within this elevational band during years when the standardized temperature proxy is neutral or positive. Note that this is not a correction for cold-air drainage, which would require a different and more complicated model. Our lack of any downward correction for cold-air drainage may positively bias our estimates slightly.

STEP 12: ALLOW FOR POSSIBLE SOIL DEPLETION OR DEGRADATION

Several soil properties are subject to change from use, potentially reducing agricultural productivity. Current literature on the subject of soil degradation is dominated by discussions of soil erosion from tilling in the course of modern farming practices (Brengle 1992; Hillel 1991; see especially contributions in Part 3 of Unger et al. 1988). Since the hand-planting restriction (step 10) already reduces production on high-slope soils, we are not particularly concerned with the effects of erosion here. Our goal in this step is to account for possible depletion of soil nutrients resulting from long-term cropping in a given model cell. Discussion of this subject in the soils literature is sparse and conflicting, dealing primarily with the effects of modern intensive farming in tropical contexts. Dry farming is also sometimes addressed (Brengle 1992; Sandor et al. 1990; Unger et al. 1988), as is floodwater farming in the Southwest (Hack 1942; Norton et al. 2003).

Soil-nutrient depletion has not been intensively studied for the semiarid soils of the upland Southwest, and we have been unable to find any technical discussion of the results of long-term planting in the region. (As this book

7. Karen Adams and Kenneth Petersen (1999:26) would apparently favor a more restrictive correction: "Weather stations in or adjacent to the study area above 2134 m (7000') average 2,000 CGDD [corn growing degree days] for May-September, too few for successful maize agriculture. Stations between 1829 and 2134 m (6000 and 7000 ') average a barely adequate 2,345 units." We think that the Hesperus data show that cutting off maize cultivation at or above 2,134 m is too restrictive, but surely Adams and Peterson are right when they state that toward the upper limits of maize production, selection of favorable aspect and position with respect to cold-air-drainage patterns is critical.

went to press, Larry Benson [2011a, 2011b] published two highly relevant papers as part of the Village Ecodynamics Project (VEP) II, but his approaches could not be taken into account here.) On the basis of decades of first-hand gardening experience, Honeycutt (1995) suggests that continued horticultural planting in the more favorable soils within our study area does not result in significant declines in soil nutrients. Studies of prehistoric farming fields of New Mexico suggested long-lasting soil degradation in the Mimbres Valley (Sandor et al. 1990), but nearby research yielded an opposite conclusion: "We did not find evidence that Zuni agricultural soils are degraded" (Homburg 2000:127).

In the Mimbres study, Jonathon Sandor and colleagues found that soils deposited above prehistoric hillside agricultural terraces supported sparse vegetation to none after centuries of lying fallow. One possible explanation for this is that the native ground cover in the area is grama grass, whose germination is very climatically sensitive. Sandor et al. (1990:77) cite Wilson and Briske (1979) as finding that blue grama (*Bouteloua gracilas*) requires conditions for seedling development that have not prevailed in the region since the early Holocene. These authors also suggest that abandoned farming plots greatly promoted soil erosion, removing the topsoil. As discussed previously here, our hand-planting soil restrictions should account for this sort of degradation, which is due to erosion.

In the Zuni region, Jeffrey Homburg's studies found various effects on soils farmed in prehispanic times. Such soils exhibit "thickened A horizons and organic coatings on grains and granular peds," and "paired cultivated soils have higher bulk densities and pH levels, and either reduced or enriched levels of N and organic C" (2000:127). Homburg concludes that "although these differences are often statistically significant, they are not great enough to indicate degradation of agricultural runoff soils" (2000:127).

Another study of traditional farming practices by older Zuni farmers reports on strategies used to assure adequate harvests. Working closely with Zuni elders, Jay Norton and colleagues (2000) note that these farmers do not consider their farming as a subsistence system, explaining that "we don't having any system because everything changes every year" in the sense that "farmers apply combinations of innovation and indigenous knowledge to ever-changing situations" (Norton et al. 2000:118) to maintain adequate agricultural production. From this indigenous perspective, we can surmise that prehispanic farmers were well aware of variability in annual agricultural productivity and adapted their planting practices to assure adequate harvests.

Given these somewhat unclear and even contradictory studies, we take the approach that soil degradation under continual planting might have occurred, and so let us see if it did. We do this by parameterizing soil degradation as a variable to be studied in our simulations. As with other parameters, we eventually will be able to determine whether simulation runs with high or low values for soil depletion yield settlement output that better fits that known for our region through archaeological survey. We explored the effects of two different parameter settings. SOILDEGRADE = 1 results in a relatively mild degradation to 70 percent of potential as estimated in the previous step. SOILDEGRADE = 2 results in a more severe degradation to 40 percent of potential. In both cases, degradation begins only when more than two of the nine potential 1-acre plots in each 4-ha cell are being farmed. If exactly two are being farmed, it is assumed that local rotation can offset local degradation. If fewer than two are being farmed, then soil slowly recuperates. When two or more plots in a cell are being farmed, soil degrades by a factor proportionate to the number of plots being farmed:

$$df_{(t+1)} = df_{(t)} - 0.05/[(PLOTS + 1) - farm\_plots]$$

where df is the degrade_factor, $PLOTS$ is the number of potential 1-acre plots in each 4-ha cell (currently set to 9), and farm_plots is the number of plots in a cell currently being farmed. Degrade_factor acts as a multiplier on the potential yield. The floor for the function is set to .7 when the parameter SOIL_DEGRADE=1; the function floor is set to .4 when SOIL_DEGRADE=2. So, when SOIL_DEGRADE=1, a fully farmed cell would lose 30 percent of its potential production after six years, but would never get worse than that; when SOIL_DEGRADE=2, a fully farmed cell would lose 60 percent of its potential production after six years, but would never get worse than that. If the number of farming plots declines to one or none, soils regain their potential productivity, though more slowly than they lost it. A newly unfarmed cell that was fully depleted when SOIL_DEGRADE=1 requires 30 years to return to full potential production. When SOIL_DEGRADE=2, the same cell would require 70 years to return to full potential production. Of course, we can run the model with no soil degradation effects (SOIL_DEGRADE=0). Plates 6.4 and 6.5 display the differing effects of SOIL_DEGRADE=1 and SOIL_DEGRADE=2 on the values for degrade_factor, as a function of the number of plots under cultivation in a cell plus the number of years of cultivation.

In VEP II we hope to draw on recent research by Benson, noted earlier in this chapter, to refine these algorithms.

SUMMARY

The potential maize production landscapes attempt to retrodict annual potential production of maize in our study area back to A.D. 600, at a spatial resolution of 4 ha. These estimates take into the account the inherent fertility, available water-holding capacity, and elevation of soils. We do not attempt to control for aspect or cold-air drainage patterns, though we assume that these both have potentially significant effects on production locally. The nature of the climatic proxies we use means that very-low-frequency climate change, on the order of several decades or centuries, is probably underrepresented, as are very-high-frequency weather patterns, on the order of days or a week or two. The probable general effect of the regression approaches we use is to underestimate extreme highs and extreme lows in production. I discuss some of these and other inherent weaknesses in more detail elsewhere (Kohler 2010).

Plate 6.6 summarizes the net result of our reconstruction across the 700 years of interest here. It differs from previous representations (Kohler et al. 2007:Figure 4.2; Varien et al. 2007:Figure 3) in that it shows the sum of the productivity across the landscape for each year, plus the contribution of each elevation band to that sum.

The Bluff elevation band—the lowest 2.6 percent of the area—does not contribute significantly to farming, with an average per-hectare yield of only 137 kg ($s=21$). To put that in human terms, an adult requiring 2,500 cal / day and getting (conservatively) 70 percent of that from maize containing 3,560 cal / kg would need to cultivate about 1.3 ha (more than three 1-acre plots) to meet his annual caloric needs—with none left over for storage. This can be compared with Bradfield's often-cited figure that the Hopi must cultivate two acres (.8 ha) per person (1971:21)—although about half of that was for storage.

The Cortez band is more important for total production, but keeping in mind that it constitutes almost half of the area, its per-hectare yields are still low, averaging 192 kg ($s=31$). Following the calculations in the previous paragraph, an adult would need to cultivate more than two 1-acre plots just for annual consumption—slightly more than what the Hopi needed for both consumption and storage. On a per-hectare basis, the Yellow Jacket band is the most productive, averaging 330 kg / ha (about 5.2 bu / acre). Its standard deviation is fairly high in absolute terms ($s=52$ kg / ha), but relative to its mean, these soils are quite constantly productive (CV=15.8, versus 15.1 for Bluff-elevation soils and 16.1 for Cortez-elevation soils). An adult would need to plant about 1.4 1-acre plots in this elevational band to meet his or her annual needs—making this about as productive as the Hopi area, keeping in mind that the 2-acre / person Hopi planting

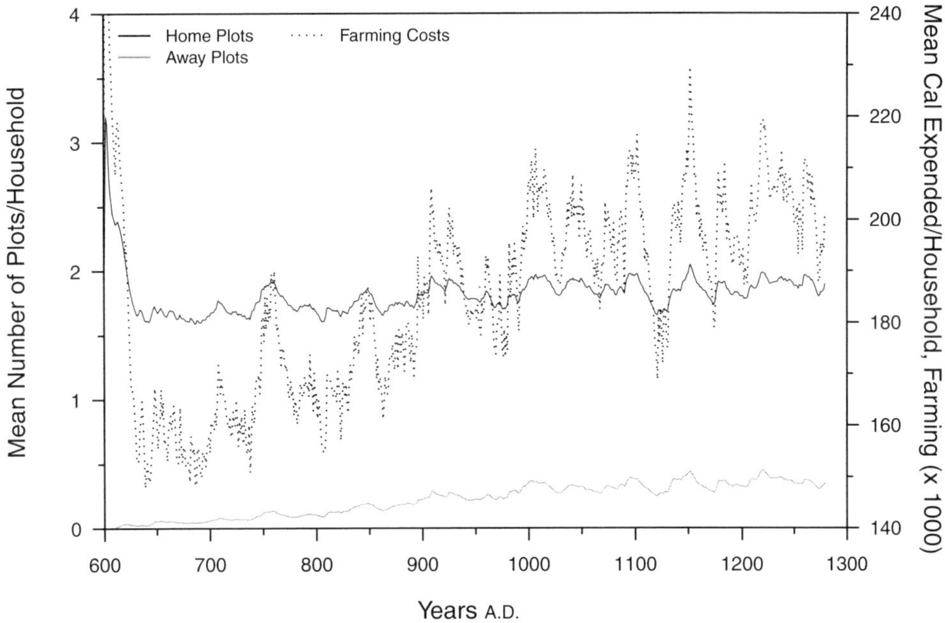

FIGURE 6.1 Average number of 1-ac plots used through time per household (512 runs) and mean calories expended per household on all farming activities. "Home" plots are located in the same 200-x-200-m cell as the residence; "Away" plots are in the 1-cell radius around the cell containing the residence.

allowed for storage. Finally, the Mesa Verde band, the highest 15.2 percent of the area, contains all the areas affected by the cold correction, so its per-hectare mean potential production is a little lower (245 kg) and much more variable ($s=88$, CV=36.3). That would mean an individual would need to plant about 1.8 1-acre plots to meet his annual caloric needs—implying production slightly lower than achieved by the Hopi.

Adding Households to the Landscape: Usage Patterns and Realized Yields

On the face of it, this seems to imply that none of the VEP area was more productive then than Hopi country was in the recent past. But Pueblo people today, and in the recent past, carefully select a few relatively small areas for planting, as did the Hopi and Zuni (as Bradfield, and Muenchrath and her co-workers, emphasize). The VEP figures just noted, on the other hand, are area-wide averages that include both unproductive and highly productive soils within each elevation band. The ancient inhabitants of this area were—and our agents are—able to find areas that exceed the overall averages by a factor of at least 3. Plate 6.7 shows how the production average across the landscape through time compares with that actually achieved in the plots sought out and used by our households, given different levels for SOIL_DEGRADE. Note that yields are slightly lower when SOIL_DEGRADE is set to 2 than when it is set to 1, as would be expected, and that both sets of yields trend downward slightly through time because of these degradation effects. Note also that the yields realized by Village farmers are much more variable than those across the landscape as a whole, because most of their plots are in the more variable, but higher-yield, Yellow Jacket and Mesa Verde elevation bands.

Finally, we want to see how much energy households in the simulation are spending on farming, and whether this is ethnographically plausible. Figure 6.1 shows the numbers of plots used, on average, through time by households. Our households tend to be smaller than those

TABLE 6.11

Estimates of Time Spent on Key Farming Activities Such as Field Clearing, Soil Preparation, Planting, and Tending and Field Processing Crops (but Also Including Time Spent on Tending Animals) Among Horticultural Societies

Society	Men (hrs / day)	Women (hrs / day)
Chittagong Hill Tracts	5.6	2.1
Mandinka	5.6	6.6
Mossi (I)[a]	3.2	5.2
Mossi (II)	4.1	2.3
Southeast Ghana[b]	5.4	4.2
Congo[b]	2.8	2.4
Banda (I)	2.5	4.0
Banda (II)	3.3	3.3
Zairians[b]	2.3	3.7
Ugandans[b]	2.9	3.0
Haya[b]	2.5	3.0
Logoli[c]	1.2	2.0
Nandi[b]	2.5	2.4
Ushi	2.9	3.3
Ifaluk Islanders[d]	.27	.45
Miyanmin	1.4	2.5
Abelam	1.4	4.3
Raiapu-Enga	2.5	4.0
Chimbu	5.3	3.7
Lufa	1.4	1.8
Tairora	3.2	4.7
Karkar Islanders	.48	.72
Orokaiva (I)	1.8	2.2
Orokaiva (II)	2.3	2.7
Oriomo	2.8	4.2
Macoita	2.8	1.5
Irapa[e]	2.5	2.0
Tatuyo	3.6	3.7
Machiguenga[e]	2.6	.78
Quechua[f]	3.9	2.7
Xavante[a]	2.5	2.3
Averages	2.8 ± 1.4	3.0 ± 1.4

NOTE: From Sackett (1996:449–462). Adult effort, averaged across men and women, is 2.9 ± 1.2 hours / day.

a. Assuming a 14-hour workday.
b. Assuming a 12-hour workday.
c. Assuming a 10-hour workday.
d. Assuming a 9-hour workday.
e. Assuming a 13-hour workday.
f. Assuming a 15-hour workday.

in the Hopi case cited earlier (see also Chapter 4), and use on average about two plots through time and across the simulations, with both of those on average being in the cell in which they are living (which minimizes their costs). The gray line in Figure 6.1 shows the total energy expended per household for the farming activities we keep track of: field clearance, planting, weeding, monitoring, and harvest (including cartage, but no food-preparation activities). These "work calories" are on top of the base calories required for mere existence, and we calculate them according to sex and age group most likely to be doing the activity (see WORK_CAL_ entries in Table 4.1). For example, we assume that children can monitor fields, which means the caloric costs for so doing are less than would be the case for the same number of monitoring hours by a man, whereas planting is done by men and weeding by women. Roughly speaking, the work calories on Figure 6.1 can be divided by 200 to estimate the mean number of hours spent by all workers in each household per year on farming chores. The grand mean of about 184,900 calories therefore represents about 925 hours of labor, which is about 231 hours a month during the four-month farming season, or 7.7 hours a day. In an average-size household of 3.3 people, then, if all of those are workers (over seven years old), each person would be spending about 2.3 hours a day on farming chores in the simulation. (These times go up sharply during strings of years with poor production as farmers add more plots to compensate for low yields.) This is reasonably close to the grand mean for time spent on farming chores by adults in horticultural societies (Table 6.11) of 2.9 hours a day, and the fact that our averages are slightly lower is possibly because the estimates in Table 6.11 include time for tending animals, where applicable, whereas our estimates do not. In Chapter 15 we will accumulate the costs for all the activities we model to examine how households' energy budgets changed through time as populations grew and climate varied.

In the next chapter Dave Johnson and I discuss how we estimate productivity for other biotic resources critical to the Pueblo farmers—browse and graze for deer, hare, and rabbits, and fuelwood—and then we briefly characterize how costs for obtaining fuels scale with population size. In Chapter 8 we discuss how our agents hunt the animals grown on these virtual landscapes.

SEVEN

Modeling Plant and Animal Productivity and Fuel Use

C. David Johnson and Timothy A. Kohler

THE SUCCESS OF OUR MODELING efforts depends on establishing plausible long-term productivity estimates for relevant aspects of the biotic environment across the Village Ecodynamics Project (VEP) study area for the seven centuries on which we are focusing. Simulating the long-term productivities of natural resources critical for human survival could not be realistically accomplished without the availability of the paleoproductivity reconstruction detailed in the previous chapter. That, in turn, relied on a substantial body of previous work on various aspects of the paleoenvironment of the upland Southwest. Similarly, this chapter builds on abundant research by a variety of scientists in soils, forestry, botany, and wildlife ecology.

The original Village Project modeling (Kohler, Kresl, Van West, Carr, and Wilshusen 2000) investigated the long-term settlement distributions of model households, using as a basis the annual maize production potential of study-area soils and proximity to water sources. The VEP, as reported in this book, greatly expands that effort, not only by extending its temporal scope by 300 years, but also by dramatically enhancing the model world in the direction of increased realism. Model agents are now presented with temporally and spatially variable supplies of what C. David Johnson (2006) terms "critical natural resources." Modeling these resources is in addition to the improvements in modeling potential maize production discussed in the previous chapter, the addition of the dynamic groundwater model described by Kenneth Kolm and Schaun Smith (Chapter 5, this book), and the ability of model households to exchange goods (Chapter 11, this book).

The critical natural resources include those required for the long-term survival of preindustrial populations. The modeled resources are the fuelwood necessary for cooking and household heating, as well as wild herbivores to supply high-quality animal protein. Fuels are modeled on the basis of the annual production of woody biomass supported by study-area soils. Meat protein is made available to model agents by modeling relevant aspects of the natural ecology of mule deer (*Odocoileus hemionus*), black-tailed jackrabbit (*Lepus californicus*), and desert cottontail (*Sylvilagus audubonii*).

The variable annual productivities of these resources are derived by combining information on the paleoproductivity of soils, as discussed in Chapter 6; on the native vegetation communities supported by each of these soils; and on the relevant ecological characteristics of the three animal species. Therefore we start here by discussing data derived from the soil surveys.

STUDY-AREA SOILS

Douglas Ramsey (2003b) describes 13 groups of soil series in the northern San Juan region. These groups are generally associated with one of three broad topographic settings, each of which roughly corresponds to an elevation range. From lowest to highest the settings are as follows: floodplains, stream terraces, and alluvial fans; hills and mesas; and mountains and hills (and some canyons). These general elevation ranges do not strictly correlate with those delineated for the modeling effort.

Two soil series groups, Mikim–Mikett and Lillings–Ramper–Fluvents, are found in the lowest settings. Midelevation settings host five groups: (1) Mack–Farb; (2) Barx–Gapmesa–Rizno; (3) Wetherill–Pulpit–Gladel; (4) Granath–Ilex–Ormiston; and (5) Morefield–Arabrab–Longburn. Finally, the highest settings support six groups: (1) Typic Torriorthents–Claysprings–Uzacols; (2) Romberg–Crosscan–Rock Outcrop; (3) Sideshow–Zigzag; (4) Wauquie–Dolcan–Rock Outcrop; (5) Sheek–Archuleta–Pramiss; and (6) Northrim–Prater–Sheek.

These groups are further divided into numerous soil components that support a wide variety of native plant species. In total, 233 soil components are described within the Mesa Verde region, and 87 of these are found in our study area. Soil map units (or soil complexes) are made up of one or more soil components, combined in ways that depend on their settings. These soil complexes present a number of different characteristics as described in the three Natural Resources Conservation Service soil surveys covering the study area. The soil surveys present a wealth of information useful to farmers, engineers, and foresters. Data pertinent to the VEP modeling effort were derived from the Animas-Dolores Area Soil Survey (Pannell 2003), the Cortez Area Soil Survey (Ramsey 2003b), and the Ute Mountain Soil Survey (Ramsey 2008).

These three soil surveys use 148 different soil map unit symbols ("musyms," in soil survey terminology) to map VEP-area soils. Many additional soil complexes are also described in those three soil surveys, some of which are identical to soils described in one or both of the adjacent area surveys (but recorded and reported using a different musym). The overlap of soil types described in the soil surveys required a method to distinguish each within the VEP model world, and this was accomplished by assigning sequential numbers to soils not included in the Cortez area survey. That is, there are 152 soil map units reported in the Cortez area survey, and although not all of those are included in the model world, those designations were retained in anticipation of increasing the size of the model world (as VEP II aims to accomplish). Soil complexes assigned different musyms in the Animas-Dolores and Ute Mountain surveys—the same complexes described as one of the 152 described in the Cortez area survey—were coded as the applicable musym from the Cortez area survey. Soils mapped by either the Animas-Dolores or Ute Mountain surveys that were not reported in the Cortez area survey were assigned VEP project-specific soil codes numbered sequentially from 153 to 193 (Table 7.1).

The VEP model world is based on 200-m-square grid cells, and at this resolution there are 139 soil complexes that dominate at least one of the 45,400 4-ha model cells in the 1,816-km^2 study area. Each of these 139 complexes has a variety of general characteristics that broadly influence the natural biotic productivity simulated by the model world. Much of the variation reported for the soil complexes

TABLE 7.1
Correlation of Soil Survey Map Unit Symbols and VEP Soil Codes

Musym	VEP Code	Musym	VEP Code
Ute Mountain Soils		*Animas Dolores Soils*	
128B	153	2	167
208C	154	12	168
390F	155	13	169
391D	156	15	170
401C	157	16	171
510C	158	17	172
511D	159	110	173
550C	160	500	174
550D	161	504	175
605D	162	506	176
605F	163	508	177
615D	164	509	178
635F	165	510	179
910	166	512	180
		527	181
		552	182
		553	183
		700	184
		801	185
		802	186
		804	187
		806	188
		813	189
		860	190
		955	191
		958	192
		959	193

NOTE: For musyms 1–152, musyms and VEP codes are the same. VEP = Village Ecodynamics Project.

derives from their constituent soil components. This discussion trends toward increased detail, moving from general characteristics of the soil complexes, through specifics of soil components, to the native vegetation supported thereon, followed by description of the resulting primary productivity and ultimately the secondary productivity used by model agents.

A number of the soil reports and tables provided in the three surveys were referenced to compile the soil characteristics needed to model the productivity of critical natural resources for the study-area soils. Foremost among those are data related to each soil component's capacity to retain soil moisture (see Johnson 2006:Appendix A for a complete tabulation). These, in turn, allow us to calculate the available water capacity for each soil complex throughout its profile.

The available water capacity varies among levels of the profile of each soil component, and for each soil complex the model input data required substantial manipulation of the available data (see Johnson 2006). First, we corrected for data missing from the soil survey tables by weighting the reported data to fill in for the missing values. There are, primarily, two kinds of missing data: unreported proportions of soil complexes, and incomplete listings of native vegetation communities. We compensated for each of these by giving additional weight to the constituents of the soil complexes and native plant communities for each soil component, which were reported so that their total proportions summed to 1.0.

Each soil complex, for example, is composed of one or more soil components. Rarely, however, do the contributions of the reported components sum to 100 percent of the soil complex to which they contribute. In the simplest cases, the complexes consist of a single component, but that reported component may account for only 90 percent of the complex, with nothing else reported to account for the missing 10 percent. In such cases, we simply multiply that percentage by the inverse of its proportion, yielding a weighting factor of 1.111 in this case, so that the complex is represented as complete.

More commonly, two or more components contribute to a soil complex, and those consistently total less than 100 percent of the soil complex to which they contribute. To make up the difference, the proportion each component contributes is divided by the sum of the proportions they all contribute. The results add up to 1.0, completing the respective soil complex. This assumes that the normal-year productions reported for each soil component, when combined, are representative of what would have been produced by the unreported proportions. This weighting was also applied to the contributing vegetation species, as discussed in more detail in the next section. Additional information on the specifics of contributing soil components is warranted before moving on to the associated vegetation.

Net Primary Productivity

The soil data most directly applicable to this chapter estimate soil productivity by weight of annual net primary productivity (NPP) per unit area. NPP (Odum 1971) is the annual production of primary biomass by each of the vegetation species supported by a particular soil complex, as reported in the soil surveys. This NPP is assumed to be produced aboveground (see Del Grosso et al. 2008 for the distinction between total and aboveground NPP), and therefore readily available for consumption as either forage by modeled herbivores or firewood by modeled households. Soil productivity varies with available soil moisture as modeled using the Palmer Drought Severity Index (Chapter 6). Each soil has other properties that contribute to productivity, such as its depth and nutrient supply. The depth data are reported in the soil surveys (see Johnson 2006:Appendix A for project-specific data). The nutrient data are more complicated and are beyond the scope of the present discussion.

Normal-year productivity is reported in the soil surveys for each soil component as pounds of new growth per acre, which is assumed to equal the NPP of each soil's native vegetation community. There are commonly different NPP values provided for each soil component in a soil complex. In all such cases, the figure for the dominant soil component within the complex is used; in a few cases this slightly increases the overall NPP of a soil complex. Although the soil surveys also report productivities for both more- and less-favorable growth years, we use the normal-year productivity figures, which are then modulated by annual estimates of stored soil moisture derived ultimately from tree rings, as described in the next section.

The values for annual NPP thus vary through time, but in raw form the normal-year productivity values (converted to kilograms per hectare by multiplying the reported pounds per acre by 1.12) for soils within the VEP study area range from 336 to 3,360 kg. This, then, represents the total new biomass expected in a normal growth year per hectare of each soil mapped within the VEP study area. In the model world, the dominant soil mapped within each 4-ha cell is taken to represent all the productive soil in that location.

Biomass productivity in each model cell is then simulated using as a basis the dominant soil in that cell, its associated native vegetation community, and the localized influence of the paleoproductivity model for a given year. In years when a particular model cell is cultivated by model households, native plant productivity is eliminated from that cell. Some additional environmental smoothing is built into the model because of a number of necessary assumptions. One of these is that the soil characteristics and supported native vegetation of the dominant soil map unit in each cell is taken to represent the entire cell. A second is that the native vegetation supported by each soil type (reported in the soil surveys) represents a steady-state climax community. The potential effects of periodic wildfires are also absent from the model, since data to accurately portray the long-term fire regime were neither readily available nor easily incorporated into the simulation. Thus no successional stages are modeled, since attempting to do so would be based on very generalized observations and would be difficult to implement in the simulation. We are considering ways to treat succession in a more realistic fashion in future research.

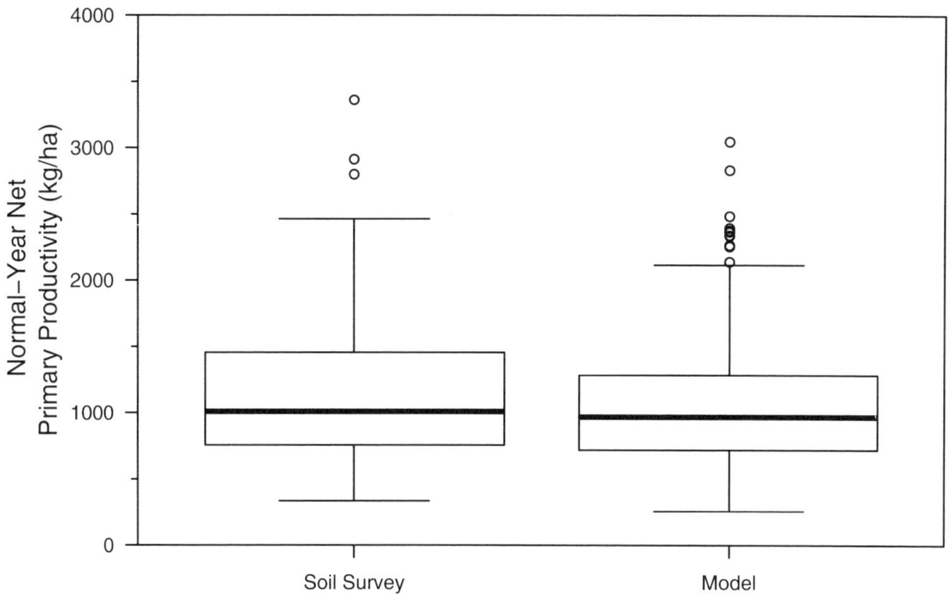

FIGURE 7.1 Comparison of normal-year net primary productivity across model cells (using modal soil complex NPP as reported by soil surveys) with model-cell mean NPP from a 700-year run of the simulation. After Johnson (2006:Figure 5.4).

Given these assumptions and simplifications, and the complexity of some of the operations leading to opportunities for error, we compared the modeled biomass production to that expected from data provided by the soil surveys. We simulated the total annual net primary productivity for the usual 700-year model run, but with no households on the landscape. This allows us to check for possible errors, and also allows us to compare the static estimates for the dominant soil in each cell provided by the soil surveys with their long-term mean production from the simulation (Figure 7.1).

This comparison shows that the values for mean, median, and upper and lower quartiles for productivity produced by the simulation are very close to those calculated from the soil survey data. The slightly lower upper value (9.4 percent) and a more notable lower low-range value (23.5 percent) for long-term model productivity are not unexpected, since there are several long periods of very unfavorable growing years during the centuries modeled. Despite these lower extreme values, the mean long-term productivity figure output by the simulation (7.3 percent) is only slightly lower than that calculated from the normal-year productivity figures provided by the soil survey data.

Figure 7.2 maps the distribution of normal-year NPP of our soil complexes as reported by the soil surveys. This differential distribution of productivity undoubtedly influenced prehistoric settlement decisions in a variety of ways, since it influences the productivity of the native vegetation that forms the basis of the real- and model-world food chains. We now proceed to a more detailed description of the biotic environment generated by the VEP simulation.

NATIVE VEGETATION COMMUNITIES

Even though the present implementation of Village does not allow the use of native plants as food for its households, native vegetation communities figure in the model for the fuelwood they produce and the animals they support.

Study-area soil components support 93 plants of four types: forbs, grasses, shrubs, and trees. In general, shrub species are the most common, followed by grasses, trees, and forbs.

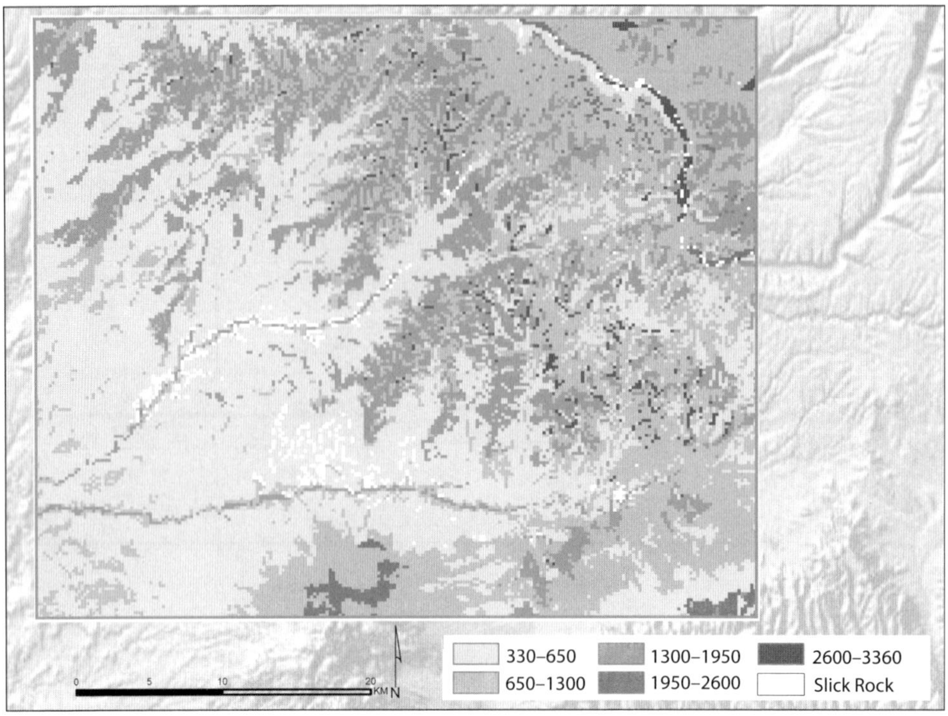

FIGURE 7.2 Map of normal-year net primary productivity of the dominant soil in each 4-ha model cell, in kg per hectare, as reported by soil surveys. After Johnson (2006:Figure 5.2).

From an ecological perspective, trees typically dominate the biomass on soils that support them, as do grasses on other soils. Forbs represent a minor component of a few vegetation communities only. Accordingly, they merit only the following brief discussion.

There are nine species of forbs in the study area. As listed in Table 7.2, only a dozen soil complexes are reported as supporting forbs, and those complexes represent a very small portion (2.36 percent) of the dominant soils in model cells. These forb species are thus effectively inconsequential to our modeling efforts to date.

By species count, grasses are the second most numerous plants in our study area's native vegetation. James Pannell (2003) and Ramsey (2003, 2008) report 35 native grass species on the soil complexes that dominate any of the 4-ha model cells (Table 7.3). More significantly, at least one grass species is supported by each of the 139 soil complexes represented on the model landscape. In lieu of extensive farming by model households, therefore, the annual NPP of at least one species of grass is available in every model cell at each annual model time step.

Grasses greatly outproduce both the shrubs and trees; across the study area, grass normal-year NPP is about 800 kg / ha, twice that of shrubs and three times higher than the trees. In the simulation, grass production supports only leporids; grasses are fully accessible to hare and rabbits because none of their aboveground grass biomass is out of reach or inedible. Moreover, grasses are reported as supplying less than 1 percent of deer food, during a short period of the early spring season each year (Krausman et al. 1997), so leporids have little competition for grasses in our model world. In the model world, therefore, we allow leporids to access 70 percent of the NPP of their preferred grass species annually. Given the wide distribution of grasses, it is therefore likely that in any given year leporids will be able to graze

TABLE 7.2
Forbs in Native Vegetation Communities of Study Area

Common Name	Scientific Name	Supporting Soils (*n*)	Model Cells (%)
Arrowleaf balsamroot	*Balsamorhiza sagittata*	2	.34
Aspen peavine	*Lathyrus lanszwertii*	1	.141
Broadleaf cattail	*Typha latifolia*	1	.269
Cinquefoil	*Potentilla* spp.	26	.176
Mule-ears	*Wyethia arizonica*	27	.165
Rocky Mountain iris	*Iris missouriensis*	367	.35
Scarlet globemallow	*Sphaeralcea coccinea*	354	.81
Skyrocket gilia	*Gilia haydenii*	1	.033
Western yarrow	*Achillea millefolium*	1	.08

within any model cell that is not completely farmed. (Modeled herbivore populations are not allowed to feed on maize, since we assume that in fact people would have done their best to prevent that.)

Grasses are more than twice as productive as the next most productive vegetation class, shrubs. Annual, normal-year NPP of grasses on all model-world soils is roughly 800 kg / ha. A major difference between the grass component of the total primary productivity and those of the shrub and tree components is, however, that grasses are neither initialized nor maintained as standing crops on the model landscape (as are shrubs and trees). Grasses are modeled as annuals, despite a strong contingent of perennial grasses within the native community.

Shrubs are the most numerous, by species count, of the native vegetation reported (Pannell 2003; Ramsey 2003, 2008) for study-area soils. Table 7.4 lists 39 species among our native plant communities. Shrubs produce substantially less NPP than do grasses; however, they have uses within the model—as fuels and as browse for deer—that grasses do not. Unlike grasses, shrubs typically have woody structures supporting foliage. Their dense woody tissue can make for excellent fuel (but is not typically very edible or nutritious), while the leaves are commonly preferred feed for herbivores. So only a portion of shrub NPP biomass is available as browse.

The actual distribution of different forms of shrub biomass is not frequently described in the botanical literature. In the VEP model, we consider all new shrub growth to be leafy biomass, and that not consumed by herbivores during a model year (minus 25 percent for leaf fall) becomes woody biomass in the following year, contributing to the standing crop of each species. Using as a basis a study of woody tissue distribution of mountain mahogany (*Cercocarpus* spp.), we consider 3.35 percent of all shrubs to be deadwood (Chojnacky 1984:Table 2). The remainder is live woody biomass that, along with the deadwood, contributes to fuel loading on the model landscape.

Shrub biomass is variably distributed across the study-area landscape and the model world. Nearly all the soil complexes represented in the model (132 of 139) support at least one shrub species. Nevertheless, some terrain in the study area—notably, canyon slopes—supports much higher levels of shrub biomass than do others (Johnson 2006:Figure 6.3).

Our model of shrub species is implemented in several steps. First we estimate a standing crop for shrub biomass based on the long-term productivity of shrubs in each modeled soil complex as modulated by the annual soil-moisture signal proxied from tree rings dis-

TABLE 7.3

Grasses in Native Vegetation Communities of Study Area

Common Name	Scientific Name	Supporting Soils (*n*)	Model Cells (%)
Alkali sacaton	*Sporobolus airoides*	28	14.96
Arizona fescue	*Festuca arizonica*	47	7.87
Baltic rush	*Juncus balticus*	2	.61
Blue grama	*Bouteloua gracilis*	12	6.24
Bluegrass	*Poa* spp.	7	.48
Bottlebrush squirreltail	*Sitanion hystrix*	49	50.91
Columbia needlegrass	*Achnatherum nelsonii nelsonii*	1	.01
Elk sedge	*Carex garberi*	4	.27
Foxtail barley	*Hordeum jubatum*	2	.83
Galleta	*Hilaria jamesii*	51	53.85
Indian ricegrass	*Oryzopsis hymenoides*	65	80.22
Inland saltgrass	*Distichlis spicata*	19	4.47
Kentucky bluegrass	*Poa pratensis* L.	4	1.51
Letterman needlegrass	*Achnatherum lettermanii*	5	1.41
Mesa dropseed	*Sporobolus flexuosus*	1	.10
Mountain brome	*Bromus carinatus*	38	7.94
Mountain muhly	*Muhlenbergia montana*	42	7.80
Muttongrass	*Poa fendleriana*	59	68.13
Needleandthread	*Stipa comata*	38	42.04
Needlegrass	*Stipa columbiana*	15	2.63
New Mexico feathergrass	*Stipa neomexican*	9	4.10
Nodding brome	*Bromus anomalus*	8	.47
Parry's oatgrass	*Danthonia parryi*	12	.89
Pine dropseed	*Blepharoneuron tricholepis*	4	.06
Pinyon ricegrass	*Piptochaetium fimbriatum*	4	24.13
Prairie junegrass	*Koeleria pyramidata*	41	10.02
Rush	*Juncus*	7	2.30
Sand dropseed	*Sporobolus cryptandrus*	4	.56
Sedge	*Carex bella*	12	3.11
Slender wheatgrass	*Agropyron trachycaulum*	13	3.38
Thurber's fescue	*Festuca thurberi* Vasey	1	.17
Tufted hairgrass	*Deschampsia caespitosa*	4	.37
Western wheatgrass	*Agropyron smithii*	108	83.18

TABLE 7.4

Shrubs in Native Vegetation Communities of Study Area

Common Name	Scientific Name	Supporting Soils (*n*)	Model Cells (%)
Antelope bitterbrush	*Purshia tridentata*	26	24.79
Basin big sagebrush	*Artemisia tridentata tridentate*	7	.99
Big sagebrush	*Artemisia tridentata*	62	57.2
Black sagebrush	*Artemisia nova*	17	9.72
Cliff fendlerbush	*Fendlera rupicola*	1	.2
Cliffrose	*Purshia* spp.	3	.85
Common chokecherry	*Prunus virginiana* spp.	4	1.30
Common snowberry	*Symphoricarpos albus*	29	28.06
Fourwing saltbush	*Atriplex canescens*	26	15.43
Gambel's oak	*Quercus gambelii*	52	13.94
Greasewood	*Sarcobatus vermiculatus*	14	4.39
Kinnikinnick	*Arctostaphylos uva-ursi*	2	.96
Mormon tea	*Ephedra viridis*	4	1.57
Mountain big sagebrush	*Artemisia tridentata vaseyana*	7	.44
Mountain mahogany	*Cercocarpus Kunth* spp.	13	27.8
Mountain snowberry	*Symphoricarpos oreophilus*	5	1.52
Oregongrape	*Berberis repens*	3	.90
Rabbitbrush	*Chrysothamnus vaseyi*	3	1.81
Rubber rabbitbrush	*Chrysothamnus nauseosus*	16	10.74
Sagebrush	*Artemisia*	2	.05
Saltbush	*Atriplex gardneri*	5	1.69
Saskatoon serviceberry	*Amelanchier alnifolia*	10	1.18
Serviceberry	*Amelanchier*	9	2.78
Shadscale saltbush	*Atriplex confertifolia*	15	11.73
Shrubby cinquefoil	*Dasiphora floribunda*	2	.18
Skunkbush sumac	*Rhus trilobata* Nutt.	1	.02
Small Douglas rabbitbrush	*Chrysothamnus viscidiflorus*	4	.97
Snakeweed	*Gutierrezia* spp.	1	.00
Snowberry	*Symphoricarpos albus*	12	3.96
Squaw apple	*Peraphyllum ramosissimum*	1	.02
True mountain mahogany	*Cercocarpus montanus*	15	14.24
Utah serviceberry	*Amalanchier utahensis*	25	29.72

(continued)

TABLE 7.4 *(continued)*

Common Name	Scientific Name	Supporting Soils (*n*)	Model Cells (%)
Utah snowberry	*Symphoricarpos oreophilus utahensis*	1	.85
Whortleleaf snowberry	*Symphoricarpos oreophilus*	2	.33
Willow	*Salix*	3	.82
Winterfat	*Ceratoides lanata*	6	5.28
Woods' rose	*Rosa woodsii*	5	.46
Wyoming big sagebrush	*Artemisia tridentata wyomingensis*	16	5.55
Yucca	*Yucca baccata*	2	.99

cussed in the previous chapter. The NPP of each soil is then apportioned to each plant species in accordance with the proportional contribution of each species to the native vegetation community in that particular soil complex. The simulation is run through the entire 700-year period, and the annual productivity of each native vegetation species in each 4-ha model cell is recorded.

This allows us to compute the average annual productivity for each cell over the centuries of interest; this number, in turn, is used to calculate a standing crop of woody biomass in each model cell. This calculation requires knowing how much of the annual NPP persists as woody biomass. The figures we use for this calculation are based on a study of pinyon-juniper forest by Joseph Howell (1941), who found annual new growth (analogous to NPP used here) to be approximately 1.2 percent of standing biomass. We use a slightly higher figure of 1.3 percent (to allow for faster-growing shrubs), inverting that to .769 as a multiplier of the average NPP output by the model run. The resulting figures for each species of shrubs and trees, when summed for each cell, represent the existing biomass, or standing crop, on the model landscape at initialization.

All the native shrubs provide some NPP that supports the secondary production of resources important to the well-being of model households. Many of these shrubs are the preferred foods for the three herbivores whose dynamic populations are simulated by the model. All shrubs also represent sources of woody biomass that model households can collect for use as fuels.

Trees are the least prevalent vegetation class in terms of the number of species represented, and as a class they also produce the least total annual NPP. Nine species of trees grow in the study area (Table 7.5). They are the longest-lived vegetation, and their accumulation of woody biomass, in the absence of fire, greatly exceeds that of all other vegetation classes (on those soils with trees) over the long term.

The distribution of biomass on individual trees is also an important factor. Like shrubs, their NPP includes both woody biomass and foliage or fruits (or both) that are consumable by herbivores. As is easily observed, live woody tissue is by far the largest proportion of biomass on living trees. The proportions of other forms of arboreal tissue are less easily distinguished. Figure 7.3 illustrates the rather similar distribution of aboveground biomass for the two most common species of trees in the study area. Standing crops of trees are initialized in the same manner as those of shrubs.

The archaeological record of the VEP area indicates that tree wood was the preferred fuel for prehispanic Pueblo peoples; however, mac-

TABLE 7.5
Trees in Native Vegetation Communities of Study Area

Common Name	Scientific Name	Supporting Soils (*n*)	Model Cells (%)
Cottonwood	*Populus fremontii*	1	.74
Douglas fir	*Pseudotsuga menziesii*	5	2.86
Narrowleaf cottonwood	*Populus angustifolia*	3	.82
Ponderosa pine	*Pinus ponderosa*	24	4.42
Quaking aspen	*Populus tremuloides*	3	.94
Rocky Mountain maple	*Acer glabrum*	2	.96
Rocky Mountain juniper	*Juniperus scopulorum*	11	2.73
Two-needle pinyon	*Pinus edulis*	24	41.07
Utah juniper	*Juniperus osteosperma*	22	45.58

robotanical remains do show that shrubs were also burned in ancient hearths (Adams and Bowyer 2002; Kohler and Matthews 1988). The fact that smaller-diameter hardwood fuels typically burn hotter and produce less ash probably contributes to the typical dearth of shrub remains in thermal features.

In our present implementation of fuel use, households utilize whatever dead woody fuels are closest first, and then chop down living wood for fuels only if deadwood is unavailable within a specified radius (in the sweep reported here, we used 10 cells, or a 2-km radius). After that point households maintain an offset of

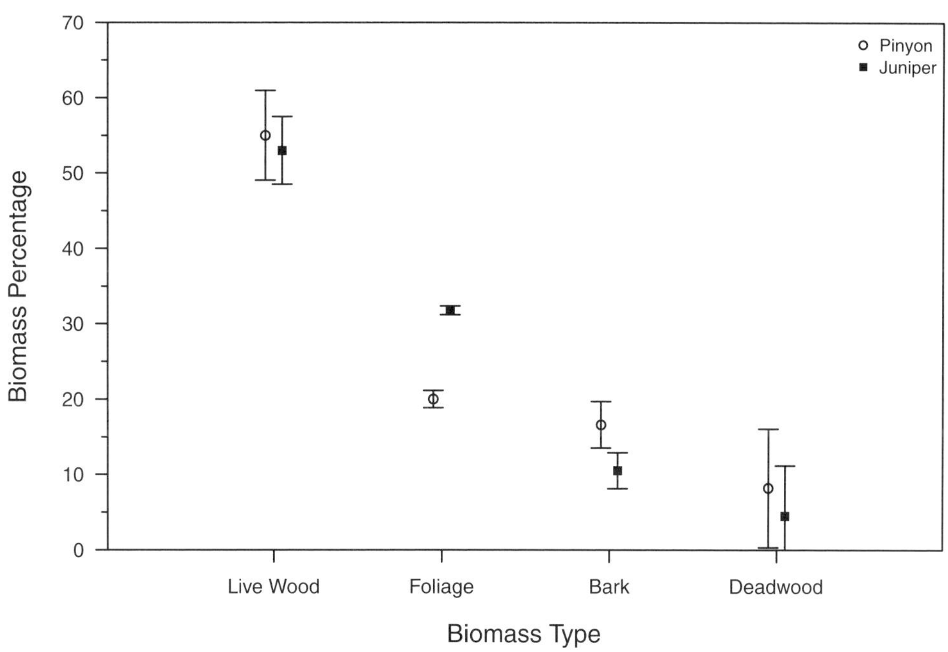

FIGURE 7.3 Distribution of biomass on the two species of trees, pinyon and juniper, dominant in the study area. After Chojnacky (1984:Table 2).

2 km in these behaviors; that is, they would be willing to travel 2.2 km for deadwood at the same time as they were cutting live wood within a .2-km radius.

The contributions of native vegetation to each soil component, like those of each soil component to the soil complexes, are incompletely reported in the soil surveys. Therefore we undertook a second series of weighting calculations to account for the unreported native vegetation data. This is similar enough to what we did for soils that we will not describe it here; readers interested in more detail should consult Johnson (2006). We now turn to explaining how our model for native plant productivity was used as the basis for modeling the herbivores most important to local Pueblo peoples.

MODELING HERBIVORES

In preindustrial societies, hunting can account for a considerable proportion of the diet, at least in environments supporting numerous game animals. The archaeofaunal record from sites in the American Southwest attests to the importance of wild game in the prehistoric diets of the region (Driver 2002; Nelson and Schollmeyer 2003; Shelley 1993; Speth and Scott 1989).

Our simulation of herbivore populations is designed to shed light on the influence of game distributions on long-term settlement patterns. Model households attempt to supply each of their members with some (parameterized) amount of meat protein. Hunting requires energy to search, capture, and transport game; the energy thus expended must be paid in calories obtained from cultivated maize. Recommended amounts of high-quality proteins vary greatly, but generally range from 30 to 90 g per person per day (United States Department of Agriculture 2005; Wing and Brown 1979; see also Chapter 8, this book). Even though prehistoric agriculturalists obtained some protein from domesticated and native plant foods, meat would have been highly prized (Speth and Scott 1989; Chapter 8, this book). The VEP simulates populations of three species of herbivores on the model landscape.

The species we model—mule deer (*Odocoileus hemionus*), black-tailed jackrabbits (*Lepus californicus*), and desert cottontail (*Sylvilagus audubonii*)—are native to the upland Southwest and represent significant proportions of most archaeofaunal assemblages in the region. The model landscape grows vegetation for them to eat, and they in turn may become prey for model households.

As explained in Chapter 8, mule deer are modeled within what we call "deer cells" that are 1 km^2, so there are 1,816 of these in the study area, each containing 25 of our model's 4-ha cells. Mule deer prefer to feed on 21 of the plant species supported by study-area soils. On average they ingest 21.9 g of vegetation per day per kilogram of body weight (Alldredge et al. 1974:Table 3). Mean body weight for adult mule deer is approximately 60 kg (Alldredge et al. 1974; Hanley and Hanley 1982; Hobbs and Swift 1985), so on average each deer consumes 1.31 kg of NPP from the model world daily.

Deer are allowed to access 50 percent of the NPP produced annually by each of their preferred native plant species; we assume that the other half is either out of reach or too costly to eat in terms of energy return. Mule deer densities therefore depend on the concentrations of appropriate native vegetation, and population levels respond to annual variations in NPP of those plants that are driven by the soil-moisture signal from tree rings.

The mule deer population and also the leporid population discussed in the paragraphs that follow are seeded across the model landscape in the same way as are the standing crops of trees and shrubs already described. Populations may then be depleted by human predation and replaced through annual reproduction at a mean maximum rate of population growth (r_{max}) of .4 (see Johnson 2006:185–194 for a thorough discussion of mule deer ecology). In the absence of hunting, the long-term deer population varies between about three and 12 individuals per deer cell (Johnson 2006:Figure 7.3). We do not model any carnivores competing with humans for deer or leporids. Our hunting algorithm, described in the Chapter 8, does not allow any of the animal species to be completely killed off in any cell, although populations may be severely depressed by predation.

Black-tailed jackrabbits are the next-largest herbivore modeled on the VEP landscape. Since these hare individually consume far less vegetation than the deer already described, they are modeled within the same 4-ha cell framework used by the rest of the simulation. Individual jackrabbits are unable to migrate from cell to cell, and in fact, like deer, these animals exist not as individuals (or agents) but as variables maintained by each of the cells. Since this species exhibits a relatively high reproductive rate, and has a relatively small home range, we do not believe that lack of diffusion is seriously unrealistic.

Black-tailed jackrabbits prefer 24 of the local plant species, consuming an average of 122 g of plant NPP per day, an amount based on an average 2.3 kg of body weight (Haskell and Reynolds 1947). Juvenile hare are weaned at the age of six weeks and grow to maturity very quickly; by 25 weeks, juveniles are eating as much as adults.

Hare are allowed to access 70 percent of the NPP produced annually by each of their preferred plant species; we assume the other 30 percent to be either out of reach or too costly in terms of energy return. Jackrabbit densities therefore depend on the concentrations of appropriate native vegetation across space, while through time their populations also respond to annual variations in the NPP of those plants.

The jackrabbit population is seeded across the model landscape in the same way as standing crops of trees and shrubs as described previously. Population levels are then drawn down by human predation and replaced through annual reproduction at a maximum population growth rate (r_{max}) of 1.75 (French et al. 1965; Haskell and Reynolds 1974; Lechleitner 1958).

In the absence of hunting, long-term hare populations vary between about two and nine individuals per 4-ha model cell (see Johnson 2006:Figure 7.6). Their density is approximately 5.5 per cell on average, which is fairly congruent with hare population densities reported from various areas of the western United States (Gross et al. 1974; Knick and Dyer 1997; Smith 1990).

Desert cottontails are the third herbivore population on the model landscape. Since these rabbits individually consume even less vegetation than the hare already described, they too are modeled within 4-ha model cells. Like the jackrabbits, desert cottontails cannot migrate between cells.

Desert cottontails prefer 30 of the plant species supported by study-area soils. They consume an average of 190 g of plant NPP per day based on an average 895 g of body weight (Ingles 1941). Rabbits are allowed to access 70 percent of the NPP produced annually by each of their preferred plants; the other 30 percent is assumed to be either out of reach or too costly in terms of energy expended. Cottontail densities depend on growth of their preferred plants, and they fluctuate in response to annual variations in the NPP of those plants, as well as in response to hunting pressure. Rabbits are seeded onto the model landscape in the same way as the species already described. Cottontails have the highest reproductive rate of the three species we model, with an r_{max} of 2.3 (Myers 1964), and they are therefore less susceptible to depression through hunting than the other two species. In the absence of hunting, the long-term rabbit population varies between about four and 14 individuals per 4-ha model cell (see Johnson 2006:Figure 7.9).

At the outset of this project, we had no idea whether these three species of herbivores would supply model households with enough meat for their long-term health and survival, or whether combinations of low-production years and high human populations would result in mild or severe depression of some or all species. The answer is detailed in Chapters 8 and 9. Here we discuss how consumption and production

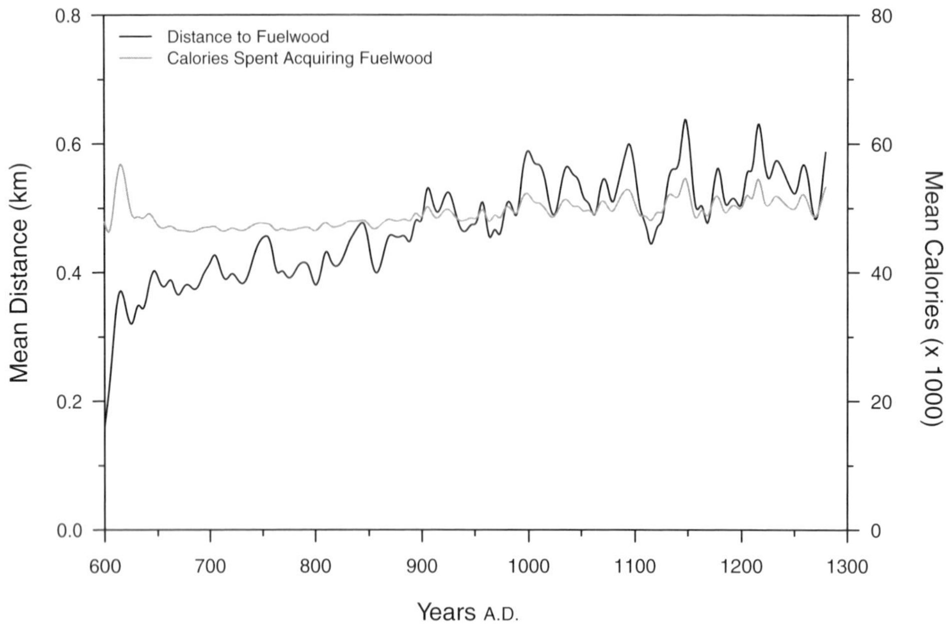

FIGURE 7.4 Average distance households traveled (left vertical axis, black line) and average caloric costs / household / year (right vertical axis, gray line) through time to acquire fuelwood in the 512 runs reported in this volume.

balance, as human populations change, in the case of fuelwood. These are of interest since some analysts have identified changes in fuel usage in Pueblo I villages in our area that seem to be attributable to relatively severe local depletion (Kohler and Matthews 1988), whereas others, examining usage at long-lived sites from our area with an important Pueblo III component (Duff et al. 2010), are more impressed by apparent stability in usage patterns.

MODELING HOUSEHOLD FUEL USE

Fuelwood is a basic necessity of preindustrial life in the upland Southwest. Consumption of maize as a staple (Decker and Tiezen 1989)—and even more, the addition of beans to the diet—required regular cooking fires. Several months a year of below-freezing nighttime temperatures would have certainly caused households to maintain fires for heat on a regular basis throughout at least the winter months.

Much recent literature discusses the fuelwood needs of households that do not have access to other energy sources (Brondeau 2001; Contreras-Hinojosa 2003; Israel 2002). Rates of household fuelwood use depend on a number of factors, including climate, elevation, housing construction, and food preparation techniques. Reported rates of annual fuelwood use by rural peoples range from 629 to 1,633 kg per person (see Johnson 2006:Table 4.1 for more details, and Johnson et al. 2005 for additional discussion).

In the simulations reported here, households use 1,130 kg of fuel per capita (Khan et al. 2001). Each simulated household must find and collect this much woody biomass for each household member every year. Each load of fuelwood weighs 25 kg and incurs both collection costs and transport costs. Costs rise if households have to chop wood as opposed to gathering deadwood. These costs are denominated in calories, paid through the consumption of maize, which households must either grow or obtain through exchange.

Figure 7.4 graphs (in black) the mean distances traveled by members of households

TABLE 7.6

Parameters that Significantly Affect Average Distance to Fuelwood[a]

Parameter (Values)	Mean Distance Traveled for Fuelwood (r / p)	Mean Calories Expended to Acquire Fuelwood (r / p)	Mean Distance Traveled for Fuelwood (r / p) Controlling for n Households	Mean Calories Expended to Acquire Fuelwood (r / p) Controlling for n Households
NEED_MEAT (0, 1)	−.51 / <.0001	−.44 / <.0001	−.28/ <.0001	−.04 / .337
SOIL_DEGRADE (1, 2)	−.47 / <.0001	−.45 / <.0001	−.11 / .012	−.038 / .393
HARVEST_ADJUST_FACTOR (.75, .95)	−.25 / <.0001	−.23 / <.0001	.24 / <.0001	.267 / <.0001
STATE_GOOD (0.1, 0.2)	.19 / <.0001	.22 / <.0001	−.20 / <.0001	−.06 / .165
PROTEIN_NEED (10, 25)	−.16 / .0004	−.1 / .023	.02 / .579	.18 / <.0001
H_2O_TYPE (1, 3)	−.14 / .001	−.16 / .0004	−.02 / .639	−.06 / .146

a. Evaluated across 512 runs at A.D. 1093, when most runs have reached their maximum populations.

TABLE 7.7

Comparative Contemporary Data on Time Spent "Managing Fuel" (Men Only)[a]

Society	Observed Energy, Calories / Year	Activity
Karkar Islanders	54,960	Chopping firewood
Mossi II	81,308	Collecting firewood
Irapa	86,113	Chopping firewood
Mossi II	98,392	Cutting wood using machete
Lufa	114,096	Chopping firewood
Chimbu	140,256	Splitting firewood
Machiguenga	142,172	Chopping firewood
Quechua	146,161	Chopping firewood using axe

a. From Sackett, who specifies that "fuel and water work involves carrying loads, chopping wood and some lifting" (1996:212).

through time in each of the 512 runs reported in this book; evaluated at around A.D. 1100, the average distance to fuelwood for the runs averaged together in Figure 7.4 ranges from about .3 km to about 1.7 km. Not surprisingly, these distances are strongly and significantly greater in runs that generate more agents ($r=.96$; $p<.0001$). Therefore, our estimates of fuelwood distances and caloric costs are likely to be conservative for the second population cycle, when even our most populous simulation runs underestimate the peak population we estimate for the archaeological record (Chapter 4). Table 7.6 reports the parameters that result in significantly greater average distances to fuelwood, by declining importance. For example, reading across the first row and down the second and third columns, the higher value for NEED_MEAT significantly decreases mean distance to firewood and mean calories expended getting it, apparently since NEED_MEAT=1 households are forced into undepleted areas. On the other hand, a higher value for STATE_GOOD is significantly correlated with greater distances to firewood and greater costs for procuring it, since the higher value for STATE_GOOD boosts agent populations.

Several of these factors are no longer significant when we control for the effects of population size across the runs using partial correlation (see the two rightmost columns, Table 7.6). After partialling population, the most important remaining parameter is NEED_MEAT which, when zero, still entrains larger distances to fuels, perhaps because under this setting agents often settle in densely populated places. The results for HARVEST_ADJUST_FACTOR and STATE_GOOD are particularly interesting because they remain significant but change in sign after controlling for population size. Thus, a higher value for HARVEST_ADJUST_FACTOR—which decreases maize harvests relative to the lower parameter value—results in greater distances to fuelwood after partialling population. Agents are also more likely to have to go farther for fuels when STATE_GOOD=.1 than when it is .2 when we control for population size. Both of these results, which we did not anticipate, may reflect some suppression of mobility under these parameter settings, relative to their other possibilities. However, it is important to remember that the amount of variation explained by these relationships is very small in relation to that explained by the total population size.

The average calories spent per household acquiring fuelwood through time in each of our 512 runs are graphed in gray in Figure 7.4. Obviously, calories and distances covary strongly ($r=.97$; $p<.0001$). The time allocation literature provides a measure of external validation for the calories our households spend on this activity. Table 7.7 shows a variety of societies for which we have been able to find data, with expenditures for obtaining firewood ranging from about 55,000 to about 146,000 calories. Except for the three societies with the highest expenditures, our simulated populations have expenditures similar to those tabulated. Since all of these societies have steel tools, we consider it likely that our caloric estimates for firewood acquisition are highly conservative, especially as local deadwood stores become depleted, since our estimates do not take into account the time-consuming preparation and maintenance of stone axes.

CONCLUSIONS

We have discussed how the VEP simulation generates some of the natural resources critical to long-term survival in the prehistoric American Southwest. We also briefly summarized patterns of resource use through time for fuelwood. Not surprisingly, higher populations lead to the need to travel greater distances for fuelwood, plus increasing caloric expenditures for its acquisition. These results—similar to those presented by Johnson et al. (2005:Figure 5) using an earlier version of the model—reinforce the possibility that some of the architectural and behavioral changes we see in Pueblo II and III from earlier times (e.g., architectural massing) might result from attempts to minimize fuel use. In Chapter 8 we provide the background needed to present parallel results for hunting in the VEP world in Chapter 9.

EIGHT

Supply, Demand, Return Rates, and Resource Depression

HUNTING IN THE VILLAGE ECODYNAMICS WORLD

Jason A. Cowan, Timothy A. Kohler, C. David Johnson, Kevin Cooper, and R. Kyle Bocinsky

MANY EXCELLENT STUDIES of hunting in small-scale agricultural societies are available in the archaeological and zooarchaeological literature, including several (e.g., Muir and Driver 2002) for the portion of the U.S. Southwest on which this chapter focuses. It is our contention, nevertheless, that such studies would be strengthened if they could document not just the "demand" side of hunting as known from the archaeological record, but also the numbers and spatial distributions of various animals in the environment from which hunters sampled—the "supply" side. The absence of models of resource supply leads to difficulties in adjudicating controversies such as that between a demand-side explanation (costly signaling-driven increases in big-game hunting) and a supply-side explanation (increased availability of big game in the environment) for changes in the faunal record of the Middle Archaic period in the Great Basin (Byers and Broughton 2004; McGuire and Hildebrandt 2005). Beyond this specific issue, some of the most innovative zooarchaeological approaches in recent years (e.g., Stiner et al. 2000; Waguespack and Surovell 2003) owe much of their power to having found ways to begin to put resource use into the context of models of resource abundance.

This chapter describes how hunting is modeled in the Village Ecodynamics Project (VEP). Our agent-based model for hunting among the small-scale agricultural societies of the late prehispanic U.S. Southwest departs from traditional approaches in several ways, most importantly by our construction of a dynamic spatial model for both the demand and the supply side of hunting. This enables us to determine whether protein derived from hunting was limiting on this landscape, and whether certain features of the archaeological record in the central Mesa Verde region can therefore be plausibly explained as means to overcome these limitations—or as failure to do so. More generally we are interested in how hunting worked in Neolithic societies. We have known for some time that dense and sedentary populations almost invariably depress populations of high-return prey (Grayson 2001:5). Is it possible to build an agent-based model that reflects this process accurately? Can such a model suggest other consequences of hunting that might not be so clearly recognized? More generally yet, this hunting model—as part of the larger VEP simulation—aims to allow us to test hypotheses about how successfully people optimized their use of the landscape, whether there were trends toward increasing optimization under competition, and (eventually) whether we can detect in these processes levels of selection above the individual.

This chapter and Chapter 9 are companions to Chapter 7 by C. David Johnson and Timothy Kohler, in which a model for wood use in this area is presented and its results explored. Here we do the same for hunting, discussing how we model the animals that are hunted, as well as the process of hunting itself. In Chapter 9, we explore the general effects of varying four parameters related to the interaction of the hunters and the hunted: (1) how far people go to hunt; (2) how much protein they attempt to get from hunting; (3) the consequences of failing to meet those targets; and (4) how much priority households give to being able to hunt successfully as they consider whether and where to move. We close that chapter with a few observations about how our results provide insight into the dynamics of Pueblo society in our study area.

THE IMPORTANCE OF HUNTING

In the U.S. Southwest, studies of the ways that variability in subsistence success affected prehispanic societies have generally emphasized the erratic reconstructed fortunes of maize production (e.g., Axtell et al. 2002; Van West and Kohler 1996). The effects of variability in hunting success for Southwestern farmers have received less consideration (but see James 2004 for some relevant examples). At its most fundamental, hunting is of course a way of providing sustenance to oneself and one's kin. This is quite likely not the whole story, however, since it has been noted in many different settings that hunters retain relatively little of what they acquire for themselves and their families (Hawkes 1993). Kristen Hawkes suggested that success in hunting brings favorable attention from the group and allows the hunter a "larger, readier pool of potential allies and mates" (1993:341), a position that has been called the "show-off hypothesis." More recently, Rebecca Bliege Bird and Eric Alden Smith (2005) have shown how this hypothesis can be encompassed within signaling theory: a successful and generous hunter is providing an honest, and costly, signal of his capabilities. Clearly hunting satisfies both subsistence and social goals, especially when garnering relatively large-bodied animals that are uncommon and can be shared widely. This helps resolve the puzzle as to why men may persist in such hunting even when average return rates are low—indeed, lower than those available from alternative activities (Hawkes 1993:348).

While many answers are possible to the question of why meat and hunting are "consistently and ubiquitously valued more than plants and gathering and farming" (Kent 1989:7), for our purposes in this chapter what matters is this value, not its causes. Among small-scale horticultural societies, hunting may be particularly important where there are no large-bodied domesticates as alternative sources of fat and protein—as was the case for the societies we discuss here. Florence Hawley remarked on the importance of meat to a mid-twentieth-century Puebloan people, the Zia, as follows:

> Meat is the food most desired by people and most difficult to obtain. The small amount of meat purchased from the trading post by even the well-to-do families is almost negligible in the diet . . . very few families can afford to buy sheep or goats . . . . The principal meat supply for the village is made up by the meat dried in the fall, supplemented by rabbits. Several times a year there are communal rabbit hunts, which last all day and in which everyone takes part. Occasionally some one [sic] kills a wild turkey, and during the deer season every man who can afford a license hopes to bring in a large animal for both its meat and for its skin. But even under the most

> favorable conditions, the people do not expect to have meat more than two or three times a week, and then more as a flavoring than as a food. They themselves say that they think they would feel better if they had meat oftener, and . . . they speak of the last date on which they enjoyed fresh meat as if they were naming a Saint's day [Hawley et al. 1943:552].

Their clinical study of 20 Zia schoolchildren indicated "insufficient nutritional assimilation of nicotinic acid, riboflavin, ascorbic acid, and, presumably, protein of high biologic value" (Hawley et al. 1943:555). Even so, they remarked that the diet in some contemporary pueblos—-e.g., Taos—was worse than that at Zia in that the Taos diet contained less animal protein and fewer green vegetables.

A persistent problem in studying hunting has been the difficulty of building models that reasonably approximate expectations for success and return. They are difficult to build because animals (unlike, e.g., tool-stone sources) are a dynamic resource whose densities depend heavily on other dynamic variables, namely past and current hunting pressure and local environmental conditions. Without such models, researchers in contemporary societies (e.g., Kaplan and Hill 1985) can observe only return rates for animals that are actually hunted, not estimate what return rates might be on animals that are not hunted.

Of course these problems are compounded for the archaeologist, since observation of return rates generally is limited to assessing the relative frequencies of faunal materials from excavations that have been subject to greatly differing recovery techniques and preservational biases. Estimates on how much time or energy might have been required to return the animals thus represented are usually limited to indirect inferences derived from patterns of representation of particular bones, the degree of fragmentation, and so forth (see discussions in Grayson 1984; Lyman 1994; O'Connell and Hawkes 1988).

Alternatively, or in addition, static models can be constructed showing the radius around a site required to satisfy the various needs of the inferred population, as recently demonstrated by Margaret Nelson and Karen Schollmeyer (2003) for the protein and fat needs of Mimbres populations in the U.S. Southwest. Steven Simms (1987:49; see also Sept 2001:93), in discussing a dynamic model he developed to help estimate encounter rates for plant resources in the Great Basin, remarks that this task is relatively easier for plants, since "plants are generally more ubiquitous and stationary, reducing search costs, while animals tend to be mobile and cryptic."

Our motivation for developing the dynamic models presented here is the fact that, ultimately, such models will allow us to build expectations about what the archaeological record should look like under any assumptions we care to make about agent behavior. (We have chosen to assume that behaviors are approximately optimal with respect to the use of four types of resources.) These expectations can be compared with the archaeological record to investigate many classes of hypotheses (Stephens and Krebs 1986:183–184), including the simple goal, pursued here, of putting zooarchaeological remains into a context that also gives them more value for cultural and historical interpretation.

We begin with a very brief overview of the logic and structure of hunting in the VEP model. We focus initially on the supply side of the model by describing how we "grow" the animals that become prey for the human predators. Modeling deer presents special problems in that we must accommodate their great mobility; this is done with diffusion equations that are also described in the supply section.

We then examine the demand side of the hunting model. First we review the literature related to the amount of protein required in the diet, and then describe how this protein is acquired by our agents. The supply and demand sides are put together by running the simulation, and our discussion in Chapter 9 concerns how hunting plays out on this landscape under some different choices for parameters, such as the amount of protein required, and the consequences to the human populations of failing to acquire those amounts.

MODELING RESOURCES: THE SUPPLY SIDE

Our model landscape provides sources of protein in the form of three nondomesticated herbivore populations: mule deer (*Odocoileus hemionus*), black-tailed jackrabbits (*Lepus californicus*), and desert cottontails (*Sylvilagus audobonii*). These populations vary in size—spatially and temporally—in response to many factors. They "grow" on a model landscape that contains 139 soil complexes, each of which supports some combination of 93 native vegetation species. Animal population sizes depend in part on how much of their preferred browse is available within local plant communities. Browse production, in turn, is controlled by the action of a climatic signal on a soil of a specific category.

The deer, jackrabbit, and cottontail populations subsist on various combinations of 23, 26, and 31 species of vegetation, respectively. The browse preferred by each herbivore is highly variable in distribution across the study area. Deer populations, for instance, are most abundant at high elevations, where their preferred browse is most common and productive (e.g., northeast of the Dolores River, in the upper right of the maps in Plate 8.1 in the color insert of this book). Jackrabbits, on the other hand, exhibit higher population levels in low-lying, drier, grassland settings. Cottontail habitat generally dominates intermediate elevations in the study area. Of course, numerous species of other animals are present on the modern landscape and in prehistoric faunal assemblages. We limit our model to these three species because of their consistent importance among the wild fauna in the zooarchaeological record throughout the Southwest. As will be discussed in Chapter 9, though, turkey, which we do not model, becomes an important domestic and wild resource toward the end of the time span we target.

Jackrabbit and cottontail populations are maintained within each 4-ha model cell, but deer populations, because of their larger body size and greater mobility, are maintained within 1-km^2 units composed of 25 4-ha cells. At model initialization, animal populations are seeded within each 4-ha cell in the case of lagomorphs, or in 1-km^2 "deer cells" in the case of artiodactyls, in proportion to the mean productivity of their preferred browse species in that space. Chapter 7 provided a description of how the annual net primary productivity (NPP) of the browse species is determined by the model. Briefly, they are governed primarily by means of a calculation of stored soil moisture that is ultimately generated from local tree-ring data calibrated with local historical weather station data.

The mean NPP, or annual new growth, of each plant species is determined by running the model for the full 700 years without human agents and recording the average production within each model cell, taking into account the proportion available as food for each species. After initialization, the annual NPP of the browse species in each cell preferred by each species of animal determines the carrying capacity (K) of that herbivore (N) according to the discrete logistic growth function (Gurney and Nisbet 1998:Eq. 3.65) given in equation 8.1:

$$N_{t+\Delta t} = \frac{KN_t}{N_t + \gamma(K - N_t)} \tag{8.1}$$

where $\gamma \equiv e^{r\Delta t}$, the model time step t is one year, and the rate of intrinsic increase r is .4 for mule deer (McCullough 1997; Medin and Anderson 1979), 1.75 for jackrabbits (Wooster 1935), and 2.3 for cottontails (Myers 1964). These rates assume mortality from a variety of factors, including wild predation and disease. When populations are below K, from either an increase in the NPP or the harvesting of animals, they grow toward K at the rate of their given r, as modulated by equation 8.1. Populations decline toward K if they exceed it due to decreases in local NPP governed by the external climate signal. This function runs in each cell in each year for each of the three modeled herbivores.

Even though deer are modeled at a larger cell size, their mobility requires that we also allow them to "move" from cell to cell. The next section explains how this is done, given that deer populations are a variable maintained by each cell rather than mobile objects, or agents, like the households.

Deer Mobility Based on a Discrete Fisher Equation

Since the deer represent a supply variable, and since they are maintained by each model deer cell rather than as individual agents, they are most easily modeled using classical methods involving partial differential equations. The standard abstract model for such problems in population dynamics is Fisher's equation (Gurney and Nisbet 1998). In two space dimensions, this is $\frac{\partial N}{\partial t} = rN(K-N) + d\Delta N$, where r represents a rate of reproduction, K the carrying capacity, and d is a diffusivity, describing the tendency of the deer to wander. Both N and each of the parameters in the equation are a function of t and both space variables. Of course, ΔN is the Laplacian of N. In the absence of information concerning populations outside the study area, periodic boundary conditions are used.

Given the discrete nature of this model, this differential equation is replaced by a discrete version. Letting N_t denote the function N evaluated at time t, the time discretization can be performed using the growth function given in equation 8.1, which is denoted by $g(N) = Kf / (N + \gamma(K-N))$. This function actually comes from an analytical solution of $\partial N / \partial t = rN(K-N)$. This, coupled with a one-year time step, allows the discrete time model to be written as $N_{t+1} = g(N_t) + d\Delta N_{t+1}$.

The diffusion term is handled by a centered finite difference discretization on the 1-km² cells, so the model equations, for cells interior to the study region, are

$$N_{t+1}^{i,j} = g(N_t^{i,j}) + (N_{t+1}^{i+1,j} + N_{t+1}^{i-1,j} + N_{t+1}^{i,j+1} + N_{t+1}^{i,j-1} - 4N_{t+1}^{i,j}) \quad (8.2)$$

for $i = 2, \ldots m-1$ and $j = 2, \ldots, n-1$. Here $N_t^{i,j}$ represents the population of deer in the i, jth cell at time t. The equations at the boundary cells are analogous, but involve superscripts that assume $N_t^{i,n+1} = N_t^{i,1}$ and $N_t^{m+1,j} = N_t^{1,j}$.

The collection of these equations may be written as a matrix problem. Let $\upsilon_t = N_t^{1,1}, N_t^{1,2}, \ldots, N_t^{1,n}, N_t^{2,1}, \ldots, N_t^{2,n}, \ldots, N_t^{m,1}, \ldots, N_t^{m,n})^T$, $G(\upsilon_t) = (g(N_t^{1,1}), \ldots, g(N_t^{m,n}))^T$, and let A denote the matrix of coefficients for the discrete Laplacian. Then the system of equations in 8.2 becomes

$$(I - dA)\upsilon_{t+1} = G(\upsilon_t) \quad (8.3)$$

Observe that this places all the nonlinear terms on the right, so that only a linear system must be solved at each time step. It should be noted that an explicit model could have been formulated in the form $\upsilon_{t+1} = G(\upsilon_t) + dA\upsilon_t$. Such a model would be computationally very simple, but its numerical stability would depend strongly on the size of the parameter. For the range of values used here, the explicit iteration would be unstable. For this reason we chose the implicit algorithm in equation 8.3, which is stable for all $d > 0$.

The use of periodic boundary conditions means that the matrix $I - dA$, while sparse, is not tightly banded. Solving the system using an LU or Cholesky decomposition would result in a great deal of fill-in, leading to excessive computation time and computer memory use. For that reason, the system was solved at each time step using a preconditioned conjugate gradient iteration (Golub and Van Loan 1983). This involves only matrix multiplications, requiring less memory and somewhat less computation.

Plate 8.1 illustrates this diffusion process in action; at A.D. 611 we artificially kill off all the deer within an area 10 km on a side. Deer are able to rebound to their previous levels within about 25 years because of a combination of regrowth and diffusion. The figure illustrates

TABLE 8.1

Average Standing Crops of Prey Species in Model in Absence of Hunting, Compared Against Contemporary Data

Modeled Species	Modeled Ind / ha Mean / sd	Comparative Information for Validation		
		Ind / ha	*Location*	*Source*
Deer	.07 / .0142	.032	NE Oregon	Ager et al. 2003
		.04 –.08	Intermtn West	Mackie 1994[a]
		.10	AZ, NM	Short et al. 1977
Jackrabbit	1.32 / .26	<.3	N Utah	Smith 1990
		.11	SW Idaho	Knick and Dyer 1997
		.2 – 1.2	Southwest U.S.	Chapman and Willner 1986[a]
Rabbit	2.1 / .413	.02 – 2.5	CO–OR	Chapman and Willner 1986[a]
		6	IL	Giuliano et al. 1994

Ind = individuals.

a. Sources we consider most applicable for validation.

how diffusion would work with and without regrowth, with no hunting and in fact no humans on the landscape to complicate the picture. Of course, in the model, regrowth always occurs, though at variable rates depending on climatic conditions.

Exercise and Initial Validation of the Supply Side

Wildlife ecologists have amassed an enormous literature on the behaviors of various species at both the individual and population levels. Of particular interest here are the potential numbers of animals available for our model households to hunt; it is important that the model generate realistic population densities for the species in the simulated world.

As noted, we model both lagomorph populations at the normal 4-ha model-cell level and the artiodactyl population at a 1-km^2 deer-cell level. The rationale for this strategy is that while both the actual and virtual landscapes of our study area are capable of supporting many cottontails, jackrabbits, or both within a given 4-ha area, neither can do so for deer. As shown in Table 8.1, even the highest counts observed by wildlife studies have less than one-half of a deer in a 4-ha area.

Our virtual landscape supports deer population densities of .28 individuals per 4-ha model cell, or .07 / ha. This figure falls within the range reported by Richard Mackie (1994) for mule deer densities in the Rocky Mountain foothills, in which our study area lies. Although higher deer densities are reported from areas south of the Four Corners by Henry Short et al. (1977), their study was limited to pinyon-juniper forest, much of which historically was heavily modified for range enhancement. Our study area, by contrast, contains a wide range of vegetation, some of which supports fewer deer than the pinyon-juniper forest.

For jackrabbits, the model generates mean population densities that are slightly higher than those reported in what we consider to be the most pertinent published comparison. For cottontails, the model generates mean population density figures that are on the high side but within the range observed today in comparable environments. There are many possible reasons for this, one likely candidate being the complete absence of hunting by humans in generating these figures. In addition, our study area is quite productive as opposed to similar areas within the region. Overall, it appears to us that animal densities generated in the model are reasonably congruent with relevant studies by wildlife biologists.

MODELING HUNTING: THE DEMAND SIDE

A basic philosophy governing all aspects of our modeling (indeed, of any model) is that we cannot hope to capture in its fullest detail all the richness of our target system. For example, we expect of our households that they satisfy 70 percent of their caloric needs through farming. We do not attempt to model the various sources of wild foods, or calculate the calories obtained from meat in hunting, that would have made up the remainder. Instead we assume that these could be obtained costlessly, in the course of other activities. In this sense, and also in the relatively generous stocking of the landscape that we allow, our model may slightly overestimate the number of households that can easily make a living in the study area. This design bias is consistent throughout the model, because if, in the end, we wish to argue that resource deficiencies are responsible for depopulation or aspects of culture change, we do not want the model to overestimate those deficiencies.

This bias is somewhat compensated for by the fact that, for hunting, the model is based on the potential productivity and harvest of only three species that together constitute the majority of most local archaeofaunal records (whether calculated by number of identifiable specimens NISP or by percent contribution to the meat protein intake). Although the amount of protein households require from hunting in the model is a tunable parameter, in most runs we keep this fairly low (e.g., 10 or 25 g / person / day) which is, as we will see, less than the amounts that nutritionists today suggest as appropriate. We do not model protein returns from hunting other species, from other activities like trapping, or the protein acquired from cultivated or gathered plants. Nor, as we noted, do we model protein returns from domesticated turkey, which becomes an increasingly important part of the local archaeofaunal records in the A.D. 1000s.

Current VEP research explores boundedly rational behaviors for our agents, including in their hunting. Agents have only limited, local information about the environment, and they have only limited computing power with which to determine the "best" decisions from that information. As a result, we make no claims that our agents create globally optimal settlement systems, although we consider the settlement patterns they generate to make a first approximation to the optimal solution. We have designed their decision algorithms for what to hunt, how much to plant, where to live, and so forth to generate cheap (from a computational perspective) approximations to optimal solutions. This may be a weakness from the perspective of rational choice theory, but it is in line with much research on how people actually make decisions (Gigerenzer 1999). Thus, the hunting model we present in the paragraphs that follow is not a strict implementation of optimal foraging theory, but its design assumptions are the same: that protein acquisition should be efficient so that diet breadth will change as resource abundance changes, and that the sites from which hunting is conducted—like those of central-place foragers (Stephens and Krebs 1986:53–60)—should be located so as to maximize the total efficiency of use for production of water, woody fuels, maize, and animals. Our algorithm for determining these central places was discussed in Chapter 4. A fundamental attraction of simulation here is the possibility of simultaneously calculating diet breadth and patch choice in an environment in which both resource and human distributions change every year.

Protein

Just as model households have caloric needs that can be satisfied only through farming, they have protein needs that can be satisfied only through hunting. The difference is that caloric needs vary according to levels of household activity, whereas protein needs, for simplicity, are assumed to be static. Each of the three herbivores we model provides a different average meat weight and protein content.

Proteins form the basis of life. All organisms require various forms of proteins, and, in fact, the simplest life forms consist entirely of proteins. Proteins are composed of carbon, hydrogen, oxygen, nitrogen, and smaller amounts of many other elements including copper, iodine, iron, phosphorus, and sulfur. Twenty "alpha" amino acids are the building blocks of all proteins, and combinations of from eight to 18 of these make up the known proteins required by living cells (Cordy 1976). Nine of these alpha amino acids are termed "essential amino acids" because the human body is unable to synthesize them; they must be provided in the diet (Cordy 1976; Wing and Brown 1979).

The human diet requires an adequate protein intake on a regular basis. "The cells of the body need a constant supply of proteins for building new tissues, for regulatory functions, to use as the precursors of enzymes, hormones, antibodies, and some vitamins, and to provide energy if needed" (Cordy 1976:8). To maintain optimal health, proteins consumed must be complete: they must supply all the essential amino acids. "Those foods containing amino acids in the proportions best utilized by the human body are said to contain 'high-quality' proteins .... Proteins from animal sources (meat, fish, milk, and eggs) are of high quality [but most] vegetable or grain proteins are low in some essential amino acid(s)" (Wing and Brown 1979:52). Of course, in combination, a variety of complementary vegetable proteins can supply adequate levels of essential amino acids.

Reports of protein requirements for people of different ages often differentiate between a minimum, or base, requirement and a recommended daily intake, or optimum amount (Wing and Brown 1979:Figures 3-2 and 3-3). Children, having much lower body weights than adults, require a *minimum* of 10 g / day. Minimum requirements slowly increase with age, leveling off at approximately 30 g / day at maturity. Recommended daily allowance values, on the other hand, vary much more widely. According to Wing and Brown (1979), the daily requirement increases from 10 to 60 g / day for males from birth to maturity. The United States Department of Agriculture (2005) recommended daily allowance for adults is 91 g / day of protein, in conjunction with a 2,000-calorie / day diet.

VEP households are required to satisfy only a portion of their protein needs from hunting animals provided by the model landscape. We explore the effects of 10 and 25 g / person / day in Chapter 9, assuming implicitly that the remainder can be obtained costlessly either as a side benefit of eating carbohydrate-rich foods and legumes, or by trapping and hunting animals—whose populations are not modeled—incidental to other activities. In modeling protein consumption by prehispanic Southwest U.S. populations, other archaeologists have also tended to suggest that relatively low levels of the recommended daily allowances were derived from hunting. Katherine Spielmann and Eric Angstadt-Leto (1996), for example, suggest 10 g / person / day as a reasonable estimate of quality protein obtained by prehistoric Southwestern peoples from hunting artiodactyls and lagomorphs. In her discussion of the diet of New Mexico's Arroyo Hondo population, Wilma Wetterstrom (1986:Table 31) computes an average protein requirement of 19.4 g / person / day. Assuming reduced net protein utilization associated with a heavy reliance on maize agriculture—e.g., in her 70 percent column—that figure increases to an average of 30.6 g. In their model of game use in the Mimbres Region of southwest New Mexico, Nelson and Schollmeyer estimate that "in a diet drawing 80% of its total calories from maize, game meat would have provided ... 48% of the protein" (2003:84). Using the Food and Drug Administration daily reference value of 50 g per person, Nelson and Schollmeyer's model thus requires 24g / person / day. In sum, our use of 10 and 25g / person / day, though low by modern standards, is consistent with the estimates used by others who have modeled the protein requirements of prehispanic Southwestern populations.

In contemporary societies, protein deficiencies are usually accompanied by caloric deficiencies. Such malnutrition "has its greatest impact on the young. Deaths from protein-calorie deficiency result from the failure of the child to thrive, with progressive weight loss and weakness, which in turn can lead to infection and disease, usually some form of gastrointestinal bacterial or parasitic disorder" (Encyclopaedia Britannica 2005). Rarer cases with adequate calories but inadequate protein can cause kwashiorkor, a term meaning "'deposed child' ('deposed' from the mother's breast by a newborn sibling) in one African dialect and 'red boy' in another dialect. The latter term comes from the reddish-orange discoloration of the hair that is characteristic of the disease. Other symptoms include dry skin and skin rash, potbelly and edema, weakness, nervous irritability, anemia, digestive disturbances such as diarrhea, and fatty infiltration of the liver" (Encyclopaedia Brittanica 2006).

The Hunting Algorithm

Hunting is a complex skill that requires knowledge of the landscape and mastery of a variety of techniques for tracking and capture. Success can be highly variable, even for expert hunters. Modeling this behavior, in turn, is a complicated process that we break down into several distinct parts (see the overview in Figure 8.1). In the model, for simplicity, hunting occurs just during the summer (-step_procure_summer). Before hunting, the household determines the amount of protein required for the year. This is calculated by multiplying the number of family members by the daily protein requirement in grams, which is set as a parameter. Once the protein requirement is known, the household checks whether it has any protein in storage from the previous year. Protein can be stored from year to year (presumably by smoking or drying) if a surplus is harvested, but we subtract 25 percent per year from the stored amount to represent spoilage or loss through nonhuman consumption. If a household has enough protein in storage, it consumes its required amount of protein and does not hunt during the year. If a household requires additional protein, it continues with the hunting process.

Search Costs

Once an agent has decided it needs to hunt, the next step in the process is to determine which cell or cells the agent will try to hunt. The hunting algorithm considers individuals from households as agents. Agents begin this reconnaissance by looking for deer, since deer supply more protein per animal than other species. The search starts in the cells closest to the agent's home and continues progressively outward by expanding a square search pattern one cell width at a time.

Once a hunter reaches a certain distance from his home, he adjusts his reconnaissance strategy to include rabbits and hare. The distance (radius) that agents search for deer before looking for other species is a parameter that, in the runs reported here, is set to 20 cells (4 km). Once this radius d is reached, the agent will search in his home cell for rabbits and hare before looking farther away for deer. During each additional expansion of the hunting distance the agent will also look for rabbits and hare within a radius of d-20 cells from home. Choice of d in any particular context by a researcher represents a belief, or calculation, that at that distance the return rate for deer would approximate the return rate for lagomorphs in the target system at distance d-20.

For example, then, when $d=20$, a hunter will look for deer at increasing distances from home in each of the cells within this radius, but no lagomorphs will be hunted until the distance from home exceeds 20 cells. If a hunter

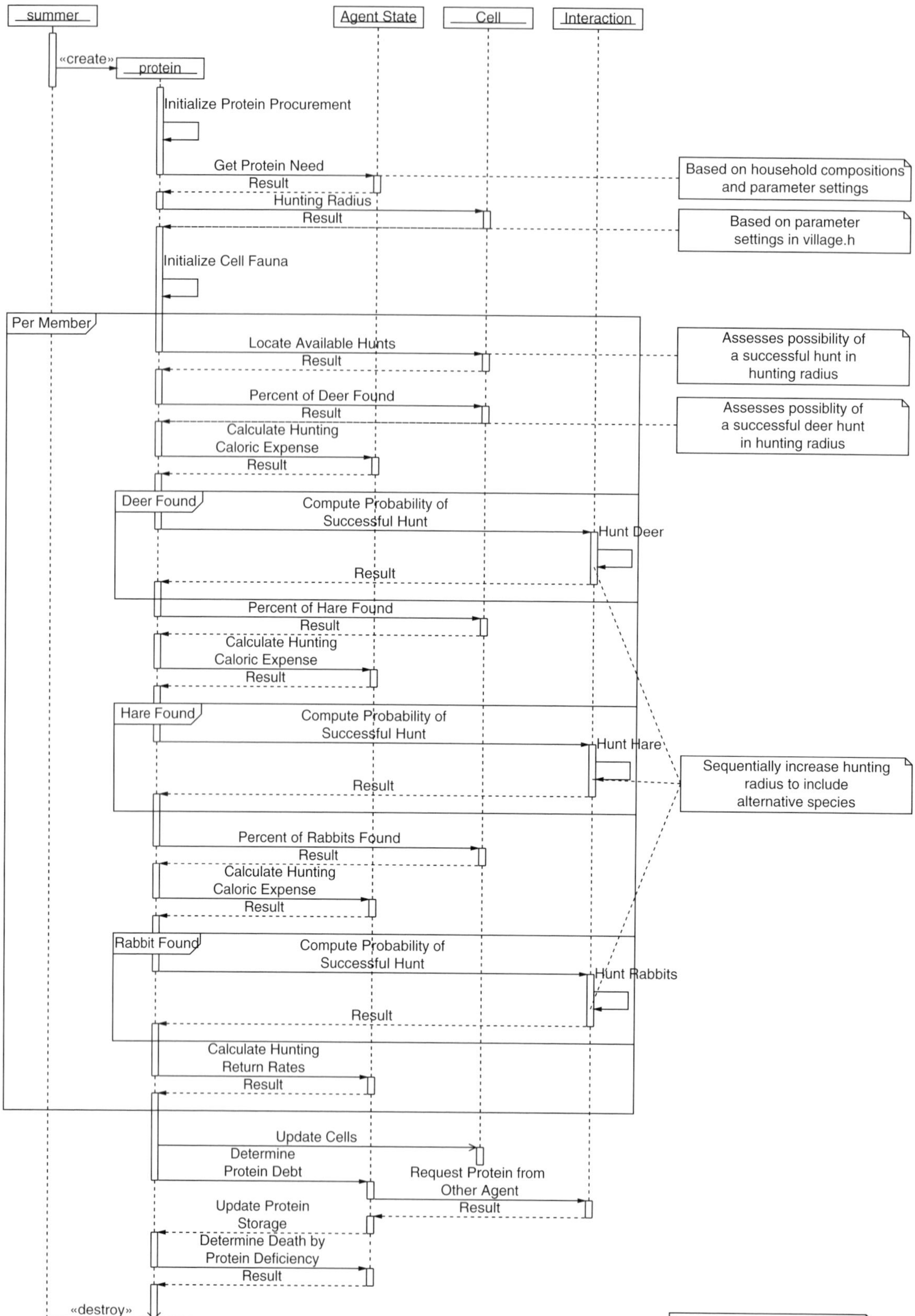

FIGURE 8.1 Overview of hunting methods.

TABLE 8.2

Example Calculation for the Number of Animals Available for Hunting in a Cell

Species	Total in Cell[a]	Agent Knowledge (%)	Animals Available for Hunting
Deer	3	55	1
Jackrabbit	9	66	6
Rabbit	10	34	5[b]

a. For deer, the cells are 1 km^2; for lagomorphs, they are 4 ha.

b. The agent finds 3.3 rabbits, which is greater than 50 percent of 5, the number necessary for hunting. This allows the agent to increase the number to the minimum needed for a single hunt.

has had to travel 25 cells from home, then he will have also looked in all cells within a 5-cell radius of home for rabbits and hare before expanding the search for deer to cells in the 26-cell band. This search continues until the household satisfies its protein needs, reaches its maximum hunting distance (also an adjustable parameter), or runs out of time for hunting. The maximum hunting distance can be set to any value. Here we use 20 or 40 cells (4 or 8 km), more or less corresponding to the small and large hunting areas discussed by Lewis Binford (2001:Table 7.13). The number of hours an agent can hunt is determined by subtracting the total number of hours it spent in the previous year's nonhunting activities (as an estimate for the current year) from the total hours an agent can work in a year, which we set rather generously at 14 hours per day for each person in the household over the age of seven.

Since animals, as noted by Simms, are "cryptic," hunters cannot find all the animals living in the cells they search: for each animal species, a hunting agent is allowed to gather knowledge of between 30 and 75 percent of the animals supported by each searched model cell during that year. This implementation of imperfect hunter knowledge is intended to approximate reality; even the most up-to-date techniques for estimating animal populations result in highly variable numbers (Langdon 2001). Knowledge of the number of animals is calculated randomly within the 30 to 75 percent range as each cell is reconnoitered by a hunting agent for each species. This is necessary since the number of animals in a cell is in a constant state of flux due (on an annual scale) to varying amounts of animal food in the cell each year, deer migration, and (on an interannual scale) hunting by multiple agents. Thus one hunter may discover no deer in a cell in the same year that another found one, either because of a different discovery rate or because the underlying population number has changed. After calculation, the figure is rounded down to the nearest whole animal to determine the number of animals available for hunting—except in the case where the number of animals found is less than the number needed to allow a hunt. In that case, if an agent finds at least 50 percent of the required number of animals for a hunt, he will assume that there are that many animals available. Table 8.2 shows how this process works. For each species there is a separate agent knowledge percentage. This means that even though there may be, for example, fewer hare than rabbits in the cell, more of them may be hunted if an agent learns of more hare than rabbits in the cell.

Deer present a special case since they do not live at the standard model-cell level; they live and grow at the deer-cell level, which at 1 km^2 is much larger than a single cell. Each deer cell comprises 25 4-ha model cells, and the productivity from all of these regular cells is combined to determine the amount of food available to deer. When hunting occurs in any model cell, the number of deer available in that 4-ha cell is the number occupying the deer cell of which the hunted cell is part, discounted by imperfect knowledge. If deer are taken from the hunted cell, then the total number of deer in the associated deer cell is decremented accordingly.

Information is not free, but in our model the partial knowledge of how many animals are in a cell is given to the agent at relatively low cost. Agents engage in, and are charged calories and time for, many activities other than hunting that take them out onto the landscape: tending crops, collecting water, and gathering fuel resources. We assume that during such trips people can accurately assess where animals are *not* present. Given the nature of our growth functions, we do not allow animal populations to become zero, and we require a minimum number of animals—one deer, two hare, or five rabbits—to be in a cell before agents can recognize that they are there. If the animal numbers are below those thresholds, we do not charge households for searching that cell (which from the perspective of the household is empty). But households are charged a caloric and time cost for going to the closest cell in which there are in fact animals above those thresholds, as well as for any additional trips to cells with animals they may need to hunt to satisfy their protein needs.

Once a hunter "knows" how many animals are in a cell, he will hunt there until he kills enough deer to satisfy his household's protein needs, he runs out of time for hunting, or he kills all known deer in the cell. A lagomorph hunter uses slightly different rules. First, the hunter must have a certain number of people in the household over the age of seven to participate in the hunt, since most lagomorph hunts are communal and the model is set up to hunt multiple lagomorphs at the same time. (In the runs reported in Chapter 9, however, the number of lagomorphs taken during a single hunt is small, so this parameter has likewise been reduced and currently requires only one person over seven years of age to hunt in this way.) If a household has enough hunters to hunt lagomorphs, the next step is to determine if there are enough animals in the cell to trigger a hunt. The number of lagomorphs needed for a hunter to consider hunting in a cell is either five rabbits or two hare. If hunting these populations causes known lagomorph numbers to fall below this threshold, the agent will move on to another cell for hunting.

Handling Costs

Once a hunter has found a cell to hunt in—one with enough known animals to hunt—he then goes on a "hunting trip." The purpose of a hunting trip is to get a chance at acquiring animal protein. During each trip a hunter has a defined chance of killing either one deer, two hare, or five rabbits. The model calculates how much time and caloric effort is expended to perform this operation. The time and costs for each hunting trip depend on the type of animal being hunted (see Table 8.3). In optimal foraging theory terms, the costs of the "hunting trip" include both pursuit time (whether or not the hunt is a success) and processing time (if the hunt is successful). Hunting deer is an individual task, but it is possible to have more hunters when hunting the other species (though this is not implemented in the runs reported here).

Each hunt incurs all the base costs regardless of whether the agent is successful in a given trip. During each hunting trip, an agent has a 30 to 75 percent chance of killing the animal or animals being hunted. This success rate for hunting is highly variable since it is meant to apply to various species in a variety of environments. Robert Kelly lists ethnographic accounts of hunting success that range from 9 to 100 percent (1995:Table 3-7). If an agent fails to harvest any animals, he will conduct additional hunting trips until successful, or until he runs out of time for hunting. If successful, the household's protein storage is adjusted according to the resources gained (see rightmost column of Table 8.3). A caloric cost is levied for meat transport, based on travel distance and 25-kg cargo weight (Lightfoot 1979). The household then recalculates its protein needs, and either continues hunting or moves on to other activities.

TABLE 8.3
Hunting-Trip Information

Type of Hunt	Time Expended	Calories Expended	Chance of Acquiring Resource (%)	Resources Gained
Deer Hunt	4 hrs	960 Kcal based on 4 hrs work by a man	30–75	1 deer 10,800 g protein
Rabbit Hunt	.5 hrs / person	120 / 100 / 46 Kcal: 120 for single person based on work by a man, additional 100 for second person based on work by a woman, and 46 extra for each additional person based on work by children 7 or older	30–75	5 rabbits 886 g protein
Hare Hunt	.5 hrs / person	120 / 100 / 46 Kcal: 120 for single person based on work by a man, additional 100 for second person based on work by a woman, and 46 extra for each additional person based on work by children 7 or older	30–75	2 hare 910.8 g protein

Exchange

If a household runs out of time or hunting area and thus fails to meet its protein target, it may use up to three exchange networks to attempt to obtain its needed protein. Our implementation of exchange allows four options. The first option excludes exchange, the second allows generalized reciprocity, the third allows both generalized and balanced reciprocity (as defined by Sahlins 1972), and the fourth allows exchange between households that are highly successful in both networks (which we call hub nodes). The Village Project's economic network was originally designed for the trade of maize between agents, but it has been extended to the exchange of protein among households; a full discussion of how this network operates is provided in Chapter 11 of this book and elsewhere (Kobti 2004; Reynolds et al. 2005). Hub-node exchange is not implemented in the runs discussed in this book.

When both generalized and balanced reciprocal exchange are enabled, agents will attempt to acquire protein resources first through their kinship network (generalized reciprocity). Failing that, they will attempt to acquire repayment from nearby agents who owe them protein (balanced reciprocity). Failing that, they attempt to go into debt with other nearby agents to acquire protein. In the runs reported in Chapter 9, we average across all nonhunting parameters in order to focus our analyses on hunting strategies.

Protein Deficiencies

A household that has exhausted all available avenues for protein acquisition and still has not met base requirements is considered protein deficient. We explore the effects of two different penalties on households reaching this state. The first affects the birth and death rates of the agent. If an agent has sufficient maize in storage, and it is growing a surplus of maize each year, the birth rate of that household is incremented by 10 or 15 percent, and the death rate decremented by 10 or 15 percent, on the basis of a standard life table drawn from Kenneth Weiss (1973:156), to reflect its thriving status. This is a

TABLE 8.4

Summary Data on Animals in Model

Attributes	Deer	Hare	Rabbits
Average Weight (kg)	60	2.3	.895
Edible meat weight (kg)	36	1.38	.537
Protein / kg (g)	300	330	330
Kcal / kg	1,580	1730	1730
Protein / animal (g)	10,800	455.4	177.2
Kcal / animal	56,880	2387.4	929.01

NOTE: Hobbs and Swift 1985 (data for deer); Haskell and Reynolds 1947 (data for hare); Ingles 1941 (data for rabbits).

parameter (STATE_GOOD) that we vary in the parameter sweep reported in this book. Such a household is in what the model considers "State 2." The first penalty for insufficient protein erases this modification—if it is in place for a particular household—by lowering the birth rate and raising the death rate to the base levels (State 1). The second, and harsher, of the penalties for protein deficiency that we implement here uses the two-step process of first applying the birth and death rate penalty to households, but, in addition, if a household continues in a protein-deficient state for DEATH-YEAR consecutive years, then it dies. Given our discussion of the effects of inadequate protein in the diet, we consider the first penalty to be somewhat more realistic, although it likely underestimates the effects of severe shortfalls over several years. The extent of the mortality induced by the second penalty depends greatly on the other parameters that are in place.

The amount of protein received by an agent when it has successfully hunted an animal is calculated by taking the base weight of the animal and multiplying this number by .6 to determine meat weight. Theodore White (1953) calculated that mule deer, hare, and jackrabbit dress out to 50 percent of their body weight. More recent anthropological researchers (e.g., Simms 1987) typically estimate that 60 percent of a deer's total weight is edible, based on Andrew Christenson's (1981) discussion of what aboriginal people consider edible. We likewise use an edible meat weight for lagomorphs of 60 percent of total body weight. This meat weight is then multiplied by the number of grams of protein per kilogram of edible meat (United States Department of Agriculture 2005) to get total protein provided by animals of each species (see Table 8.4). Calories per animal are calculated in the same manner.

Agent Movement

Considerations of access to protein also affect the decisions a household makes to determine where to go when relocating. Households move when they are using more calories than they can expect to produce locally and when their maize storage cannot be expected to make up the difference. When this happens, a household will search the landscape within its radius of "vision" for a location that maximizes caloric (maize) production and minimizes caloric costs for hunting collection of water and fuel collection. Using this procedure, households might relocate to a cell with a high maize production value even if other needed resources were locally rare. Because it seems in some sense "unfair" that a household might perish from inadequate protein when it did not in fact put a priority on protein in its locational decisions, we created a parameter—NEED_MEAT—within the movement algorithm that optionally forces households to consider cells for relocation only if local hunting is possible. This parameter may be set to either zero, in which case the household operates under strict rules for maximizing calories, or one, in which case the household attempts to maximize calories within the constraint of moving to a place where some hunting is possible.

SUMMARY AND CONCLUSIONS

In this chapter, we have described how agents acquire protein resources in the VEP world; our model attempts to account for both the supply and demand sides of hunting. Protein resources in the form of deer, hare, and rabbit are generated across the landscape, and their population sizes are affected by environmental dynamics and by hunting. Hunting is performed by agents to meet their (parameterized) protein needs; agents make bounded rational decisions about where to live, and what to hunt from their home cells, by employing the energy-minimizing strategies with which they were encoded. The next chapter explores model output with varying hunting parameters in the model.

A few final points need to be made about our assumptions in this part of the model. First, we do not explicitly include wild animal predation on our modeled prey, though it should be noted that predation is assumed in our population-growth algorithms for the prey species that we model. Second, we do not allow for differentiation of hunting skill among agents; hunting success is determined stochastically. Skills and techniques are not heritable or transmissible, and one's hunting success in the past does not improve hunting success in the future. Finally, agents are unable to strategically devote more resources to either hunting or exchange—as we expect real people might in response to their circumstances—nor are agents allowed to exchange between currencies (i.e., meat for maize). It is quite possible that networks of intercurrency exchange occurred in the past; enabling such exchange in the Village world would potentially allow for economic specialization and increase systemic resilience to productivity depressions that were subregional in scale. We are exploring several of these issues in VEP II.

NINE

How Hunting Changes the VEP World, and How the VEP World Changes Hunting

R. Kyle Bocinsky, Jason A. Cowan, Timothy A. Kohler, and C. David Johnson

THIS CHAPTER EMPHASIZES those results from running Village, the simulation of the Village Ecodynamics Project (VEP), that help us understand how the agents behave as we vary parameters that directly affect their hunting, as described in Chapter 8. The sweep reported here consisted of 512 runs in which we vary nine parameters with two states each. In this chapter we "compress" these 512 runs into 16, each representing a unique combination of values for the four parameters explained in the preceding chapter (see also Table 4.1) that directly affect hunting. These are (1) the grams of protein needed per person per day (PROTEIN_NEED); (2) the penalty that households suffer if they fail to acquire this minimum (PROTEIN_PENALTY); (3) the radius within which households could hunt (HUNT_SRADIUS); and (4) the priority given to protein in relocation decisions (NEED_MEAT). This compression was done by averaging. For example, results from all 32 runs with the specific combination of hunting parameters shown in the top row of Table 9.1 were averaged to graph the demographic history for that specific combination of parameters in Plate 9.1 (see color insert in this book). Two composite runs that differ by only their value for one parameter will be referred to as a pair of runs. For example, composite runs 2 and 4 (Table 9.1) differ only in that run 2 has a 10-g protein requirement whereas run 4 has a 25-g protein requirement. Table 4.1 lists other parameters averaged across the runs reported here.

Other chapters in this book also address questions relevant to understanding hunting in the Village simulation. Chapter 10, for example, discusses which combinations of hunting parameters result in spatial behavior that most closely approximates those known from the archaeological record (see also Johnson 2006). Here we concentrate on how the hunting parameters we varied affect human and animal population sizes. These populations are dynamically linked, of course, since more people hunt more animals, thereby generally lowering animal populations (though only to the extent explained later here). We also discuss changes in the return rates (generated by the model) on hunting each

TABLE 9.1

Parameters Varied in Composite Runs Reported in this Chapter

Composite Run	PROTEIN_NEED (g / person / day)	PROTEIN_PENALTY	HUNT_SRADIUS (cells)	NEED_MEAT
1	10	0[a]	20	0[c]
2	10	1[b]	20	0
3	25	0	20	0
4	25	1	20	0
5	10	0	20	1[d]
6	10	1	20	1
7	25	0	20	1
8	25	1	20	1
9	10	0	40	0
10	10	1	40	0
11	25	0	40	0
12	25	1	40	0
13	10	0	40	1
14	10	1	40	1
15	25	0	40	1
16	25	1	40	1

a. No birth or mortality restriction.

b. Birth or mortality restriction.

c. May move to protein-depleted area if costs are otherwise low.

d. May not move to protein-depleted area, regardless of other costs.

animal species over time, and we compare those to rates found in the hunting literature on these species. This helps us to understand the behavior of this model and to assess its realism.

GENERAL OBSERVATIONS

Except during the population trough from the tenth to mid-eleventh centuries, all these runs underestimate the number of households in the study area (Plate 9.1). In general, the simulated populations are less subject to large abrupt changes in size than were the real populations—although this is potentially misleading since the real human populations can be estimated only within discrete periods, leading to abrupt changes at period boundaries in some cases.

In considering the reasons for these differences, and particularly for the typical underestimation of the real populations by the simulations, it is important to remember that—unlike the real occupants of our study area—our agents are incapable of any intensifications other than switching among deer, rabbits, and hare, and simply working longer. More subtle forms of intensification, including domestication of turkey, are not within their repertoire. And yet, we know from the archaeological record that turkey become important by the middle of the A.D. 1000s (Badenhorst and Driver 2009), precisely as the real populations in Plate 9.1 begin to surpass most of the populations generated during these simulations.

Population growth in the model levels off during the eleventh century and holds steady

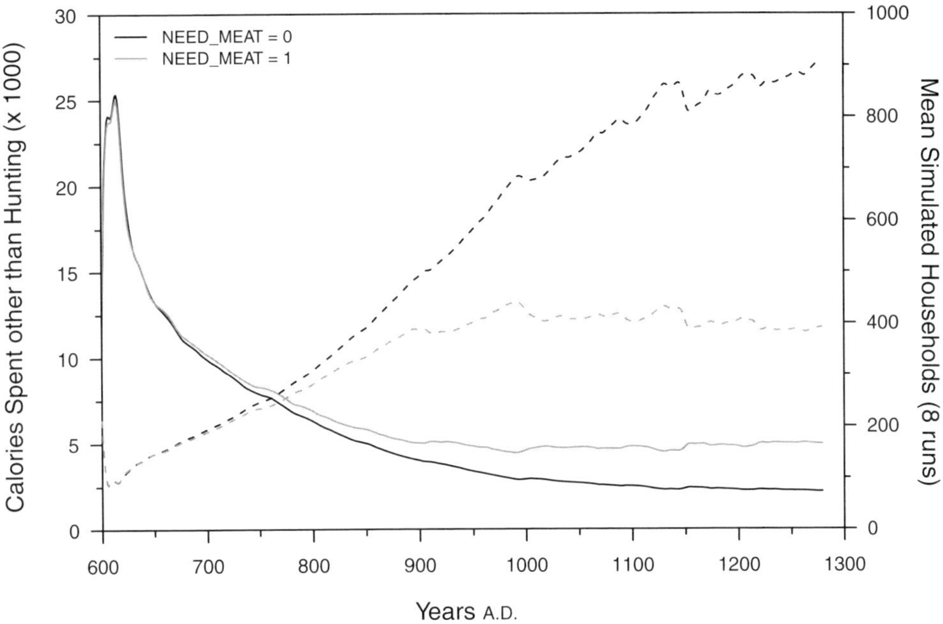

FIGURE 9.1 Modeled calories spent on non-hunting tasks (solid lines), compared to the number of simulated agents on the landscape (dashed lines), for runs with contrasting values for NEED_MEAT.

until the very end of the simulation. We postulate that this is primarily due to protein depletion. Deer populations decline in conjunction with rising human populations, and they also flatten off at around the same time as do the human populations (Plate 9.2). Protein limits on agent population growth are expected when PROTEIN_PENALTY=1 or NEED_MEAT=1, but what is the mechanism for this when those are set to 0? When PROTEIN_PENALTY and NEED_MEAT are zero, agents spend a great deal of effort (calories) on hunting even though it is not paying off. Here, as elsewhere, the model honors the prediction from life history theory that resources allocated to somatic effort—in this case, hunting—detract from those available for reproductive effort.

Also, as noted earlier, differences between simulated and estimated population levels may be due to factors that affected the archaeological record but are not present in the model, such as turkey use and domestication. Remembering also that all decisions in the simulations are made atomistically by individual households, a fertile area for future work will be to consider the extent to which the large increases and decreases in the real populations, as opposed to those of the simulated populations, can be attributed to highly conformist decision making either above the level of the household or at the household level.

VARYING THE PRIORITY GIVEN TO PROTEIN AVAILABILITY IN DECISIONS TO RELOCATE

Of the four hunting parameters, varying the priority given to protein in relocation has the greatest impact on the simulated numbers of households: populations are consistently lower when agents are required to move to areas where there are local protein resources (when NEED_MEAT=1) (Figure 9.1). Agents decide to move their homes when they become unable to sustain themselves at their current location. When this parameter is toggled on (= 1), agents typically are forced to move to a cell with fewer attractive returns to effort, overall; this depresses global populations, eventually. (If they are unable to find a new home near protein resources—because all such sites in the range are at their population capacity, or because no such sites exist within the agent's search radius—they relocate to the highest-ranked cell that is available; see Chapter 4). As deer, hare, and rabbit populations become depleted in the best areas, the locations chosen are increasingly expensive calorically (increasingly out of the optimal niche). This difference becomes more pronounced as agent populations increase.

VARYING PROTEIN REQUIREMENTS AND PROTEIN PENALTIES

Other parameters being equal, the runs in which household members each need only 10 g of protein daily (composite runs 1, 2, 5, 6, 9, 10, 13, and 14; Table 9.1) generate more households than runs in which households seek 25 g (Plate 9.2). This is expected, since a higher protein requirement means that agents attempt to hunt for more game, and in so doing they expend more calories. Since hunting in the model returns only protein and not calories to the agent, more effort spent on hunting leads to lower viability and fertility in the agents. In the same manner, runs in which PROTEIN_PENALTY=0 (no penalty to birthrates for not reaching protein needs) have higher populations than runs with the PROTEIN_PENALTY=1, which returns agents' birthrates and death rates to the base rates if they fail to get adequate protein even if they are otherwise thriving (in state 2). Not surprisingly, PROTEIN_PENALTY=1 imposes a dramatically greater impact on population size on runs where PROTEIN_NEED=25—such as runs 5 and 13—than on runs with lower protein requirements, since agents are much less likely to be able to attain the higher amount of protein, especially late in the simulation. This nicely illustrates what is in fact very common: the effects of these parameter settings are frequently not simply additive; more commonly, they harbor nonlinearities and threshold effects that we contend are also common in the reference system of the Pueblo past.

VARYING HUNTING RADIUS

For runs where agents are required to relocate to cells with nearby protein resources, human populations tend to be higher when agents are allowed to hunt at a greater distance, since the definition of "nearby" is expanded by the hunt-radius parameter. When NEED_MEAT=0, however, a wider hunting radius is detrimental to agents so long as PROTEIN_PENALTY=0 (Plate 9.2). This is because a smaller hunting radius reduces the total costs the agents accrue since the agents will stop attempting to fulfill protein requirements sooner—and at a lower cost to their calorie budget. This combination of parameters is probably not very realistic, however.

EFFECTS OF HUNTING ON ANIMAL POPULATIONS

Deer

All runs show decreases in the deer populations (Plate 9.2), though the greatest depletion is clearly correlated with the higher value for PROTEIN_NEED. Deer populations bottom out at around 2,000. They are not completely wiped out because when deer populations are this low, the average deer per deer cell (a 100-ha area composed of 5 × 5 cells) is less than the density needed for hunters to be able to "observe" that deer are present. As soon as there are enough deer in a cell to hunt, however, those deer are taken, keeping the total number of deer close to this minimum. In most runs, deer reach this nadir around A.D. 1000. The arrival at this point is fairly uniform across all runs, with the exception of runs 7 and 8, which reach their minimum around A.D. 750. These runs are characterized by the higher protein requirement (25 g / person / day), smaller hunting radius (20 cells), and the requirement to relocate to protein-rich areas. This may explain their extreme depletion, because relocation decisions early in the simulation favor areas rich in deer. In these runs, agents effectively chase deer across the landscape, allowing for their more rapid depletion. Also, the combination of lower search radius and higher protein need may lead to less access to leporid populations, thereby placing an even greater burden on the deer. Other runs with NEED_MEAT=1 also begin significantly depleting deer earlier in the simulation, though they reach their lowest points in tandem with the majority.

Leporids

Plates 9.3 and 9.4 shows rabbit and hare populations over time for each of the 16 composite runs. For both species, the runs fall into three categories, exhibiting low, moderate, or high depletion. High-depletion runs include runs 3, 4, 11, and 12; medium-depletion runs are 1, 2, 9, 10, 13, 14, 15, and 16; runs 5, 6, 7, and 8 all result in low leporid depletion. It is not surprising that the degree of depletion is roughly correlated with the total number of households, but there are some other interesting correlations as well. The parameter with the least effect on the degree of leporid depletion is the penalty for not getting the required protein. This is not very surprising, because this does not alter protein-acquisition effort. When the other parameters are held constant (here, by averaging their effects), all pairs of runs differing only by their value for PROTEIN_PENALTY group together in terms of depletion severity. The amount of protein needed by each agent is slightly more influential in determining the degree of leporid depletion, though pairs of runs differing by only PROTEIN_NEED and PROTEIN_PENALTY are usually adjacent.

Hunt radius has a much larger impact on leporid populations. When agents are able to hunt at greater distances, they depress these populations more consistently. The need to move to areas with protein resources, however, is the best predictor for the degree of leporid depression. It is interesting that the runs where this is *not* required (NEED_MEAT=0) are those that show the greatest suppression of leporid populations. The requirement to relocate near protein resources generally puts agents into areas where deer populations have not yet been depleted, where they hunt fewer leporids; moreover, on average there are fewer agents in these runs. Model runs with the more restricted hunting distance (20 cells) and whose agents are required to move to where there are protein resources are in the group with the least depletion. Runs with the larger hunting radius, no protein consideration in relocation decisions, and higher protein requirements are in the high-depression group, at least in part because these runs generate high human populations. All other runs fall in the middle of the depletion range.

Hunting Return Rates

Hunting return rates measure the potential caloric gain from eating prey in relation to the expenses of hunting. In the simulation, we calculate return rates by *dividing* the calories gained from consuming one animal by the amount of time (in hours) it took to find that animal on the landscape. This computation differs from some of the literature on human behavioral ecology—which calculates return rates by *subtracting* calories spent by calories gained—but aligns with the calculation of encounter rates generated by Steven Simms (1987) using direct field experiments hunting deer, rabbit, and hare in the Great Basin of the western United States. The amount of time it takes to hunt an animal in the simulation is a function of how far from the home cell that animal was found.

Comparing deer return rates across all 16 of these composite runs (Plate 9.5), we see a change through time in which of the parameters most affect these return rates. All runs show a profound decrease in return rates from the outset of the model. Initially, the amount of protein needed by each person segregates runs into two clear groups, with those requiring less protein (10 g / person / day) enjoying higher return rates. This is correlated with the degree of deer depression, as one might expect. Around

TABLE 9.2
Summary Data on Animals in Model

Attributes	Deer	Hare	Rabbits
Average Weight (kg)	60	2.3	.895
Edible Meat Weight (kg)	36	1.38	.537
Protein / kg (g)	300	330	330
Cal / kg	1,580	1,730	1,730
Protein / animal (g)	10,800	455.4	177.2
Cal / animal	56,880	2,387.4	929.01

NOTE: Hobbs and Swift 1985 (data for deer); Haskell and Reynolds 1947 (data for hare); Ingles 1941 (data for rabbits).

A.D. 850, runs with the protein-based move requirement (NEED_MEAT=1) begin to generate higher return rates than runs where NEED_MEAT=0. The NEED_MEAT requirement enforces an "equilibrium" in terms of deer return rate; while return rates for PROTEIN_NEED=25 runs continue to decline throughout the simulation, average return rate is maintained at higher levels when NEED_MEAT=1 and even grows modestly from about A.D. 900 through the conclusion of the simulation.

Because return rates integrate a great deal of information on both the supply and demand sides of protein procurement, the comparison of the return rates for hunting generated in these models to those found in the literature provide an attractive summary measure that allows us to evaluate aspects of the realism of these models. These comparisons must be made in the same currency. The model uses calories as the currency for all costs, though protein gain is assessed in terms of grams of protein, not calories. Most of the relevant literature uses a caloric currency for all costs and returns, so protein amounts gained by agents in the model are converted to calories for comparison, using the conversions in Table 9.2.

Plate 9.6 graphs the average household return rates for each species hunted in a sample run; because household levels for composite run 1 were the highest and thus closest to empirical estimates (Plate 9.1), we take that run for these analyses. It is significant that this composite run has NEED_MEAT=0, and that all runs with this parameter state tend to track estimated population levels with greater fidelity. Return rates for deer decline rapidly as these runs begin, and they continue downward, settling at about 3,500 cal / hr after A.D. 1000. Hare and rabbit return rates increase from zero only after the first 50 years of the simulation—agents prefer to hunt deer and will take leporids only when deer become scarce—and then closely approach 800 and 400 cal / hr, respectively, and decline slightly through time. The initial increase in leporid rates is an artifact of the way we calculate average return rates. When agents are able to capture a deer before they hunt for leporids—and if, as a consequence, they do not hunt leporids—their return rate for leporids is calculated as a null value since no leporids were taken. If agents spend at least a little time hunting for leporids, but none are taken, their return rate will be zero. Remember that once an agent is a sufficient distance from his home cell (*d*=20 in the runs discussed here), he will begin searching for leporids automatically, even while still preferentially hunting deer. This lowers the average return rates for leporids at the beginning of the model as some of the agents cross this distance threshold.

Too literal an interpretation of Plate 9.6 would suggest that no hunting of leporids

TABLE 9.3
Resource Return Rates from Composite Run 1, in cal / hr

	Literature	Model (Year)	
Taxon	Simms (1987:74)	(650)	(1250)
Deer	316–632	5400	3625
Hare	1038–8477	382	682
Rabbit	726.6–1262.9	199	345

should take place, since return rates for deer appear to be always higher. It is important to remember that these rates summarize the experience of a large number of households in very different circumstances, rather than some nonexistent "average household." When deer can be hunted, these hunts do generate higher return rates, as we model the process.

Comparing the A.D. 1250 return rates (Table 9.3) to those provided by Simms (1987), we see that our rates for deer are far higher than the range he reports (3,625 cal / hr in the simulation, versus 245.3–496.5 cal / hr by Simms's estimate), whereas our return rates for rabbits are much lower than his (345 versus 3,122.5–3,578.7). For hare our return rates are within the range given by Simms, though toward the bottom of it (682 vs. 656.5–4,025.5). Our results are affected by the problem, already noted here, that if a deer hunter gets to the end of the permissible radius without having found any deer, the wasted energy does not decrease his overall deer-hunting return rate; that is, his return rate will be zero regardless of whether he failed to obtain a deer within 20 or 40 cells.

There are additional reasons why our hunting routine may be giving a much high return rate on deer than Simms's values. First, Simms was computing encounter rates for the Great Basin, which—as he and others note (Simms 1987:56)—supports a lower density of mule deer than the surrounding mountainous regions. His estimates are based on deer densities of 5 to 20 deer per square mile (Simms 1987:59), whereas our landscape can support local deer densities an order of magnitude higher than that (Johnson 2006:Figure 7.1). When averaging across our landscape, however, our densities are on the upper end of the range Simms reports (roughly 18 deer per square mile; Johnson 2006:190).

Another factor that might be generating high and admittedly misleading return rates in our simulation is that we rather strictly cap the amount of time agents may hunt. Our accounting of "average" return rates is literally the arithmetic mean of each agent's individual annual return rate, and not the net calories gained by all agents divided by the net hours spent by all agents. Agents have a maximum hunting radius (HUNT_SRADIUS) of either 20 or 40 cells. If a simulated hunter satisfies his protein needs via hare, and stops hunting at radius 35, his deer return rate will be zero. A simulated hunter who has no success meeting his needs and stops hunting at radius 40 will also report a deer return rate of zero, even though these two agents clearly expended different amounts of effort deer hunting. Similarly, if a simulated hunter is successful in killing a deer in the most distant cell when HUNT_SRADIUS = 40, he will enjoy a return rate of 5,925 cal / hr, while an agent capturing a deer in the cell adjacent to his home plot enjoys a gratuitous return rate of 237,000 cal / hr. That the average return rates in the runs reported here are much lower than this reflects the reality that many agents are unsuccessful at hunting deer within HUNT_SRADIUS and thus report a return rate of zero. Placing more realistic limitations on agent hunting, and designing return rate reporting methods that consider both encounter rates as well as caloric costs of hunting, will allow for more satisfying comparisons with the ecological and ethnographic literature.

DISCUSSION AND CONCLUSIONS

Perhaps the greatest revelation from our work on hunting in the VEP world is the ease with which humans deplete deer on this landscape. Plate 9.2 shows that by the time human populations reach about 800 households, marked depression of deer is well under way in most scenarios.

Local archaeofaunal data also record declining proportions of deer through this temporal sequence. Shaw Badenhorst and Jonathan Driver (2009) demonstrate that turkey and leporids begin to replace artiodactyls in Pueblo II period (A.D. 900–1140) faunal assemblages, and that in later assemblages deer are very scarce. Our results strengthen the hypothesis that this replacement is induced by depression of deer populations accompanied by declining return rates for deer hunting. We can also hypothesize that—and have begun investigating through simulation whether (Bocinsky 2011)—return rates for turkey, even if they are fed maize, become higher than those we generate for deer in the later portions of our simulations.

TEN

Exercising the Model

ASSESSING CHANGES IN SETTLEMENT LOCATION AND EFFICIENCY

Timothy A. Kohler, R. Kyle Bocinsky, Stefani Crabtree, and Ben Ford

SO FAR WE HAVE EMPHASIZED how the Village model was constructed, how resources are produced and consumed, and how our parameter choices affect simulated population sizes and time allocations over the course of the occupation.

But of course the model "solves" not only for population size, given specific parameters, but also for household location. So we begin this chapter by providing an overview of how our parameter choices affect the ways households locate within the model world. We then consider how well the simulated settlement distributions fit those known from archaeological research—including the Village Ecodynamics Project (VEP) fieldwork discussed in Chapters 2 and 14—as those have been distilled using methods reported by Scott Ortman et al. (2007, and Chapter 2, this book). This turns out to be a way of asking how efficient the real settlement system was in each period, with respect to the household-level access it provided to the specific resources we model. Then we develop a way to determine the relative importance in each period of the various parameters we chose to examine in the sweep reported here. These last two analyses provide an important basis for interpreting the culture history of our area, in this chapter and in those that remain.

HOW WELL DO THE SIMULATED HOUSEHOLD LOCATIONS MATCH THOSE KNOWN FROM THE ARCHAEOLOGICAL RECORD?

> A particular model can be tested either by a direct test of its assumptions, or by comparing its predictions with observation.
>
> MAYNARD SMITH (1978:5)

Although it is important to know how the various parameters we used affect the spatial relationships among the simulated households, we are at least as interested in how well the various runs fit the known record, as well as how this changes through time. That is the subject of this section. Then we will see how specific parameter values increase or decrease this goodness-of-fit, as a novel approach to understanding what factors affect locational practices through time.

To first summarize the most relevant aspects of the Village simulation from the previous chapters, our model households are central-place foragers (Orians and Pearson 1979) who, on relocation, attempt to place themselves where they will minimize their caloric costs for meeting their requirements for calories, protein, water, and fuel, satisfying the constraints of (1) living no further than approximately 300 m from any of their fields, (2) not working more than the available labor in each household allows, and (3) deciding among places only within their specified search radius, set to 20 cells (4 km) in the simulations reported here. We do not model the field-house strategy that is common in aggregation, in which residents place seasonally used structures adjacent to their fields but possibly quite distant from their permanent residence. Although it would be desirable to do so, it is a complication that we did not want to attempt. This is probably most significant as a distortion of real locational practices during periods 9 and 14 through 18, when field houses were most common. Households stay where they are as long as they anticipate being able to satisfy their needs at their current location over the next year. Thus agent placement might be viewed as combining aspects of optimization (on relocation) and "satisficing." (Satisficing is an approach to bounded rationality that emphasizes "good-enough" solutions rather than optimal ones, which make extraordinary demands on cognition [Simon 1991:256].)

Of course, even if real people were perfect central-place foragers, there are many reasons why this *model* will not perfectly predict household placement. These reasons fall into three major categories. The first includes errors and oversimplifications in the resource domain. The second includes errors and oversimplifications in the behavioral and cultural domains—including our particular assumptions about decision making, cognitive capacity, social relations, and political organization. Finally, even the most devoted defenders of the adaptationist paradigm (e.g., Mayr 1983) do not expect perfect adaptation in specific traits. Biologists expect organisms to be only as perfect as, or slightly more so than, the organisms with which they compete (selection is for relative fitness). Individual biological traits are not selected individually; the target of selection is the whole individual, effectively constraining the "perfection" of the integrated parts. Moreover, chance events have some role in selection. These considerations also apply to optimization of cultural practices, especially one such as placement of settlements that integrates many considerations and is therefore not cleanly divisible from other practices.

In the resource domain, households use many resources, such as stone for tools and in some periods for building, clay for ceramics, and wood for construction, and so forth, whose locations, and costs to obtain, we do not model. Of course, the models for the resources that our agents do consider are probably not perfectly accurate, even at our chosen spatial and temporal granularities, which may not themselves correspond to the granularities used in decision making. Also, we do no checking to see if cells are in fact habitable before allowing households to locate there, so that cells offering nothing but steep slickrock, for example, might be chosen in the model but presumably would have been avoided by real households.

In the second domain, we expect slippage between the model and the real world for many reasons. For example, the social or temporal scales at which locational decisions are made might be crucial. In the model, these decisions are made by the household, use an extraordinary amount of very precise information, and are reevaluated annually. In the real world it is possible that such decisions are made at the community level (or perhaps some lower but still suprahousehold level such as the clan segment), and they are perhaps evaluated rather infrequently and possibly on the basis of much less information or by reference to simple rules of thumb. Finally, there are important social processes that we omit entirely from the model—e.g., conflict—that likely had effects on Pueblo settlement decisions, as they do on those of nonhuman central-place foragers (Thomson et al. 2006).

Results from 512 Runs of Village

So with these caveats, how good is the fit between our model households and those in the archaeological record? This is of much more interest to us than would be the simple test, recommended by John Maynard Smith in the epigraph, that implies a Popperian rejection or failure to reject. We might assess this fit along many different dimensions, including total population, frequency of household movement (within our ability to discern this in the archaeological record), global levels of aggregation, and so forth—but in this chapter we deal in site location. Specifically, we examine the real and simulated records in ways that account simultaneously for the locations *and* numbers of their households. Before beginning what is unfortunately a lengthy discussion, we want to emphasize that the measures of goodness-of-fit that we develop do not count unsurveyed cells with no sites in the real world as contributing to these measures in any way: they are treated as missing values. It is possible that this results in a subtle bias, because outside of the block-survey areas what we know about are mostly the larger sites, whereas inside the block-survey areas we know about both the large and small sites. In effect, to an unknown extent, large sites contribute more to our assessments (see the paragraphs that follow) of goodness-of-fit than do small sites. It would be possible to evaluate the extent of this bias by limiting our calculations to just the block surveys, but given how complicated our discussion already is, and given the reduction in sample size that would entail, we did not think that that was desirable here. We believe this bias to be small, in any case.

Evaluating Spatial Goodness-of-Fit

For both the archaeological record (within the limits outlined in Chapter 2) and the simulation, we know the spatial locations and numbers of households in each 200-m cell on our landscape, for each of our 14 periods. Our ideal measure of spatial goodness-of-fit, therefore, would simultaneously assess both the numbers of households at each location and their spatial locations. We use two complementary methods to achieve this. First, we calculate Pearson product-moment correlation coefficients between the number of household-years accumulated in each of the 45,400 cells in our model world and in each of 14 periods in the archaeological record. This use of the correlation coefficient requires us to consider the two data planes being compared as variables, with each of the locations (cells) at which the comparison takes place being observations, or cases. This is an extremely local measure, since it compares household-years at each pair of cells having precisely the same location in the real and simulated data planes, and therefore we also smooth the distribution of households in the empirical record slightly using a uniform smoothing within a 3 × 3 cell window. (Such a smoothing, for example, would change an isolated cell with nine household-years into counts of one household-year each in the target and the eight surrounding cells of the Moore neighborhood.) This is still an extremely high spatial resolution test, since it yields positive correlations overall only if on average the simulation places households within 200 m of a cell that was actually occupied, in a given period.

Table 10.1 shows that in periods with low, dispersed populations, the comparison using the unsmoothed landscape of real sites yields a higher proportion of positive correlation coefficients, whereas when population is more aggregated, the comparison using the smoothed landscape of real sites yields a higher proportion of positive correlation coefficients. To minimize these effects, we average these two sets of proportions in what follows.

It is not clear how these proportions of positive *r* values should be interpreted: do the values we calculate indicate a good fit between the simulated and real distributions, or not? To

TABLE 10.1

Comparison of Real and Simulated Site Sizes and Locations Using Pearson Product-Moment Correlation Coefficients[a]

Period	Date (A.D.)	Unsmoothed *p* Positive[b]	Highest *r*, Unsmoothed	Smoothed *p* Positive[c]	Highest *r*, Smoothed	Average *p* Positive	Average *p* Positive, rand1[d]	Averag *p* Positiv rand2[e]
6	600–720	.02	.007	0	—	.01	.497	.286
7	720–800	.184	.023	.156	.046	.17	.395	.403
8	800–840	.35	.03	.457	.09	.403	.352	.433
9	840–880	.334	.03	.633	.076	.483	.351	.411
10	880–920	.578	.034	.815	.074	.696	.308	.447
11	920–980	.43	.048	.134	.03	.283	.404	.254
12	980–1020	.078	.024	.012	.006	.045	.445	.251
13	1020–1060	.061	.046	.006	.015	.033	.446	.332
14	1060–1100	.031	.016	.549	.074	.29	.323	.206
15	1100–1140	.059	.034	.635	.075	.347	.272	.134
16	1140–1180	.047	.042	.631	.091	.34	.304	.13
17	1180–1225	.049	.019	.578	.063	.314	.302	.152
18	1225–1260	.066	.024	.465	.065	.266	.396	.127
19	1260–1280	.037	.044	.5	.059	.277	.328	.133

a. For the real-site data planes, cells outside of block survey areas with no sites are considered missing values and do not enter into these comparisons. Thus the sample sizes for the comparisons between the real and simulated data planes are different in each period.

b. Proportion of the 512 correlation coefficients comparing the unsmoothed real- and simulated-site data planes that is positive.

c. Proportion of the 512 correlation coefficients comparing the uniform-smoothed real- and simulated-site data planes that is positive.

d. Average proportion of positive correlation coefficients between the uniform-smoothed real-site and the unsmoothed real-site data planes, wi a randomization of the total *n* of households on the simulated-site data planes.

e. Average proportion of positive correlation coefficients between the uniform-smoothed real-site and the unsmoothed real-site data planes, wi a randomization of the cells from the simulated-site data planes.

help us interpret these results, we also calculated the correlations between the real sites and the simulated households, in each period, after randomizations of the simulated households. We applied two different randomization schemes to the simulations. In the first (referred to as "rand1"), we took the average number of households in each period in each of the 512 runs of the simulation and redistributed that same number of households randomly on 512 new landscapes for each period. Then, by period, we calculated correlation coefficients between each of these 512 randomized landscapes and each of the two (smoothed and unsmoothed) real landscapes. The average proportion of positive *r* values from this comparison is reported in Table 10.1, second column from the right.

The black line in Plate 10.1 (see color insert in this book) shows the difference between the average proportion of positive *r* values in the real versus simulated comparisons, and the average proportion of positive *r* values in the real versus random comparison, when the random set is calculated using our first method. This method retains the total household counts in each simulation in each period—and therefore the mean number of households per cell in each simulation in each period—but does not, except by chance, reproduce the standard deviation, skewness, and kurtosis of those counts from the simulations. The black line in Plate 10.1

then reports the difference between columns 7 and 8 in Table 10.1.

This difference (between the proportion of positive correlations between the real-site locations and the simulated site locations in each period, on one hand, and the proportion of positive correlations between the real-site locations and the "rand1" version of the simulated data planes, on the other hand) is positive when the model predicts the locations and numbers of the real households better than would this null model. We will interpret these differences in more detail later, but for the moment we can observe that the null model—constructed in this way—outperforms the simulation when the black line in Plate 10.1 is below the vertical reference line at zero on the left vertical axis. Remembering that greater similarities between the real and simulated distributions suggest greater efficiency of household location for the real sites, we can posit a general trend within both occupation cycles (which is stronger in the first) for greater efficiency of location as populations increase.

We will now consider a second way ("rand2") of constructing a random, null model for comparison. We can retain higher moments of the simulated site distribution in each period (as well as the mean number of households per cell) by randomly permuting the locations of the cells (retaining their counts of households) in each simulated run in each period. The rightmost column in Table 10.1 presents, for each period, the average proportion of positive correlation coefficients between each of the 512 simulated landscapes randomized in this way, as well as each of the two (smoothed and unsmoothed) real landscapes. The red line in Plate 10.1 displays the goodness-of-fit, relative to this random expectation, between real and simulated site locations and sizes. Specifically, it reports the difference between columns 7 and 9 in Table 10.1.

In our interpretations of the measures we will emphasize upward and downward trends. Therefore we do not have to decide which of the randomizations provides the better null model, since their trends coincide even though their absolute values differ.

These two approaches based on *r* give us very local measures of goodness-of-fit, which is not necessarily bad. However, they put us in the rather odd position of viewing our data planes as variables rather than objects (i.e., they provide *r*-mode comparisons [Rummel 1970:192–202]; see Cowgill [1990] for possible pitfalls).[1] We now turn to a more traditional *q*-mode analysis that not only allows us to implement a slightly broader spatial comparison but also provides a check on the validity of the patterns in Plate 10.1. Specifically, we calculate the similarity coefficient *s* (Research Institute for Knowledge Systems 2009:24), averaged across all the cells in the two maps being compared:

$$s(a,b) = 1 - ((|a-b|) \,/\, \max(a,b)),$$

where *a* is the number of household-years in a specific cell in the real-site distribution for a specific period and *b* is the number of household-years in the same cell, for a simulated distribution in the same period. The comparison follows a smoothing using a 1-cell-(200-m-) radius neighborhood. Table 10.2 shows the similarities, calculated in this fashion, among six simple 5 × 5 cell maps. In contrast to the previous comparison, we now smooth *both* the empirical and the simulated data planes, with the result that the spatial comparison is slightly more general but still quite local. Moreover, the similarity coefficient causes matches with zeroes in both cells to contribute to the overall similarity. For these two reasons it is not surprising that this approach yields much higher-appearing similarities between the real and simulated sites (Table 10.3) than the correlation coefficient approach reported in Table 10.1 and Plate 10.1. Once again, however, we need to

1. While we take Cowgill's points seriously, three elements in our approach obviate the problems he raises. First, we are not using closed arrays (percentages) in our comparisons. Second, we use only the proportion of positive correlation coefficients and not their absolute values. And third, our random comparisons using the same statistic provide adequate control.

TABLE 10.2

Illustration of Plasmode Data Sets of Computed Values for s, *Utilizing Same Settings (Radius 1 Smoothing) Used for Comparisons Between Real- and Simulated- (and Real vs. Random) Site Locations*

Pattern (maps)	0 0 0 0 0 0 0 9 0 0 0 0 0 0 0 0 0 0 0 0	0 0 0 0 0 0 0 0 0 0 0 0 9 0 0 0 0 0 0 0	0 0 0 0 0 0 0 3 0 0 0 0 0 0 0 0 0 0 0 0	0 1 1 1 0 0 1 1 1 0 0 1 1 1 0 0 0 0 0 0	1 0 0 0 1 1 0 0 0 1 1 0 0 0 1 1 1 1 1 1	0 0 0 0 0 0 0 0 0 0 0 0 0 0 0 0 0 0 0 0
0 0 0 0 0 0 0 9 0 0 0 0 0 0 0 0 0 0 0 0	100	52.5	70.0	53.1	14.4	55.0
0 0 0 0 0 0 0 0 0 0 0 0 9 0 0 0 0 0 0 0		100	52.5	42.9	23.9	55.0
0 0 0 0 0 0 0 3 0 0 0 0 0 0 0 0 0 0 0 0			100	36.3	24.3	55.0
0 1 1 1 0 0 1 1 1 0 0 1 1 1 0 0 0 0 0 0				100	20.0	20.0
1 0 0 0 1 1 0 0 0 1 1 0 0 0 1 1 1 1 1 1					100	10.0
0 0 0 0 0 0 0 0 0 0 0 0 0 0 0 0 0 0 0 0						100

calibrate our interpretation of these results by comparing them with the same two randomizations of the simulated maps already explained here. This is done in Plate 10.2, whose left axis shows the mean *s* for the real versus simulated comparisons from which we have subtracted a null model (either that provided by the first randomization, in black, or that provided by the second randomization, in red).

Discussion of Goodness-of-Fit

The largest differences between these metrics for goodness-of-fit are present during the population trough between A.D. 920 and A.D. 1060, when *s* reports relatively better fits than does *r*. We attribute this to the positive value that *s*, but not *r*, accords to the matching zeroes that are most common on our maps in periods of low population. Another difference is that *s* finds greatest settlement efficiencies at the end of the second population cycle, whereas *r* finds them at the end of the first cycle. In general we prefer the *r*-based measure, and we use it as a basis for most of the discussion that follows, though many of the results are visible in both.

Plate 10.1 reveals two peaks to settlement efficiency which, with important exceptions,

TABLE 10.3

Comparison of Real and Simulated Site Sizes and Locations Using Similarity Coefficient s[a]

Period	Date (A.D.)	Mean *s* Real vs. Simulated[b]	Std *s* Real vs. Simulated[c]	Mean *s* Real vs. rand1	Std *s* Real vs. rand1[c]	Mean *s* Real vs. rand2	Std *s* Real vs. rand2[c]
6	600–720	.814	.001	.819	.001	.817	.002
7	720–800	.920	.001	.921	.002	.920	.003
8	800–840	.944	.001	.943	.003	.943	.003
9	840–880	.928	.002	.928	.002	.928	.003
10	880–920	.985	.002	.982	.004	.985	.004
11	920–980	.958	.002	.955	.003	.956	.005
12	980–1020	.954	.003	.950	.004	.953	.005
13	1020–1060	.938	.003	.934	.004	.938	.004
14	1060–1100	.900	.003	.899	.003	.900	.004
15	1100–1140	.916	.003	.912	.004	.914	.005
16	1140–1180	.908	.003	.906	.003	.908	.004
17	1180–1225	.904	.003	.905	.003	.905	.004
18	1225–1260	.908	.003	.906	.003	.908	.004
19	1260–1280	.954	.004	.947	.004	.953	.005

a. For the real-site data planes, cells outside of block survey areas with no sites are considered missing values and do not enter into these comparisons. Thus the number of cells across which the mean *s* for each comparison is calculated is different in each period.

b. Mean of 512 similarity coefficients *s*, one for each comparison between the smoothed real- and simulated-site data planes.

c. Standard deviation for the mean *s* from the previous column.

track the population size. This is precisely what we would expect if, as Robert Foley (1985:230) has noted, pressures toward optimality are always strongest when competition is most severe. Until after A.D. 800 in the first cycle (the Basketmaker III and first part of the Pueblo I periods), settlements were well outside what we might call the optimal niche as estimated by the agent-based model. Given that these early settlements represent the first regional use of these deep upland soils for farming, this may reflect not only the absence of competition, but also a lag (perhaps analogous to the biological concept of phylogenetic inertia) in learning how to best exploit this landscape for farming. (This 200-year lag in turn suggests that farming practices were socially learned, possibly through a conservative mechanism like vertical transmission, with little individual learning or experimentation.) Despite this, the success of these pioneers—judged from their growth in numbers—clearly demonstrates that it was possible to live and even thrive outside the optimal niche as defined by our agents so long as regional populations remained relatively low, especially since the climate was fairly forgiving for most of this period.

This first population cycle came to an end with large-scale emigration in response to cold and dry conditions exacerbated by some depression of large game (Chapter 9) and fuelwood (Chapter 7). During the initial depopulation from A.D. 880 through A.D. 920, settlements that remained in place and retained their size or grew tended to be more optimally placed than those that shrank or were abandoned. Although this is not surprising, it does suggest that the end of the first population cycle was precipitated at least in part by resource stress that winnowed inefficient locational strategies.

The two measures disagree on whether the immigration between A.D. 1060 and A.D. 1100, coincident with the appearance in this area of Chaco-related manifestations such as great houses, increased locational efficiency. This question warrants further investigation, since we would like to know whether apparent Chacoan desires to maximize factors such as visibility (of great houses to each other, of great houses from their surroundings, or of sacred topographic features from great houses, Lekson et al. 2006:73), or perhaps control over populations, took precedence over efficient access to resources.[2] Both measures, however, suggest some slight declines in locational efficiency during the period between A.D. 1140 and A.D. 1225. (Although settlement location was relatively constant during this second cycle, our measures will be sensitive to whatever small changes there were, as well as to reshuffling of households among existing sites.) We suspect that the high levels of violence in these two periods that Sarah Cole documents in Chapter 13 prohibited households from arranging themselves in a way that would be maximally efficient in relation to resources.

Our *r*-based measure (but not the *s*-based metric) also records slight declines in efficiency during the Pueblo III population peak, when many small community centers appeared in the western portions of our study area (Chapter 14): these late arrivals may have been restricted to less-than-optimal locations. As at the end of the first population cycle, both measures reflect an increase in efficiency as the second depopulation begins. Once again it appears that the less optimal locations were abandoned first during the events that led to the complete depopulation of our study area and in fact the entire northern Southwest. Even though we cannot know how strong the attraction to other areas might have been, this is the pattern we would expect if VEP households were all making rough calculations about whether they would do better someplace else with some given degree of attraction, and then leaving when they believed that to be the case based on their local circumstances.

MODEL SELECTION: WHICH NICHE?

The logic behind the approach so far is that the ensemble of runs undertaken in this parameter sweep characterizes alternative optimal niches, though in the previous analysis we do not care about the ways in which different parameter settings cause simulation runs to be different. Different combinations of parameters do identify somewhat different locations as optimal, and to the extent that these do not overlap, the empirical record cannot fit all of them simultaneously.

Recognizing this, another use to which we can put our parameter sweep is to investigate which combinations of parameters result in the best fits in each period, and we can analyze the changes in the best-fitting parameters through time as an indication of the directions in which locational practices were tending. Appendix A presents the parameters used in each of the 512 runs analyzed here, and Table 4.1 explains how these parameter choices affect the running of the simulation.

Methods and Results

To identify how the parameter values in each run interact with the goodness-of-fit in each run, we begin with a database having 7,168 observations, one for each combination of the 14 periods and 512 runs. For each observation, we record the values for the nine parameters used by that run, and the correlation coefficient between the site sizes and locations generated by that run and the uniform-smoothed distribution of real sites. These correlation

2. In VEP II we hope to apply location-allocation models developed in geography to the settlement distributions of the second population cycle to assess the degree of centralized political control represented by the great houses. These techniques have been fruitfully applied to ancient Egyptian settlement patterns (Church and Bell 1988).

TABLE 10.4

Eigenvectors on the First Principal Component, by Period, for Each Parameter in the Sweep

Period	SOIL_ DEGRADE	COOP	NEED_ MEAT	PROTEIN_ NEED	PROTEIN_ PENALTY	HUNT_ SRADIUS	HARVEST_ AD-JUST_ FACTOR	STATE_ GOOD	H_2O_ TYPE
6	−.57	−.04	.01	−.01	−.01	−.01	.33	−.24	.07
7	.32	−.11	.05	−.37	−.18	−.45	.07	.08	.04
8	.22	−.06	−.25	−.18	.01	.09	−.56	.17	−.04
9	.032	.14	−.19	−.36	.17	−.14	−.33	−.17	.35
10	.10	.31	−.38	−.34	.09	−.09	−.11	−.22	.24
11	.22	−.14	.01	−.18	.06	−.38	.23	−.45	−.03
12	−.05	−.04	.64	.18	−.05	−.18	.03	−.14	−.03
13	.11	.01	.66	.11	.01	−.14	.02	−.10	.09
14	−.52	−.03	.38	.21	.03	.02	−.03	.09	−.18
15	−.54	.06	.38	.17	.02	.05	−.03	.06	−.11
16	−.48	.02	.47	.19	.02	.02	−.06	.07	.01
17	−.53	−.06	.43	.12	.04	.08	.00	.09	.01
18	−.46	−.03	.50	.09	.04	.055	−.11	.11	.04
19	−.46	.03	.51	.08	.03	.01	−.10	.05	.07

coefficients are unsquared, and therefore they take on both positive and negative values. By their sign and their size they measure the extent to which those particular parameter choices reproduce the actual settlement pattern in each period.

Now we need a multivariate approach to determine the relationship, across all the 512 runs in each period, between the parameter values and the associated measure of goodness-of-fit. We selected the simplest and most direct approach, a principal components analysis (PCA; Khattree and Naik 2000), performed separately for each period. For each period, the PCA extracted 10 orthogonal (i.e., uncorrelated) eigenvectors (or components). The PCA was performed on the correlation matrix, so each of the 10 variables (nine parameter values and one correlation coefficient) has the same weight in the analysis. The first component extracted is by definition the most important, and in these data sets it accounts for between 11 and 16 percent of the total variation in the data. Table 10.4—read along the rows—reports the loadings (or elements of the eigenvectors) for each parameter in each period, on the first principal component. The loading for the measure of goodness-of-fit is not reported in Table 10.2, since it has a constant, high positive value for each period (.71), and a loading of zero on components 2 through 9.

When a parameter has a *positive* loading, therefore, it indicates that *high* values for that parameter tend to increase the goodness-of-fit across all the runs in that period. A *negative* loading means that *low* values for that parameter tend to increase the goodness-of-fit across all the runs for that period. The higher the absolute value of the loading, the more important the effect, since these loadings represent the correlation coefficients between the component and the original variable. There is no strict rule as to when loadings are high enough to be worth interpreting, but a commonly used rule of thumb is that values greater than |.4| are of particular interest.

Plate 10.3 graphs the loadings for each parameter through time from Table 10.4, except that two parameters (COOP and PROTEIN_PENALTY), whose loadings never exceed |.18|, are excluded to make the graph more legible. Parameters whose loadings exceed the horizontal reference lines at |.4| in particular periods are especially important in increasing the goodness-of-fit between the simulations and the real settlement practices in those periods.

Discussion

Even on quick inspection, Plate 10.3 reveals two distinct regimes of settlement practice, one beginning in A.D. 1060 and one before that. The later regime is characterized by high values for NEED_MEAT (in black) and low values for SOIL_DEGRADE (red). (This will be our shorthand way of saying that runs in which the parameter value for NEED_MEAT is 1 rather than zero fit much better after A.D. 1060, as do runs in which SOIL_DEGRADE is zero rather than 1.) The earlier regime is most obviously characterized by relatively rapid fluctuation in the controlling parameters. Our interpretation of these parameters is admittedly ad hoc and exploratory; we are unaware of any models for this sort of analysis in either archaeology or geography.

Organizing our discussion by parameters, the initially low value for SOIL_DEGRADE is unexpected but may reflect the continuation of an agricultural tradition based on overbank flood farming or other relatively spatially stable Basketmaker II / III farming systems. (In the simulation, a value for 1 instead of 2 for SOIL_DEGRADE results in relative stability of fields and therefore settlements.) Positive values for SOIL_DEGRADE from the mid-A.D. 700s through the mid-A.D. 900s perhaps reflect the dominance of a shifting farming regime during this period (see Kohler and Matthews 1988). Finally, low values for SOIL_DEGRADE beginning in the period between A.D. 1060 and A.D. 1100 reflect the very high residential stability of the second population cycle. They also underscore the possibility that techniques for slowing soil erosion or retaining fertility might have been important for achieving that stability.

The role of HARVEST_ADJUST_FACTOR is somewhat less certain. High values for this parameter result in globally low values for maize production, and vice versa. Therefore, when the simulation has a high value for this parameter, population growth rates (and the exploration of new locations for settlement) tend to be low, as was indeed the case in the first period. The low value for this parameter in the early A.D. 800s may suggest that the areas into which people were moving at that time (principally the Dolores River valley) were attractive (or perhaps plausible) for settlement only when production was quite high. In the first occupational cycle there is negative correlation between this parameter and the low-frequency temperature proxy (Wright 2006, and Chapter 3, this book). High values for HARVEST_ADJUST_FACTOR (which induce low production in the model) in the A.D. 600s and A.D. 900s correspond to short summers in the VEP area. This relationship—if it even exists—breaks down, however, in the second population cycle, when values for this parameter change little from the warm A.D. 1100s to the cold A.D. 1200s.

Along with the low values for SOIL_DEGRADE, the second cycle is defined by high values for NEED_MEAT, which in fact usher in the last 300 years of the occupation in an interesting way. During these years the runs that fit best are those in which agents are constrained to move to cells from which hunting is possible within HUNT_SRADIUS, even if that is not the least costly alternative. This seems to have two rather contradictory effects. It can prevent agents from moving at all, if they find no place that is better than where they are presently located. However, if they do move, it is to someplace with better hunting prospects. Note also that PROTEIN_NEED is roughly correlated with NEED_MEAT; both tend to be low in the first cycle and high in the second. Both of these parameters suggest that access to protein is relatively easy in the first cycle and relatively difficult—and a high priority—in the second. The HUNT_

SRADIUS may also be illuminating in this regard; except for the anomalous A.D. 600s, it tends to be lowest when population levels are low (which makes sense) and has a peak in the mid-A.D. 800s, when Timothy Kohler and Charles Reed (2011) suggested that there was a strong cultural priority on deer hunting. However, its neutral values throughout the second population cycle are a little puzzling, given the evidence already presented here, unless it merely reflects the dense population packing during that period.

Finally, we might note that STATE_GOOD tends to be increasing when populations are increasing, and vice versa, and that H_2O_TYPE correctly identifies the period in the late A.D. 800s, when settlement is maximally associated with the Dolores River, the only type 3 water source in our area.

HOW MUCH EXPLANATORY CAPACITY IS LOST WITH MODEL SIMPLIFICATION?

Before leaving this discussion of how various parameters affect goodness-of-fit, an obvious question to ask, of this or any model, is whether it could be simpler and still perform as well (in our case, for predicting site size and location).

We do not want to try to evaluate this question comprehensively here, but we will present some exploratory results. The approach is to select a run that fits reasonably well and then progressively relax the criteria on which agents make their locational decisions until the only remaining requirement is that they get adequate maize. We cannot remove that requirement and still maintain living agents on the landscape—it is programmed too deeply into the simulation. At each step in the simplification we evaluate the goodness-of-fit in relation to the settlement pattern. This is conceptually similar to knockout analysis of genes, a common tool for localizing and assessing function in analysis of complex biological systems.

As suggested, the locational practices of our real households are different enough in the two population cycles that no single run fits well in each cycle. We selected one of the best-fitting runs for the second population cycle, run 230, which as shown in Appendix A and Table 4.1 is characterized by SOIL_DEGRADE 1, COOP 4, PROTEIN_NEED 25, PROTEIN_PENALTY 1, HUNT_SRADIUS 20, HARVEST_ADJUST_FACTOR .75, NEED_MEAT 1, STATE_GOOD .1, H2O_TYPE 1, WOOD_NEED 1130 kg / person / year, and WATER_NEED 3650 liters / person / year. We constructed seven simplifications of this run, exploring the consequences of setting water need, wood need, and protein need to zero, in all possible combinations.

We judged the performance of these as follows. First, we constructed 10 "rand1" and 10 "rand2" variants for that output from run 230 and each of its variants in which the number of household-years for each period was accumulated in each of the 45,400 cells. Then, for the original unrandomized simulation output for each variant, and each of the additional 20 randomizations for each variant, we computed correlation coefficients, as already described, with the unsmoothed and the radius-1-smoothed real-site distributions. We then averaged these correlation coefficients by run and period.

Plate 10.4 graphs the difference between the correlations obtained for each of the simulated outputs, versus the randomizations that are specific to each of them (and that thus control for the different population sizes that each generates). Therefore, for example, the height of the solid black line shows how much better run 230 performs than the average of its 20 randomizations in each period. When the solid black line is above the reference line at zero, the simulation performs better, on average, than its randomizations. The location of the solid black line in relation to the other lines tells us how run 230 performs in relation to its randomizations and allows us to compare it against its simplifications in relation to their own randomizations.

In the second population cycle, run 230 fits better than any of its simplifications, suggesting that all those parameters are needed. It also fits better than its randomizations. During most of the first population cycle, on the other hand, run 230 is outperformed by a simplification in which agents need no water (and thus water location does not enter into their locational decisions), and moreover, it does not fit as well as do its randomizations, on average.

It is probably not a good idea to read too much into results from one run. A comprehensive look at this question would require repeating this analysis across all the runs in our 512-run sweep and analyzing the ensemble. But on the basis of the results here we predict that runs in which households are required to consider all four resource categories in their locational decisions will ordinarily perform the best in any period. Deviations from that—if any—will contain useful information. For example, should the deletion of the water requirement generally provide a better fit throughout the first population cycle, as it does here, then we would have to consider what aspects of the cultural traditions or social organization of the Basketmaker III and Pueblo I societies allowed the value of the labor for getting water, which is normally woman's work, to be discounted.

Moreover, there is value to tracking resource use even if it could be shown that these resources did not contribute to locational decisions, since this allows us to estimate the calories and time needed across the spectrum of modeled resource use, as we will do in the final chapter.

CONCLUSIONS

When the senior author began this work years ago, he did not know for sure whether we could actually build a model for settlement practices that accounted for the dynamic placements of both resources and households, providing a usable set of expectations through time as to what settlement patterns should look like if they were formed by some well-characterizable process such as minimization of household-level energy. We believe the evidence presented in this chapter validates this risky experiment.

Nor did we know whether such a model would produce behaviors we could compare with those known from the archaeological record to recover meaningful information about actual settlement practices. We have shown that by examining the parameter values that increase the goodness-of-fit between simulated and actual household locations, we can reproduce some things that were otherwise already known about this record and suggest other things worthy of further examination. Our success here is not complete—e.g., we would like to have gotten more definitive results from the exchange parameter (Table 10.4), which we discuss in Chapter 11. It is nevertheless extremely encouraging, since it suggests that we may be able to identify other relatively subtle influences on settlement practices so long as we can parameterize them meaningfully. Obviously this approach is not immune to the dangers of equifinality. It is possible, for example, that practices in the second population cycle that we connected to considerations akin to the NEED_MEAT and SOIL_DEGRADE parameters in the model may reflect some other reality—such as population packing, investments in facilities, or accreted property ownership—that similarly constrain mobility.

The computational approaches pioneered here give us a set of tools beyond those traditionally available to archaeologists. These tools appear to yield powerful and subtle results, but they do not obviate the need for good sense, creative thinking, and the examination of explanations from a variety of angles. In the next chapter we try to apply these techniques to recovering information on behaviors suspected to lie at the very heart of human cooperation, which nevertheless have been difficult or impossible for archaeologists to document.

ELEVEN

Simulating Household Exchange with Cultural Algorithms

Ziad Kobti

SO FAR IN THIS BOOK we have summarized key points about the archaeological record in the Village Ecodynamics Project I (VEP I) study area (Chapter 2) and presented an overview of how the Village simulation works (Chapter 4), with special attention to how we model the supply and use of various resources (Chapters 5 through 9) and how various parameter settings alter the choices agents make about where to live (Chapter 10).

With the minor exception that newlyweds start their households relatively close to that of one of their parents, nothing in the modeled interactions of households that we have yet discussed is social. It has always been our intention to leave most problems involving how groups beyond those generated by close kinship define themselves and collaborate on projects of mutual interest to our next phase (VEP II). But transfers of food resources are so basic to all societies—and likely so fundamental to the critical problems of how groups form and maintain themselves (Winterhalder 1996)—that we built our first small steps toward studying the social in VEP I by modeling resource transfers among related, and unrelated, households.

DEVELOPMENT OF RESOURCE TRANSFERS IN VILLAGE

Our motivation for introducing social connections among households is to generate a social fabric in an otherwise autonomous agent model. Along with these social ties, exchanges of resources and knowledge can be enabled, since communication is undoubtedly critical in the evolution of cooperation and cultures. A key hypothesis by Timothy Kohler and Carla Van West (1996) argues that Pueblo I villages in the northern Southwest in particular are organizations that formed in order make their exchanges efficient. Trading efficiency naturally leads to a better quality of life for village inhabitants, and from the perspective of the model it enables an underlying social support system essential for survival.

Stimulus-Response (SR) agents are a classical form of artificial entity that respond to their environment as a stimulus arises. In the early versions of the Village model, agents could respond to low productivity only by increasing the number of farmed plots for the following year or by relocating to more productive areas nearby (Kohler, Kresl, Van West, Carr, and Wilshusen 2000). There was no explicit movement of information or resources within the model. Each agent (or household) acted individually, and the decision to move was solely a household's response to productivity levels of the current location. While it enabled a successful examination of the degree of environmental impact, as a first attempt at creating the artificial model the simplified formula provided by these SR agents falls short in explaining what other reasons (i.e., if not environmental) might have led to the eventual exodus of the settlers. What social system did the population develop, and what might in turn have been its impact on population decisions as well as on individual household choices?

Ziad Kobti (2004) proposed that with the rise of a stronger system of social welfare, individuals became increasingly dependent on the more productive households; over time, the productive households rewired the social fabric and exchange practices to maximize their own benefits. But when system resources became limited, the sudden reduction in the resource production could lead to the collapse of the households that had sponsored most of the social connections. Perhaps the social system was stressed to the point where a large-scale social collapse, starting with the loss of the key connecting households, could ripple through the system, initiated by the breakdown of the "hubs"—the important and highly frequented exchange partners—of the social network, leading to chaos in the social welfare and trade system.

HISTORY OF EXCHANGE IN VILLAGE

Kobti and colleagues (2003) extended the early completely autonomous household model to include a preliminary state model for each household. This state model is the basis on which various models of resource transfers among the households have since been built. It allows the flow of resources between households having different states—e.g., between those with excess productivity and those needing food. We provided three different frameworks for such flows among households related by descent or marriage. In the first case, those who are in need ask their relatives for assistance. In the second case, individuals with excess food poll nearby relatives to see if they need assistance. The third approach combines both of these activities within the same network. We consider all these to model aspects of what Marshall Sahlins (1972:192–193) called generalized reciprocity. We suggested that the combined (third) approach, as opposed to the condition of no exchange, synergistically enhanced the resilience of households in dealing with food shortfalls. We refer to the network generated by the combined approach as the generalized reciprocity network (GRN). This network is based on kinship. Communication, and hence exchange, can be established between related households. Since each agent is a household composed of a husband, wife, and children, now with the social connection we enable the concept of an extended household. In the model, households connected on the GRN resemble extended households to the extent that they are in very close proximity and are affected by the same environmental conditions. Their cooperation through exchange, forming in effect one virtual workforce, allows them to produce a crop yield that could sustain the related households, thus allowing an initial form of suprahousehold group to emerge.

At first, the currency for exchange was defined in terms of calories, and maize was the only source of food. The simulation was later enhanced to enable meat to be another food source. Meat must be acquired by hunting deer, jackrabbit, and rabbit, which are "grown" by the landscape (see Chapters 7 and 8). A household must acquire both maize and meat in order to meet its minimal requirements of calories and proteins to survive—though the way we handle protein shortfalls depends on the current setting for the parameter PROTEIN_PENALTY (see Chapters 4 and 9). One kilogram of maize generates 3,560 calories, and each member of the household (men, women, and children) has distinct minimum requirements to ensure individual survival. The first exchange scenario within the GRN is initiated by a request from a relative based on need. The requesting agent tries a few relatives a preset maximum number of times, relaying the number of calories required for its survival; if that amount is not met, the agent falls short of the minimum caloric level that it needs to survive and may die and be removed from the system, depending on the severity of the shortfall and whether the option of moving to a better place is available. If, on the other hand, the caloric need is met by one or more relatives, then the household would accumulate sufficient maize to survive another year. The second GRN scenario arises from an abundance of goods. Even a household with plentiful maize may not store over a certain maximum amount because of the physical limitations of storage. Instead of discarding the maize, or failing to harvest what one has grown, a household can become philanthropic. In this case the donor triggers the exchange event, a scenario that differs from the first one in which a needy household requests food. This establishes a framework whereby successful households might be able to increase their reputations or put other households into a debtor relationship, or both, even while they help related households that are less successful.

In the GRN, then—modeled after the compassionate and human response of social beings—agents may ask their relatives for food in time of need or donate their surplus to relatives during prosperous times. These exchanges are activated by the requestor, the donor, or both. Each version is potentially reciprocal; the only difference is in terms of who provides the information that triggers the exchange. The current approach to exchange implements a more refined version of Sahlins's (1972) "generalized reciprocity" than that represented previously by Kohler and Lorene Yap (2003), since the exchanges here are indeed limited to kin, and present as possibilities both asymmetric and symmetric exchanges. Our use of these terms focuses on whether an exchange can be initiated only by the donor or alternatively only by the receiver (asymmetric) or by either (symmetric).

Unlike "trade" between agents as we will define it later in the chapter, the GRN model of exchange does not keep a record of debts owed by particular agents. In the initial exchange, model donors are not remembered for their generosity, and a recipient has no recollection of the exchange that saved its life, nor is it obliged to repay its debts. The model was consequently enhanced to enable learning and simulate memory of whom the household exchanged with. Again following Sahlins (1972:194–195), we refer to these transfers as "balanced reciprocal exchanges"; the network they form is abbreviated as BRN. Furthermore, in this BRN each household recalls, for each partner, the quality of the exchange—whether or not the trade was successful. This in turn makes it possible for agents to learn over time who would be a good future partner for trade. This added memory becomes a critical component in formulating a population evolutionary algorithm to guide the simulation.

In the BRN, exchanges are tracked, and the generosity—and hence reputation—of donors is noted in order to produce influence and generate status. Balanced reciprocal exchanges are initiated between nearby—but not necessarily related—households. Forming this network requires that we track and capture social intelligence, and we adopted the cultural algorithm framework (Reynolds 1979) for modeling changes in behavior and cultural belief systems over the course of the historical sequence that we model.

Cultural algorithms (CAs) are population evolutionary models for implementing artificial learning and consist of a social population and a belief space (Reynolds 1979). Selected individuals from the population space contribute to the cultural knowledge by means of the acceptance function. The knowledge resides in the belief space, where it is stored and manipulated depending on individual experiences and their successes or failures. In turn, the knowledge controls the evolution of the population by means of an influence function. A CA thereby provides a framework in which to accumulate and communicate knowledge so as to allow adaptation both at the individual and at the population levels.

While an individual, or local, memory is added within each agent to enable it to recall its own history of exchanges, the CA works at a more generalized level, accumulating the collective experiences of the population in driving its evolution. Knowledge from population experiences over time are accumulated and maintained in the belief space, so that the collected knowledge reflects the beliefs and attributes of the society that are based on the underlying population experience. The belief space is updated over time, and the outcome influences the population as it evolves toward optimal use of the knowledge to enhance survivability. The economic network further teaches the individual additional rules of interaction, generated from common beliefs and trading patterns. This is done by the influence function of the CA, in which a household, just before it selects a particular exchange partner, checks the belief space for a contributing or influencing input. An example of an influence from the belief space is an exemplar, or good pattern, stored in the global belief space, that could be selected by the individual household agent to align its own behavior more closely to that of the exemplar. Thus the belief space represents a repository of popular beliefs, in this case selection strategies of partners for potential exchange, shared among individual households.

Before explaining the BRN system, it is worth noting that learning can be achieved at the GRN layer. The selection strategy that a household would employ to request a resource exchange from another household is critical to the household's success and ultimately its survival. When a household is nearly out of food, it has to economize its caloric expenditures. Consequently, travel distance and other communication costs have to be accounted for when an exchange request is initiated. To that effect, a parameter is defined in order to limit the maximum distance a household can search for potential exchange partners. If the request is futile, the energy invested was wasted and the household may not have enough energy to initiate another request with a different household in its quest for food. This defining moment in an agent's life is selected for the implementation of the learning strategy. A genetic algorithm (Holland 1992) provides a framework for learning at the individual level. Driven by the Darwinian notion of the "survival of the fittest," a successful experience is reinforced, whereas a failed experienced is penalized or even eliminated. However, the experiences of one individual household with an exchange partner are not shared with other households.

A population evolution algorithm—in this case, the CA because of its unique characteristics in capturing the human experience over more realistic evolutionary periods—enables the specific maintenance of plans for the interaction with other agents. Rather than randomly interacting, agents can decide with whom they wish to interact; this selection is learned first by means of reinforcement at the individual level, and if the selection has high impact, it is accepted into the belief space of the cultural algorithm and shared by the population. Households can maintain plans concerning whom they prefer to interact with at the individual level, and they can produce generalization in the belief space that indicates the type of individuals that are best to exchange with. Thus, every individual household possesses and maintains a local strategy for selecting one of its kin for exchange.

Year after year, individuals learn to bias the selection process toward exchange partners with whom they are more likely to interact successfully. The selected partner is asked for food on the basis of the requesting household's need. The selected kin's response is noted by the requesting agent. If the request is successful, the requesting agent will maintain the request plan; if unsuccessful, the requesting agent will modify it using the genetic operators. The goal is to allow the agent to rank its kin based on experience in an effort to enhance its selection and chances to choose one who is more likely to donate or agree to an exchange request. From an individual's perspective, when the agent reaches a critical stage and is facing death from starvation, selecting the right donor for food is essential. A parameter is defined in the model to set the maximum number of attempts an agent can make within its exchange radius to find suitable exchange partners and satisfy its needs. Should the maximum number of attempts be reached without the agent being able to satisfy its ratio for survival, the agent could subsequently die.

Given the CA, individual households in the population can learn to modify their plan concerning whom they will interact with so that their interactions are more successful. Learning can also take place at the cultural level in the belief space. In the belief space, generalization about classes of kin that are most likely to respond positively to exchange requests can be produced. What one individual knows can be communicated to others. If an individual discovers that a wealthy relative is consistently able to cooperate and provide food on request, other related individuals may pick up that strategy by means of the influence function in the CA. The belief space keeps track of the likelihood that certain kin are more likely to respond to requests than others; this is represented in the belief space as an instance of situational knowledge. Although higher productivity, or crop yields, would in fact be a better indicator of a household's ability to give in response to a request, that knowledge is not publicly available. The global, or cultural, knowledge presents one household with a belief about the likelihood that another household will cooperate, and therefore it directs the household-level experiences. This global knowledge is based on the collective experience of the top performers, or exemplars, defined as the richest in terms of the amount of maize or protein (meat) they possess.

To clarify, the CA's acceptance function is responsible for recording the success of individuals and maintaining their strategy in the belief space. When an agent in the population is about to select its plans about who to request maize from, the information in the belief space is used to condition the probability of choice by making the agent more likely to select previously successful categories of relatives in order to formulate its new plan. The influence of the CA plays this latter role, simulating the impact of popular beliefs on household-level decision making.

The variability and unpredictability of the environment in this region over the long term necessitates interactions between individuals. One farmer may experience a surplus while another experiences a deficit during a given year. The current model allows the kinship and neighborhood networks, the GRN and BRN, to evolve as necessary to reflect the relations between surviving households. The overall social network framework takes a hierarchical form derived from the combination of the two base networks, the GRN and BRN.

Within a social network, which we can think of as a community, the notions of status, trust, and beneficence play a role in enhancing a household's reputation. This reputation can be an important factor in the organization of new social networks at the population and regional level. In the BRN, a measure of quality is maintained with each exchange. These learned values enable the agent to utilize its higher status to become a member of a network of "hub nodes," or high-performing households, as discussed in Kohler et al. (2007: Table 4.4). These hub-node exchanges are not, however, implemented in the exchanges discussed here, which are limited to those shown in Table 11.1.

The BRN provides an economic framework supporting the exchange of maize, or calories

TABLE 11.1

Description of the Exchange Strategies (Values for COOP Parameter)

Cooperation Method	Description
0	There is no exchange of food between households.
1	When an agent requires food, it is allowed to select and request food from within its kinship network in order to survive.
2	When an agent has excess food (above a determined threshold amount), it is allowed to select one or more individuals from its kinship network and donate some of the excess.
3	Both methods 1 and 2 are enabled (GRN).
4	There is full cooperation across the kinship and economic networks (GRN and BRN simultaneously).

BRN = balanced reciprocal exchange network; GRN = generalized reciprocity network.

as a general unit, between neighboring agents. In a balanced reciprocal transaction, the donor expects payback of an equivalent amount relatively soon. Such exchanges are localized to enforce the physical constraints of travel distance. This distance constraint is consistent with what was implemented in the GRN. Each agent maintains a set of trading partners who are not necessarily associated with the kinship network. A trading partner can be any agent within a given radius from the agent, as parameterized in "VILLAGE.H," with the selection guided by the individual and population learning mechanisms previously discussed.

In our most recent versions of the model, as reported here, we have added protein from hunted game as a resource that is subject to exchange. Depending on the parameter set (see Chapters 4, 8, and 9), households require a minimum caloric level as well as a minimum protein level for survival. From a social system perspective, a society can exchange maize and meat across its dynamic exchange networks.

Social networks change not only through learning, but also in response to the creation of new households or the deaths of households, as well as to alterations in distance among households, due to mobility. Overall, the GRN and BRN form two base networks capable of maintaining reciprocal communication and exchange of resources between individuals, thus creating the fundamental fabric of a society.

EXCHANGE IMPLEMENTATION AND OBSERVATIONS

The main trigger initiating an exchange in the model occurs when an individual is in a needy state in terms of resources. After updating their networks, they first try to satisfy their resource need by calling in debts (if any) from their neighbors through the BRN. If they are not successful, they request aid from their relatives through the GRN. If they still are deficient in terms of resources, they go back to the economic network (BRN) (Kobti et al. 2003).

Under the CA framework, agents learn to adjust their plans at the individual level on the basis of their experiences, and at the cultural level in the belief space. Belief-space knowledge is used to condition the changes individuals make to their plans as the result of failures to interact. In a series of experiments, we progressively introduce one layer at a time and observe its effect on population counts, as well as the volume of the social network, in reference to the established number of exchange links between individuals. These results were reported in previous work (Kobti 2004; Kobti et al. 2003).

Initially we observe that without the presence of a social network—i.e., with households acting individually in response to the environmental changes—a dip in population is observed during drought years. It comes as no surprise, therefore, to observe an enhanced resilience of the population after the addition of the social networks and exchange acting as a social welfare system for hungry households. This helps validate the complex model as agents respond to the dynamic environment.

At every step in the simulation, the agent performs a deterministic set of actions specific to exchange. Here we summarize the exchange of maize:

1. Update the GRN.
2. Update the BRN.
    - 2.1. Remove dead (and inactive, out-of-region, or expired) partners.
    - 2.2. Search each neighboring cell within a trade radius and get its settlers list, and add new ones to the trade list up to a MAX_TRADE_LIST.
3. Request payback of debt from BRN partners.
4. If HUNGRY/CRITICAL,
    - 4.1. Request food from the GRN (no payback).
5. If HUNGRY/CRITICAL,
    - 5.1. Request food from the BRN (with payback promise).
6. If CRITICAL,
    - 6.1. The agent is DEAD and is removed.
7. If PHILANTHROPIC/FULL,
    - 7.1. Donate the surplus into the GRN.
    - 7.2. Pay back debt owed into the BRN.

For completion, we show the impact of protein on exchange steps, simplified here to illustrate an example of how resource procurement and exchange are integrated into the model for each major resource.

1. Calculate how much protein a household needs for the year to survive, based on amount set by parameter PROTEIN_NEED.
2. Hunt for deer, rabbit, and/or hare to fulfill the protein need.
3. The household consumes the protein it gathered from its hunting activity.
4. If the household did not satisfy its protein requirements, then
    - 4.1. The household requests the needed protein from its relatives, uses the CA to learn whom to ask, and updates its knowledge accordingly.
    - 4.2. Calls in the debt by asking the trading partners to pay back their owed balance, and updates the quality criterion with each partner accordingly.
    - 4.3. If the household is still in need, it will request a protein trade from known partners on the BRN. It will also seek out new partners on the BRN if needed.
5. If the household did not meet its protein requirements, it may die, depending on how the parameter PROTEIN_PENALTY is set.

The outcomes of the exchange strategies are sensitive to the model parameters defined in VILLAGE.H. For the purpose of illustrating exchange in general, we will employ output from run number 134, whose parameters are given in Appendix A. This run generates a spatial distribution of agents that best fits the known distribution of households between A.D. 1020 and A.D. 1060, and again between A.D. 1260 and A.D. 1280. However, it generates relatively few agents overall (mean = 494 households; minimum, 113 households in year A.D. 612; maximum, 672 households in year A.D. 1214), considerably fewer than we estimate from the archaeological record.

In Figure 11.1 we isolate the number of households involved in actual exchange. This and all the simulations discussed in this book

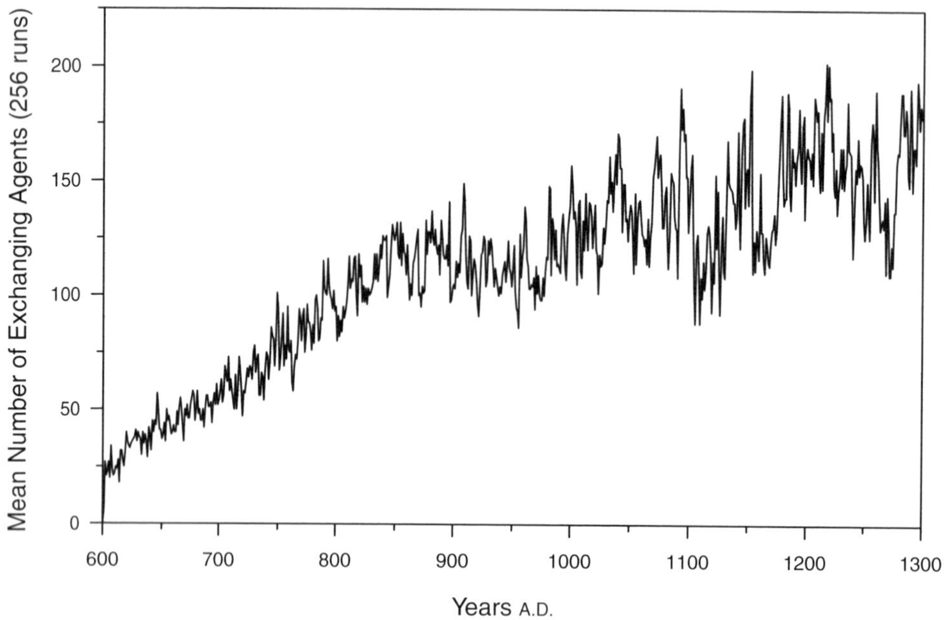

FIGURE 11.1 The number of households trading each year, averaged across 256 runs of the simulation.

begin with 200 agents, among whom approximately 25, a mere eighth of the founding population, invoke exchange. The growing number of trading households is apparent with the duration of the simulation, with the highest numbers—and greatest fluctuations in numbers—appearing in the final 200 years. Comparing this graph with the mean maize production through time (Plates 6.6, 6.7), we can see that exchange peaks during production valleys. For example, the sharp peak in the number of trading agents just before A.D. 1110 in Figure 11.1 corresponds to a series of below-mean maize production years between A.D. 1089 and A.D. 1097, in which A.D. 1090 was particularly disastrous.

Another way to view exchange is to examine the number of trades taking place. We can break down this view in terms of the type of resource being exchanged (i.e., protein or maize) shown in Plate 11.1 (in the color insert of this book); the type of network where the trading activity is taking place (i.e., GRN or BRN), shown in Plate 11.2; or the type of exchange (i.e., a donation, payback, or request) as shown in Plate 11.3.

In Plate 11.1 we see the protein trade grow and then suddenly level off around A.D. 900, when it is surpassed by maize, which in turn grows throughout the rest of the simulation years. Since the vast majority of the exchanges are donations in the GRN, it is reasonable to suppose (in conjunction with the data on deer depression in Chapter 9) that this represents an important threshold for large-game depletion in this simulation, after which households rarely have excess protein to gift. (Gifting of meat from hare and rabbits will be rare, since households hunting those species will ordinarily take roughly what they need without having much excess.) Plate 11.2 confirms that the GRN is the most active network throughout the simulation, again with the largest fluctuations in the final 200 years, which—as Plate 11.3 reveals—is due to high fluctuations in donations as households move between situations of maize excess and shortage at higher frequencies than was previously the case. In addition to counting the exchange frequencies between agents, we can examine the trade volumes. Volume here is defined as the amount of resource in kilograms exchanged by all individuals in a

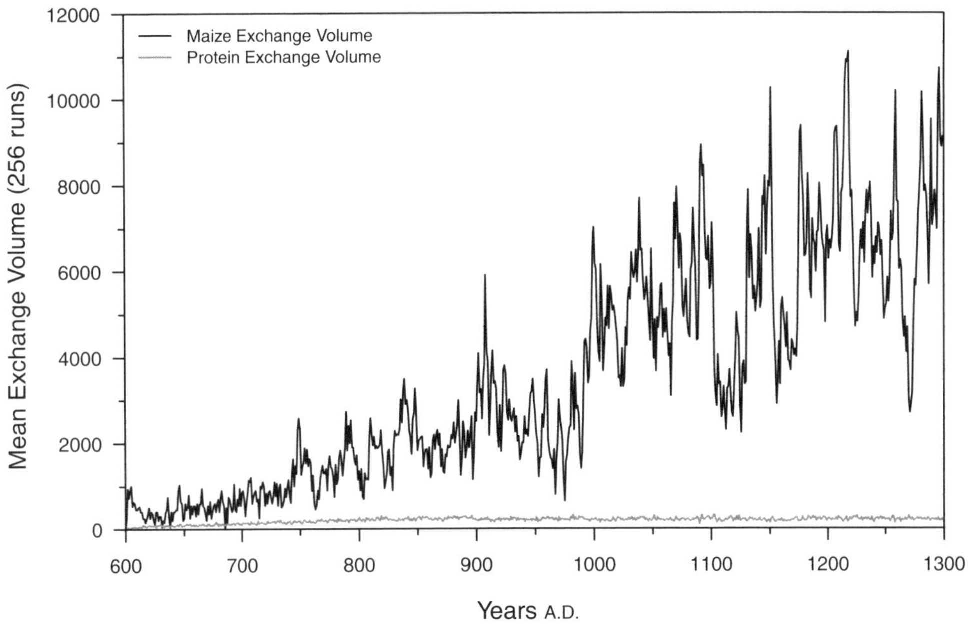

FIGURE 11.2 The amount of protein vs. maize exchanged over time in Kg.

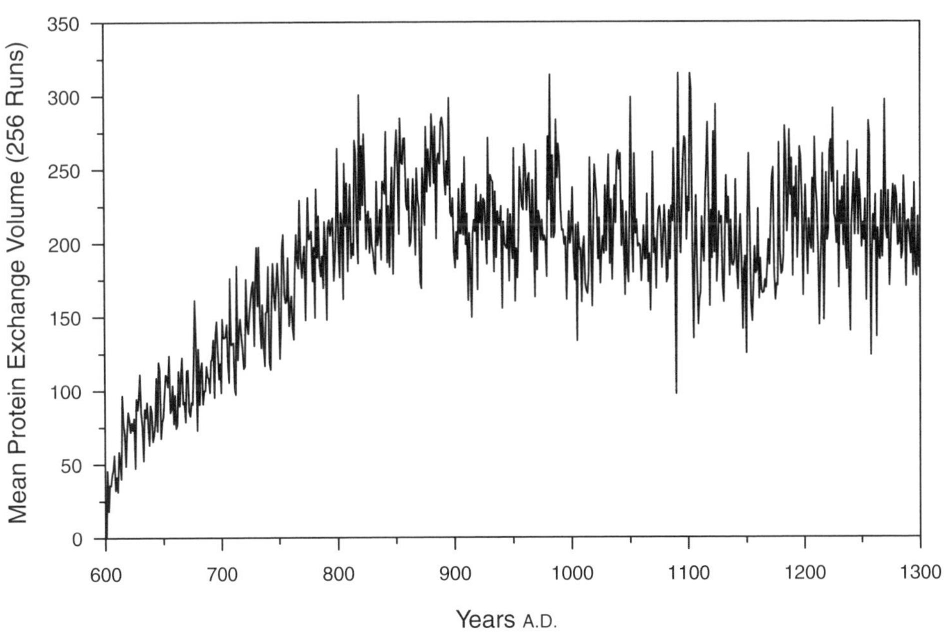

FIGURE 11.3 The amount of protein exchanged over time in Kg.

given year. In Figure 11.2 we observe the amounts of protein and maize exchanged over time, noting the significant amount of maize traded. Maize exchanges are curtailed in the drought years of the A.D. 900s, then again around A.D. 1100, with the largest fluctuations in the later centuries. This may suggest a high demand and a serious stress on the resource as the population size increases in the latter years. When we examine the protein exchanges in Figure 11.3, we see that after an initial increase their volume remains fairly stable. If we break down the trade according to its type, in Plate 11.4 we see that in later centuries paybacks are very limited. The volume of requests remains at its high levels, yet donated amounts do not appear to keep up with the demand.

Further studies are needed to examine the underlying social networks responsible for the trades. An offline network viewer was designed to track the nodes (households) over time. An edge in the graph represents a relation or social link between two households. The GRN and BRN are displayed in different colors and can be selected separately. While this tool allows us to visualize the networks and examine the distribution of the relations as well as the ability to track exchanges from an individual household's point of view, much work remains to be done to examine the mathematical properties of these dynamic social networks. This research is planned for VEP II.

CONCLUDING REMARKS

Communication between individuals is a key to forming societies. Exchange networks implemented in the model enable, thus far, two types of interactions between households: one between kin related by blood or marriage, another built on proximity and the concept of neighborhood. Households can specify messages triggered by various actions reflected by their current state. For instance, a hungry household can ask for maize or protein, and a wealthy household then donates. Overall, the model is transitioning from an SR-based simulation to a set of planning agents that are more proactive and that can not only communicate their needs but also pick and track their trade partners and trade qualities. Such is the case with the BRN.

The design of planning agents is far from over. Households have beliefs, desires, and intentions, hidden or known. As in any exchange system, resource-rich households and successful traders may gain higher status in society. They may become movers and shakers and may have an advantage in competition for leadership positions. Work has already been started to identify these "hub" nodes, with a particular interest on examining their numbers, noting their success and downfall, and examining the extent to which their locations through time match those of the community centers discussed in the next chapter. After all, the collapse of leaders can have a negative impact on the connectivity of the graph and the success of the trades. The next development phase in VEP II can now embark on simulating with more realism households that have some degree of control over their trade networks and some ability to manipulate them to their advantage.

TWELVE

Tool-Stone Procurement in the Mesa Verde Core Region Through Time

Fumiyasu Arakawa

THE VILLAGE ECODYNAMICS PROJECT (VEP) was designed to integrate a computer simulation with archaeological analyses to better understand the long-term interaction between humans and their environment and to clarify general evolutionary processes. One goal of these integrated studies was to better understand settlement patterns, including the episodic formation of aggregated villages. By modeling the exchange of maize and meat among households, the computer simulation illustrates how exchange helps us understand settlement patterns and the formation of villages. But it is difficult—perhaps impossible—to assess the local exchange of maize, of the sort we model, in the archaeological record. Therefore, the VEP needed to carry out studies to examine exchange in a medium that would be visible in the archaeological record, in order to complement the results of the simulations. This chapter reports on the investigation of how the use of stone raw material changed through time in and around the VEP area; what this reveals about exchange; how those changes are related to settlement patterns; and whether any boundaries indicative of communities or alliances of communities can be identified. I use estimates of energy expenditure for tool-stone procurement across space, and through time, as my major metric for studying these phenomena. I show that tool-stone procurement informs on a wide range of human behavior at least as well as the more traditional study of functional and morphological characteristics of stone tools.

STUDY AREA AND TIME PERIODS

The Mesa Verde region is an ideal area to study lithic materials to investigate both exchange and how settlement patterns and community organization change through time. The area has been the subject of numerous archaeological projects, with abundant and well-documented data sets (Chapter 2, this book; Adler 1990; Breternitz et al. 1986; Kohler, Kresle, Van West, Carr, and Wilshuen 2000; Lipe et al. 1999; Lister 1968; Rohn 1977; Varien 1999a; Varien and Wilshusen 2002). This chapter puts the VEP study area in a larger

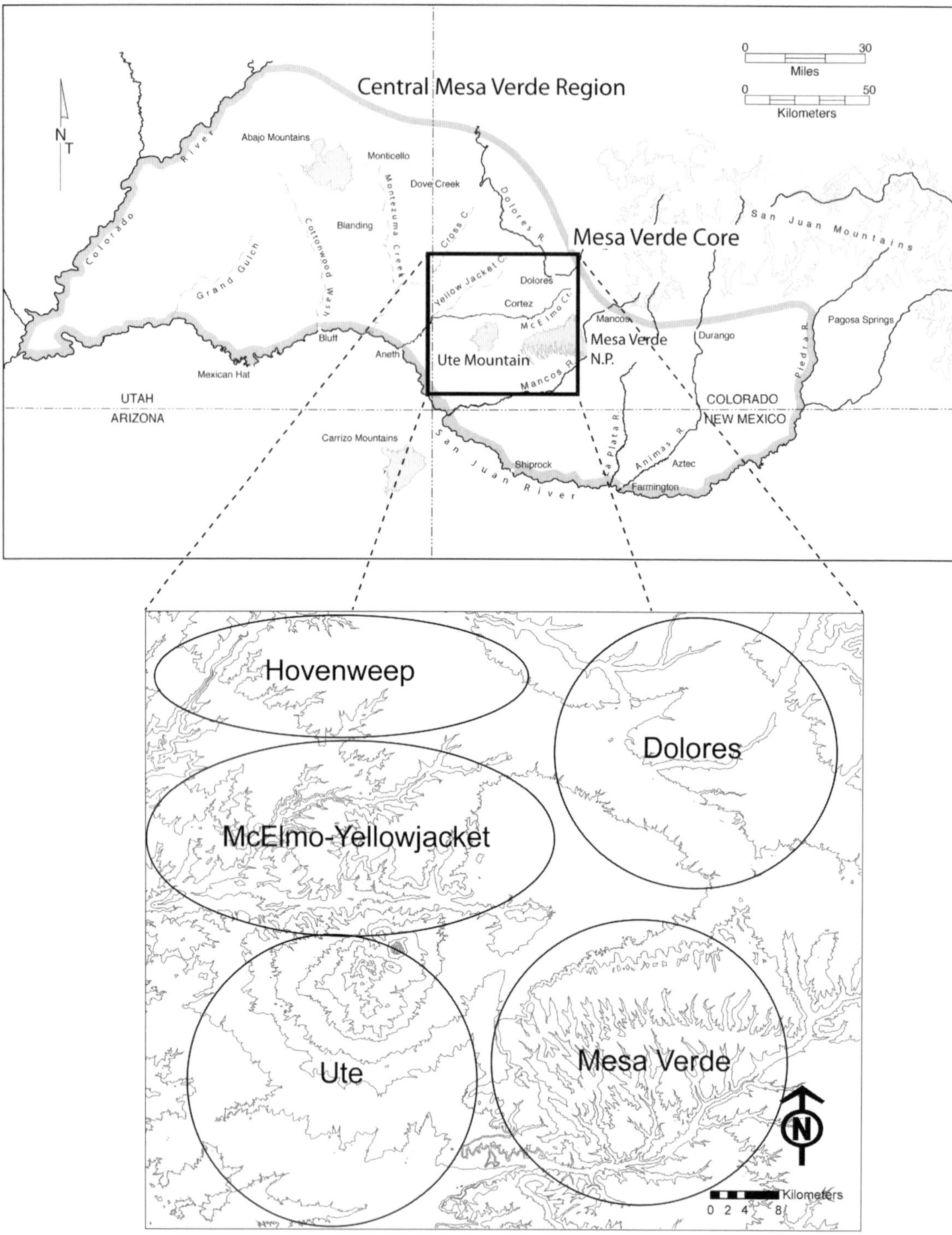

FIGURE 12.1 (Upper panel) central Mesa Verde region; (lower panel) localities in the Mesa Verde core area.

TABLE 12.1
Periods Used in this Chapter (Subdivisions of Traditional Pecos Classification) and Their Correspondence to Calendar Dates and VEP Chronology

Periods	Dates (A.D.) Used in This Chapter	VEP Periods[a]
Basketmaker III	600–725	6
Pueblo I	725–920	7–9
Early Pueblo II	920–1060	10–13
Late Pueblo II	1060–1140	14–15
Early Pueblo III	1140–1225	16–17
Late Pueblo III	1225–1280	18–19

a. Ortman et al. (2007) and Chapter 2, this book.

regional context by examining adjacent areas to the south and east, whose addition almost triples the size of the VEP I area. This enlarged study area (the upper panel of Figure 12.1) is similar to what Mark Varien and Richard Wilshusen (2002:4) defined as the central Mesa Verde region, but since my study area is not identical to theirs, I will call my study area the Mesa Verde core region. This includes the Dolores, Mesa Verde, and McElmo-Yellow Jacket districts as defined by Varien and others (Varien et al. 1996).

This chapter employs four scales of analysis: sites, communities, localities, and the region (Varien 1999a; Willey and Phillips 1958). Habitation sites—e.g., the Duckfoot site (5MT3868), Shields Pueblo (5MT3807), and Yellow Jacket Pueblo (5MT5)—represent the smallest units. A community is defined as many households that regularly engage in face-to-face interaction and share social and natural resources. Even during periods when aggregated villages form, communities include multiple sites (Kolb and Snead 1997; Murdock 1949; Varien 1999a:19).

A locality is defined as an area larger than a community but smaller than a region (Varien 1999a:23; Willey and Phillips 1958:18). It is generally identified by physical boundaries such as rivers, canyons, and mountains. Here, my localities are defined by canyons, and they include the McElmo-Yellow Jacket, Hovenweep, Dolores, Mesa Verde, and Ute Mountain areas (Figure 12.1). The largest scale I discuss will be the region as thus defined.

It was not possible to use the 14 modeling periods employed by the VEP because there would not be adequate well-dated samples for each of the 14 periods. Instead, I use the six periods shown in Table 12.1. These were defined to isolate settlement patterns that were relatively homogeneous internally so that breaks between periods coincide with discontinuities in settlement pattern (Ortman et al. 2007). By considering a 2,500-sq-km study area and a 700-year time span, I can recognize changes in settlement patterns and the development of territoriality, as measured by changing costs of tool-stone procurement. Before discussing these, however, it is important to distinguish between the concepts of territoriality and the land-tenure system, and to provide a general economic perspective on both.

LAND-TENURE SYSTEMS AND TERRITORIALITY

Some scholars have used the concepts of territoriality and land-tenure systems interchangeably, but their definitions differ slightly (Netting 1982). Land-tenure systems are defined only for human societies. Michael Adler (1996:338) defined these as "the systems of rights and privileges that human groups use to protect their resources and resource areas from outsiders." Studies of land-tenure systems generally focus on how exploitive uses of local resources within a community are buffered from similar use by other communities. In general, land-tenure systems become more salient under competitive conditions that include high population densities (Adler 1996; Kohler 1992a; Smith 1988).

Eric Smith (1988:244–245) outlined a continuum in the development of land-tenure systems among hunter-gatherer societies in which five states can be identified: (1) commons (common property); (2) reciprocal access (communal property); (3) territoriality (ownership by local group); (4) private property (ownership by kin group); and (5) private property (individual ownership). Common property does not involve strict enforcement of access; people regularly develop permission by consensus that allows land use and access by others. The second system allows groups to reciprocally access land controlled by another group. In this system, groups can easily negotiate with each other to move in, use resources, or both. The third system provides more strict controls on land access, limiting it to local members. Systems four and five indicate a strong control of lands through ownership by kin group or by individuals. Although two different land-owning groups may negotiate reciprocal access, the chances of obtaining it are reduced under these private-ownership arrangements. Smith (1988:245–246) recognized that societies might employ different systems for different resources.

In archaeology, territoriality is often defined through empirical inference of social, economic, or ecological restrictions on resource use that are imposed by individuals or groups on other individuals and groups, such as lineages, clans, moieties, or communities. Archaeologists have generally employed land-use and catchment patterns to understand and reconstruct territories in hunter-gatherer and agricultural societies (Bettinger 1989; Gendel 1984; Varien 1999a). Study of land use focuses on the modification, manipulation, and utilization of ecofacts and artifacts—such as plants, fauna, soils, water, and lithics—in a large landscape. The study of catchment patterns concentrates on exploitation of resources from a central place (e.g., residential sites and community centers) and generally considers the distance, cost, or both of procuring and transporting resources to understand the nature of mobility or degree of accessibility. M.R. Jarman and colleagues (1982:38) rightly differentiated between catchment and territory; they defined the former as "an empirical statement of observations concerning the geographic relationship between an archaeological site and its constituents, and whether these arrived there through geological, meteorological, human, or other biological agencies." On the other hand, territory is understood not only through empirical observation but also through more theoretical expressions of what we believe to be restrictions on access by individuals or groups given specific social, economic, or ecological contexts (Jarman et al. 1982:38).

To build a theoretical expression of territories, Eric Higgs and Jarman (1972) and Jarman et al. (1982) emphasized the importance of time-distance factors. Jarman et al. (1982:32) estimated that it usually takes people about one and two hours to walk 5 to 10 km on flat terrain. Topography and geographic features such as canyons and rivers can affect travel times considerably.

In archaeology, catchments are often defined arbitrarily as artificial radii of some set distance from a feature, such as a habitation site. Even when these radii are computed in terms of travel time, the study of arbitrary catchments provides an analytical context, and this is distinct from territorial boundaries generated in the systemic context. Although catchment analysis can provide data that contribute to inferences about human use of a landscape, it may not provide direct information about land-tenure systems or territories.

In this respect, land-tenure systems are different from catchments but similar to territories; land-tenure systems are the human way of being territorial. Biologists and ecologists have investigated the relationship between spatial organization and ecological niche among animals, but they use the terms "territory" or "home range" instead of "land tenure" (e.g., Revilla and Palomares 2002). Ethnographers often refer to land-tenure systems in describing indigenous peoples' economic and sociopolitical organization *within* their territories (Cashdan 1983; Netting 1982). To attempt to reconstruct and interpret the complexity and variability of territories in the prehispanic Mesa Verde core region, I focus on the economic concept of territoriality derived from human behavioral ecology (HBE).

Human behavioral ecologists (Bird and O'Connell 2006; Emlen 1987; Smith and Winterhalder 1992; Winterhalder and Smith 2000) assume that human behavior and the content of decisions are strongly influenced by their natural and social contexts—including the predictability and abundance of resources such as food, water, and mating partners. According to Bruce Winterhalder and Smith (1992:6), "variation in social organization between and within species [can] be analyzed as evolutionary responses to local social and ecological conditions." Human behavioral ecologists have developed neo-Darwinian explanations using relatively simple mathematical models (particularly optimization and game-theoretical approaches) to account for human organization and behavior. The neo-Darwinian perspective emphasizes the role of natural selection as formative in organisms' survival and reproduction, leading over evolutionary time to "adaptive design" for behaviors as well as for morphology (Williams 1966).

Thomas Malthus (1803) played a key role in developing the basis for such evolutionary arguments by illustrating the power of exponential growth; increasing population density strongly influences resource availability and opportunities for mobility, since rising agricultural production cannot, in general, keep up with human population growth. This effect inevitably leads to competition for resources. Following Malthus, many scholars (Boserup 1965; Brown and Podolefsky 1976; Dyson-Hudson and Smith 1978; Netting 1969) suggested that territories emerge as a typical response to such competition, and they developed cost-benefit arguments to understand the conditions under which such systems could be expected.

A great deal of work within the HBE tradition employs optimal foraging theory, defined by Douglas Bird and James O'Connell as productively employing "the assumption that maximizing the rate of nutrient acquisition enhances fitness, either by increasing nutrient intake or by reaching some intake threshold more quickly, thereby freeing time to pursue other fitness-related activities" (2006:146).

Natural selection should favor behavior that maximizes net acquisition rates and minimizes the cost of obtaining resources, or increases efficiency in these tasks. I employ the term "cost" as synonymous with "time and energy" through the remainder of this chapter. Such costs are also a fundamental category in analyses by neoclassical economics, to which the HBE approach is allied. For instance, on the basis of considerations of time allocation, we expect that when people increase their time and energy expenditures for activities such as food production, feasting, or defense, they must reduce the time and energy spent in other activities (Jeske 1992:469). That is, trade-offs can be expected in time and energy budgets. Additionally, we assume that when individuals invest much time in obtaining a transportable item in a landscape, they may encounter scheduling conflicts with food-production tasks, tool manufacture, or other activities (Jones and Madsen 1989:533).

Following this logic, I assume that humans tend to minimize the costs (whether measured in time or energy) of traveling. As Jarman et al. (1982:26) note, "The more distant the resource to be exploited, the more expensive it is in energy costs, and the more its exploiter is exposed to predation and competition." Many archaeologists have used the cost-minimization concept, particularly as measured by distance, in their research (Findlow and Bolognese 1984; Morrow and Jefferies 1989:30). Studies of agricultural societies, in particular, support the idea that people minimize their travel cost. When population density increases in a community or village, activities become more constrained by social, economic, and political involvements within the society. Various archaeologists (Crown and Wills 1995; Jeske 1992; Parry and Kelly 1987) have discussed how time allocation interacts with degree of sedentism. Agriculturalists tend to limit their mobility, and in so doing, reduce their costs of traveling to procure resources on a daily basis.

Ethnographic studies of catchment patterns suggest that agriculturalists generally conduct their activities within a primary area where most resources needed on a daily basis—such as water—are found. Jarman et al. (1982) define this "primary area" as being 1 km in radius; Varien (1999a) uses 2 km for a similar concept. There also exists a larger "exploitative zone" where most fuel or hunted animals, or both, are located (7 km in Arnold [1985], 5–10 km in Jarman et al. [1982], and 7 km in Varien [1999a]). Beyond that is a nonexploitative zone where other materials, including clay and lithic sources, may be located (more than 10 km in Jarman et al. [1982] and 18 km in Varien [1999a]). All these studies suggest that agriculturalists concentrate their activities within the primary area, and that their exploitation radius decreases with an increase in agricultural intensification. These studies are therefore based on a logic consonant with that of HBE: humans minimize the costs of traveling to obtain their resources within their landscape.

Economic Defensibility Model

Rada Dyson-Hudson and Smith (1978) investigated the relationships between human spatial organization, resource density, and resource predictability. They contrasted the conditions under which four spatial organizations should develop: (1) high mobility, information-sharing, spatiotemporal territories; (2) increased dispersion and mobility; (3) geographically stable territories; and (4) home-range systems (Figure 12.2, upper panel; Dyson-Hudson and Smith 1978:26). We expect high mobility, information-sharing, spatiotemporal territories when resource density is high but resource predictability is low. When both resource density and predictability are low, we expect increased dispersion and mobility: people frequently move to acquire resources. When resources are both plentiful and predictable, we expect geographically stable territories, and in these cases it may be economically advantageous to spend more time and energy defending such territories, especially when the gains from such defense are greater than the costs of losing access to the resources. Finally, we expect home-range systems when resource predictability is high but resource density is low. The mammalogist W. H. Burt (1943:351) defines the home range as "the area, usually around a home site, over which the animal normally travels in search of food. Territory is the protected part of the home range, be it the entire home range or only the nest."

We can use Burt's distinction between home range and territory to understand the economic defensibility model. Suppose people from one community in the Dolores valley always found deer in that valley, but perhaps people from other communities in the Mesa Verde core region also went hunting in that area. Deer tend to be fairly low in density, even in favored areas, and they are not very predictable in their distribution. Thus, although the Dolores inhabitants might have wished to control and restrict access to their territory by others, the size of the territory that would have to be defended would have been very large, and the costs of so doing might well have been greater than the benefits. Thus, home-range systems define territories, but these may not be defended in their entirety.

Dyson-Hudson and Smith's economic defensibility model was developed for foraging societies but can be applied, with some changes, to

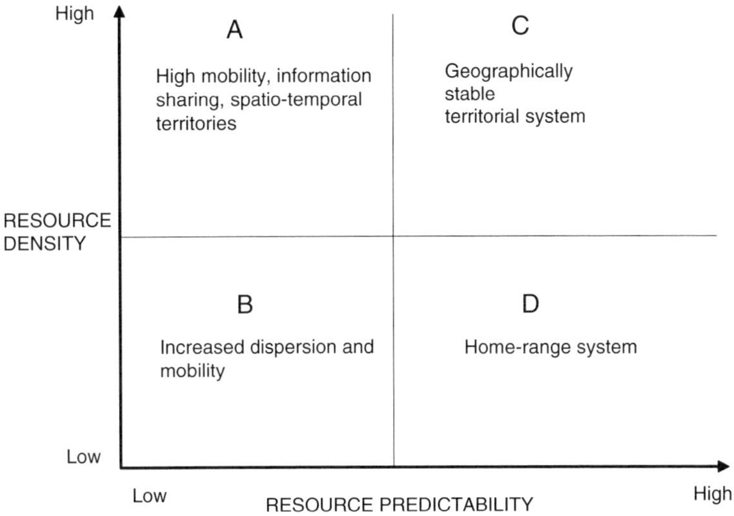

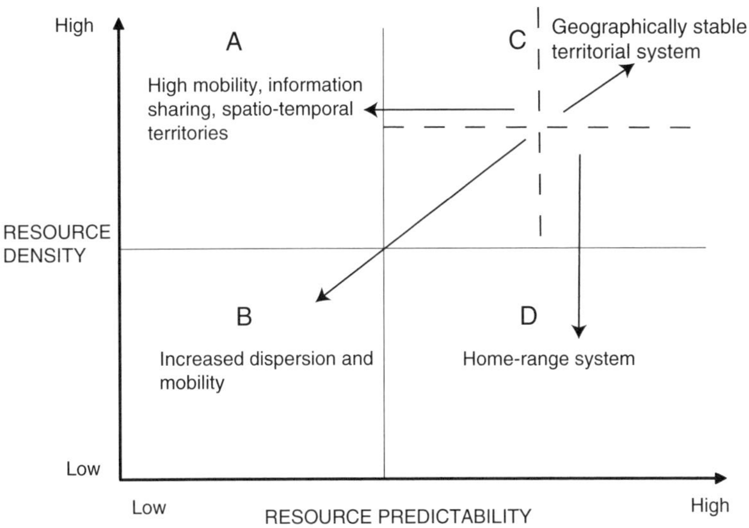

FIGURE 12.2 Resource predictability versus resource density. Top panel: Dyson-Hudson and Smith's original model. Bottom panel: Dyson-Hudson and Smith's model as modified in this chapter.

agricultural societies. Through domestication, people produce resources that are denser and more predictable than most food resources used by foragers. Agricultural systems, then, will tend to gravitate toward the upper right-hand corner of Dyson-Hudson and Smith's model—the geographically stable territorial systems (Figure 12.2, lower panel), but this may apply only to domesticated resources and not to the wild and natural resources on which agriculturalists rely, and for these resources the territories may resemble home-range systems. In terms of domesticated resources in the Mesa Verde core region, agricultural production increased with rising population density, and the importance of domesticated turkey also increased beginning in the mid-A.D. 1000s (Driver 2002). The increasing importance of domesticated resources and the other effects of population density affect where people can be in the diagram. Together, high population densities and local patches enriched through human labor will tend to generate geographically stable territories. Therefore I argue that the agricultural societies we are dealing with in this book tend to develop geographically

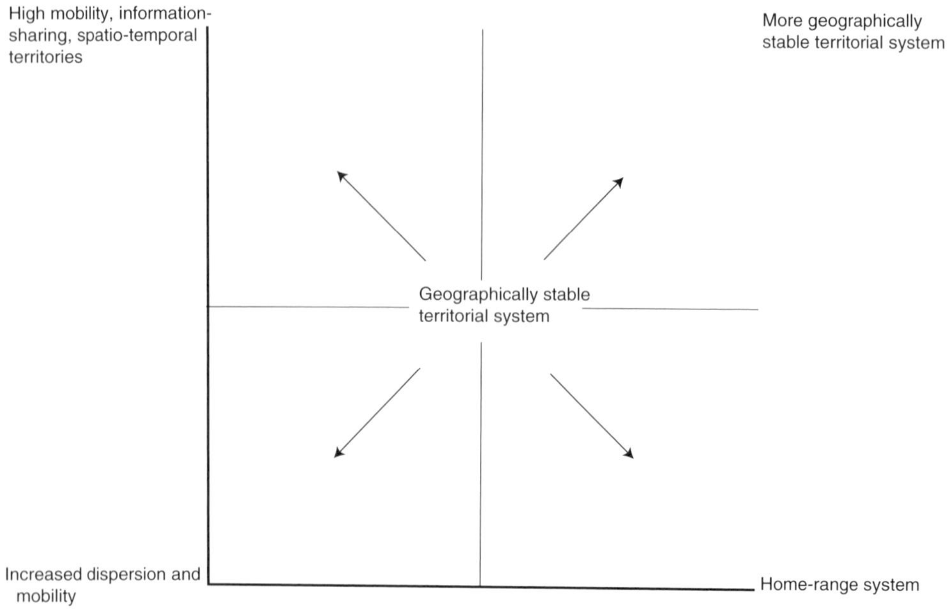

FIGURE 12.3 The portions of Dyson-Hudson and Smith's territorial system model relevant to this chapter, expanded.

stable territories (Figure 12.3), though I identify periods in which circumstances moved them slightly into one of the other economic defensibility categories.

Tool-Stone Procurement Patterns

Tool-stone procurement patterns reflect important aspects of territoriality by suggesting how far and how often people traveled to obtain raw materials. Many archaeologists have focused on catchments and resource procurement to understand and reconstruct territoriality and accessibility in hunter-gatherer and agricultural societies (Arnold 1985; Bettinger 1982; Harro 1997; Walsh 1998). Robert Bettinger's (1982) use of Julian Steward's (1937) family-band model to study land use and territoriality in Owens Valley is analogous to the approach I develop here. Using linear-regression analysis of the distributions of obsidian frequency in the valley, Bettinger expected that the frequency of obsidian would be high close to a quarry and relatively low in more distant areas if specific groups controlled or dominated that quarry. Bettinger (1982:121) put distance on the *x*-axis as the independent variable and the percentage of obsidian on the *y*-axis, as dependent on distance. Obsidian frequencies above the regression line were found mainly within 15 km of the source, while sites below the regression line were mostly within 15 to 25 km. Bettinger also examined the spatial distributions of obsidian projectile points during three periods in Owens Valley and discovered a major shift toward higher frequencies of this tool type within the falloff zone (as defined by Colin Renfrew's [1977] distance-decay model) during his late period (A.D. 1300–historic). Bettinger concluded that this change in tool-stone distribution was probably due to trade. Overall, his study suggested that there was a territorial boundary within the supply zone (within 15 km) that impeded direct access to the lithic source for those living beyond this distance threshold, but strong trading networks offset the decline in frequencies of finished tools with greater distances during the late period in his study area.

My general approach is also indebted to Robin Torrence's (1989) book, in which many of the contributors use time and energy as currencies for understanding how humans use lithic materials, starting from the assumption that humans have been shaped by natural selection to find more or less optimal solutions to problems, including those posed by lithic technological organization. According to Michael Jochim, "The study of stone tools has entered an exciting stage because of theoretical developments linking technology to the overall organization of behavior within a framework of economics and evolutionary ecology" (1989:111).

In this chapter I follow Jochim's lead and use distance-decay models and cost distances for obtaining tool stone to understand the relationship between settlement placement, patterns of resource productivity, and population size, and to shed light on the changes in territoriality through time. Previous research (including analyses in Chapter 2) and generalized application of HBE ideas both suggest that mobility was relatively unrestricted from the Basketmaker III to the early Pueblo II periods because of fairly low population density coupled with greater reliance on hunting than in later periods. The possible exception may be the period of peak population and aggregation achieved during the late Pueblo I period (ca. A.D. 840–880; Varien et al. 2007). During the late Pueblo II and III periods, mobility was more restricted, and defended social boundaries or territories emerged in this region.

I show how variation in the proportions of debitage raw materials reveals these differences in mobility and access to these raw materials through time. When access was relatively unrestricted, I expect to see use of materials that are best suited to the performance requirements of particular tools or tasks, as well as variation in raw material use otherwise determined by cost distance. When access was restricted by defended territories and reduced mobility, I expect to see more use of local materials, with variation in raw material use determined more by social accessibility and less by cost distance. Therefore, the distance-decay model for tool-stone procurement should help document changes in the types and scales of territoriality during the prehispanic occupation of the Mesa Verde core region. Douglas Harro (1997) provides a general example of this methodology, as does Michael Walsh (1998) for the development of ethnic boundaries in the northern Rio Grande.

Methods and Expectations[1]

To calculate costs (energy expenditure) for tool-stone procurement, I implemented cost-weight analysis based on slopes, using ArcGIS. The cost-weight analysis is a method for calculating approximate energy-expenditure values by weighting the distance from point *x* to *y*. These energy-expenditure values (or cost distances) provide a better estimate of walking costs in this mesa-and-canyon landscape than would straight-line distances.

I use these cost-weight estimates in three different analyses. First, I use the "high-resolution" data from the excavated sites and employ a modified distance-decay model to determine which sites within a locality deviate from the predicted model. In this analysis, I define the percentage of each raw material type across all sites as a dependent variable, and the cost distance for that material from each site as the independent variable. I use linear regression, correlation coefficients, and analysis of residuals to examine these data (Arakawa 2006).

In the second approach, I build energy-expenditure maps for the entire study area for each period defined in Table 12.1. These maps display the spatial distribution of energy expenditure for tool-stone procurement in each period. For this analysis I added other local materials—red jasper, indurated shale, silicified mudstone, and metaquartzite—to the

1. The assemblages selected for this research come from sites with a single component as well as some sites deposited over two or more periods, where periods can be distinguished using tree-ring-dated contexts. I analyzed debitage from 76 excavated site collections, which were either not reported or not previously reported in enough detail to be used for this analysis, to add to the excavated and well-reported assemblages (see Arakawa 2006: Appendix B, for a complete tabulation of the assemblages used).

high-resolution data from the excavated sites to provide as even a spread of data points across the study area as possible in each period. These maps can be "differenced" to show the areas of largest change from one period to the next. The changes highlighted by this approach provide a regional context for the high-resolution data already described here and may reveal regional patterns not obvious from more local analyses.

Finally, I use the results of the distance-decay analysis and the correlation coefficients it generates to investigate whether residents of the Mesa Verde core region developed increasingly restricted territories through time. This will lead to a discussion of how other factors—environmental, demographic, or sociopolitical—can be adduced to help explain the extremely interesting trajectories of change in lithic material distributions through time that I identify in the Mesa Verde core region.

Expectations for the Linear-Regression Analysis

All things being equal, I expect people who live closer to a lithic source to use a larger proportion of that raw material. Thus, the regression of percentages of raw materials against distance from source should show a negative slope. If this is not the case, then a high degree of trade, other interaction, or frequent mobility must be inferred. Anomalies may also be interpretable as resulting from patterns of alliance among communities in the region.

Second, as Robert Neily (1983) argued, I expect that through time inhabitants of the study area should tend to rely increasingly on materials closer to their habitation sites if the population increase that generally coincided with aggregation also resulted in the development of more restricted territories. This would lead to increasingly negative slopes in the regressions for these materials through time. If this is not true (i.e., if there is no change over time or if people living farther away from a quarry had a larger proportion of certain raw materials), this might suggest that some people had more open or unrestricted mobility (or both) and territory, or had more access to trade, than did others. Dramatic change in subsistence patterns could also affect this analysis. For example, if Pueblo peoples were forced to reduce their hunting distances because of territorial considerations (Muir and Driver 2002), then they could not embed procurement of high-quality materials into long-distance hunts. The many factors affecting lithic procurement make this analysis more difficult, on one hand, but also enrich the range of interpretations that can be offered.

The slopes and r^2 values from these analyses also help us understand how Mesa Verde peoples used their landscapes. When the regression line has a steep negative slope, people had a marked tendency to use material available near their habitations. In this case they are very "sensitive" to the cost-distance relationship for that material. People might appear to be cost sensitive because they actually were, because they were quite confined to their local areas by population packing, or because they could obtain everything they needed locally (as might be the case, for example, if they lived in an area in which game was not depressed). Exchange of lithic materials, on the other hand, can lead to relatively more shallow negative slopes, or even positive slopes, in the cost-distance relationship.

The size of the r^2 values indicates variation in procurement patterns across the landscape. When everyone is similarly sensitive to the cost-distance relationship, the r^2 values will tend to be high, and the relationship between the percentage of raw materials and distances appears significant because all sites fall near the regression line. If inhabitants in some sites are exchanging and others are not, or if territories are strongly marked on some portions of the landscape and not on others, the cost-distance relationship will be weakened and the r^2 values will be lower.

If Mesa Verde Pueblo people developed more restricted territories and employed less logistic mobility over time, I expect the slopes of these linear regressions to become steeper and the correlation coefficients stronger. If resources are distributed nonrandomly and some people were prevented from using richer patches, such conditions might also be accompanied by a less even distribution of social power within this region through time—as was, in fact, inferred by William Lipe (2002; see also Kohler and Varien 2010). On the other hand, slopes that are less negative through time might indicate more frequent logistic mobility, less restricted territories, more exchange, or all of these.

Energy Expenditure and Exchange: Model and Expectations

To help understand changes in tool-stone procurement patterns through time from another angle, I also estimated the energy required to obtain the mix of raw materials present in each site in each period. I assume that the (typically very small) proportion of nonlocal materials, such as obsidian and Narbona Pass chert, came from quite distant sources and were probably acquired through regional exchange, and so I excluded such materials from this analysis. I make the calculation on the assumption that local raw materials were procured directly, not through trade or exchange. Energy (E) is measured as the sum, across all materials, of the proportion of debitage (by weight) in each site, multiplied by the distance from that site to the nearest quarry, for each of the 10 represented materials:

$$E = \sum_{i=1}^{10} p_i d_i \qquad (12.1)$$

where, for each raw material type i, d is the cost to travel from the site to its nearest quarry, and p is its proportion in the debitage assemblage by weight (for both flakes and angular shatter). I calculate proportions by weight instead of by size or count since it generally provides a more suitable measure for discriminating reduction stages (Ammerman and Andrefsky 1982; Andrefsky 1998) and provides results similar to those obtained by measuring debitage length and width (Amick et al. 1988; Ammerman and Andrefsky 1982; Magne and Pokotylo 1981; Mauldin and Amick 1989:77; Shott 1994), but less costly to generate.

Interpolated Cost Maps

Finally, I use the summed energy expenditures to characterize variability in lithic procurement strategies among four localities. For this analysis, I use 10 local or semilocal materials discussed in more detail later in this chapter: chalcedony, Cretaceous Dakota–Burro Canyon quartzite (Kdbq), Cretaceous Burro Canyon chert (Kbc), Morrison, Jurassic Morrison Brushy Basin chert (Jmbc), and igneous materials (Ign)—which are frequent in our assemblages (Table 12.2; Arakawa 2000). Three other raw materials—indurated shale, silicified mudstone, and metaquartzite—are less common and variably distributed in this region, but are included in the analysis, as is red jasper, for which only one source was identified during the quarry survey (Arakawa 2006). Including those four additional materials provides a more reliable map for the total energy expenditure through time. I argue that specific localities showing high energy expenditures for their lithic assemblages either employed high logistical mobility or engaged in a great deal of exchange. I will use kriging (a geostatistical method for interpolating a smooth surface for some value, using known points) to interpolate these cost maps, to reveal patterns in a regional context.

Assumptions

The energy expenditure calculation (equation 12.1) is a relatively simple modification of Renfrew's (1977) distance-decay model that sums the products of the proportions of 10 raw material types multiplied by their cost distances. This approach is possible only because the

TABLE 12.2
Material Types, Locations, and Quality

Material Types	Names Used for This Paper	Local, Semilocal, or Nonlocal	Quality
Cretaceous Burro Canyon / Jurassic Morrison Brushy Basin chert and silicified mudstone	Morrison	Local	Low
Cretaceous Dakota / Burro Canyon quartzite	Kdbq	Local or semilocal	High
Cretaceous Dakota / Burro Canyon chert	Kbc	Local or semilocal	High
Jurassic Morrison Brushy Basin chert	Jmbc	Local or semilocal	Medium
Chalcedony	Chalcedony	Local or semilocal	High
Igneous	Igneous	Local or semilocal	Medium
Red jasper	Red jasper	Local or semilocal	High
Indurated shale	Indurated shale	Local or semilocal	Low
Silicified mudstone	Silicified mudstone	Local	Low
Metaquartzite	Metaquartzite	Local	Low

Mesa Verde core region has fairly well-known quarry locations (Arakawa 2006). In using this equation, I make three main assumptions:

1. Debitage found at a site was not produced by craft specialists.
2. Debitage found at a site resulted from in situ production.
3. Direct procurement strategies (including embedded procurement) were employed.

These assumptions appear to be met in this study. Because most stone tools found at these sites are expedient, and even projectile points seem to be expediently manufactured, craft specialists probably did not produce these tools. This is important because if specialization for manufacturing tools were common, then these tools may have been traded some distance and we could not measure energy expenditure for the inhabitants of a site with a calculation of this form. For similar reasons, I assume that debris generated from stone-tool production in one community was not carried and dumped into middens of other communities. Finally, whether tool-stone procurement was direct or traded is not a significant issue when energy expenditure is considered because this calculation focuses on the total energy embedded in an assemblage, not on how or what strategies individuals used to obtain raw materials. Whether a piece of debitage was obtained during a hunting trip, was the result of direct procurement, or was brought by someone from other communities, in this analysis it will represent the same amount of embedded energy. In fact, the extent to which local exchange affects the settlement patterns and social organization in our study area seems to be rather small, at least as suggested by the simulations reported in Chapters 11 and 15. Exchange of materials indicates that there were cooperative relationships between inhabitants who lived closer to a quarry and people who lived farther away from this source. I follow Renfrew's (1969:152) definition of trade as "reciprocal traffic, exchange or movement of materials or goods through *peaceful* human agency" (emphasis added). Tracking these relationships is an important goal of the VEP research. Although it is difficult to distinguish between the movements of lithic raw materials by direct procurement versus exchange, I examine the case for each in the interpretations that follow.

LITHIC RAW MATERIALS

Correct raw material classifications and identifications of the source areas are crucial to this study. I now briefly discuss the six most common materials by three broad classes of similar knapping quality (Table 12.2). Together, these six account for more than 90 percent of the lithic assemblages in this region.

High-Quality Materials

Three high-quality materials—chalcedony, Kdbq, and Kbc—may be especially sensitive to changes in access that might accompany growing community territoriality. These materials were used in making formal tools, and their quarries are well known, recorded, and relatively ubiquitous in this study area. Quarries for these materials fall within 18 km of most villages in the Mesa Verde core region (Arakawa 2000). For instance, Shields Pueblo is approximately 3 km from the nearest Kdbq quarry, while the nearest quarry for this material is approximately 12 km from Yellow Jacket Pueblo (Arakawa 2000). Because these materials are frequent choices for manufacturing formal tools, and since the projectile points and bifacial tools that are the most common formal tools are generally associated with activities likely dominated by males (such as deer hunting), changes in the proportions of these materials through time might indicate, in part, changes in male hunting territories.

Low-Quality Materials

Morrison materials are of a lower quality and therefore provide another interesting medium for investigating changes in tool-stone procurement patterns. Neily (1983) and Fumiyasu Arakawa and Andrew Duff (2002) recognized that the central Mesa Verde residents used a larger proportion of these low-quality materials during the later Pueblo periods. If, as population increased and communities became more aggregated, communities exercised greater control over their immediate territories—and if it was more difficult to obtain materials from the territories inhabited by other communities—then most people would have come to rely on resources near their communities. Accessibility may have been reduced by hostilities between communities or by other pressures not to encroach on other groups' territories. As population increases, people might try to expand their territories to procure adequate resources, but this may not be possible in an increasingly populous landscape.

Jmbc and Ign

Finally, both Jmbc and Ign are especially interesting materials for this study because the sources for both of these medium-quality materials can be pinpointed (Arakawa 2006). Jmbc is from the Morrison Formation, but this material type shows distinctive attributes, such as multiple colors and hard textures. One use of Jmbc in our region was to make an elongated, flat stone tool called a Tchamahia. Although this tool might have been used for digging or cutting purposes (or both), some archaeologists (Brew 1946:241–242; Voth 1903:286) have proposed that Tchamahias were used for ceremonial purposes. The sources of Jmbc are mostly limited to the southwestern portion of our region. Ign (both intrusive and extrusive) are also found in its southern portions. Arakawa and Kimberlee Gerhardt (2007, 2009) reported some aphanetic minette (a fine-grain, extrusive material) source areas on and around Mesa Verde National Park and Ute Mountain Ute Tribal Park in Colorado (Figure 12.1). If my approaches reveal spatial patterns in the frequencies of these materials that are unrelated to distance to quarries, consideration must be given to interactions on the supracommunity scale. For example, if some households or

TABLE 12.3

Summary of Relationships Between stb *(Standardized Slope Estimates) and* r^2 *Values from Basketmaker III to Late Pueblo III Periods*

STB VALUES WITH CORRELATION COEFFICIENTS (r^2) > .3 SHOWN IN BOLD.

Material			Periods			
Types	*BMIII*	*PI*	*EPII*	*LPII*	*EPIII*	*LPIII*
Chalcedony	–.244	–.233	–.223	–.279	**–.374**	–.226
Kdbq	–.255	–.068	**–.418**	–.176	**–.523**	**–.415**
Kbc	**–.608**	–.323	–.232	**–.426**	**–.328**	–.263
Morrison	**–.424**	–.292	**–.816**	**–.765**	**–.900**	**–.360**
Jmbc	**.475**	–.207	–.262	**.456**	**.441**	**.567**
Igneous	**–.861**	**–.661**	**–.954**	**–.860**	**–.765**	**–.795**

BM = Basketmaker; EP = early Pueblo; LP = late Pueblo; P = Pueblo.

communities in the study area have relatively high frequencies of Jmbc and Ign even though they are relatively distant from these quarries, alliances with communities closer to the quarries might be inferred.

RESULTS

Standardized Regression Slopes (*stb*) and r^2

The standard slopes for the regressions are the slopes calculated after standardizing both the independent and the dependent variables. A value of –.5, for example, means that the percentage by weight (on the *y*-axis) declines by half a standard deviation unit as the cost distances (on the *x*-axis) increase by one standard deviation unit (Shennan 1988). Table 12.3 summarizes the relationships between the *stb* values and the r^2 values through time; *stb* values in boldface have strong correlation coefficients. Strong negative slopes indicate use of a raw material close to home, either because of constraints imposed by competitive processes among dense populations or simply because people were able to procure all raw materials they needed locally, perhaps because they lived close to abundant raw materials with appropriate characteristics.

The size of the r^2 value indicates both the variability in tool-stone procurement patterns across the landscape and the degree to which the slope is nonzero. Holding slope constant, when everyone is similarly sensitive to the cost-distance relationship, the r^2 value tends to be high because all sites fall near the regression line. If inhabitants in some sites exchange and others do not, the cost-distance relationship will be weakened. If everyone exchanges raw materials across the landscape, the r^2 values become unpredictable, though these cases would probably be identified by low negative or even positive slopes in the cost-distance relationship. This would be an interesting topic for simulation.

One problem for this analysis needs to be clearly stated: I was not able to obtain an early Pueblo III lithic assemblage from the Mesa Verde locality. Because residents in the Mesa Verde locality were some distance from many lithic sources, these missing data in the percentage versus cost-distance model for the early Pueblo III period makes them somewhat incompatible with those of different periods. To investigate the consequences of this problem, I reinvestigated assemblages from other periods without the Mesa Verde locality to determine whether the early Pueblo III assemblages would then be comparable with respect to the calculated r^2 and slopes. Except for Jmbc, which was probably procured and utilized for symbolic or special purposes, the late Pueblo II and late Pueblo III assemblages showed only small differences in r^2 and slopes when I analyzed them without the Mesa Verde assemblages. On the other hand, without the Mesa Verde assemblages, Basketmaker III, Pueblo I, and early Pueblo II assemblages showed quite different values for r^2 and slope. This occurred because habitations in those early periods are located in mostly the Dolores and Hovenweep localities, far away from the Mesa Verde locality. The absence of assemblages from the Mesa Verde locality makes the percentage versus. cost-distance models for those early assemblages fit more poorly (specifically, made the slopes flatter). In general, I think the unfortunate fact that early Pueblo III assemblages do not include data from the Mesa Verde locality does not constitute a major confounding factor for analyzing and comparing the percentage versus cost-distance and energy expenditure models of this time period with those of other periods.

Figure 12.4 shows that the early Pueblo III assemblages tend to exhibit the most strongly negative *stb* values and strongest correlation coefficients, especially for high-quality materials of chalcedony and Kdbq. Local early Pueblo III people tended to acquire these high-quality materials close to their residence. This implies that their movements across the landscape were more restricted, by territoriality or other factors, than in other periods.

The early Pueblo III assemblages also show a strong correlation between the percentage by weight and the cost distance for Morrison materials (Figure 12.4). The early Pueblo III residents had a tendency to use Morrison materials when these were close to their habitations. This sensitivity to cost distance for Morrison materials presumably also reflects increasingly restricted territories due to increasing population and the development of large villages. The high r^2 values in the early Pueblo III likewise suggest similar procurement patterns across this landscape, though this may be affected to some extent by the absence of any Mesa Verde locality sites in this period, since procurement behaviors on Mesa Verde appear to be different from elsewhere in this region.

Strong interaction among localities is indicated by the tool-stone procurement pattern for Jmbc (Figure 12.4). Assemblages in the Basketmaker III, late Pueblo II, and early and late Pueblo III periods show relatively strong correlation coefficients and steep *positive* slopes for the relationship between the percentage by weight of Jmbc and its cost distance. Of course, a positive slope is very surprising, since it means that more distant locations used more of this material. This might make sense if part of the value of a material lay in its being exotic; it may be that the Mesa Verde core inhabitants procured and utilized this material for symbolic or ceremonial purposes. Even people living far from its sources managed to obtain relatively large amounts of Jmbc, particularly during the late Pueblo periods (Arakawa 2006; Arakawa and Gerhardt 2007; Glowacki 2006; Pierce et al. 2002; Robinson 2005), even though this material was not suitable for manufacturing formal tools, such as projectile points and knives. It is curious that the expected negative relationship between abundance and distance does hold for the Pueblo I and early Pueblo II periods.

The tool-stone procurement pattern for the medium-quality category of Ign also conforms closely to the pattern predicted for increasing territoriality over time (Figure 12.4). The slopes and r^2 values for Ign show a consistently strong negative relationship between weight and cost distance through time. Only people living close to the sources acquired and utilized these materials in any abundance, though this tendency is least pronounced in the Pueblo I period and most pronounced in the early Pueblo II period. These relationships may suggest that Pueblo I people ranged widely across the landscape, or that some Pueblo I people obtained this material through exchange. On the other hand, the early Pueblo II inhabitants, especially in the Mesa Verde and Ute localities, procured Ign only if they were close to their habitations.

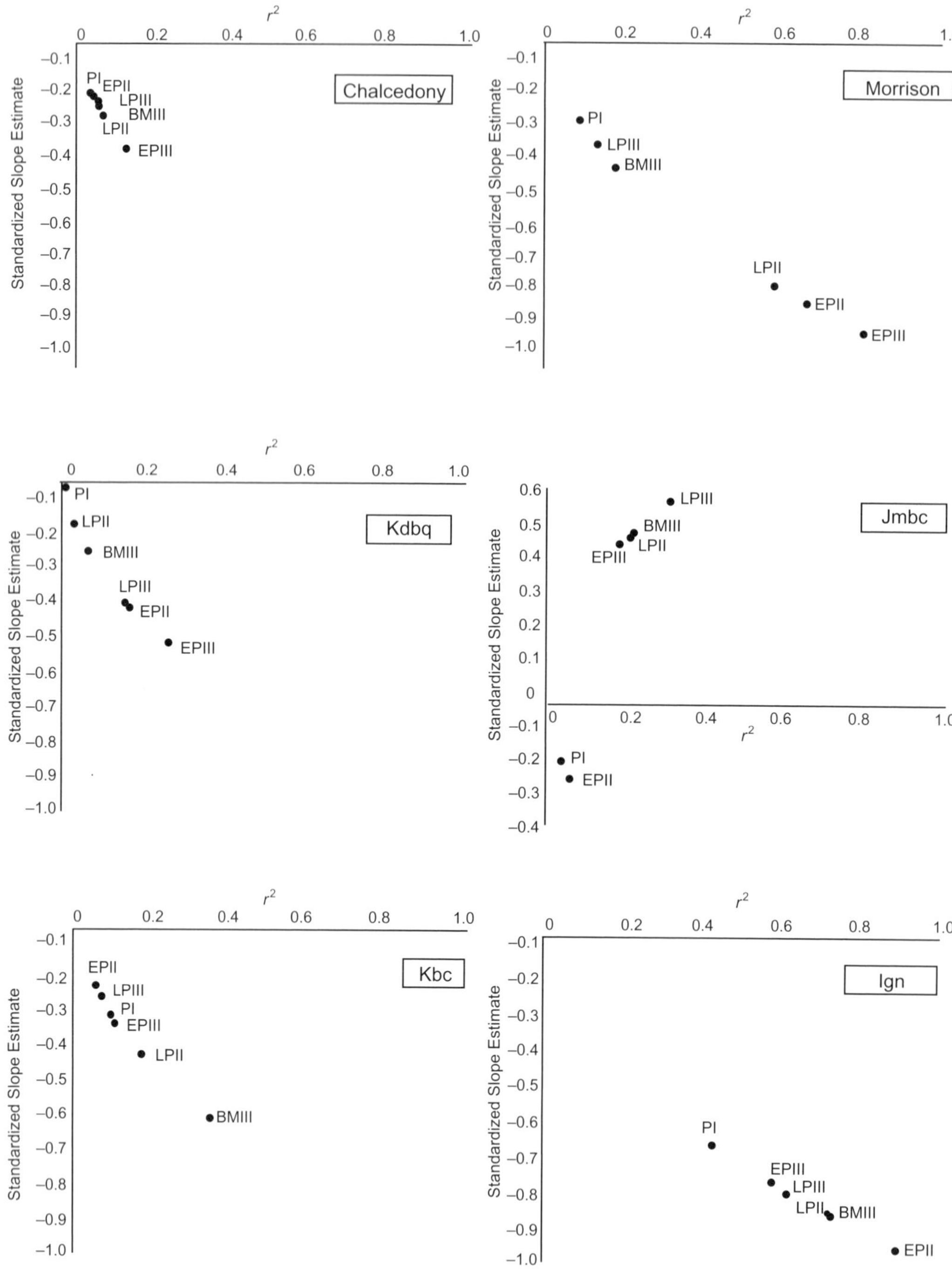

FIGURE 12.4 The relationship between standardized slope estimates and r^2 values for six raw materials through time.

Energy Expenditures

To visualize where and how people expended energy in tool-stone procurement, I interpolated energy-expenditure maps for each period. Figure 12.5 shows the estimated energy expenditures (derived from equation 12.1) for acquiring the lithic assemblages seen at each site in my sample, interpolated across the landscape using kriging (Arakawa 2006). A prediction that can be derived from HBE is that energy expenditures calculated over all significant activities should be approximately equal across the settled portions of the landscape (Smith 1979). We anticipate, therefore, that areas requiring only low energy for tool-stone procurement might have high energy expenditures for other activities, and vice versa, although temporary disequilibria could exist. Moreover, if not all voices in a community have equal say in where settlements are located, more powerful voices may be more successful in achieving their locational goals (i.e., those that minimize their own energy expenditures) than are weaker voices.

During the Basketmaker III period, high amounts of energy were expended to procure tool stone first in the Mesa Verde and Ute localities and then in the northern portion of the Hovenweep locality. This might suggest the existence of other attractions for living in these areas; for example, if these high expenditures are a proxy for high travel costs for hunting, perhaps these areas had other, offsetting advantages. During the Pueblo I period, study-area residents continued to expend a high amount of energy to procuring lithic materials in the Mesa Verde locality, whereas these costs were considerably lower in the Dolores locality, which was also a favorable area for deer hunting (Kohler and Reed 2011).

Figure 12.5 shows no dark colors across the landscape during the early Pueblo II period; perhaps the population trough in early Pueblo II (see Chapter 2) enabled all residents to expend less energy in procuring raw materials. If so, the contrast with Basketmaker III, which also enjoyed low population sizes but exhibits some areas of high expenditure for lithic materials, is notable. Early Pueblo II residents in the Ute Mountain, Mesa Verde, and portions of McElmo-Yellow Jacket localities did expend slightly more energy in tool-stone procurement, whereas residents of the Dolores locality, as in the two earlier periods, expended less. During the late Pueblo II period, inhabitants expended more energy in acquiring raw materials in the Mesa Verde locality than did residents in the previous periods. This means that materials, on average, were traveling farther across the landscape during this period. On the other hand, the people in the Hovenweep locality expended less energy in acquiring raw materials during the late Pueblo II period.

During the early Pueblo III period, residents of 5MT3778 (Casa de Sueños) in the Dolores locality and inhabitants in the Ute locality expended relatively high amounts of energy to procure lithic materials. In general, though, the energy expenditures for tool-stone procurement in the early Pueblo III period were less than those of the late Pueblo II residents. I suggest that this was strongly related to the development of restricted territories in the Mesa Verde core region during the early Pueblo III period, which in turn affected the radii used for hunting, thereby limiting the embedded collection of distant lithic materials. During the late Pueblo III period, inhabitants of the Mesa Verde locality expended higher amounts of energy to acquire raw materials. Energy expenditures were also relatively high in the Sand Canyon area in the McElmo-Yellow Jacket locality. Interestingly, energy-expenditure values for the community centers of Yellow Jacket Pueblo, Shields Pueblo, Sand Canyon Pueblo, and Castle Rock Pueblo were lower than for the small sites in the Sand Canyon area during the late Pueblo III period. Overall it can be concluded that the late Pueblo III inhabitants participated in strong interactions—either increased mobility or increased exchange—especially between the McElmo-Yellow Jacket and Mesa Verde localities, which caused tool stone to move large distances within the study

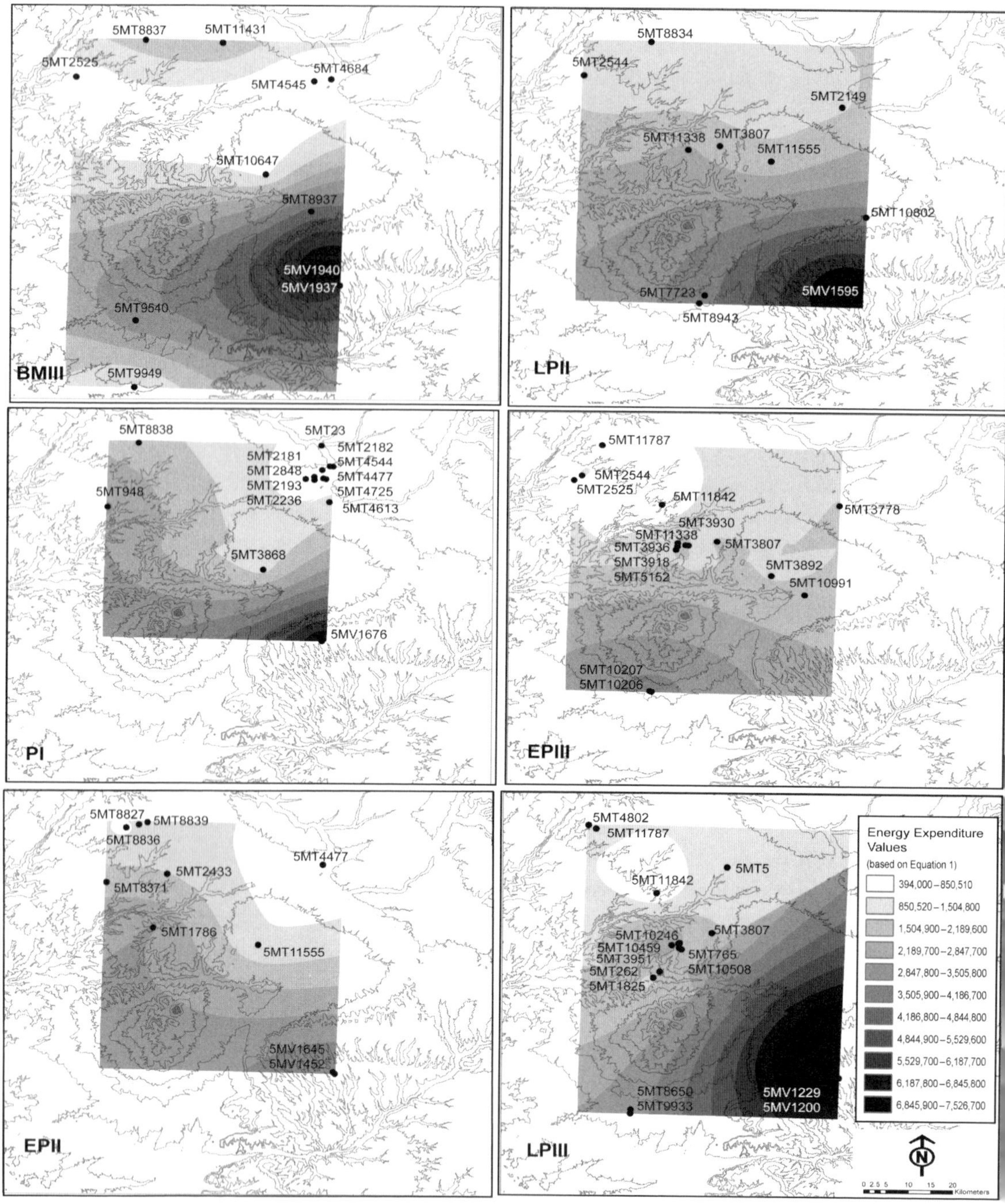

FIGURE 12.5 Maps of energy expenditure for lithic materials, interpolated using kriging, for six time periods in the central Mesa Verde region.

area (Arakawa and Gerhardt 2007). This is interesting because some studies indicate that the Mesa Verde core region was isolated from other regions by this time (Neily 1983), as evidenced by a dramatic decline in long-distance exchange (Lipe 1995). However, my data show that the exchange or movement of local materials *within* the Mesa Verde core area intensified during the final period of occupation.

In general, we expect that a geographically stable territorial system (Dyson-Hudson and Smith 1978) would have emerged in the late Pueblo III period because there were more aggregated settlements then, a larger population, more evidence of violence than in most periods (though less than in the early Pueblo III), and presumably more attention devoted to defending resources and territories (Kohler and Turner 2006; Kuckelman 2002; Kuckelman et al. 2000). Lithic data, however, seem to suggest maximum territoriality in this landscape in the early Pueblo III period. In the next section, I compare these lithic data with the Dyson-Hudson and Smith model to interpret the relationship between agricultural resources and landscape use.

Discussion: Two Different Regimes

To summarize tool-stone procurement patterns through time, I created the upper panel of Figure 12.6, showing the values for the *stb* and r^2 values from Figure 12.4, averaged across the six tool-stone categories shown in Figure 12.4. (Note that the *y*-axis is inverted). Periods falling in the upper right-hand quadrant exhibit strong negative slopes for the relationship between abundance and distance, as well as high coefficients of determination (all sites near the regression line and little variability across space). Periods in the lower left-hand quadrant exhibit weak negative slopes and at least potentially more variability across the landscape in degree of territorial restriction.

Figure 12.6 reveals the existence of two different regimes of lithic procurement for these six periods, all of which fall more or less within Dyson-Hudson and Smith's compartment C. Basketmaker III, Pueblo I, and—surprisingly—the late Pueblo III periods constitute one regime, whereas the consecutive early Pueblo II, late Pueblo II, and early Pueblo III periods constitute another. In the first (and generally earlier) of these regimes, people apparently were engaged in more mobility or more exchange; in the second regime people, procured their tool stone close to their residences much more consistently.

The bottom panel of Figure 12.6 makes the time trajectory for these two regimes explicit. For the most part, each population cycle in the VEP area, as defined in Chapter 2, is within a single regime. In the first demographic cycle, the Basketmaker III inhabitants tended to procure many raw materials locally, but they also participated in long-distance activities such as hunting large- and medium-size animals (Chapters 8 and 9, this book). The Pueblo I inhabitants procured varied raw material types, some located far away from their residences. Pueblo I residents were the first on this landscape to experiment with village life, yet they still ranged widely to procure tool stone.

This first demographic cycle ended—and a completely different mobility and tool-procurement regime began—with the early Pueblo II period, in which population sizes were small and settlement practices returned to a nonaggregated stance. The early Pueblo II inhabitants relied heavily on resources close to their habitations for tool stone (and, by extension, for acquiring other resources, to the extent that tool-stone procurement was embedded in these other activities). This is an unexpected result because population density was so low during this period. Sarah Cole's study of warfare (Chapter 13, this book) may offer a clue to interpreting this puzzle. She finds that the early Pueblo II period had a higher-than-expected level of warfare, given the low population density. Kohler and Varien (2010) suggest that this may have resulted from local populations attempting to resist the intrusion of Chacoan influence during this time. My data

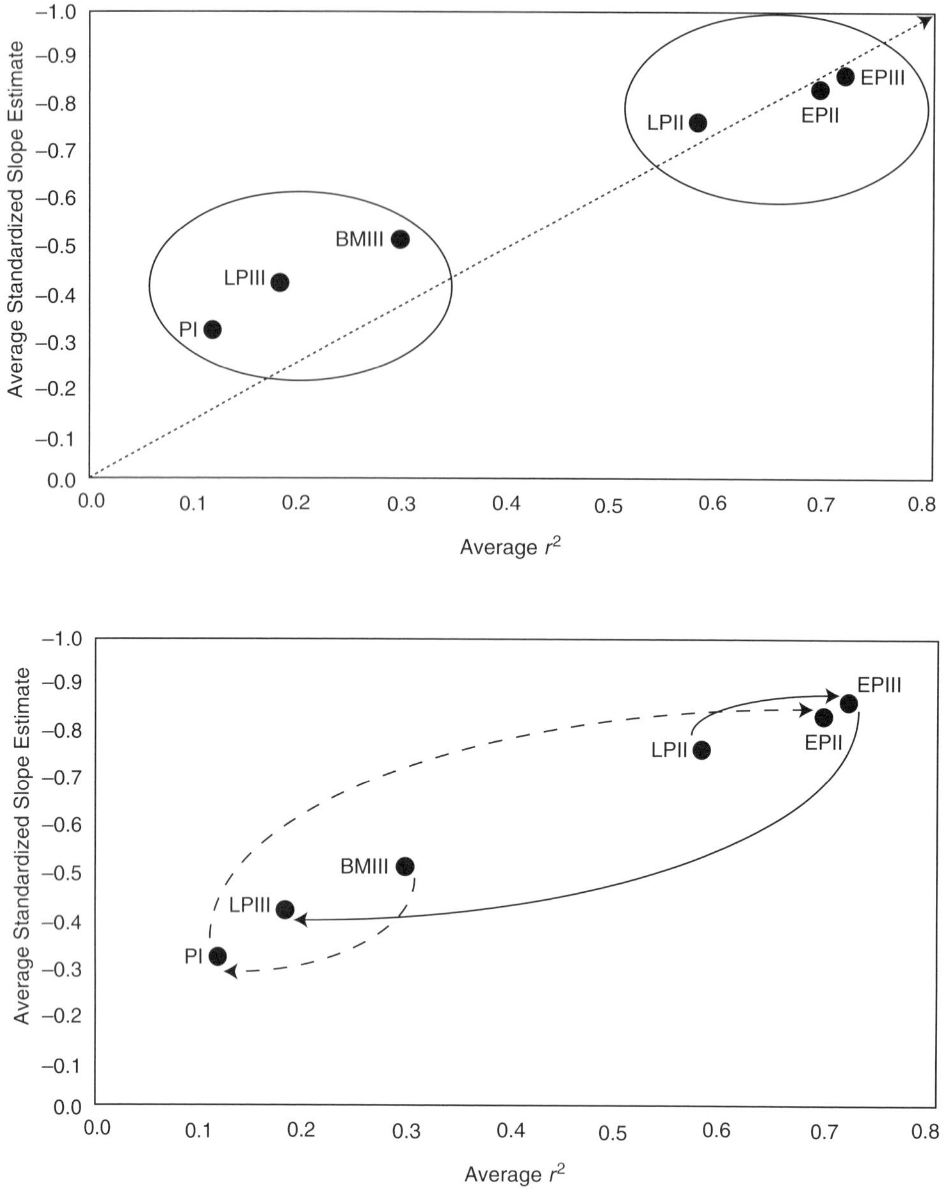

FIGURE 12.6 Top panel: two different regimes in the phase space defined by the average stb values and r^2 values for each period. Bottom panel: these regimes in the context of their time trajectories from the Basketmaker III through the late Pueblo III periods.

suggest that despite the low population density, communities developed geographically stable territorial systems in this period. Since this result is unanticipated by the Dyson-Hudson and Smith model, the factors limiting early Pueblo II mobility deserve further consideration.

In the late Pueblo II period (A.D. 1060–1140), population increased through immigration (Chapter 2, this book). These new inhabitants continued the earlier Pueblo II reliance on tool stone close to their habitations. These behaviors probably reflect both the difficulty of ranging widely across this landscape—which by now had more inhabitants than it had ever had—and also the suddenly increasing dependence on domesticated turkey for protein, rather than on deer or other medium or large game animals (Driver 2002; Chapters 8 and 9, this book). These tendencies became even more marked in the early Pueblo III period, when people appear to have relied on very restricted territories because of growing population size, a higher proportion of people living in aggregated community centers, and continued increases in reliance on domesticated turkey rather than deer.

During the first part of the late Pueblo III period, population continued to increase and aggregate, but at some point in this period migration began, and it continued until the regional depopulation was complete in the A.D. 1280s. My analyses of the lithic data suggest a return to a tool-stone procurement regime that had more in common with that of Basketmaker III and Pueblo I than with that of late Pueblo II and early Pueblo III. This is a completely unanticipated result.

I suggest that this return to an earlier regime was not due primarily to increased logistical mobility in order to directly procure resources (although Kristin Kuckelman [2010a, 2010b] suggests an attempted return to foraging in the final years of the Pueblo III period). Instead, I think this pattern reflects stronger interactions between the Mesa Verde and McElmo-Yellow Jacket localities (Arakawa and Gerhardt 2007). I believe that formation of large villages with public architecture increased the episodic movement of people between these centers—e.g., movement to attend ceremonies that took place in the public buildings at these centers—and that exchange intensified with this episodic movement. The development of this network of villages with public architecture in turn may have stimulated (or may simply reflect) the development of confederacies among the aggregated communities in our area during this period, as suggested by Kohler and Varien (2010). This model is supported by neutron activation analysis of black-on-white bowl sherds from Sand Canyon Pueblo and Castle Rock Pueblo in the McElmo-Yellow Jacket locality and the Mesa Verde locality in the late Pueblo III period, and this also suggests strong interaction among communities at this time (Glowacki et al. 1998; Pierce et al. 2002). It is interesting that the Pueblo I and late Pueblo III periods of maximum aggregation in the Mesa Verde core area, as defined in Chapter 15, both also resulted in greater movement of lithic materials, albeit through possibly different mechanisms—a result that is more complex and interesting than the original model that guided this research.

CONCLUSIONS

Tool-stone procurement patterns are very useful for understanding and reconstructing human behaviors because they are related to many other aspects of life in the past. To understand tool-stone procurement in the Mesa Verde core region, we need to consider not only resource productivity and predictability as examined by the VEP simulation, but also changing community sociopolitical organization, territoriality and land-tenure systems, interaction (exchange), and subsistence emphases.

Changes through time in these tool-stone procurement patterns allow me to infer that local ancestral Pueblo peoples developed well-marked economic territories during the early Pueblo II period, which became even more pronounced by the early Pueblo III period. The ancestral Puebloans not only experienced relatively scarce and unpredictable resource productivity during these years—including very difficult climatic conditions and low resource productivity during the drought between A.D. 1130 and A.D. 1180 (Van West and Dean 2000:37)—but they also had to cope with continued population increase.

The energy expenditure model also shows that, except in the central McElmo-Yellow Jacket locality during the early Pueblo III period, the ancestral Puebloans expended less energy for lithic procurement in most of the Mesa Verde core region during the second population cycle, because they were increasingly constrained to their local territories. The grouping of early

Pueblo II with later Pueblo II and early Pueblo III is unanticipated because of the low population during the early Pueblo II period. Therefore my analyses may provide independent support for Cole's interpretation of higher-than-anticipated degrees of conflict during the early Pueblo II period.

The grouping of the high-population late Pueblo III with the low-population Basketmaker III and Pueblo I populations is likewise unanticipated. I suggest that this can be explained by either increased logistical mobility or greater exchange in these two periods, whereas Basketmaker III is a member of this regime because these early farmers ranged broadly to procure their lithic raw materials directly in landscape characterized by low population density and the absence of defended community territories. By contrast, Pueblo I and late Pueblo III were periods of maximum village aggregation. I suggest that the presence of large villages with public architecture increased movement between villages so that residents could attend ceremonies and participate in other activities, and that this resulted in greater movement of raw materials, probably primarily through exchange.

A final conclusion of this work is to point out the anomalous positive correlation of the relative abundance of Jmbc with distance from its source for some periods. I suggest that this is explained by the greater exchange value of this specific material. One advantage of the sort of research reported in this chapter is that it allows us to clearly identify such anomalies as deserving of more consideration.

THIRTEEN

Population Dynamics and Warfare in the Central Mesa Verde Region

Sarah M. Cole

THE VILLAGE SIMULATION (Chapter 4, this book) is a model of functional regularity. Agents never so much as argue with each other. To bash someone on the head, or steal a wife or some maize, is quite literally unthinkable.

Of course, there was a time when archaeologists harbored an almost equally pacific vision of prehispanic Pueblo societies. The last two decades of archaeological research in the Southwest, however, have been as unkind to that stereotype as they have been to the ideal of the ecological Indian (Krech 1999). But the causes of conflict in the ancient Southwest remain elusive. In this chapter I explore the power, and the limits, of a simple explanatory model of warfare as a lagged consequence of increases in population size. Such a model assumes that people will be unable or unwilling to intensify production to keep pace with population growth, and that therefore disputes over resources will arise. As we will see, this model has two virtues; in fact, though very simple, it has considerable explanatory power for the first 400 years of our area's occupation. Later in the sequence, where it fails, it helps us identify the probable sources of the different sort of violence in the later Pueblo II and Pueblo III periods.

For a long time anthropologists have been aware of an important relationship between an increase in population size and subsistence intensification (Boserup 1965), increasingly hierarchical sociopolitical structure (e.g., Johnson and Earle 2000; Morgan 1965), and higher rates of technological change (Henrich 2004). At the same time, the relationship between population size or density and warfare has been the subject of much rather contradictory analysis (e.g., Cowgill 1975; Kang 2000; Keeley 1996:117–121; Redmond 1994; Vayda 1974; Wright and Johnson 1975).

Recently, Peter Turchin and Andrey Korotayev (2006) have raised the possibility that these two variables have a dynamic, time-lagged interaction that was not typically taken into account by previous analyses. In this chapter I examine this possibility, using population size data from the Village Ecodynamics Project (VEP) (Chapter 2, this book) and data on sociopolitical instability (or warfare) for the central Mesa Verde region that I collected during thesis research (Cole 2007).

The time span covered here, from A.D. 600 through A.D. 1280, is subdivided into the 14 modeling periods established by the VEP, as discussed in Chapter 2. This finely divided but lengthy span of almost 700 years allows for as precise an examination of the relationship between population and interpersonal violence (or warfare) as we can currently hope to achieve with an archaeological record for a nonliterate society.

WARFARE IN THE AMERICAN SOUTHWEST

Archaeologists have used several proxies for precontact conflict in the American Southwest, including the presence of stockades, towers, site-enclosing walls, burned sites and structures, iconography, settlement aggregation, and defensive site location (Chenault et al. 2002; Ellis 1991; Haas and Creamer 1993; Holmes 1878; Jackson 1876a, 1876b; Kuckelman 2002; Lancaster and Pinkley 1954; LeBlanc 1999; Mackey and Green 1979; Morley and Kidder 1917; Wilcox and Haas 1994). Of course, most of these classes of evidence are subject to multiple interpretations and might be attributed to ritualistic behavior (Cameron 1990; Creel and Anyon 2003; Darling 1999; Lancaster and Pinkley 1954; Lightfoot 1994; Rohn 1971, 1989; Walker 1998) or other behaviors unrelated to conflict (Chenault et al. 2002; Fewkes 1916; Green 1962; Hibben 1948; Johnson 2003; Kenzle 1993; Rohn 1989; Walt 1985).

Violent trauma recorded on skeletal remains is a more direct index of interpersonal strife so long as other potential sources of osteological damage, such as accidents, can be reasonably controlled for. Not surprisingly, more attention has recently been placed on this line of evidence. Several cases have been identified in which prehistoric cultural modification and disarticulation of human remains in the central Mesa Verde region suggest anthropophagy or extreme violence (Baker 1990; Billman et al. 2000; Dice 1993; Fink 1989; Kuckelman 2002; Kuckelman et al. 2002; Lambert 1999; Lightfoot and Kuckelman 2001; Luebben and Nickens 1982; Malville 1989; Turner and Turner 1992, 1999; White 1991). Although a largely definitional debate remains about whether this evidence indicates warfare, nearly all researchers agree that these cases represent some sort of violent encounter between people.

I will follow R. Brian Ferguson (1984:5) in defining warfare as the "organized, purposeful group action, directed against another group that may or may not be organized for similar action, involving the actual or potential application of lethal force." This definition requires "no particular level of organizational complexity or use of specific military tactics" (Kuckelman 2002:234) and thus includes raiding, as well as conflict between groups that are not necessarily politically independent, while excluding interpersonal conflict such as homicide or domestic violence.

A study of warfare as part of a dynamic system, coupled with population, allows us to form a new perspective on the generally agreed-upon fact that the final depopulation of the VEP study area occurred *amid* conflict (Lipe 1995). In fact, warfare can *cause* depopulation. For example, warfare can reduce the range of options available to settled populations by limiting residential mobility, which in turn may result in local overexploitation of resources (Wilcox and Haas 1994) and constrain populations to distribute themselves in ways that may be suboptimal (see Chapter 10, this book), leading to negative effects on rates of birth and mortality (Turchin and Korotayev 2006:121). Finally, there is the phenomenon—unfortunately, still well known today—of refugees fleeing zones of conflict for safety.

Many researchers consider competition for scarce resources, resulting from increases in population size, to be among the primary causes of violence between groups (Billman et al. 2000; Carneiro 1972; Ember 1982; Ferguson 1984, 1990; Haas 1990; Harris 1977, 1984; Johnson and Earle 2000:25; Kantner 1999a;

Kuckelman 2002; LeBlanc 1999; Meggit 1977; Price 1984; Rappaport 1968; Shankman 1991; Stone and Downum 1999; Varien et al. 1996; Webster 1975). These researchers argue that in the absence of technological or social innovations to proportionally increase resource availability as population increases, resources such as arable land, water, or hunting territories become scarce and conflicts over resource access or control can develop between groups. Of course, resource scarcity can also be induced by environmental degradation. This is most likely to be significant, however "in the face of high population densities" (Haas 1990:183); otherwise mobility can buffer such scarcities.

Others suggest that at least in some cases there is no relationship between population increase and conflict between groups (Billman 1997; Billman et al. 2000; Chagnon 1983, 1988; Cowgill 1975; Ember and Ember 1992; Helms 1994; Johnson and Earle 2000:20; Kang 2000; Keeley 1996:117–121; Lekson 2002; Maschner 1997; Patton 2000; Redmond 1994; Vayda 1974; Wright and Johnson 1975). These authors suggest various alternative explanations for conflict:

1. As a corrective response to environmental stress without population increase, because of "sudden and unpredicted climatic change over a short period, including drought, heavy rain, typhoon, frost, unusual temperature fluctuation, volcanic eruption, earthquake, and snow" (Kang 2000:878)
2. As a response to underpopulation when the groups are so small that they are subject to fluctuations in size, sex ratio, and age distributions, and they compensate for these fluctuations by taking captives belonging to appropriate age and sex categories (Oberg 1955)
3. Competition over increasing wealth
4. Striving for status
5. Revenge
6. As a result of socialization for fear or intimidation
7. In conjunction with competition for mates

A DYNAMIC APPROACH TO POPULATION AND WARFARE

One reason most researchers have downplayed a link between population size and warfare may be that tabulations across societies of concurrent measures for each often fail to identify a positive correlation. For example, Lawrence Keeley (1996:117–121) compared two cross-cultural samples of societies (from Murdock and Wilson [1972] and Ross [1983]) and concluded that there was no significant positive correlation between the frequency of warfare and the density of human population. He argued that groups with densities of less than one person per square mile are just as likely to engage in warfare as groups with densities 100 or more times higher. Although Keeley acknowledges that using population density by itself—without being able to take into account other factors, such as carrying capacity—is too simplistic, he argues that population density is more strongly correlated with sociocultural evolution than with warfare. He concludes that if there is a relationship between population size and warfare, it is either very complex, very weak, or both.

But perhaps this is the wrong approach. If population size and warfare are *dynamic* variables, *dynamically* linked, they both change with time and affect each other only with a lag. "Population and warfare are two aspects of a nonlinear dynamical system, in which population growth leads to increased warfare, but increased warfare in turn causes population numbers to decline" (Turchin and Korotayev 2006:113–114), though with a time lag. If these the two dynamic variables cycle with the same period but are phase-shifted (or lagged) with respect to one another (Turchin 2003), then cross-sectional correlation analysis will not be fruitful.[1]

1. Some previous archaeological research has acknowledged the possibility that population's effects on contemporary practices may be lagged. Kohler et al. (2004:299–302) found that a multiple regression model that included two predictors—population lagged one period, and a

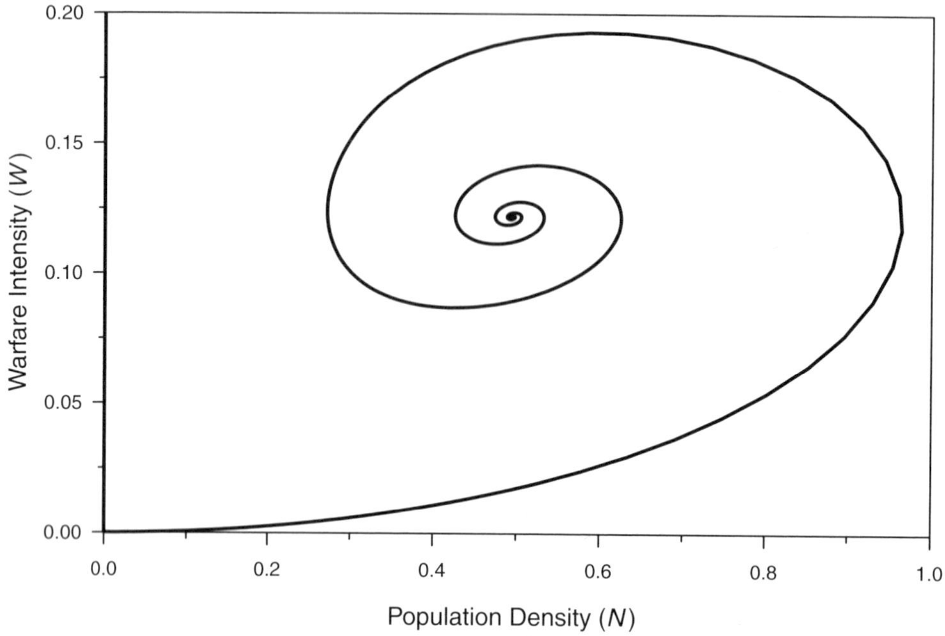

FIGURE 13.1 Relationship of population density (*N*) and warfare intensity (*W*); time begins at the periphery of the spiral and moves towards the equilibrium point at the center. Parameter values: $r = .13$, $a = .02$, $b = .04$, $c = 1$, $K = 8$; initial conditions $N = 5$, $W = .3$.

Figure 13.1 graphically depicts the relationship suggested by the Turchin-Korotayev (T-K) model between population (*N*) and warfare intensity (*W*), to be formalized in the paragraphs that follow. In the absence of changes in carrying capacity (*K*), *W*, and *N* eventually reach a stable equilibrium.

This type of nonlinear dynamic approach is widely used in the ecological literature, for example, to model predator-prey interactions (Lotka 1925; Volterra 1926). Turchin and Korotayev (2006) do not suggest that the relationship between warfare and population size in human societies is the same as in the predator-prey model; they do propose that the basic concept of a nonlinear, lagged dynamic approach could be useful.

One way to understand the relationships they propose is to focus on the rates of change of the structural variables (*N* and *W*) rather than the structural variables themselves. The rate of population change is negatively affected by the intensity of warfare, and the rate of change in warfare is positively affected by population density. In such a system, the peaks of warfare intensity will follow population peaks. In other words, they will be phase-shifted (or lagged) with respect to one another (Figure 13.2).

Turchin and Korotayev (2006) focus on internal warfare—within or between culturally similar groups—because this type of conflict in small-scale societies will be most directly affected by population dynamics. External warfare is an exogenous phenomenon driven by factors outside of their modeling framework (Turchin and Korotayev 2006:115).

current estimate of agricultural production (the Palmer Drought Severity Index)—provided a good fit to the history of aggregation from A.D. 1150 through A.D. 1325 in Bandelier National Monument. This finding is all the more relevant here, since some scholars have considered the degree of aggregation to be a useful index for the degree of conflict.

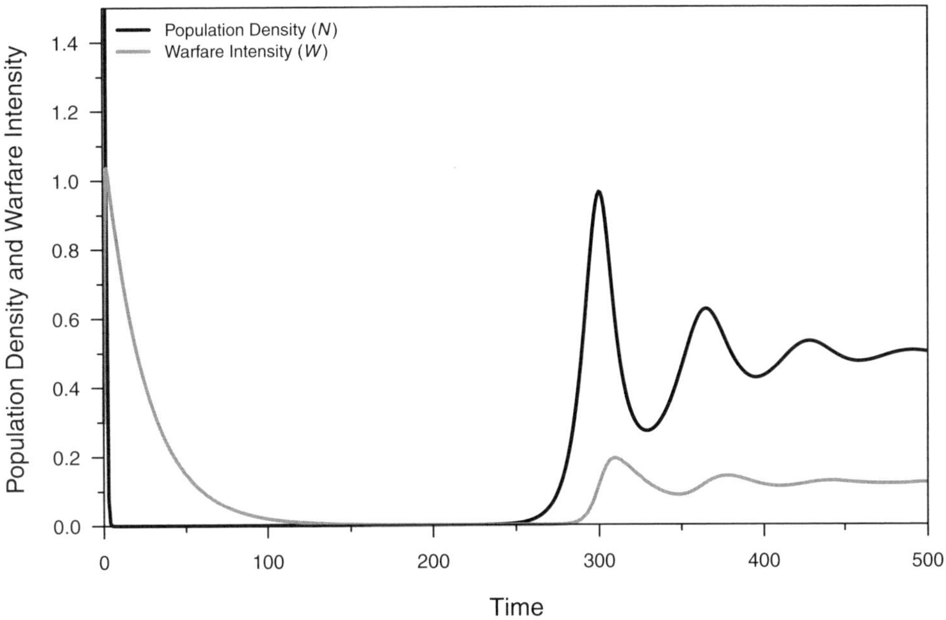

FIGURE 13.2 The relationship of population density (N, black line) and warfare intensity (W, gray line) through time. Parameters and initial conditions as in Figure 13.1.

The Turchin-Korotayev Model: Assumptions and Application

To construct the equation for the rate of change of population density (N), Turchin and Korotayev (2006) begin with the widely used logistic model of population growth

$$\frac{dn}{dt} = rN\left(1 - \frac{N}{K}\right), \tag{13.1}$$

where r is the rate of population growth when there is no scarcity (the Malthusian parameter), and K is the carrying capacity. The rate of population growth depends on population density and is restricted by carrying capacity. Therefore, as population increases the growth rate declines, resulting in an s-shaped (logistic) curve for population growth through time.

If we further assume that some deaths are due to warfare, and that the death rate due to warfare is directly proportional to warfare intensity, then equation 13.1 becomes

$$\frac{dN}{dt} = rN\left(1 - \frac{N}{K}\right) - cWN. \tag{13.2}$$

Timothy Kohler, Sarah Cole, and Stanca Ciupe (2009) added the constant c while fitting the model to empirical data; it is not part of the model Turchin and Korotayev published. This new term calculates the rate at which warfare leads to additional deaths in the population. Although the effects of resource acquisition and external factors such as climate change are known to have an effect on population size and growth rates (Boone 2002; Shennan 2002:118–119), the T-K model is the first that I am aware of to include warfare intensity as a factor in population fluctuation.

The instantaneous coefficient of population growth (r) in equations (1) and (2) is the growth rate of population (or population density, holding the area considered constant) when there is no scarcity. A minimum estimate for this value, in our case, can be calculated from the growth rates we estimate for our study area. The N of households in the VEP study area grew from 304 around A.D. 663 to 1030 (after smoothing) by A.D. 860 (this uses the middle-range population estimated for these periods,

TABLE 13.1

VEP Modeling Periods with Chronology, Population, and Carrying Capacity

Village Modeling Period	Begin (A.D.)	End (A.D.)	Midpoint (A.D.)	Total No. Momentary Households	Flux	Total No. Momentary Households (Natural Log)	Carrying Capacity Estimate (*K*)
–	600			0			—
6	600	725	663	304		5.717	1,580
7	725	800	763	326		5.787	1,580
8	800	840	820	836	180	6.729	1,580
9	840	880	860	1,030		6.937	1,580
10	880	920	900	370	–477	5.914	1,580
11	920	980	950	289		5.666	1,515
12	980	1020	1000	653	92	6.482	1,345
13	1020	1060	1040	671		6.509	1,430
14	1060	1100	1080	1,385	110	7.233	1,635
15	1100	1140	1120	1,940		7.570	3,234
16	1140	1180	1160	2,077		7.639	3,234
17	1180	1225	1203	2,326		7.752	3,234
18	1225	1260	1243	3,234		8.081	3,234
19	1260	1280	1270	1,770	–	7.479	1,770
–		1300		0			1,400

and the midpoints for these periods; see Chapter 2, this book), implying an annual growth rate (*r*) of .006 (see also Chapter 4). In fitting the T-K model to the data, values for *r* were used that ranged from .007 to .014 (see Kohler, Cole, and Ciupe 2009). Paul Ehrlich and Anne Ehrlich (1970) argue that the maximum attainable *r* for human populations may be as high as .02. Fluxes (instantaneous additions of population) were also added to the model where growth was identified "in excess of that achievable through in-place processes" (Kohler, Cole, and Ciupe 2009:285) (see Table 13.1). The population densities (*N*) shown in Table 13.1 are also based on data compiled by the VEP using the methods discussed in Chapter 2 of this book and illustrated in Table 2.2 as the total number of momentary households for the VEP study area.

Estimating carrying capacity for human populations is difficult and complex, and attempts to do so cause contention. Carrying capacity can be affected by subsistence intensification, which might include working longer or developing new techniques (Graber 1997:264), but beyond that it can be affected by an almost endless list of qualifications. Taken together, these imply that carrying capacity is not necessarily a fixed value (Sanders 1997:383). Humans use space in a flexible manner and switch patches to maximize return if they are not limited by ownership considerations or distance. Farmers' tendency to be territorial (see Chapter 12) may cause late arrivers to be forced into patches that are less desirable. Humans must also satisfy their need for water and fuel while meeting their needs for subsistence, which can put some areas off limits for exploitation. If buffer zones are created between groups because of conflict, they also limit the amount of land available for exploitation. If two or more groups can be incorporated into a new larger group, then buffer zones are not needed between them and the lands becomes usable. This, in turn, can rapidly increase carrying capacity (LeBlanc 2006:446). Because of food taboos or food that is difficult to process or low in density, humans are also limited by their intake of calories and protein. Furthermore, it is difficult to estimate carrying capacity when interannual climatic variability can have a profound effect on temperature and precipitation. And last, humans can degrade the environment, and—depending on the circumstances—this can have either temporary or permanent impact (Rappaport 1968:88–90). Since environmental degradation can be considered a continuous process, there is no threshold at which degradation begins (Dewar 1984), further complicating estimation of carrying capacity. Given all these factors, it is quite understandable why some researchers have suggested that carrying capacity "cannot be calculated reliably for any archaeological sequence" (Dewar 1984:610).

For the purposes of this analysis, carrying capacity estimates (see Table 13.1) are taken from the agent-based simulations (see Kohler, Cole, and Ciupe 2009 for more discussion). Although the carrying capacity values used in this analysis are admittedly rough estimates (and in fact are based on an earlier version of the simulation than the one discussed in this book), agent-based models have great potential to provide methods for estimation that overcome at least some of the criticisms usually mounted against the concept. In any case, these estimates of carrying capacity affect only the fitting exercise reported in Kohler, Cole, and Ciupe (2009) and otherwise do not enter into the analysis.

The state equation for intensity of warfare (W) assumes that the hostile encounter rate between individuals belonging to different groups (a) will be directly "proportional to the product of population density and warfare intensity" (Turchin and Korotayev 2006:118).[2] The equation also assumes that there is a gradual decline in warfare intensity, at an exponential rate b, which reflects the probability that groups are willing to forgive and forget past hostilities (if no hostilities have recently occurred). This decline is proportional to W, because "high warfare intensity causes greater degree of warfare fatigue, and therefore, great willingness to de-escalate conflict" (Turchin and Korotayev 2006:118). Together, these assumptions result in the following expression for the rate of change of W:

$$\frac{dW}{dt} = aWN - bW. \tag{13.3}$$

Calculating the Warfare Index

From 88 sites located within the central Mesa Verde region, 621 individuals were selected for analysis. Half of these are within the spatial boundaries of the VEP (Figure 13.3), and the rest are nearby. Although the central Mesa Verde region is larger than the VEP area, I assume that the cultural processes and population fluctuations were much alike—an assumption supported by the similarity of material remains throughout the region. The use of this larger sample creates a more stable analysis, which in my view takes precedence over the possibility that we encompass subareas with slightly different patterns of violence.

The human remains selected for analysis did not solely include formal burials: some were determined during excavation to be in a wide variety of natural and cultural depositional contexts. Not all human remains encountered during excavation are included in this analysis, however. I included only individuals with enough skeletal elements to determine the presence or absence of trauma.

In turn, only certain types of trauma point to the sort of violence likely in intergroup conflict.

2. Turchin and Korotayev (2006) consider two forms for the state equation for W. Initial testing of both for goodness-of-fit by Kohler, Cole, and Ciupe (2009) caused us to settle on the form presented in equation 13.3.

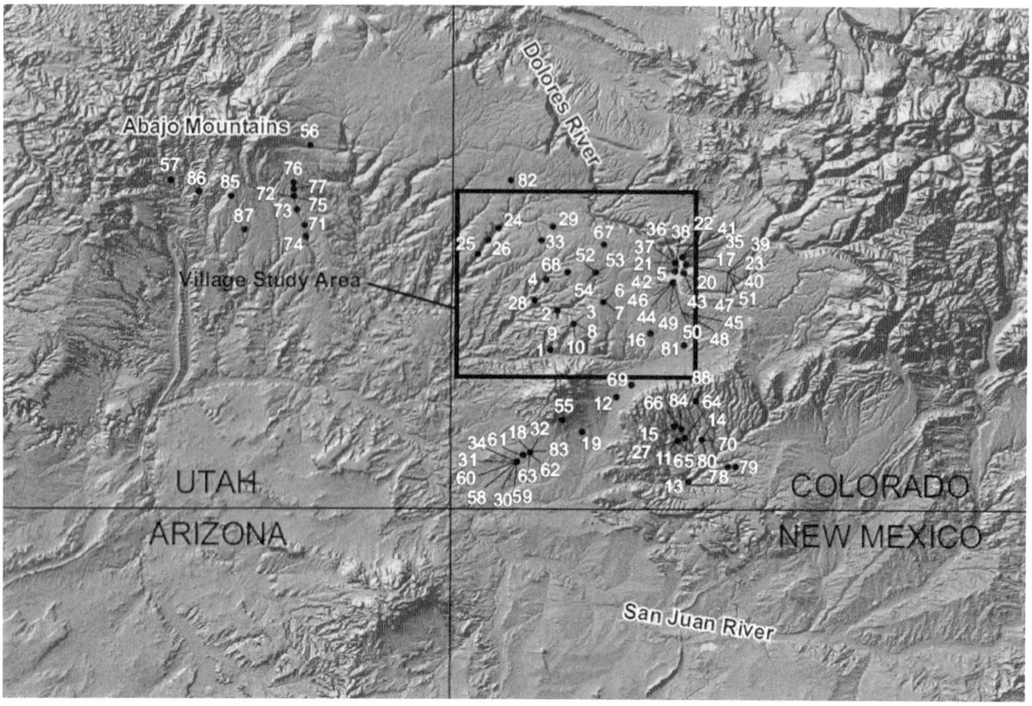

FIGURE 13.3 Map of the sites selected for analysis in this study relative to the approximate VEP boundary. Numbered sites are identified in Table 13.3.

I recorded fractures to the radius, ulna, or both since such trauma most likely resulted from a blow to the arm raised in defense (Angel 1974; Lahren and Berryman 1984; Wood-Jones 1910). The eight adults (six females) in this analysis with fractures to the radius, ulna, or both did not exhibit evidence of cranial trauma (see Cole 2007:Appendix). On females, fractures to the radius, ulna, or both without evidence of cranial trauma are consistent with intergroup rather than interpersonal (female-directed) violence (Martin et al. 1993; Martin and Akins 1994; Shermis 1982 / 1984; Wilkinson and Van Wagenen 1993).

Certain cranial fractures were recorded as well. Cranial trauma from accidental falls is more common among the children or the elderly (Hussain et al. 1994). Similarly, facial fractures to the nasal bone or orbital region are likely to occur during spontaneous interpersonal violence and may not indicate intergroup conflict (Walker 1997). Thus, such fractures are assumed to be from accidental falls or interpersonal violence and are not included in the computation of *W*.

Cultural modification and disarticulation (as might result from anthropophagy) were recorded as warfare and contribute to *W*. These include instances of what Kristin Kuckelman and colleagues (2000) define as "extreme processing," in which bodies were intentionally dismembered and bones were broken into small fragments.

Although some fragmentary remains were the result of human modification and disarticulation, this was not always the case. With such instances, I needed enough skeletal elements to determine the presence or absence of appropriate trauma. At a minimum, the skeletal elements must include the presence of a cranium, radius, or ulna, which must be complete enough to determine the presence or absence of trauma. Table 13.2 displays the types of evidence recorded on skeletal remains that I considered as indicating warfare. I also used additional information to alter my coding in cases where overwhelming evidence

TABLE 13.2

Skeletal Trauma Inferred as Evidence of Warfare

Type of Trauma	Count[a]
Fractures of the radius	5
Fractures of the ulna	10
Nonfacial cranial trauma in subadults and nonelderly adults	49
Human modification and disarticulation in the form of extreme processing	136

a. Some of the human remains exhibited more than one piece of evidence of violent trauma. Therefore, the total number shown here will be more than the total number shown as *x* in Table 13.4.

from the investigator suggested that strict adherence to my rules was not sensible.

All the information on human remains data was obtained through library research in site reports and "gray" literature and not through direct examination of the remains. Although direct examination would have been desirable, since different analysts report different data and make different inferences from these data, it was not logistically feasible; for example, some of the remains have been reburied. Regardless, the literature consulted included detailed descriptions of the pathological conditions, the position (i.e., flexed, semiflexed, and so forth), and location (i.e., midden, kiva, pistructure, and so forth) of interment of each individual, and any interpretations made by the author. See Cole (2007) and Table 13.3 for more details on the data set used to calculate *W*.

The data set for warfare intensity is of course not a complete record of all individuals who lived and died from A.D. 600 to A.D. 1280 in the central Mesa Verde region—a situation common in other types of archaeological data sets as well (including the VEP population estimates discussed in Chapter 2 of this book). Empirical Bayesian methods can help provide a more meaningful estimate of *W* for the entire region, one based on the compiled population. The Bayesian method of parameter estimation used here follows techniques discussed by Ian Robertson (1999), who in turn was inspired by Miriam Chernoff (1982), George Cowgill (1968, 1974, 1994), and Cowgill et al. (1984). This method, developed with the assistance of Scott Ortman of the Crow Canyon Archaeological Center, is useful in dealing with samples that are small and variable in size, as it obviates the need to arbitrarily aggregate or eliminate small samples. Table 13.4 displays the *n* of skeletal remains for each period, the number of individuals with conflict-related trauma (*x*), and the resulting warfare intensity (*W*), slightly refined using Bayesian techniques. The final version of *W* in the farthest right-hand column of Table 13.4 is calculated by rounding the raw warfare index to four decimal places and converting negative values to zero.

EVALUATING THE T-K MODEL

This model can be evaluated in various ways. Following Turchin and Korotayev's approach (2006), the time series data for *N* and *W* are first smoothed using 40-year running means to obtain a continuous measurement. Figure 13.4 displays the time series data for *N* and *W* before and after smoothing.

These smoothed data series can also be superimposed, after standardizing both variables, to simultaneously display the trajectories of *N* and *W* through time (Figure 13.5). Figure 13.2 displayed the dynamic relationship predicted by the T-K model. Thus, population growth should lead to increased warfare, and increased warfare in turn should cause population numbers to decline. Thus, the response of each variable to changes in the other should be lagged, although the period of lag is unknown.

Excluding the first period with its slightly higher than expected value for *W*, the first cycle

TABLE 13.3

Data Set of Human Remains Used to Calculate Warfare Index (W)

No.[a]	Site Number	Site Name	Period	Total No. Human Remains	Total No. Contributing to Warfare Index (*W*)	References
1	5MT1825	Castle Rock	19	41	41	CCAC 2001, 2004; Kuckelman et al. 2002
2	5MT3	N / A	13 15	18 16	15 3	Malville 1989; Swedlund 1969; Turner and Turner 1999; Yunker 2001
3	5MT765	Sand Canyon	19	33	8	CCAC 2003, 2004; Kuckelman et al. 2002
4	5MT11842	Woods Canyon	18	11	0	Churchill 2002;
5	5MT2235	Marshview Hamlet	15	6	6	Gross and Kane 1983; Stodder 1987; Turner and Turner 1999
6	5MT11338	G and G Hamlet	14	1	0	Varien, ed. 1999
7	5MT3930	Roy's Ruin	17	0	0	Varien, ed. 1999
8	5MT3951	Troy's Tower	17	1	1	Varien, ed. 1999
9	5MT10246	Lester's Site	19	1	0	Varien, ed. 1999
10	5MT10459	Lookout House	19	1	0	Varien, ed. 1999
11	5MV1200	Long House	19	39	1	Cattanach 1980
12	Unknown	Grinnell	16	7	7	Luebben 1983; Luebben and Nickens 1982; Turner and Turner 1999
13	5MTUMR2346	Mancos	15	29	29	Turner and Turner 1999; White 1991
14	5MV499	N / A	15	10	1	Lister 1964; Turner and Turner 1999
15	5MV1229	Mug House	18	46	0	Miles 1975; Rohn 1971
16	5MT3868	Duckfoot	9	14	2	Lightfoot and Etzkorn 1992
17	5MT23	Grass Mesa Village	9 10	1 2	0 0	Lightfoot et al. 1988; Stodder 1987
18	5MT10207	N / A	16	14	13	Dice 1993; Errickson 1993; Turner and Turner 1999
19	5MT7723	N / A	14	4	1	Dice 1993; Errickson 1993; Turner and Turner 1999

(continued)

TABLE 13.3 *(continued)*

No.[a]	Site Number	Site Name	Period	Total No. Human Remains	Total No. Contributing to Warfare Index (*W*)	References
20	5MT2336	Kin Tl'iish	10	2	0	Kane et al. 1986; Stodder 1987
21	5MT4545	Tres Bobos Hamlet	7	1	0	Kane and Gross 1986; Stodder 1987
22	5MT2858	Apricot Hamlet	7	1	0	Kane and Gross 1986
23	5MT2182	Rio Vista Village	9	2	1	Stodder 1987; Kane and Robinson 1986
24	5MT2519	Herren House	14	3	0	Morris 1991
25	5MT2525	Knobby Knee Stockade	17	2	0	Morris 1991
26	5MT2544	Roundtree Pueblo	14 17	1 1	0 0	Morris 1991
27	5MV1544	N / A	6	1	0	Birkedal 1976
28	5MT2433	Aulston Pueblo	13	1	1	Morris 1986
29	5MT8827	Dobbins Stockade	13	1	0	Kuckelman and Morris 1988
30	5MT8651	N / A	18	1	0	Billman 2003; Lambert 1999
31	5MT9942	N / A	15	3	0	Billman 2003; Lambert 1999
32	5MT10010	N / A	16	11	8	Billman 2003; Lambert 1999
33	Unknown	Ackmen 1-1937	11	1	0	Martin 1938
34	5MT9943	12	15 18	7 5	3 5	Billman 2003; Lambert 1999
35	5MT2181	Hamlet de la Olla	8	1	0	Kohler et al. 1985; Stodder 1987
36	5MT2192	Pheasant View Hamlet	9	2	0	Stodder 1987; Yarnell 1982
37	5MT2236	Horsefly Hamlet	7	1	0	Stodder 1987; Kane and Chenault 1982
38	5MT2378	Poco Tiempo	7	2	0	Kohler et al. 1985; Stodder 1987
39	5MT2848	Rusty Ridge Hamlet	7	1	0	Kane and Gross 1986; Stodder 1987
40	5MT2853	Deer Hunter Hamlet	8	1	0	Kane and Gross 1986; Stodder 1987

(continued)

TABLE 13.3 *(continued)*

No.[a]	Site Number	Site Name	Period	Total No. Human Remains	Total No. Contributing to Warfare Index (*W*)	References
41	5MT4671	Periman Hamlet	8	2	0	Stodder 1987; Yarnell 1983
42	5MT5108	Golondrinas Oriental	10	3	2	Stodder 1987; Kuckelman 1984a
43	5MT2320	House Creek Village	8 9	1 2	0 0	Kane and Robinson 1986; Stodder 1987
44	5MT4475	McPhee Village	10 11	4 2	0 0	Stodder 1987; Kane and Robinson 1988
45	5MT4477	Masa Negra	10	1	0	Stodder 1987; Kuckelman 1984b
46	5MT4480	Rabbitbrush Pueblo	9	1	0	Stodder 1987; Kuckelman and Harriman 1984
47	5MT4684	Chindi Hamlet	6 7	2 3	0 0	Stodder 1987; Tucker 1983
48	5MT4725	Tres Chapulines Pueblo	9	3	0	Stodder 1987; Chenault 1983
49	5MT5106	Weasel Pueblo	9	4	0	Stodder 1987; Morris 1984
50	5MT5107	Pueblo de las Golondrinas	9	8	0	Stodder 1987; Kane and Robinson 1988
51	5MT5985	Standing Pipe Hamlet	9	1	0	Stodder 1987; Nelson 1985
52	5MT9168	Rabbit Site	14	1	0	Chenault et al. 2002
53	5MT9343	Dancing Man Hamlet	6	2	1	Chenault et al. 2002; Underwood 2004
54	5MT7522	Humming-bird Pueblo	9 13	5 3	0 0	Chenault et al. 2002
55	5MT11884	Conejos Camp	14	2	0	Chenault et al. 2002
56	42SA3724	N / A	12	1	1	Fetterman and Honeycutt 1990; Fink 1989
57	42SA12209	Cottonwood Wash	10	4	4	Fetterman et al. 1988
58	5MT8651	N / A	14 15	1 1	1 0	Billman 2003; Lambert 1999
59	5MT9541	N / A	18	1	1	Billman 2003; Lambert 1999

(continued)

TABLE 13.3 *(continued)*

No.[a]	Site Number	Site Name	Period	Total No. Human Remains	Total No. Contributing to Warfare Index (*W*)	References
60	5MT9924	N / A	15	11	1	Billman 2003; Lambert 1999
61	5MT9933	N / A	18	1	0	Billman 2003; Lambert 1999
62	5MT7704	N / A	16	1	0	Dice 1993; Errickson 1993
63	5MT8943	N / A	14	7	2	Dice 1993; Errickson 1993
64	5MV1676	N / A	9	7	0	Hayes and Lancaster 1975
65	5MV1452	Badger House	10	2	0	Hayes and Lancaster 1975
			11	10	0	
			12	2	0	
			13	9	0	
			17	2	0	
			19	1	0	
66	5MV1595	Big Juniper House	15	20	0	Swannack 1969
			6	6	1	
67	5MT1	Porter-Stevenson	12	4	4	Malville 1989; Swedlund 1969; Turner and Turner 1999; Yunker 2001
			14	5	0	
			15	1	0	
			17	3	0	
68	5MT5	Yellow Jacket	19	5	1	Kuckelman 2003
69	5MT10967	N / A	15	1	1	Nickens and Mabry 2000
70	5MV866	N / A	13	13	0	Lister 1966
71	Unknown	Alkali Ridge 5	13	12	1	Brew 1946
72	Unknown	Alkali Ridge 13	7	1	1	Brew 1946; Turner and Turner 1999
			13	1	0	
			15	3	0	
			17	2	2	
73	42SA1	Alkali Ridge	17	1	0	Brew 1946
74	Unknown	Alkali Ridge 3	13	4	0	Brew 1946
75	Unknown	Alkali Ridge 7	13	1	0	Brew 1946
76	Unknown	Alkali Ridge 8	13	2	0	Brew 1946

(continued)

TABLE 13.3 *(continued)*

No.[a]	Site Number	Site Name	Period	Total No. Human Remains	Total No. Contributing to Warfare Index (*W*)	References
77	Unknown	Alkali Ridge 12	13	3	0	Brew 1946
78	5MT2831	N / A	8	1	0	Reed et al. 1985
79	5MT2826	N / A	9	5	0	Reed et al. 1985
80	5MV875	N / A	13	1	0	Lister 1965
81	5MT6970	Wallace Ruin	18	10	0	Bradley 1988
82	5DL975	N / A	17	1	1	France 1988; O'Neil 1984
83	5MT10206	N / A	16	2	2	Dice 1993; Turner and Turner 1999
84	5MV1645	Two Raven House	12	13	0	Hayes 1998; Miles 1975
85	42SA18434	Rattlesnake Ruin	14	20	20	Baker 1990
86	42SA6396	N / A	13	1	0	Davis 1985
87	42SA7660	Happy Salamander	17	1	0	Firor et al. 1998
88	5MV34	Soda Canyon Pueblo	17	19	0	O'Bryan 1950

CCAC = Crow Canyon Archaeological Center.

a. As referenced in Figure 13.3.

in Figure 13.5 shows *N* peaking before *W*, then *N* decreases as *W* peaks. As *W* decreases, *N* begins to increase once again. This cycle is as the model predicts. The cycle from about A.D. 1000 to A.D. 1200, however, does not follow the model, since *W* increases *before* a corresponding increase in *N*.

Another way to evaluate the model is to graph the trajectory of *N* and *W* in their phase space. Figure 13.1 displayed the trajectory of these two variables in the phase space as predicted by the T-K model. Thus, an increase in *N* should lead to an increase in *W*, which eventually results in a decreasing *N*; this results in a counterclockwise trajectory through the phase space. As is apparent in Figure 13.6, the A.D. 600s through the middle of the A.D. 900s and the A.D. 1200s do behave as the model predicts (although in different regions of the phase space), but the three period midpoints between A.D. 1080 and A.D. 1160 strongly contrast with the model's predictions.

A closer look at the A.D. 600s through the A.D. 900s indicates no linear nonlagged relationship between *N* and *W* (Figure 13.7). This result is suggested by the model, since a growth in *N* leads to increased *W* and increased *W* in turn causes *N* to decline (Figure 13.8); dynamic systems theory gives us no reasons to expect a clear correlation between these two variables when they are not lagged (Turchin and Korotayev 2006). In fact, one might see a weak positive, a weak negative, or even no correlation between the two variables. The deficiencies of such an approach should by now be obvious.

TABLE 13.4

Final Data Used in Calculating the Warfare Index (W)

VEP Modeling Period	Midpoint	*x*	*n*	*x* / *n*	Raw Warfare Index[a]	*W* (Final Version)
6	663	2	11	.1818	.10375	.1038
7	763	1	10	.1000	–.01622	.0000
8	820	0	6	.0000	–.32309	.0000
9	860	3	55	.0545	.03581	.0358
10	900	6	18	.3333	.31537	.3154
11	950	0	13	.0000	–.10708	.0000
12	1000	5	20	.2500	.22202	.2220
13	1040	17	70	.2429	.23535	.2353
14	1080	24	45	.5333	.53870	.5387
15	1120	44	108	.4074	.40655	.4065
16	1160	30	35	.8571	.88939	.8894
17	1203	4	34	.1176	.09148	.0915
18	1243	6	75	.0800	.06732	.0673
19	1270	51	121	.4215	.42102	.4210

a. See Cole (2007:85–89) for a detailed discussion of the Bayesian techniques used to calculate *W*.

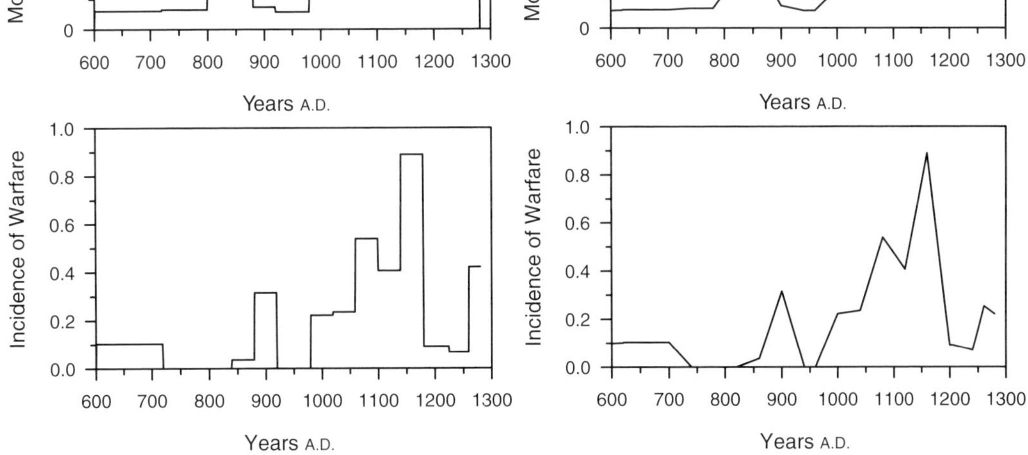

FIGURE 13.4 Time series for population size and and warfare index. Top: unsmoothed (left) and smoothed (right) versions of population (*N*) by number of households. Bottom: unsmoothed (left) and smoothed (right) versions of warfare intensity (*W*). *W* is the proportion of individuals in each period with evidence of warfare-related injury, slightly refined using Bayesian techniques (see Cole [2007]).

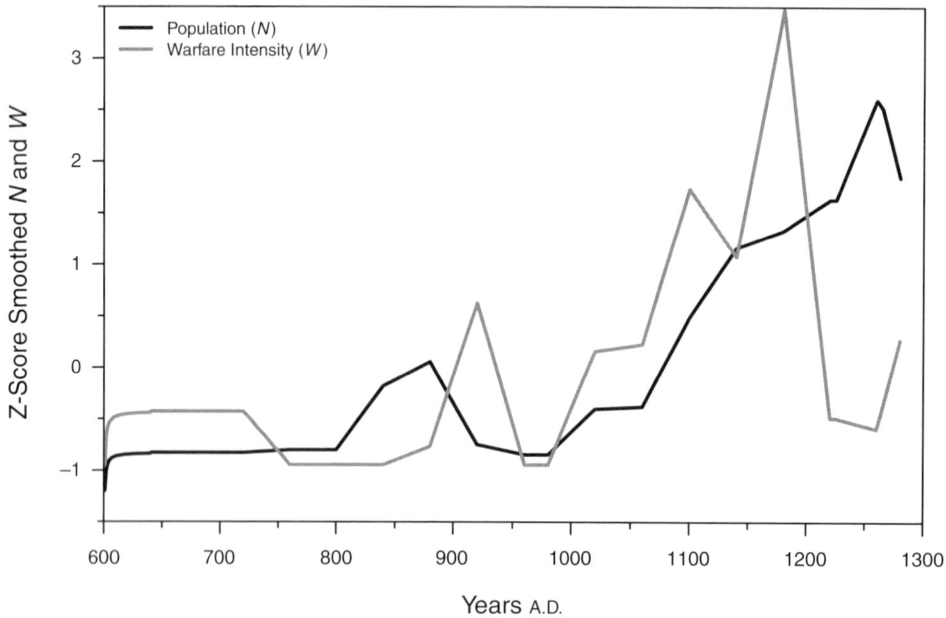

FIGURE 13.5 Dynamics of smoothed population (*N*) (black line) and smoothed warfare intensity (*W*) (gray line) converted to z-scores.

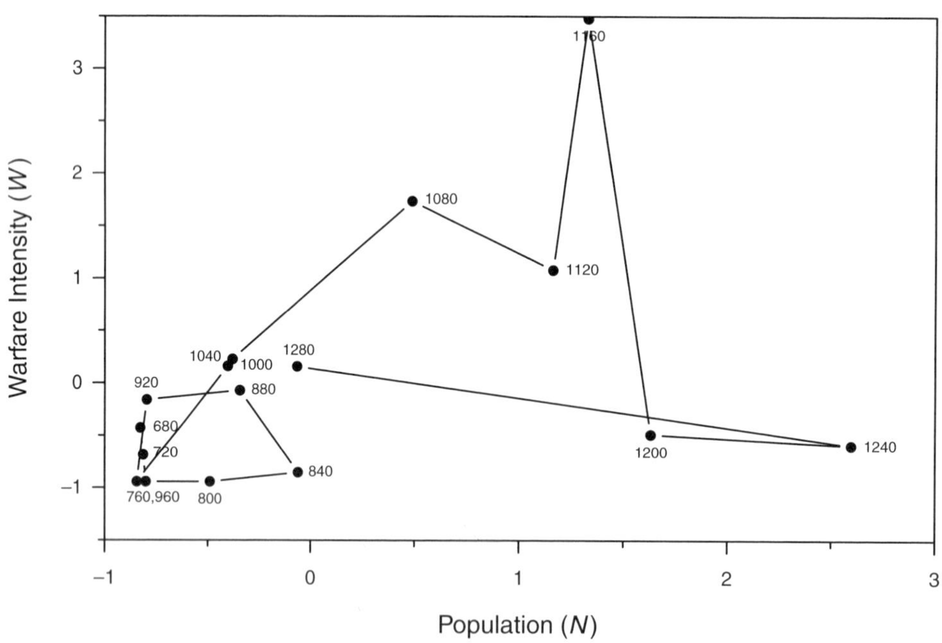

FIGURE 13.6 The population-warfare trajectory in the phase space (smoothed and converted to z-scores).

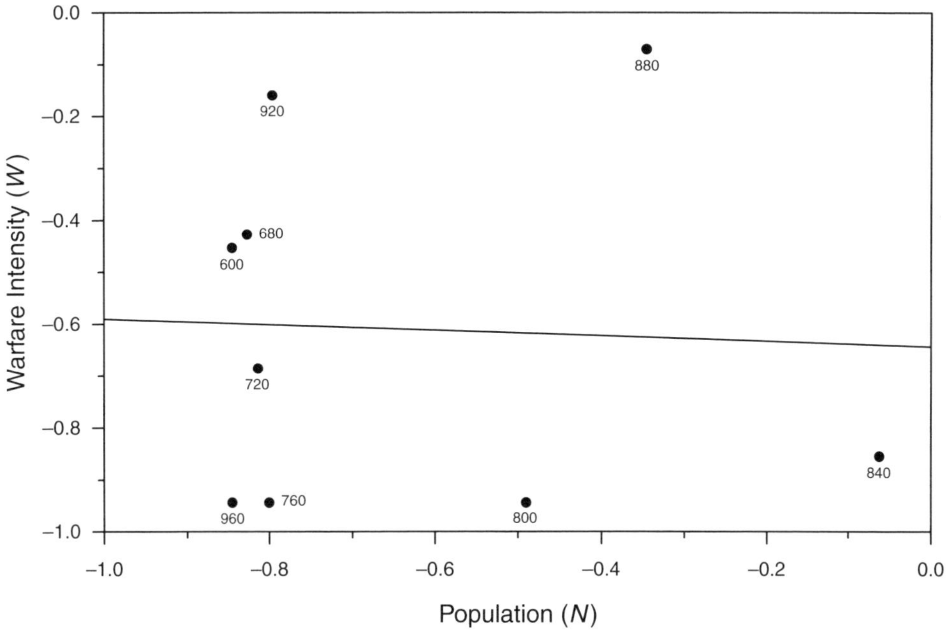

FIGURE 13.7 Linear regression of warfare intensity (W) on population (N) for the A.D. 600s through the 900s (smoothed and converted to z-scores; $r^2 = 0.002$, $p = 0.91$.

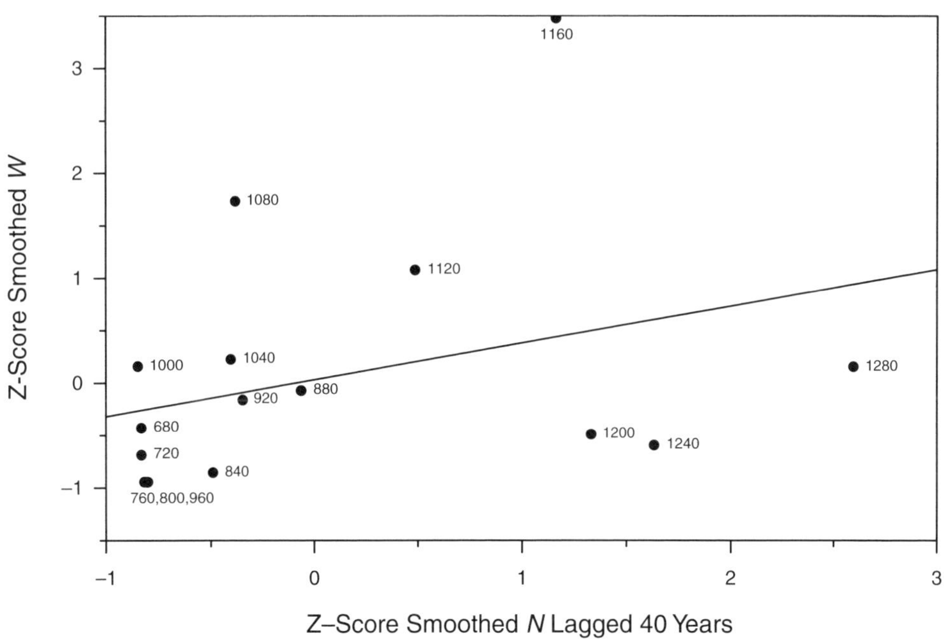

FIGURE 13.8 Linear regression of Warfare Intensity (W[t]) against Population (N[t–1]) (smoothed and converted to z-scores); $r^2 = 0.102$, $p = 0.2278$.

A better approach to evaluating the model is to plot *W* against lagged values of *N*. If the lag is about 40 years in length, then a strong positive relationship should be observed between *N*—lagged 40 years—and *W*. The results in Figure 13.8 suggest that there is a positive, but not significant, relationship between the two variables when all periods are included. The three data points representing the periods from A.D. 1080 through A.D. 1160 are strong positive outliers and appear to represent a consistent pattern, but a pattern very different from that between the A.D. 600s and the early A.D. 1000s. The final three periods are negative outliers.

Finally, the most rigorous method for evaluating the T-K model is to fit the model's values for *a*, *b*, and *c*, using the empirically derived values for *N* and *W*, and then assessing the goodness-of-fit. This analysis was conducted by Ciupe, then of the Santa Fe Institute, as reported in Kohler, Cole, and Ciupe (2009) and Cole (2007); here I merely relay its conclusions.

We found that although the model fits reasonably well in both the beginning and end of this sequence, the somewhat high *N* and very high *W* in the A.D. 1000s and A.D. 1100s and the very high *N* and moderate or low *W* in the A.D. 1200s cannot both be accommodated by the model. In short, three different regimes for the relationship between *N* and *W* are apparent: one from the beginning of the sequence until the early A.D. 1000s; another for the middle A.D. 1000s through the middle or late A.D. 1100s; and a third for the final 100 years or so of the occupation.

In summary, the T-K model fits well during the first population cycle, from A.D. 600 through the early A.D. 1000s, when exogenous factors appear to be weak enough not to influence the dynamic relationship between *N* and *W*. The lack of a significant relationship, overall, may be due to the high degree of endogenicity required by the model. This is difficult to obtain in a nonclosed, prestate system.

The first portions of the second population cycle, especially the two centuries from about A.D. 1000 to A.D. 1200, do not fit the model's predictions. Increases in *W* actually *precede* increases in *N*, directly in contrast to what the model predicts. I propose that this is due to exogenous factors that had a significant impact in the study area (see the discussion that follows). Finally, in the A.D. 1200s the predicted relationship reappears; however, values for *W* are much lower than one would expect on the basis of the relationship between *W* and *N* in the first population cycle.

It is important that the T-K model is supported during the first population cycle from A.D. 600 through the A.D. 900s, and during the A.D. 1200s, but it is equally important that the second cycle does not support the model. Both the link between *N* and *W* in the first cycle (and in the A.D. 1200s) and its disconnection in the second cycle require further explanation.

DISCUSSION

The portion of the Basketmaker III period (A.D. 600–725) included in this analysis suggests a slightly higher than expected level of *W*. Kuckelman et al. (2000) have stated that there are no known instances of violent deaths that date to this period, but their analysis focused on incidents of human modification and disarticulation and did not examine the evidence for conflict used here. The independent evidence of partially or fully stockaded settlements in this period (Chenault 2002; Desruisseaux et al. 2004; Fuller and Morris 1991; Phillips and Chenault 2004; Rohn 1974, 1975) supports the inference that violence was present and higher than expected during this period (see also Chapter 1). Chenault et al. (2002) argue that defense against human aggression was the primary purpose for the construction of stockades, with protection from the elements, privacy, and definition of space being secondary functions. Furthermore, architectural features at numerous Basketmaker III sites had burned at or near the time of abandonment. Chenault et al. (2002) have concluded that there appears to be a correlation between Basketmaker III sites that have burned and the presence of stockades. Thus, it is possible that the slightly higher than expected value for *W* calculated here may be an accurate reflection of conflict during this period, which in turn perhaps reflects an earlier cycle of population growth—presumably outside our area—terminating as the immigration into our area began.

The first cycle of *N* and *W* includes the entire Pueblo I period and the first part of the Pueblo II period. The relationships between *N* and *W* in this cycle are as predicted by the T-K model. *N* peaks before *W*, then decreases as *W* peaks. As *W* decreases, *N* begins to increase again. This cycle falls within a period recognized as having "rapid demographic and organizational change" (Wilshusen 1999:200). It has also been suggested that population changes during this time were outside the reasonable mortality and natural increase rates for societies of similar size and social organization (Schlanger 1988; Wilshusen and Ortman 1999). Migration in as well as out of the region has been considered important in understanding these population changes (Chenault et al. 2002 Schlanger and Wilshusen 1993; Wilshusen 1999:234; Wilshusen and Ortman 1999).

Explaining why aggregation occurred at this time (in the famous Dolores Pueblo I villages) has been difficult. Some researchers interpret the need to aggregate as a defensive (and offensive [Kuckelman 2002]) measure (Haas 1982, 1989; Hunter-Anderson 1979; Kohler 1989; LeBlanc 1999:283; Rice 2001; Solometo 2004, 2006; Wilcox and Haas 1994), but the results of the T-K model for the first population cycle in this region suggest otherwise. According to the data presented here—and as predicted by the model—values for *W* were low when aggregated sites first appeared in the middle to late A.D. 700s (as noted by Orcutt et al. [1990:211]). *W* then increased sharply beginning in the A.D. 800s, and local population began to decline by the late A.D. 800s. Wilshusen (1999:233) and Wilshusen and Ortman (1999) have suggested that the population decline at this time was a large-scale emigration to northwestern New Mexico that eventually contributed to the development of the nascent Chaco system.

What were the causes of this emigration? Some have argued that the abandonment of the aggregated villages was due to the series of drought years beginning in the early A.D. 880s (Schlanger and Wilshusen 1993; Varien et al. 1996; Wilshusen 1999:231). According to their analyses, the people of the central Mesa Verde region became locked into a system in which high-risk, unsustainable methods of farming—in combination with climatic fluctuations and population pressure—resulted in village abandonment and emigration. Others (e.g., Kohler and Matthews 1988) have suggested that deforestation, and possibly other human impact on the environment, may have contributed to this depopulation. Considering the fit of the T-K model, however, one might add that high *W* will also have a "negative effect on demographic rates (birth, mortality, and emigration)" (Turchin and Korotayev 2006:121). Most researchers concerned with population dynamics in the central Mesa Verde region have not fully considered the importance of the presence of warfare during this period's population decline (see LeBlanc 1999:145–146 for a notable exception—although his interpretation refers to rather ambiguous data in citing burned sites and structures and unburied bodies). It is likely that the *coincidence* of mounting human impact (see also Chapters 7 and 10 of this book), unfavorable climate (see also Chapter 3 herein), plus a desire to escape conflict all contributed to the depopulation of the region as the sociopolitical transformations at this time were unable to mitigate these adverse impacts.

In the second cycle, from about A.D. 1000 to A.D. 1200, relations between *N* and *W* do not proceed as predicted by the T-K model; instead, increases in *W* appear to lead to increases in *N*, reversing the direction in which their relationship through time traverses the phase space (Figure 13.6). To understand why the model fails during this time, we must take into account exogenous factors (outside of the T-K modeling framework)—specifically, the developing system in Chaco Canyon to the south and its expansion into the VEP study area. Although scholars (e.g., Kantner 1999a, 1999b; Kuckelman et al. 2000; Lekson 2002; Lipe 2006; Turner and Turner 1999) differ as to what sort of influence the Chacoan system had on the inhabitants of the central Mesa Verde region, most agree that various violent behaviors occurred during these two centuries. Whether such incidents should be classified as anthropophagy (Turner and Turner 1999) or we should follow Kuckelman et al.'s (2000) definition of extreme processing (Kantner 1999a; Lekson 2002) is still debatable.

Lipe (2006) and Kuckleman et al. (2000) have asserted that the time-space distribution of Chaco-style great houses in the central Mesa Verde region is the best indication of Chacoan influence. They further suggest that most of the skeletal evidence of human modification and disarticulation occurred outside the period of great house construction, with many falling well outside this period. Some of the best-documented and well-dated examples of extreme violence from the region fall within a period between A.D. 1130 and A.D. 1160, after great house construction in Chaco Canyon had stopped. By the A.D. 1140s, the Chacoan system—as originally constituted around Chaco Canyon, at least—appears to be falling apart. It is highly likely, as Lipe (2006) and Kuckelman et al. (2000) suggest, that such instances of violence were correlated with the collapse of the Chacoan system, possibly through score settling or because of crop failure or both during the long drought in the middle A.D. 1100s.[3]

Two elements that are missing from such theories, however, are (1) an explanation for the instances of extreme violence that predate great house construction in the central Mesa Verde region, and (2) an explanation for violent behavior that is not represented by human modification and disarticulation. Construction of great houses began in the central Mesa Verde region around A.D. 1080 (Lipe 2006:292). Some direct evidence of conflict, which includes not solely human modification and disarticulation, but also defensive wounds and cranial trauma, occurred before great house construction and also appears to be unrelated to population size in the way predicted by the T-K model.

One way to explain the anomaly of the second cycle is to suggest that these architectural indicators may not be the earliest evidence of Chacoan influence. Instead, "we [may] need to look for external influences before they become obvious as Chacoan-style architecture" (Kohler, Cole, and Ciupe 2009:289). Rising levels of violence and stockaded sites before A.D. 1080 may indicate resistance on the part of the local inhabitants to the first attempts at Chacoan expansion. Partly or completely enclosed stockaded sites in the mid-A.D. 1000s, predating the construction of great houses in this region, include Dobbins stockade (Kuckelman 1998), Dripping Springs stockade (Harriman and Morris 1991), the Ewing Site (Hill 1985), and Two Raven House (Hayes 1998). These sites (and possibly more that remain undetected) could represent such resistance, especially as they seem to accompany contemporaneous increases in violence, documented osteologically.

Stockaded sites were also present during the period of great house construction (Mustoe Site [Gould 1982], Yellow Jacket [Lange et al. 1986], Casa Bisecada [Morris 1988], and Roundtree Pueblo [Morris 1991]). Osteological evidence suggests that conflict was present during this time as well, and at a much greater frequency than before great house construction.

Regardless of when the Chacoan expansion began, the result of the T-K model for this cycle clearly indicates that the Chacoan influence,

3. It is unfortunate that one of our period boundaries—between periods 15 and 16—falls at A.D. 1140, which is probably when many or most of our extreme processing events occurred. Some of these events may get artificially separated into two different periods, as a result. We should also keep in mind that some of the sites assigned to particular periods have more secure dating than others. This is of most concern for sites with large assemblages, like Rattlesnake Ruin, assigned here to period 14, although period 15 is also a possibility (Scott Ortman, personal communication, January 20, 2011).

expansion, and resistance were profound enough to override the earlier dynamic relationship between population size and warfare. In fact, whatever happened was so significant that it changed the direction of the time trend in phase space from clockwise to counterclockwise—in effect, reversing history. One wonders whether another impact of warfare on the inhabitants of the VEP study area might have been to reduce their locational efficiency vis-à-vis resources (see Chapter 10, this book).

The relationship between *N* and *W* in the A.D. 1200s once again falls within the model's predictions. *N* peaks before *W*, then decreases as *W* increases. The shift to canyon-oriented aggregated settlements after about A.D. 1225 has been attributed to a rise in conflict (Kuckelman et al. 2000; Lambert 2002; LeBlanc 1999; Lipe 2006; Lipe and Varien 1999b; Wilcox and Haas 1994). The osteological index of conflict developed here, however, suggests that the inhabitants of the VEP study area moved to aggregated villages before the conflict intensified. As noted in Chapter 14 of this book, the shift to canyon-oriented settlements may have begun as early as A.D. 1140 to A.D 1180, when conflict was very high by the osteological index, so the relationship of these settlement changes to conflict remains a little unclear.

For the A.D 1200s, the direct evidence most often used to infer conflict consists of the villagewide massacres at Castle Rock and Sand Canyon Pueblos, which occurred near or at the time of general depopulation in the late part of that century. Direct evidence of conflict in the early to middle A.D. 1200s is rather limited, though Kuckelman et al. (2002) point out that many of the people who died in the late 1200s at Castle Rock and Sand Canyon had healed skeletal trauma, suggesting that conflict had been occurring for some time. Possibly the formation of aggregated and often walled settlements, was, for a time at least, successful in preventing the use of strategies that would have resulted in elevated levels of *W*. Perhaps, as Tainter and Tainter (1991) suggest, aggregated settlements may have initially been visible symbols of community strength designed to be impressive to those both inside and outside the community.

In any case, given the nature of the relationship between population size and warfare in the first population cycle, warfare as measured by this index is much lower for the entire thirteenth century than anticipated by the model. Kohler, Cole, and Ciupe (2009) suggest that this reflects the increasing spatial scale of sociopolitical organization in this period. If by this time the VEP study area contained a single confederacy of communities, this formation may have been better able to promote integration and conflict resolution than the multiple competing community-level organizations likely present during the first population cycle.

Other factors, however, may explain (at least in part) the apparent low level of *W* in the thirteenth century. First, few large sites that date to the second half of the thirteenth century have been systematically excavated and reported in detail (Lipe and Varien 1999b). Nordenskiold (1973[1893]) and other early researchers reported finding human remains that were not formally buried, but such descriptions were too vague to be included in the database developed here. The few sites that have been reported, such as Long House (Cattanach 1980) and Mug House (Rohn 1971), do not appear to have the patterns of violence reported for Sand Canyon and Castle Rock Pueblos, though both of these sites suffered from a great deal of removal of materials and disturbance before their excavation. If earlier excavations also had midden-testing components that were less significant, they might have missed some osteological evidence of violence, since many people were buried in such locations.

Secondly, Kohler and Kathryn Kramer Turner (2006) have suggested that inhabitants of the Totah region to the south raided settlements in the central Mesa Verde region and took women as captives in the A.D. 1200s. Since these "missing women" have no possibility of contributing to the index for *W* developed here, as they did in earlier periods (see Cole

[2007:Figures 7.1 and 7.2]), the lower value calculated for this index in the 1200s may underestimate the true scope of conflict. The creation of a buffer zone between the central Mesa Verde and Totah regions may also be indirect evidence that supports the inference of a hostile relationship between the two areas at this time (Lipe 2006; Lipe and Varien 1999b).

Finally, when calculating *W* for the T-K model, I interpreted the different sorts of evidence of violence equally: evidence for raiding, human modification and disarticulation, and village-level massacres were all coded in the same way. Although parry fractures, skull trauma, and extreme processing events all reflect intergroup violence, a single set of causal factors may not underlie all these different behaviors—yet this is what the T-K model suggests. Unfortunately, examining various types of violence separately would result in extremely small samples for each. Moreover, such diverse types of violence could affect groups differently at a psychological level as well, in turn affecting emigration differently, as well as differentially influencing birth and death rates. Village-level massacres (such as experienced by Castle Rock Pueblo) would be psychologically devastating to any communities having ties to Castle Rock (Keeley 1996:68). Therefore, the low levels of *W* in the A.D. 1200s, unanticipated by the model given the population sizes at this time, are likely due to a combination of the small number of large sites systematically excavated, especially in the first half of the 1200s, a small possible loss of women to raiders from the Totah region, and an underestimation of the psychological impact of village-level massacres. Despite these caveats, warfare was increasing rapidly in the late A.D. 1200s and emigration followed, and this linkage is as proposed by the model.

SUMMARY AND CONCLUSIONS

The Turchin-Korotayev model contributes to defining the dynamic relationship between population and violent conflict in the prehistoric central Mesa Verde region. Examination of model fit suggests new avenues for explaining regional depopulations and points to the need for explanations for initial population aggregation that do not depend—in this area, at least—on warfare. The three regimes for the relationship between population size and warfare noted here almost certainly reflect different types and scales of sociopolitical organization (see also Kohler and Varien 2010 and Chapters 12 and 14 of this book). Modeling warfare is not currently a part of the VEP simulations, but the results of the T-K model demonstrate that simulating the causes and effects of warfare on the VEP agents would add significantly to our understanding of prehistoric lifeways in the central Mesa Verde region and beyond.

FOURTEEN

Characterizing Community-Center (Village) Formation in the VEP Study Area, A.D. 600–1280

Donna M. Glowacki and Scott G. Ortman

LARGE VILLAGES AND ASSOCIATED civic architecture played instrumental roles in structuring social life throughout the ancestral Pueblo occupation of the northern San Juan region (Brown et al. 2008; Cattanach 1980; Glowacki 2010; Hurst and Till 2002; Lipe and Ortman 2000; Lipe and Varien 1999a; McKenna and Toll 1992; Nordby 2001; Toll 1993; Varien 1999a; Varien et al. 2007). The highest density of these sites is found in the canyons and mesas formed by the McElmo and Monument drainages (Glowacki 2006). The Village Ecodynamics Project (VEP) study area is in the heart of this locale (Figure 1.1). Consequently, the VEP provides an important opportunity for a detailed examination of this settlement type in a key area within the northern San Juan region.

A small number ($n=92$) of the more than 3,000 habitation sites identified in the VEP study area are either markedly larger than the rest or contain distinctive architecture associated with political and ceremonial life. These sites have been termed "community centers" because typically each constitutes the largest settlement in its vicinity, often contains civic-ceremonial architecture, and tends to have a longer occupational history than the typical small habitation site (Varien 1999a:202–207). The distinctive size, architecture, and histories of community centers imply that they played an important role in structuring not only settlement organization, but also social, economic, and political interactions within and between communities (Adler and Varien 1994; Lipe and Varien 1999b:345; Ortman and Varien 2007; Varien 1999a; Varien et al. 1996). As loci of ritual and political activities that were not expressed materially at smaller sites (Driver 1996, 2002; Lipe 2002; Muir and Driver 2002, 2004; Ortman and Bradley 2002; Varien et al. 2007), community centers are critical for understanding the social and political organization of ancestral Pueblo communities in the northern San Juan area. Our purpose here is to characterize the histories of community centers and explore their role in structuring the social and natural landscape in the VEP study area. The results of our research also have implications for general understandings of settlement aggregation as a social process (see also Chapter 2).

As discussed in Chapter 2, the overall trend in settlement patterns in the VEP study area between A.D. 600 and A.D. 1280 is characterized by two broad cycles of aggregation: an early cycle spanning A.D. 600 through A.D. 980 and a later one from A.D. 980 through A.D. 1280. (Mark Varien and colleagues [2007] and Richard Wilshusen [2002] also identify two cycles, but they use slightly different dates). How and when community centers formed and the scale of aggregation are distinctive in each cycle. Here we present an analysis of the histories of community centers in the VEP study area during these two cycles by (1) examining when and where community centers were established; (2) assessing the types of civic-ceremonial architecture present; (3) evaluating the percentage of the population living in community centers; and (4) comparing the size of centers with the estimated agricultural productivity of their surrounding catchments to assess the degree to which agricultural potential influenced the growth and decline of community centers. We also integrate these analyses with previous evaluations of rank-size relationships (Lipe 2002) to explore regional sociopolitical organization during the ancestral Pueblo occupation.

THE VEP COMMUNITY CENTER SURVEY, 2003–2004

The VEP, which involves both computer simulation and archaeological analysis, presents the most recent opportunity to further synthesize and refine data on community centers. Computer simulation is used to understand the formation of villages by identifying when and where autonomous households cluster on the landscape given certain parameters. To the extent that village formation requires social processes beyond simple interhousehold exchange, the simulation will not model village formation, although it may predict where villages should form if the minimization of resource-acquisition costs is important (see Chapters 10 and 15). Thus, as emphasized in Chapter 2, an important complementary goal of the VEP is to build a comprehensive, archaeological database of all villages that actually formed in the study area between A.D. 600 and A.D. 1280, and to analyze these data to better understand village formation, aggregation, and dissolution.

The objective of the Community Center Survey was to augment existing archaeological data for poorly documented community centers within the VEP study area to obtain the best results possible from our analyses of settlement and population. We focused on the largest sites because errors in the interpretation of settlement histories for these would have a greater impact on our overall results than would errors for small sites. Community centers are also a challenging settlement type to model on the basis of surface evidence because many of these sites are multicomponent and the latest occupation "swamps" material evidence of earlier occupations. Since data from testing or excavation are not available for most of these sites, the best chance for detecting evidence of early occupation was careful assessment of surface remains to identify what might be rare instances of early pottery types and architectural features. For these reasons, our goal was to revisit as many of these large sites as possible during our two field seasons to bring the site size and occupation data for all such sites up to a common standard.

Previous research in the VEP study area had identified more than 80 sites as community centers (Lipe and Varien 1999b; Varien 1999a; Wilshusen 1991). The first step in planning the survey was to revise and expand this list, using as a basis published and unpublished literature, data archived at the Colorado Office of Archaeology and Historic Preservation, and conversations with local archaeologists and landowners. Following Varien and others, we defined a community center as a site or group of adjacent sites containing at least one of the following: (1) 50 or more total structures; (2) nine or more pit structure depressions; and (3) public or civic-ceremonial architecture (Adler and Varien 1994; Lipe 2002; Lipe and Varien 1999b; Varien 1999a:141–143; Varien et al. 2007). These

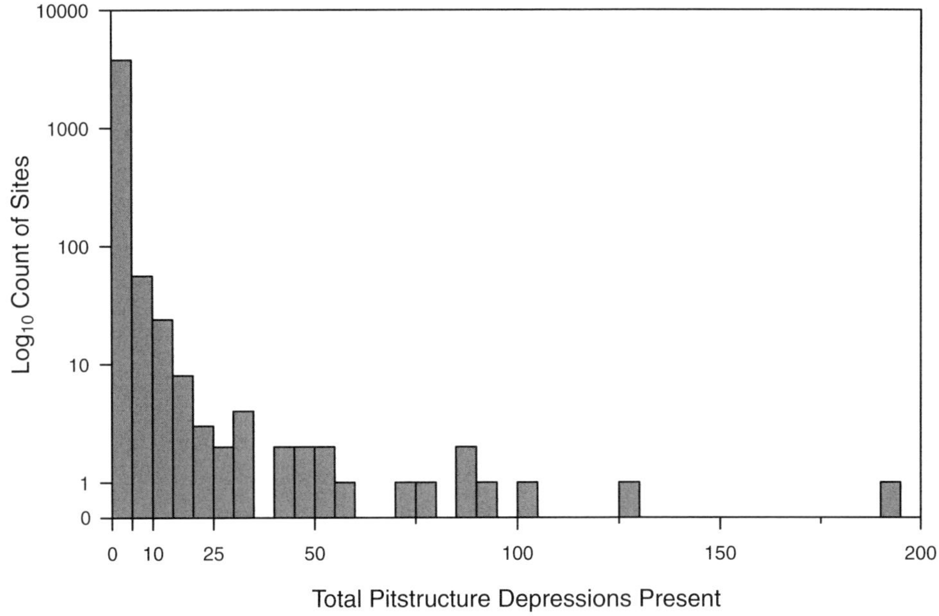

FIGURE 14.1 Histogram of pit structure counts for all recorded habitation sites in the VEP study area.

criteria are slightly modified from previous studies in the use of nine or more pit structures rather than six to eight, as suggested by William Lipe and Varien (1999b:345). The presence of nine pit structures suggests there were nine or more households at the peak of the occupation, representing roughly 45 to 50 people and thus similar to the total population expected if one used the traditional rough estimate of one person per room. Figure 14.1 suggests that nine pit structures also constitute a reasonable break point because the number corresponds to the beginning of the extended upper tail of pit structure counts for all recorded sites in the VEP study area.

These criteria provided a flexible definition that allowed us to account for variation in recording practices and changes in the layout and architecture of community centers over time, and thus to produce a more inclusive list of potential community centers.[1] Once we had created this expanded list of potential centers, we examined their existing records to identify those that lacked adequate maps and pottery tallies, then built a prioritized list of sites that we were able to visit during the survey. Centers that rose to the top of the list were those that (1) qualified as community centers but were insufficiently documented, or (2) were borderline cases that may have been community centers but required additional documentation to know for sure. During the summers of 2003 and 2004, we visited and improved the documentation for 58 of the sites on this list. As the community-center database was revised and reevaluated, some sites dropped off the list and others were added on the basis of site form reviews and site visits. Through our efforts, we added 20 sites to the list of confirmed and potential centers, for a total of 106 sites (86 community centers with resident populations, six isolated great kivas, and 14 borderline cases yet to be evaluated). The following analysis for the VEP study area is based on the 92

1. For example, some surveyors split contemporaneous room blocks within a single settlement into different sites, whereas others included all such room blocks in a single site. Also, Chaco-era great houses in the VEP study area tend to have relatively few pit structures, but they include numerous rooms and were important locations of civic-ceremonial activity.

A

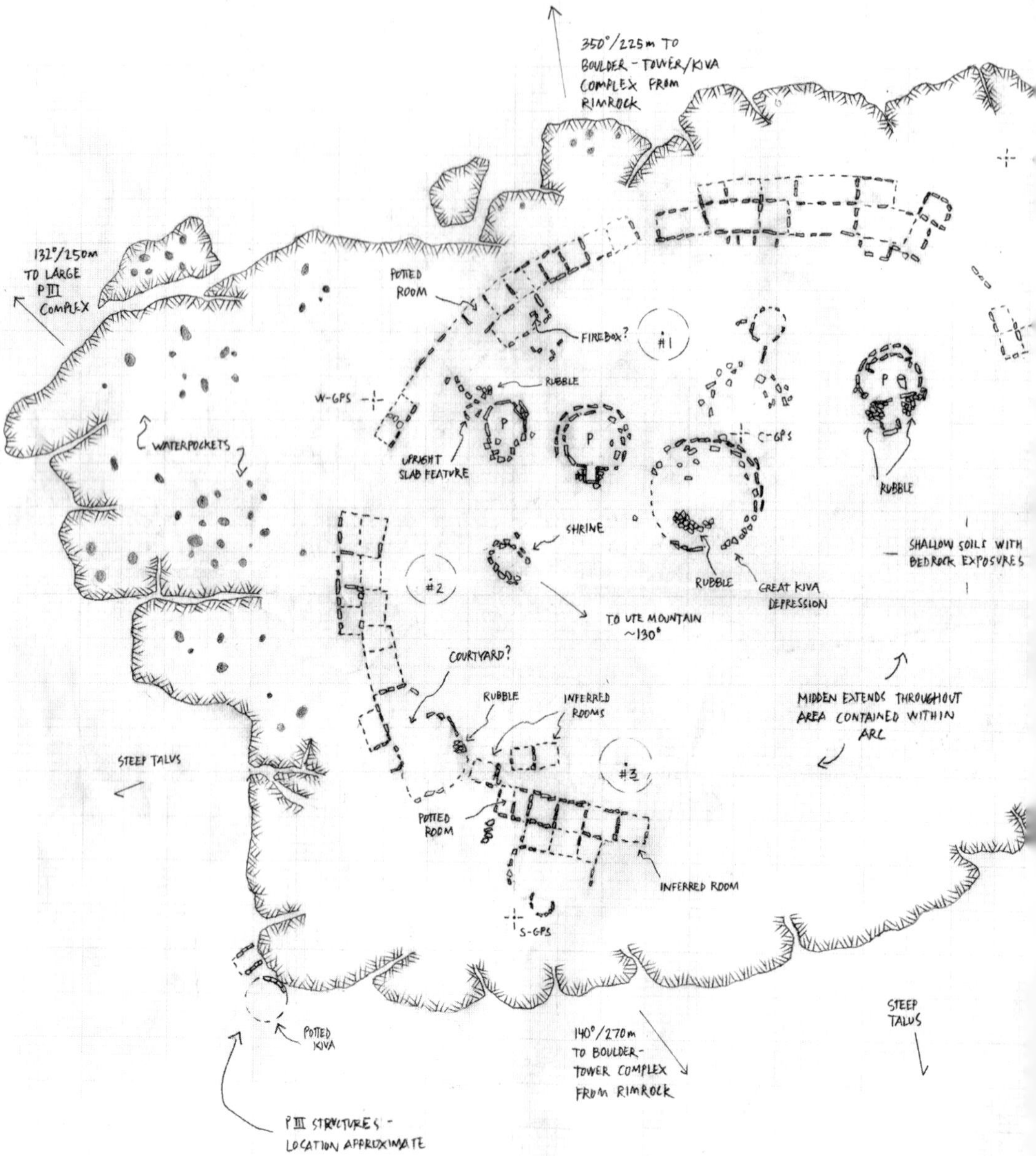

FIGURE 14.2 Maps from the CCS illustrating community centers. A) Fortress Spur (5MT296): example of a community center during the early aggregation cycle. Field drawing by CCS field crew 2003. Map B is on the page spread that follows this one.

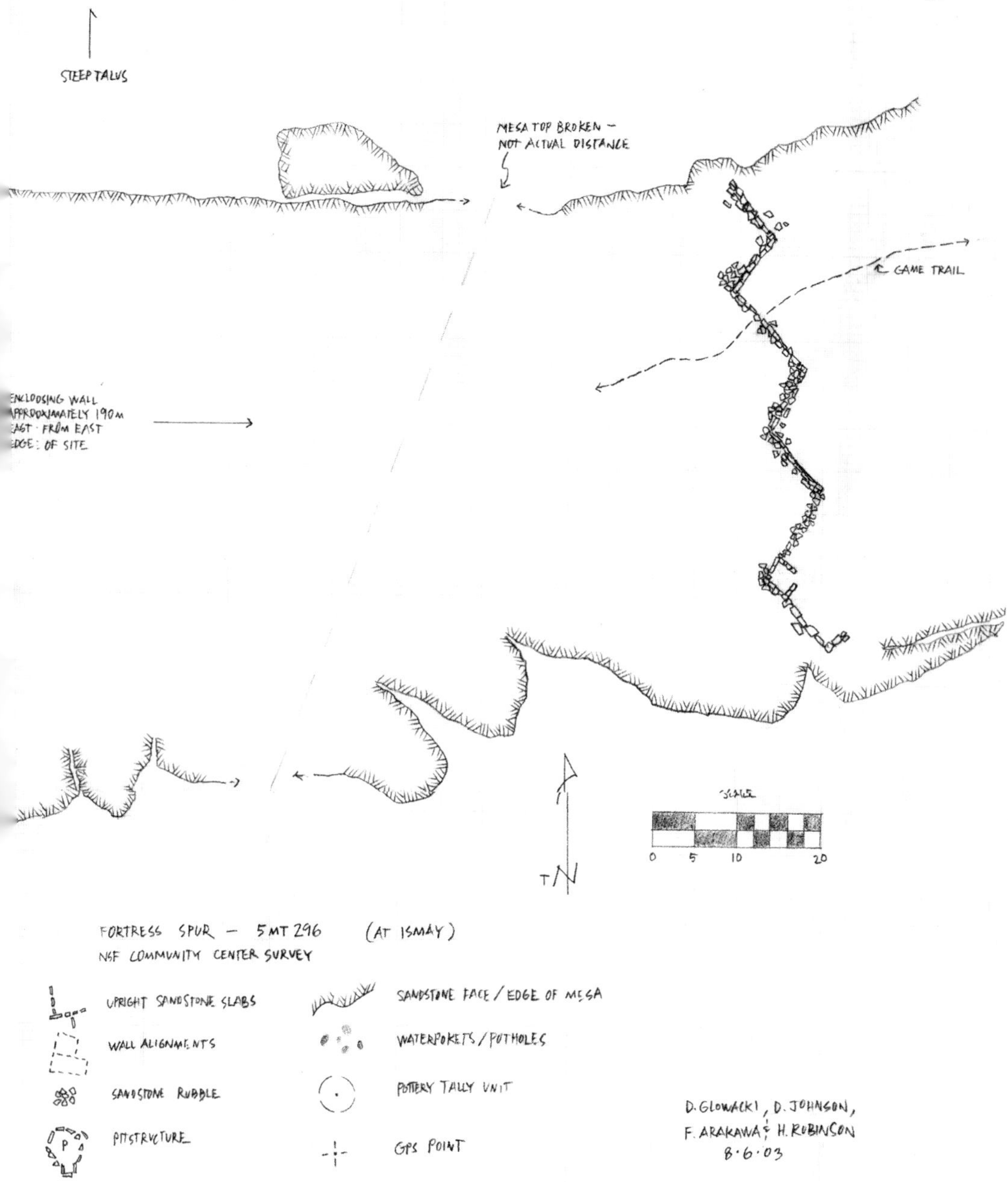
STEEP TALUS
MESA TOP BROKEN — NOT ACTUAL DISTANCE
GAME TRAIL
ENCLOSING WALL
PPROXIMATELY 190m
AST FROM EAST
DGE OF SITE
SCALE
0 5 10 20
TN
FORTRESS SPUR — 5MT296 (AT ISMAY)
NSF COMMUNITY CENTER SURVEY
UPRIGHT SANDSTONE SLABS
WALL ALIGNMENTS
SANDSTONE RUBBLE
PITSTRUCTURE
SANDSTONE FACE / EDGE OF MESA
WATERPOKETS / POTHOLES
POTTERY TALLY UNIT
GPS POINT
D. GLOWACKI, D. JOHNSON,
F. ARAKAWA & H. ROBINSON
8·6·03

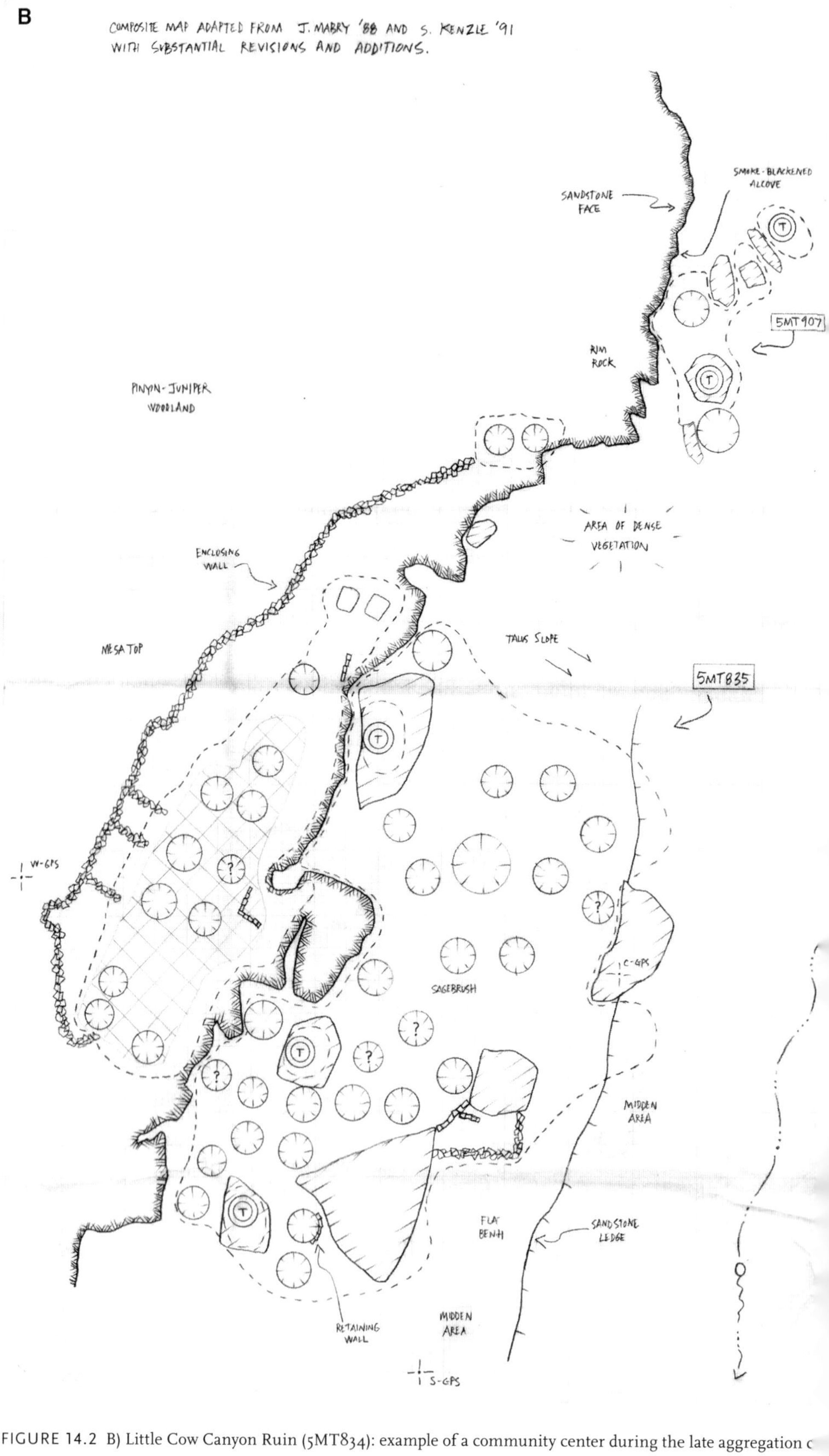

FIGURE 14.2 B) Little Cow Canyon Ruin (5MT834): example of a community center during the late aggregation c

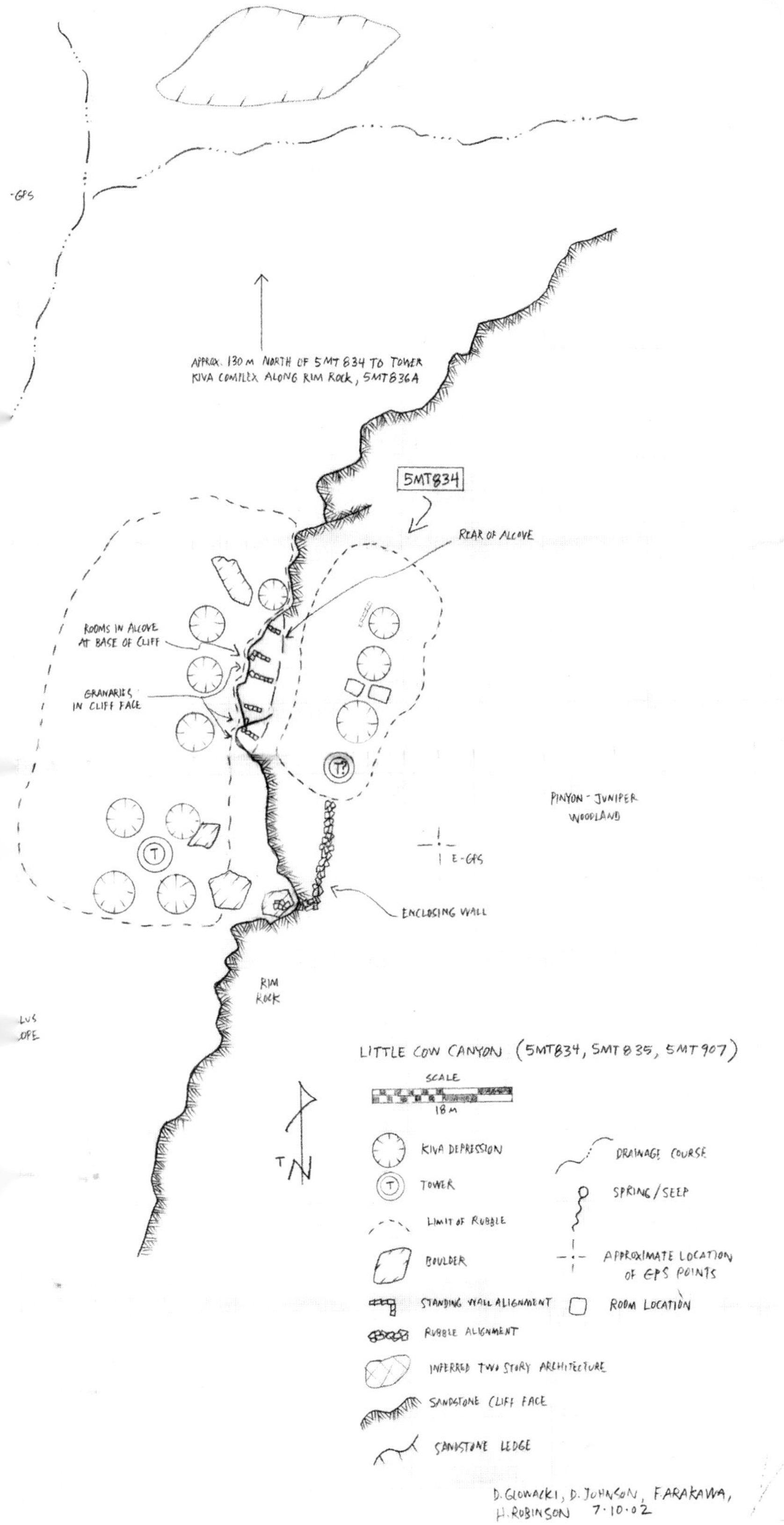
APPROX. 130 M NORTH OF 5MT834 TO TOWER KIVA COMPLEX ALONG RIM ROCK, 5MT836A
5MT834
REAR OF ALCOVE
ROOMS IN ALCOVE AT BASE OF CLIFF
GRANARIES IN CLIFF FACE
T?
T
PINYON - JUNIPER WOODLAND
E-GPS
ENCLOSING WALL
RIM ROCK
LITTLE COW CANYON (5MT834, 5MT835, 5MT907)
SCALE
18 M
TN
KIVA DEPRESSION
TOWER
LIMIT OF RUBBLE
BOULDER
STANDING WALL ALIGNMENT
RUBBLE ALIGNMENT
INFERRED TWO STORY ARCHITECTURE
SANDSTONE CLIFF FACE
SANDSTONE LEDGE
DRAINAGE COURSE
SPRING / SEEP
APPROXIMATE LOCATION OF GPS POINTS
ROOM LOCATION
D. GLOWACKI, D. JOHNSON, F. ARAKAWA, H. ROBINSON 7·10·02

confirmed community centers (see Appendix B for a list of community centers).

During site visits, our first priority was to produce a complete map. If a site was previously mapped, we compared the existing map to the remains visible on the surface of the site and if we identified discrepancies we modified it as needed. If the existing map was insufficient to assess site size, or if there was no existing map, we mapped the site using a laser range-finder and compass. Features recorded included room blocks, wall alignments, enclosing walls, towers, multiwalled structures, plazas, berms, reservoirs, and roads. We made new maps for 35 sites and redrafted or annotated 23 maps to clarify relationships among previously documented features and to add new features observed during our visit. Figure 14.2 illustrates the level of detail in our maps and provides examples of community centers from the early and late aggregation cycles. Using these maps, we estimated the total number of household residences using as a basis room block and pit structure counts, total site area, and room block area.[2]

Our second priority was to conduct infield pottery analysis to obtain quantitative data on the range of pottery types and their proportions at each site. Depending on the nature of the surface assemblage, we used either 3-m dogleash units or tallied all sherds visible on the surface. When there were distinct middens associated with individual room blocks, we made a tally from at least three middens, and we tried to make tallies in as many middens as possible. A complete site tally was used when overall sherd density was low and discrete middens were not present. In these instances, we walked transects across the site and recorded all visible sherds. In total, we classified 36,790 sherds during the survey.

The most important and long-lasting contributions of our survey are the good maps we produced and our documentation of surface remains for 58 of the largest sites in the central Mesa Verde region, performed following consistent procedures. We also supplied updated information on over 20 sites within the newly created Canyon of the Ancients National Monument that will aid resource specialists with site management and preservation. As a result, the Community Center Database (a subset of the McElmo-Yellow Jacket settlement database, version 5.3) now contains data at a consistent level of detail for all large ancestral Pueblo sites dating to all periods in the VEP study area, and thus provides a strong basis for evaluating the role of community centers in Pueblo history in the northern U.S. Southwest. In the remainder of this chapter, we explore some of the patterning in this rich data set.

CHARACTERIZING COMMUNITY CENTERS

Much of the recent research on community centers has focused on determining where they are located and on understanding their architectural characteristics and occupational histories. The term "community center" is predicated on the idea that Mesa Verde communities, which typically consisted of many distinct habitation sites, generally were situated around a central settlement. Although there has been some focus on the identification of community settlement clusters and of interactions among community centers (Adler 2002; Adler and Varien 1994; Lipe 2002; Mahoney et al. 2000; Ortman and Varien 2007; Pierce et al. 2002; Varien 1999a; Varien et al. 1996), we know relatively little about the internal organization of the communities themselves, and the types of social and political networks that formed between them. The cumulative effort of researchers over the years has produced a data set that allows us to address these issues, but first we need a thorough understanding of the breadth of the architectural variation accommodated by the term

2. See Glowacki and Varien (2002, 2003) for details regarding the field protocols for the survey. Ortman and colleagues (2007:261–262; see also Chapter 2) explain how we translated these data into estimates of the total number of pit structures and thus the number of household residences that were constructed over the course of occupation at each community-center site.

"community center." We begin to unpack some of this variability here.

Elsewhere, we described the two broad cycles of population growth and decline in the VEP area. Here, we examine the role of community centers during five periods within these two cycles. Because few community centers are known for the initial VEP modeling period, A.D. 600 through A.D. 725, we consider the first cycle of community-center growth and decline as beginning in A.D. 725 and ending by A.D. 980, when the majority of existing community centers had been abandoned.[3] The second interval, between A.D. 980 and A.D. 1060, was one in which population density remained low and community centers were rare (Lipe and Varien 1999b; Schlanger and Wilshusen 1993; Varien 1999a:145–146; Varien et al. 2007). The third interval, between A.D. 1060 and A.D. 1140, represents the late Pueblo II period, a period of population growth and influence from the Chaco regional system. The fourth interval, between A.D. 1140 and A.D. 1225, corresponds to the early Pueblo III or post-Chaco period. Finally, A.D. 1225 through A.D. 1280 represents the late Pueblo III period, during which a distinctive type of community center known as the canyon-rim village became dominant and the regional population peaked and then declined to zero.

To examine how and when community centers were established and to identify factors that may have contributed to their growth and decline, we summarize several categories of data for each of these periods. Most of these summaries are based on the number of momentary households at each community center during each occupation phase. These estimates account for the use-lives of structures and the lengths of modeling periods, and they specify the number of households present at any given moment during the occupation of the site (Ortman et al. 2007; Varien et al. 2007).[4]

Physiographic Settings

Community centers are found in a variety of settings, from upland ridges and mesa tops to river valley floodplains. Table 14.1 summarizes the locations of community centers founded during each of the four periods in which community centers were common. The table shows that during the early settlement cycle, community centers developed most frequently in upland settings or on terraces above water drainages. Over time, however, canyons came to be favored over uplands as settings for new centers. These data raise several points. First, although a number of well-known early villages were concentrated in the Dolores River Valley, in fact most early villages in the VEP study area were constructed in upland settings, either adjacent to the Dolores River Valley (Windy Ruin, Cline Crest, Holder) or in areas to the west (e.g., Ackmen Site I and II, Shields Pueblo, Radio Tower Site). Second, upland settings were also favored as locations for new community-center construction during the period of Chaco influence. Third, although it has long been known that the last community centers constructed in the central Mesa Verde region were located in canyon settings (Varien et al. 1996), the VEP community-center database suggests this move to the canyons began before A.D. 1225, during the post-Chaco period, and intensified during the final decades of Pueblo occupation. This finding is similar to the results of a ceramic seriation of 107 sites in Upper Cliff Canyon near Cliff Palace in Mesa Verde National Park, where the shift from mesa top villages to

3. By "abandoned" we mean that village occupants left their homes and moved elsewhere, leaving the structures vacant. It does not imply that these places were no longer important or central to Pueblo people (both in the past and today), but rather that the people were no longer physically living in these spaces on a year-round basis.

4. Twenty of the 93 confirmed community centers have fewer than nine pit structures observable on the modern ground surface. These qualified as community centers because they had some type of civic-ceremonial architecture, contained 50 or more total rooms, or the total site and room block areas both yielded higher estimates of the total number of pit structures than the actual observed pit structures (e.g., Crossroads–Radio Tower site, Smoot No.1). In the latter case, we assumed there were additional pit structures that are no longer visible on the modern ground surface (Ortman et al. 2007).

TABLE 14.1

Attributes of Community-Center Settings Through Time

	A.D. 725–980		A.D. 1060–1140		A.D. 1140–1225		A.D. 1225–1280	
	New Centers Established							
	23		17		18		27	
	Attributes of Setting[a]							
	N	%	*N*	%	*N*	%	*N*	%
Upland / Mesa top	16	70	13	76	8	44	5	19
Canyon head	0	0	1	6	7	39	18	67
Spring	2	9	4	24	9	50	17	63
Canyon rim	2	9	1	6	6	33	13	48
Talus slope	0	0	2	12	6	33	10	37
Bench / terrace	4	17	1	6	3	17	4	15
Bottomland / floodplain	3	13	0	0	3	17	4	15

a. Rows represent attributes of the physiographic settings of community centers established during each period. Centers may be associated with multiple attributes.

canyon rims takes place as early as A.D. 1140 to A.D. 1180 (Kleidon et al. 2007:217–223).

Community-Center Population

Figure 14.3 presents box plots of the distribution of community-center momentary populations, in households, for five modeling periods that represent important transition points in the history of Mesa Verde societies. The period of initial community-center construction is A.D. 725 through A.D. 800; A.D. 840 through A.D. 880 is the peak of early village aggregation and population; A.D. 1100 through A.D. 1140 is the period of maximum Chaco influence; A.D. 1225 through A.D. 1260 is the late-cycle population peak; and A.D. 1260 through A.D. 1280 is the final period of ancestral Pueblo occupation. This figure shows that the median population of community centers increased over time within each cycle, and that the largest median center population size was reached during the final modeling period, immediately preceding the complete depopulation of the region. The total number of centers and the size of the largest centers also increased consistently through time.

Very few community centers in the VEP study area ever grew to more than around 50 households, or about 300 persons, at any given moment. This may represent a threshold for the number of households that could easily provision themselves from a central settlement, given the ecology of maize agriculture in the VEP study area, or it may represent a threshold at which difficulties of coordination exceed the benefits, as mentioned in Chapter 1. The few centers that exceeded this threshold—particularly Yellow Jacket Pueblo, Goodman Point Pueblo, and Sand Canyon Pueblo—were either located in especially favorable areas for maize agriculture or developed in response to sociopolitical factors. From this perspective, it may be appropriate to view Yellow Jacket Pueblo as a primate center for most of the second settlement cycle in the VEP

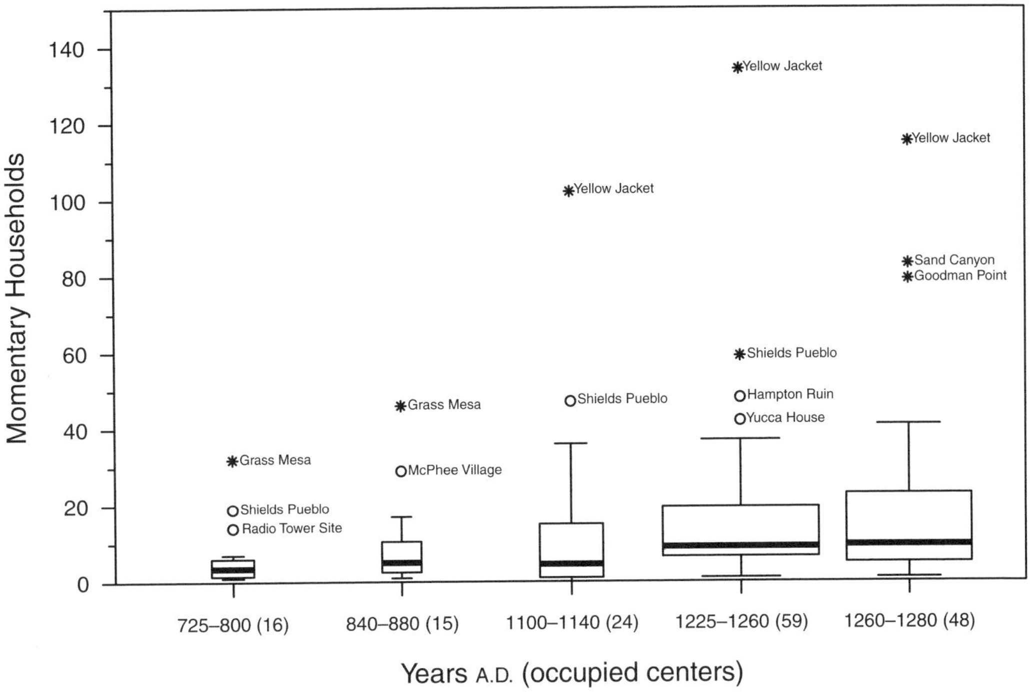

FIGURE 14.3 Box plots of community-center populations during five modeling periods. Thick lines indicate median size, box lengths indicate the middle 50 percent of cases, and whiskers represent 90 percent of cases. Outliers are indicated by open circles, and extremes by asterisks. Box widths are scaled according to the number of centers occupied during each period.

study area. In this scenario, the construction of Sand Canyon and Goodman Point Pueblos on the McElmo Dome during the A.D. 1260s might reflect the development of two more primate centers that competed with Yellow Jacket for power and influence in the region (see Kohler and Varien 2010; Lipe 2002). We explore this possibility later in this chapter through an analysis of agricultural catchments.

Civic-Ceremonial Architecture

Architecture designed specifically for civic or ceremonial purposes is present at most community centers in the study area. Table 14.2 summarizes the presence or absence of such features at centers established during various periods. In constructing this table, we have assumed that all such features were built as the center was established. We know that this was not always the case, but in the absence of independent dating estimates for each feature, we have no alternative but to follow a general and plausible rule.

As a heuristic for examining civic-ceremonial architecture through time, we categorized this architecture into three classes: group-assembly features, restricted-use features, and controlled-access features. "Group-assembly features" include great kivas and plazas, both of which were large enough to host periodic gatherings of large groups. Such features occur in most community centers dating to the first settlement cycle, as well as in centers dating to the Chaco period, A.D. 1060 through A.D. 1140, in the late cycle. After A.D. 1140, the proportion of new centers containing group-assembly features declines somewhat. During the A.D. 1200s, group-assembly features were more common at the larger centers. Among these, the average peak population of centers with group-assembly

TABLE 14.2

Percent of Community Centers with Civic-Ceremonial Architecture in the VEP Study Area

	A.D. 725–980		A.D. 1060–1140		A.D. 1140–1225		A.D. 1225–1280	
	No. of centers established							
	23		17		18		27	
	No. of centers with							
	N	%	*N*	%	*N*	%	*N*	%
Great kiva	12	52	12	71	4	22	5	19
Plaza	10	43	2	12	6	33	12	44
Oversize pit structure	7	30	6	35	2	11	6	22
Great house[a]	0	0	8	47	1	6	0	0
Multiwalled structure	0	0	1	6	4	22	7	26
Enclosing walls	0	0	1	6	12	67	21	78
Towers	3	13	5	29	15	83	24	89
Group-assembly features	19	83	12	71	9	50	12	44
Restricted-use features	8	35	9	53	5	28	13	48
Controlled-access features	3	13	5	29	16	89	26	96

a. Shields Pueblo, a center occupied during both the early and late cycles, has a great house. It is clear that this structure is associated with the late-cycle occupation in this case, and thus it is tabulated for the A.D. 1060 through A.D. 1140 period, not the initial period of occupation, A.D. 725 through A.D. 980.

features was 29 households (standard deviation = 31, $n = 27$), whereas the average for those lacking group-assembly features was 13 households (standard deviation = 8, $n = 28$). This suggests that the inhabitants of smaller centers did not require spaces for this purpose, or that they traveled to larger centers for such assemblies. The absence of group-assembly features at many smaller late centers also calls into question the extent to which these smaller centers were politically autonomous. Perhaps the inhabitants of smaller centers participated in larger polities that encompassed multiple centers, as Lipe (2002) suggested.

"Restricted-use features" are structures that were likely used for private ceremonies attended by a select few individuals. During the early settlement cycle, pit structures that were significantly larger than a typical domestic pit structure, and that often prove to contain elaborate ritual features when excavated (Wilshusen 1989), were the only type of restricted-use feature present. During the Chaco period (A.D. 1060–1140), however, great houses consisting of massive room blocks with oversize, interconnected rooms surrounding a central pit structure or kiva replaced oversize pit structures as the dominant type of restricted-use feature. This trend culminated in the development of multiwalled structures consisting of arcs of oversize, interconnected rooms surrounding one or more central kivas during the A.D. 1200s. Overall, the number of restricted-use features appears to have increased during the Pueblo III period, but because oversize pit structures are often difficult to identify on the basis of surface evidence, restricted-use features may have been more common in early centers than our data suggest.

Finally, "controlled-access features" are architectural features such as towers and enclosing walls that governed access to all or portions of a settlement. Such features are entirely absent from early-cycle centers but appeared more frequently over the course of the second cycle, to the point that nearly all centers in the A.D. 1200s included them.[5] This trend suggests that defining community boundaries, monitoring access, defensive posturing, and defense against actual attacks were important functions of community centers during the A.D. 1200s.

When viewed in terms of our classification, the dominant trend in civic-ceremonial architecture over time was a decline in the construction of group-assembly features and an increase in the construction of controlled-access features during the second settlement cycle. We think these two trends are connected. The dramatic increase in controlled-access features during the second cycle suggests that competition and conflict among communities increased, perhaps substantially, over this period. We would expect this conflict to have encouraged the development of intercommunity alliances and confederacies as a way to increase security. One means of reinforcing such alliances would be for the most influential villages in an alliance to host periodic group assemblies involving ceremony and feasting. Thus, the overall decline in the frequency of group-assembly features, and their growing association with larger centers, may reflect an expanding scale of civic-ceremonial activity associated with alliance formation, such that periodic group assemblies took place only in larger centers as opposed to every center.

5. The function of towers, found at both large and small sites, has long been debated. Suggested functions have ranged from defense and protection (Haas and Creamer 1993; Kuckelman 2002) to communication (Morley and Kidder 1917), resource monitoring (Johnson 2003), and ceremonial and symbolic functions (Lipe and Ortman 2000; Van Dyke and King 2010). Here, we view towers as a means of controlling village access because towers are strongly correlated with enclosing walls, which were clearly sociophysical boundaries (Glowacki and Varien 2003; Kenzle 1997; Kuckelman 2000, 2002). In the VEP study area, towers were present at 37 of the 42 community centers with enclosing walls (Glowacki and Varien 2003).

Episodes of Construction, Occupation, and Abandonment of Community Centers

To obtain a more nuanced understanding of village formation and depopulation through time, we explore the changing episodes of construction, occupation, and abandonment among community centers in the VEP study area. Figure 14.4 illustrates these episodes by summarizing the number of newly constructed centers (solid line), the number of centers that continue to be occupied into the subsequent period (dashed line), and the number of centers that are abandoned (dotted line) for each of the 14 modeling periods into which we have divided the ancestral Pueblo sequence. The overall trend in these data is that once centers were established, they continued to be occupied until periods of significant regional population decline at the end of each aggregation cycle. These data also suggest that the construction and abandonment of centers was related to the history of migration into and out of the study area. Specifically, two of the three periods during which there was a considerable increase in new village construction (A.D. 1060–1100 and A.D. 1225–1260) correspond to periods of significant population growth that likely included in-migration, and the two periods of marked village abandonment (A.D. 880–980 and A.D. 1260–1280) correspond to periods of significant population decline. The fact that community-center construction and abandonment were generally correlated with the history of movement into and out of the study area suggests that these population changes may have involved a variety of scales including multiple households, single households, and individuals, as discussed by Ortman and Cameron (2011).

Percentage of Population Living in Centers

Figure 14.5 summarizes the proportion of households that lived in community centers during each modeling period, based on settlement data from block surveys in the VEP study area (also see Varien et al. 2007:Figure 5A, C). These data indicate that, with the exception of

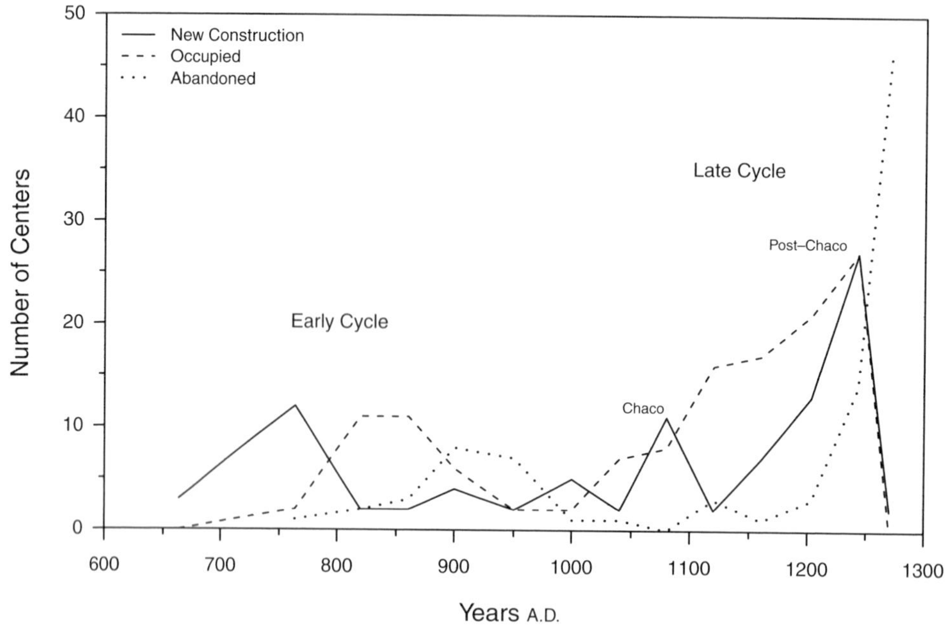

FIGURE 14.4 The number of centers that were newly constructed (solid line), continued to be occupied (dashed line), and abandoned (dotted line) during each modeling period.

the initial modeling period (A.D. 600–725), there were always at least some people living in both community centers and smaller settlements. In addition, the proportion of population living in centers generally increased throughout each settlement cycle, including the final period of each cycle when the regional population began to decline. During these two periods of peak aggregation, more than half of the study-area population lived in community centers (see also Glowacki et al. 2003; Varien et al. 2007). This, in turn, suggests either that the first people to leave the VEP study area in the late A.D. 800s and late A.D. 1200s left smaller settlements in disproportionate numbers or that the social conditions encouraging emigration also encouraged further aggregation of the remaining population into community centers.

These data also suggest that, in general, socionatural conditions throughout the ancestral Pueblo occupation promoted the gradual aggregation of population into centers, somewhat independently of climate variation (see Kohler et al. 2000; Varien et al. 2007). A number of factors were likely responsible for this. For example, one would expect the earliest centers established during each cycle to have been located in some of the most advantageous locations, allowing the populations of such communities to grow over time as a result of economic and reproductive success. As the regional population grew during each cycle, one would also expect competition over access to the best resource patches to have intensified (Adler 1992, 1994; Varien et al. 2000). Aggregation to protect people and food surpluses, especially for communities occupying the most productive and desirable patches, would have been an appropriate response to this competition. The analyses of agricultural catchments in the next section provide further evidence supporting this scenario.

Catchment Analysis

Varien and colleagues (2007; also see Chapter 6, this book) describe how the VEP produced a model of potential paleoproductivity for each

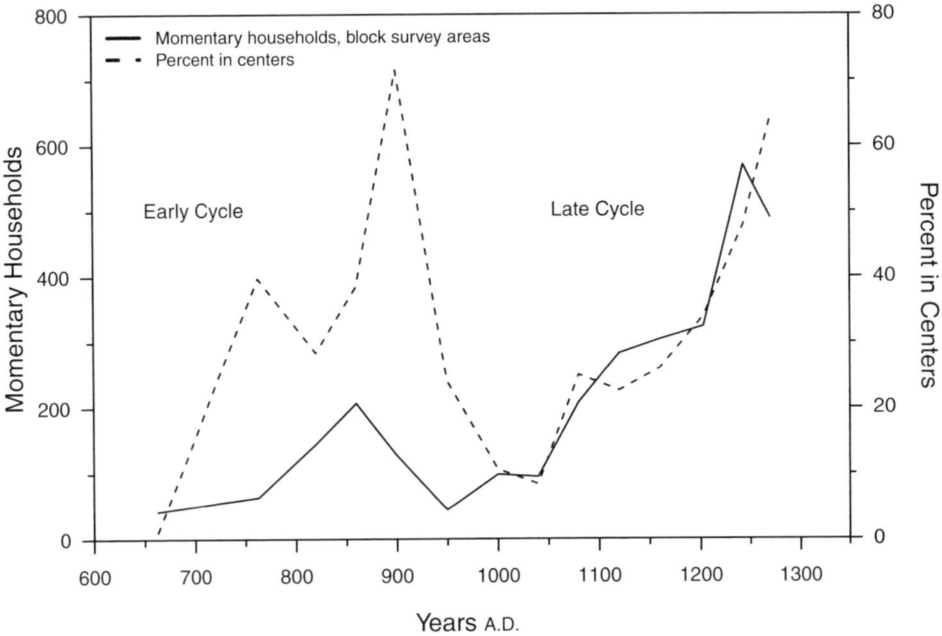

FIGURE 14.5 Momentary household population (solid line) compared to percentage of population living in community centers in block survey areas (dashed line), by modeling period.

4-ha cell within the study area. We use our reconstruction here to continue the line of research initiated by Varien and colleagues (2000) regarding the degree to which local agricultural resources, as measured by the agricultural potential of the surrounding land, influenced the formation and growth of community centers. The first question we ask is whether the eventual size of community centers was governed by the inherent agricultural potential of the land surrounding each center. Figure 14.6 presents a scatterplot of the peak momentary population against the mean potential paleoproductivity, across all modeling periods, of the lands within 2 km of each center. We chose a 2-km radius as the size of a community-center catchment on the basis of (1) ethnographic research that suggests this radius is a reasonable estimate for the area within which community members would have interacted most regularly and farmed most intensively (Varien 1999a:153–155; Varien et al. 2000:51–52), and (2) archaeological research showing that this distance corresponds to the typical area covered by dispersed community site clusters in full-coverage survey blocks within the VEP study area (Adler and Varien 1994; Kendrick and Judge 2000; Ortman and Varien 2007).

These data show that, in general, community-center peak populations are only weakly correlated with the mean agricultural potential of the surrounding land. Although the four largest centers (Yellow Jacket, Goodman Point, Shields, and Sand Canyon) are all associated with above-average catchments, many of the centers with above-average catchments did not become correspondingly large, and many centers with below-average catchments did (e.g., Easter Ruin, with 60 pit structures). In the case of the latter, these sites are typically in canyon-rim environments where check dams may have been used to enhance agricultural yields, in ways not accounted for in our current paleoproductivity models. Another possible explanation for this pattern is that the resident populations of most centers did not need to put all the surrounding land into production to provision themselves. As a result, centers could have

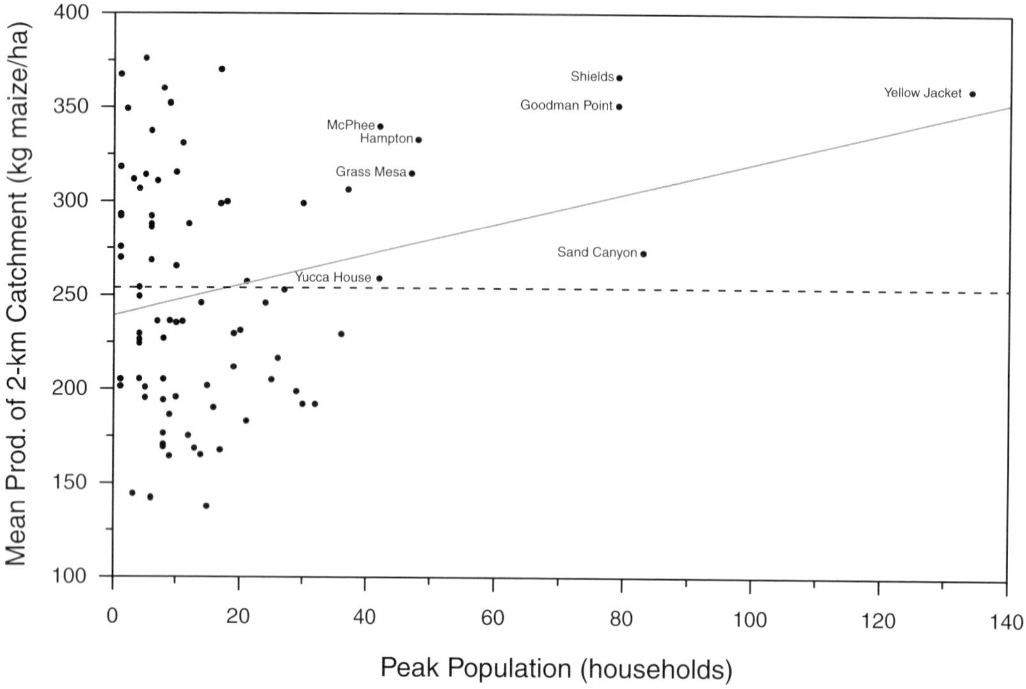

FIGURE 14.6 Peak populations and agricultural potential of community centers. The solid line is mean productivity regressed on peak population ($r^2 = 0.07$, $p = 0.015$) and the dashed line represents the global mean potential paleoproductivity of the entire study area.

developed in locations, such as canyon-rim settings, where only a portion of the surrounding catchment was suitable for farming, and the average productivity of this catchment was correspondingly lower. However, it is also important to remember that agricultural potential apparently did help the largest centers to form and maintain their centrality during both population cycles.

Figure 14.7 explores another dimension of catchment productivity by comparing the peak populations of community centers with the proportion of land in a 2-km radius surrounding the center that would need to have been cultivated to reliably provision these populations. Note, however, that we have not taken into account community members who did not live in the community center itself. Also, we have assumed that each person required 160 kg of maize per year for adequate nutrition, and that farmers generally attempted to produce a two-year surplus so as to balance out years of low production (Van West 1994:124–125).

These calculations show that, for a majority of centers, less than 40 percent of the land within 2 km would have needed to be in cultivation to reliably provision the resident population of that center. However, from 50 to 85 percent of the total land within a 2-km radius would have needed to be in production to reliably provision the peak populations of the four largest centers in the study area. This seems to be an unrealistically high percentage, since these calculations do not account for lands in fallow, which are assumed to have been necessary to maintain soil fertility over the long term, or for the significant number of people who lived within 2 km of a center but not in the center itself. Thus, these largest settlements may have been vulnerable to food shortages during extended periods of low production if they did not cultivate larger catchments or par-

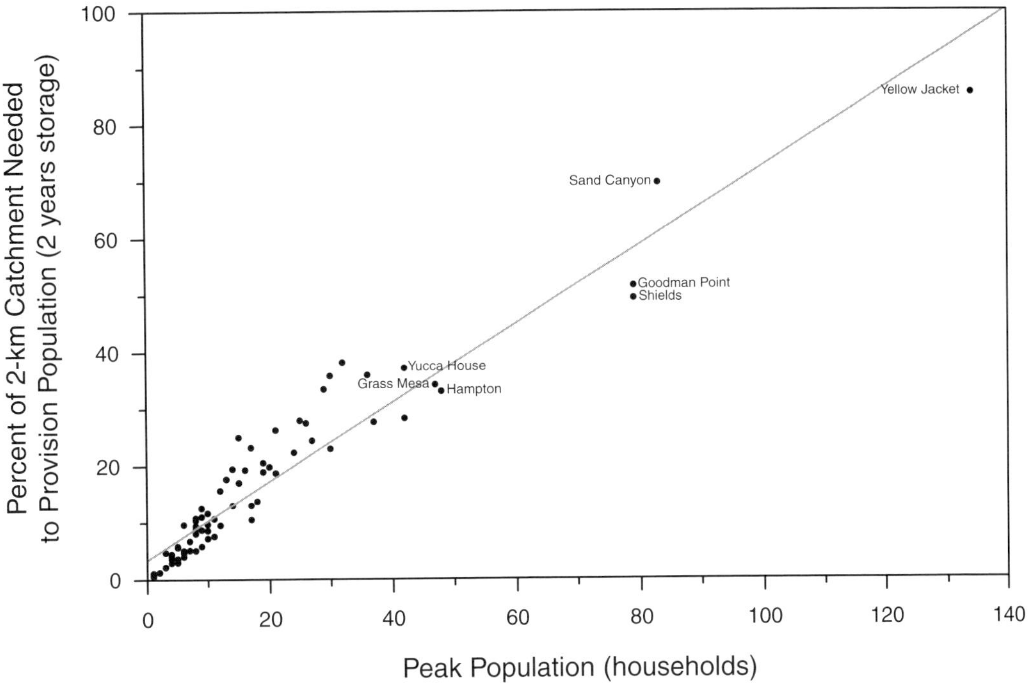

FIGURE 14.7 Peak populations of community centers vs. the percentage of land within 2-km needed to provision this population in the VEP study area. The solid line is the percent of 2-km catchment area needed regressed on peak population ($r^2 = 0.91$, $p < 0.001$).

ticipate in some mechanism of food redistribution, such as ceremonial feasting or as tribute (see Hirth 1984; Steponaitis 1981, 1984).

Even if residents of most community centers could have provisioned themselves reliably by farming a reasonable proportion of the surrounding land, it does not necessarily mean that prime agricultural land was always available for all who might have wanted it. Figure 14.8 summarizes the potential paleoproductivity of the 2-km agricultural catchments surrounding community centers founded during five different periods. These data show that, during the early settlement cycle, nearly all community-center catchments contained farmland that was above average in relation to the study area as a whole. This is not surprising, as one would expect the earliest villages in a sparsely populated landscape to emerge in highly productive catchments. Most of the new centers established early in the second cycle were also situated on above-average land, but as the second cycle progressed centers were increasingly established on below-average catchments.

Several factors may explain the declining dry-farming potential of new center catchments during the late cycle. First, new centers founded relatively early in this cycle continued to occupy their high-productivity catchments, so these were not available to subsequent communities. Second, the population density of the study area reached unprecedented heights during the final century of occupation, and this may have forced people to build new centers on lands that were less optimal. Third, it is important to note that the VEP paleoproductivity model is based on dry farming, whereas the inhabitants of new centers in canyon environments probably constructed terraces in side-drainages to capture soil and water for increased yields relative to unimproved dry farming (see Varien 1999a:214–216;

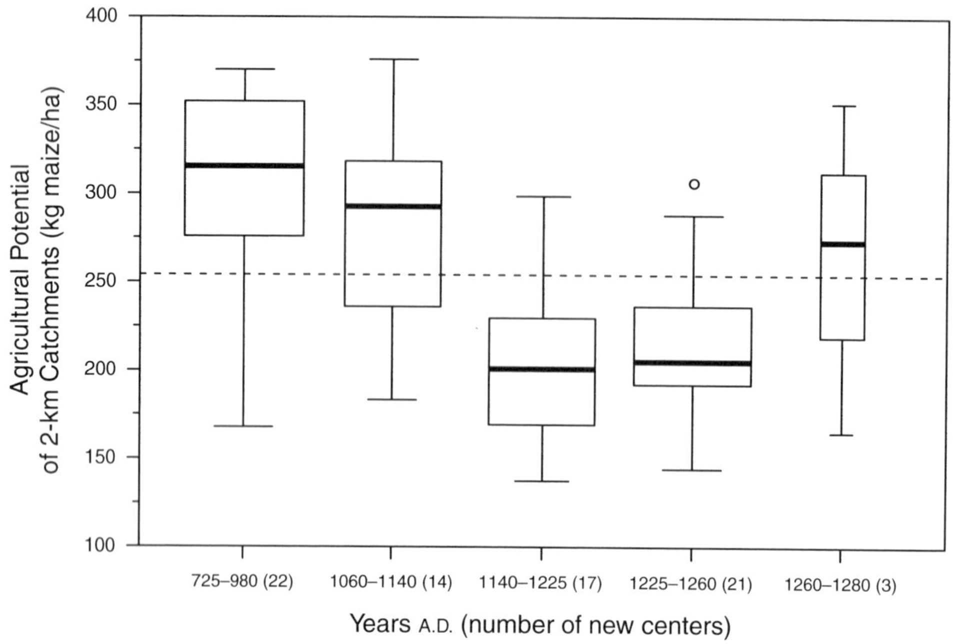

FIGURE 14.8 Agricultural potential of new community-center catchments through time. Thick horizontal lines indicate median size, box lengths indicate the middle 50 percent of cases, and whiskers represent 90 percent of cases. Outliers are indicated by open circles. Box widths are scaled according to sample size. The dashed line indicates the global mean potential paleoproductivity of the study area.

Huckleberry and Billman 1998). As a result, the productivity of catchments surrounding these new centers may not have been as low as is suggested by Figure 14.8, but canyon-based farmers would have needed to invest more labor per plot than inhabitants of the uplands who farmed the deep loess. Fourth, it may be that over time land-tenure systems were codified to the point that people no longer had to maintain a residence near their fields, thus enabling people to cultivate fields more than 2 km distant (Adler 1996; Varien 1999a).

Finally, it is possible that agricultural potential came to have a lower priority than other factors in choosing center locations. For example, centers founded early in the late cycle, such as the Yellow Jacket and Albert Porter Pueblos, were located on and adjacent to deep soils that were ideal for farming, whereas centers founded later in the cycle, such as the Sand Canyon and Woods Canyon Pueblos, were in canyon settings close to domestic water, building stone, and construction timber, but further from the best agricultural soils (Varien et al. 1996). One can also characterize these new center locations as being more defensible than earlier locations (e.g., Kuckelman 2000).

Regardless of the underlying reasons, this trend does suggest that, as the study area population grew during the second settlement cycle, increasing numbers of people lived and farmed in less-than-optimal locations. It also indicates that many people who had initially lived immediately adjacent to their fields came to live on or near their water, stone, and timber resources and walked to their fields. If, in fact, communities were forced to establish centers on less desirable land as the regional population increased, it may be that land pressure became an issue for ancestral Pueblo people in the VEP study area. A potential consequence may have been increased competition among communities, regardless of how close these communities actually were to the margins of viability.

THE HISTORY OF COMMUNITY CENTERS IN THE VEP AREA

Early Community-Center Growth and Decline (A.D. 725–980)

Early community centers were established after a period of initial colonization between A.D. 600 and A.D. 725 (Varien et al. 2007), and a total of 26 centers developed between A.D. 725 and A.D. 980. During their periods of peak occupation, these centers had an average of eight occupied room blocks (range=1 to 23; mode=7) and 16 pit structures (range=2 to 130; mode=11).[6] They were located in upland or mesa top settings or in the wide canyon bottoms of the Dolores and McElmo drainage systems in the eastern half of the study area (Figure 2.5; Table 14.1). These locations suggest that ready access to arable land and water may have been a factor in village location.

Based on our analysis, more than half of the early community centers ($n=15$) were established during the initial phase (A.D. 725–800). The early A.D. 700s were generally favorable, with above-average potential agricultural production (Varien et al. 2007:Figure 3), which could have aided the establishment and initial growth of villages, particularly since this period followed what was possibly the worst period for agricultural production in the 700-year ancestral Pueblo occupation of the study area (Varien et al. 2007:280, Figure 3).

Population levels from A.D. 725 to A.D. 800 were comparable to those of the previous period (see Varien et al. 2007:Table 3, 103 versus 134 total momentary households); thus, it seems that population pressure was not fundamental to the formation of these early villages. If it was not the number of people living in the study area per se, then cultural developments such as changes in food production, the increased tendency to settle in clusters, or both, in conjunction with climatic concerns, may have induced the aggregation noted at this time. We see evidence of both of these developments during the initial phase of the early aggregation cycle. Figure 14.8 shows that this cycle had the highest median agricultural potential of the entire occupation of the study area, which may have allowed the production of surplus crops, particularly during the early A.D. 700s and middle to late A.D. 800s, when agricultural yields appear to have been above average (Varien et al. 2007:Figure 3). This situation may have allowed food sharing networks to expand (Hegmon 1996; Kohler and Van West 1996), and this would have made it easier for larger groups of people to live together.[7]

Population more than doubled in the interval between A.D. 800 and A.D. 840, despite lower than average agricultural potential (Varien et al. 2007:Table 3). This population increase was greater than one would expect if it resulted solely from in situ growth; thus, immigration to the study area must have taken place (Schlanger 1988; Varien et al. 2007:Figure 5B).[8] Notably, our model suggests that no new community centers were established during this period. Thus, ancestral Pueblo people coming into the

6. These calculations exclude Mitchell Springs (30 room blocks, 47 pit structures) and Goodman Point Complex (58 room blocks, 239 pit structures), since most of the features at these sites were constructed during the later periods.

7. In the VEP study area, the interval between A.D. 725 and A.D. 800 also marks an important architectural transition from residing solely in pit houses to adding surface room pueblos to residential arrangements (Wilshusen et al. 2011) The reasons for this marked change in settlement pattern have long been debated, with explanations drawing on factors such as increased sedentism resulting from greater dependence on agriculture, increased population density, changes in food storage practices, and community reorganization (Feinman et al. 2000; Gilman 1987; Hegmon 1996; Rocek 1995). Our data suggest that whatever the impetus, the addition of surface pueblos to pit house villages was widespread; was not necessarily population driven; was enabled, at least initially, by a period of favorable agricultural productivity; and was associated with the formation of the early community centers.

8. Sarah Schlanger (1988) suggests this immigration occurred in the A.D. 840s. However, our analysis suggests it may have taken place in the early A.D. 800s during a period of low agricultural productivity (Varien et al. 2007:Figure 3).

study area were either moving into small habitation sites (Varien et al. 2007:Table 3) or into the existing community centers. Population during the early aggregation cycle peaked during the interval between A.D. 840 and A.D. 880, and correspondingly community-center size increased (Figure 14.3). After A.D. 840, 10 more community centers were established: two during the A.D. 840 through A.D. 880 period (Ackmen No.1, May Canyon), five during the A.D. 880 through A.D. 920 interval (Windy Ruin, Cline Crest Ruin, Singing Shelter, Carol's Ruin, Hartman Draw), and three during the last decades of the early aggregation cycle (Stix and Leaves, Kristie's, and Griffey).

Along with the cultural shifts that came with living in large villages was the development of village-level social institutions to accommodate larger coresident groups. The formation of these early large villages appears to have brought together disparate cultural groups that may have formed multiethnic and thus multihistoried communities (Potter and Chuipka 2007; Wilshusen and Ortman 1999). New or modified systems of organization may have been necessary to facilitate group cohesion among the diverse populations that may have constituted these early community centers. We see some support for this possibility in the VEP area as analyses of civic-ceremonial architecture[9] suggest significant organizational variation among early centers. In fact, in the VEP area, there appears to be even more variation than Wilshusen and Scott Ortman (1999) previously noted for the Dolores Archaeological Program area.

In the VEP area, there were two main types of architecture associated with early center organization: (1) great kivas, and (2) U-shaped room blocks associated with oversize pit structures. The use of these structures centralized aspects of ceremonialism at key locales and different scales within the early community centers (Blinman 1989; Potter 2000:466–467; Schachner 2001). They were not mutually exclusive.

The U-shaped room block and oversize pit structure complexes have received much attention because of their presence in the well-studied Dolores Archaeological Program area (Schachner 2001; Wilshusen 1986b 1989; Wilshusen et al. 2011). Oversize pit structures and U-shaped room blocks were new architectural forms that may have been imported from the west, where the earliest oversize pit structures occur (e.g., Brew 1946). These complexes have been interpreted as proto-great houses because they have well-constructed rooms with large storage areas, ritual features associated with the oversize pit structures, and ceramic and faunal evidence indicative of feasting and ritual (Schachner 2001; Wilshusen and Van Dyke 2006; Wilshusen et al. 2011; Windes 2004). These attributes have led researchers to suggest that the U-shaped room block and oversize pit structure complex were part of an intensification of ritual in conjunction with the formation of aggregated villages (Potter 1997; Schachner 2001).

In the VEP area, however, the U-shaped room block and oversize pit structure complex occurs at only a handful of early centers. The most common civic-ceremonial feature in the early community centers was the great kiva. These are found at more than half of early centers ($n = 14$), either alone or in combination with other forms of civic-ceremonial architecture. Their use for villagewide gatherings indicates a different scale of social and ritual organization than that represented by the U-shaped room block and oversize pit structure complexes. The use of great kivas in the northern San Juan was well established among local populations and began as early as the A.D. 600s (Churchill et al. 1998; Lipe 2006:267). Thus, it is not surprising that they were common during the early aggregation cycle.

Although great kivas and U-shaped room block complexes were prevalent, there does not appear to have been orthodoxy in village-level organization during the early settlement cycle. Table 14.3 shows the various configurations of

9. Note that when controlled-access architecture was present in the early community centers, it was typically found at sites with long-term occupations that extended into later periods.

TABLE 14.3

Civic-Ceremonial Architecture at Early Community Centers (A.D. 725–980, 26 Centers Total)

Architectural Attributes	*n*	%
Great kiva	7	27
Great kiva and plaza	2	8
Great kiva and oversize pit structure	2	8
Great kiva and U-shaped room block	2	8
Plaza	2	8
Oversize pit structure and plaza	1	4
Oversize pit structure and U-shaped room block	1	4
Oversize pit structure, U-shaped room block, and plaza	2	8
U-shaped room block	1	4
U-shaped room block and plaza	2	8
All four features	1	4
No civic-ceremonial architecture	3	12
Total with great kiva present	14	54
Total with oversized pit structure present	7	27
Total with plaza present	10	38

civic-ceremonial architecture present at these centers. The degree of variation suggests there may have been experimentation with various forms of community organization to attain village cohesion and stability. This variation is also evident across the greater northern San Juan region. For example, the Sacred Ridge site, near modern-day Durango east of the study area, represents an alternative form of ritual and political centralization with multiple oversize pit structures and a ridge-top complex that includes a two-story adobe tower and an adjacent, dome-covered structure surrounded by a palisade (Potter and Chuipka 2007). Despite this variation, early community centers seem to share evidence of multiscalar organization that involved both small-scale and village-scale gatherings and ritual practice. Our data suggest that community centers established between A.D. 880 and A.D. 980 remained variable, with U-shaped room block, oversize pit structure complexes, and great kivas all continuing to be constructed.

By A.D. 880, roughly half the estimated total population emigrated from the VEP study area (Varien et al. 2007:Table 3). Yet, the number of momentary households in centers remained about the same throughout the interval between A.D. 880 and A.D. 920, suggesting that most of the emigration during this time was from small sites. When the early community centers were first established, nearly 40 percent of the study-area population was living in them, and the remainder in smaller settlements (Figure 14.5). By the end of the period (A.D. 880–980), nearly 70 percent of the population was in community centers. This trend may indicate increased aggregation even as the overall population was in decline (see Varien et al. 2007:284).

It is difficult to know whether community centers were attracting new people from small sites during the late 800s or were simply more able to maintain their existing populations. The social networks of large villages may have buffered the effects of short growing seasons, drought, and resource depletion (Adams and Peterson 1999; Kohler and Matthews 1988; Schlanger 1988; Varien et al. 2007) to a greater extent than smaller settlements, at least for a time.

Our data and models suggest that the breakup of early community centers was a long-term process that extended from A.D. 880 through A.D. 980. Nine centers were depopulated during the interval between A.D. 880 and A.D. 920, and eight were vacated between A.D. 920 and A.D. 980, when the total estimated momentary population in centers decreased from 108 to 29 households (Figure 14.3; Varien et al. 2007:Table 3). The inhabitants of community centers may have been struggling as early as A.D. 840: three sites (Holder, Cirque, Ackmen No.2) were abandoned between A.D. 840 and A.D. 880 (Figure 14.3). The long-term nature of the depopulation at the end of the early aggregation cycle suggests that a variety of factors was involved—and not necessarily the same ones for each center.

Although each center had a unique history that shaped the options and decisions made by their inhabitants, climatic conditions were broadly shared. The potential paleoproductivity reconstruction presented in Chapter 6 suggests that agricultural productivity was below average between A.D. 800 and A.D. 840, above average between A.D. 840 and A.D. 880, and again below average between A.D. 880 and A.D. 920 (also see Varien et al. 2007:Table 3). Research by Sarah Schlanger (1988) and Timothy Kohler and Meredith Mathews (1988) suggests that shorter growing seasons and resource depletion were significant problems for inhabitants of the Dolores area in the late A.D. 800s. These conditions, particularly in conjunction with other social factors, would have presented challenging circumstances that could have strained interactions within and between the community centers. On the basis of an evident shift to the use of field houses to mark fields, Kohler (1992a) argues there were changes in land-tenure practices that led to the development of exclusive, long-term access. Gregson Schachner (2001) suggests that ritual control was linked to land tenure because changes in ritual organization correspond with these shifts in agricultural practices. The difficult climatic conditions of the early to middle A.D. 800s may have provided an impetus for this change, but as the systems became entrenched and were faced with social and climatic challenges, it seems these village-level institutions may not have been flexible enough to maintain village cohesion. This may have been particularly true for the centers depopulated in the period between A.D. 920 and A.D. 980, such as McPhee Village, where there is evidence of burned pit structures and a failed attempt to construct a great kiva (Orcutt et al. 1990; Schachner 2001; Wilshusen and Ortman 1999). There may have been resistance to the emerging ritual and political system at this center, or a loss of faith in village leadership. As there is also evidence of violence at other villages, the failure of village-level institutions may have been widespread (Wilshusen et al. 2011).

The A.D. 980 Through A.D. 1060 Interval

Although overall population density was relatively low during this period (Lipe and Varien 1999b), VEP data show that more people were living in the study area than previously thought (Varien et al. 2007). However, most of this remnant population lived in small habitation sites (Figure 14.5; Varien et al. 2007:Table 3). Thus, it seems the inhabitants of the study area were slow to reestablish large, aggregated villages, even after above-average climatic conditions (based on the VEP high-frequency reconstruction) returned in the middle A.D. 900s (Varien et al. 2007:Table 3). This is not to say that no one was living in villages during the interval between A.D. 980 and A.D. 1060, as 12 community centers contained between 10 to 15 percent of the total study area population at this time. At their peak occupation, these centers had an average of ten room blocks (range = 1 to 30; mode = 8) and 15 pit structures (range = 3 to 47; mode = 11.5). However, none of the centers were at peak occupation during the interval between A.D. 980 and A.D. 1060.

As seen in Figure 14.4, seven new community centers were established between A.D. 980 and A.D. 1020, four early-cycle centers continued to be occupied, and one small room block was occupied during the period between A.D. 1020 and A.D. 1060 at Mud Springs, eventually the location of a large, late-cycle center. The establishment of new centers corresponds with climatic fluctuations between A.D. 980 and A.D. 1060, with most of the new centers being established in the late A.D. 900s when agricultural productivity was above average. Only one center was established in the early to middle A.D. 1000s, when there was not only below-average productivity but also a pattern of increasing violence (Kohler, Cole, and Ciupe 2009; Chapter 13, this book). As with the early centers, large villages established during the time between A.D. 980 and A.D. 1060 were located in either upland or mesa top settings or at the wide bottoms of canyon drainages that were primarily in the southeastern portion of the study area (Table 4.1; Figure 2.7).

Most of the community centers established between A.D. 980 and A.D. 1060 were multicomponent and were occupied well into the A.D. 1200s. The long occupation histories of these sites make it difficult to generalize about the nature of village organization between the early and late settlement cycles. What can be said is that all community-center sites occupied during the interval between A.D. 980 and A.D. 1060 were associated with civic-ceremonial architecture. Village organization also exhibits continuities with earlier periods: all but three of the centers occupied during this period had great kivas. Those without great kivas had neither oversize pit structures nor plazas. Oversize pit structures were documented at three sites only. At Stix and Leaves, an oversize pit structure is associated with one of the largest room blocks at the site, but this room block does not have the U-shape characteristic of early-cycle centers. At the Lowry complex and the Lakeview group, oversize pit structures are associated with great houses that may have been constructed during this period but that clearly continued to be remodeled and used in subsequent periods. Thus, it is not possible to generalize about these structures for the period between A.D. 980 and 1060.

Many of the centers established between A.D. 980 and A.D. 1060 are associated with forms of civic-ceremonial architecture—including great houses, towers, and multiwalled structures—that would become widespread during the second settlement cycle. It is possible that the earliest examples of these features were constructed during this period. However, it seems more likely that these features were constructed in subsequent periods of occupation at these sites, when these civic-ceremonial architectural forms were common in new centers as well. Only one center was abandoned at the end of this period; the rest continued to be occupied into the subsequent period and increased in size. Given this continuity, it is tempting to characterize the A.D. 980 through A.D. 1060 interval as the beginning of the second aggregation cycle, and in some ways it was. However, for the most part, people were living in small habitation sites, and even the centers were small during this period. Thus, it appears that ancestral Pueblo people living in the study area during the interval between A.D. 980 and A.D. 1060 practiced extensive settlement and agriculture in the aftermath of the environmental downturn of the early A.D. 900s that was perpetuated in subsequent downturns of the early A.D. 1000s.

Second Cycle of Community-Center Growth and Decline (A.D. 1060–1280)

The time between A.D. 1060 and A.D. 1100 marks the beginning of a second extended period of population growth and aggregation in the VEP study area that continued until the total depopulation of the northern San Juan region toward the end of the A.D. 1200s. Our data suggest that there was a Chacoan (A.D. 1060–1140) and then a post-Chacoan (A.D. 1140–1280) phase of community-center construction within this extended cycle of aggregation.

Chacoan-Phase Community Centers (A.D. 1060–1140)

As with the early aggregation cycle, the beginning of this cycle follows a period of drought and below-average agricultural productivity (Varien et al. 2007:Figure 3), which may have engendered a predisposition to living in large, aggregated villages as conditions improved. The 22 community centers occupied during this period had an average of 13 room blocks (range = 1 to 58; mode = 11) and 41 pit structures (range = 1 to 239; mode = 15) at their peak. These community centers are dispersed across the northeastern half of the study area, primarily in the Dolores area and the upper reaches of the McElmo, Yellow Jacket, Cross, and Cahone canyons (Figure 2.8). Upland and mesa top settings were still favored, but not canyon bottoms (Table 14.1). There were also a few centers located on canyon rims. Figure 14.3 indicates that the median size of centers during this phase was slightly lower than in the previous period, but there was greater variation in sizes.

More than half of these centers ($n = 13$) were constructed during the interval between A.D. 1060 and A.D. 1100. One of these, Yellow Jacket Pueblo, was to become the largest ancestral Pueblo settlement that ever existed in the study area. Seven centers continued to be occupied from previous periods and two were established between A.D. 1100 and A.D. 1140 (Figure 14.4). Although arguably more aggregated than in the previous period, only 20 to 30 percent of the population was living in community centers at this time (Figure 14.5). Climate appears to have been generally favorable during most of this phase (Van West and Dean 2000:35–37), and agricultural productivity is modeled as being above average (Plate 6.6). However, there was a decade of long-term drought in both winter and summer from A.D. 1090 through A.D. 1100 (Wright 2010) and a short but severe downturn at A.D. 1100 (Varien et al. 2007:Table 3) that may have affected settlement trends in the early 1100s.

The community centers established during the period between A.D. 1060 and A.D. 1140 differed from those established later in the post-Chacoan phase in that they conformed to a linear, pueblo configuration (Glowacki 2010; Lipe 2006, 2010). These centers consisted of multiple, linear room blocks that ran east to west with doorways facing south, and with a row of associated pit structures, also to the south. By far the dominant forms of civic-ceremonial architecture constructed at new centers during this phase were great kivas—found at 70 percent of the centers—and great houses, which were constructed at nearly half of them (Table 14.2). This finding is not all that surprising, because this phase in community-center construction corresponds with the peak of Chaco Canyon's regional influence. Half of the new community centers lacked great houses, however, suggesting there was variation in the degree of participation in the Chaco system within the study area. All but one of these community centers had 17 or fewer pit structures (the one exception being Hampton Ruin). As with the interval between A.D. 980 and A.D. 1060, the majority of community centers occupied during this phase were multicomponent. Only two centers were abandoned during the interval between A.D. 1100 and A.D. 1140. Thus, most community centers established during this period continued to be occupied until the regional depopulation of the late A.D. 1200s.

Post-Chacoan-Phase Community Centers (A.D. 1140–1280)

There were significant changes in the nature of community centers during the post-Chacoan phase of the late aggregation cycle. The number and concentration of centers was much higher than in previous periods (Varien 1999a). The 65 community centers occupied during this phase had an average of eight room blocks (range = 1 to 58; mode = 4) and 30 pit structures (range = 1 to 239; mode = 15). These late community centers appear to have had fewer, but larger, room blocks than earlier centers. The median size of community centers was larger than at any other point in the occupation of the study area (Figure 14.3). There was also greater variation in size, which may have influenced social and political relationships among centers. For example, Yellow Jacket Pueblo, Hampton Pueblo, Yucca House, Shields Pueblo, Goodman Point Pueblo, and Sand Canyon Pueblo had significantly larger resident populations than the rest (Figure 14.3). Of these, Yellow Jacket Pueblo was by far the largest center throughout the late cycle. If size reflects importance, then these centers may have been among the most influential settlements in the study area, with Yellow Jacket Pueblo being the most influential of all.

There were also widespread changes in the local settings and regional distribution of community centers. Although centers persisted in the uplands, the interval between A.D. 1140 and A.D. 1180 marks the beginning of a shift toward canyon-rim settings (Table 14.1). With this timing, the drought context of the middle to late A.D. 1100s likely provided the initial impetus for the shift to canyon settings because the severity of this period would have caused increased concern over both secure access to water and defensive positioning. From A.D. 1225 through A.D. 1280, all new centers were built in canyon settings, and most were associated with canyon heads and springs (Table 14.1), further supporting the notion that control over water sources was of rising importance. The general shift to canyon rims also influenced the distribution of community centers within the study area, as canyon systems are more prevalent in the western portion of the study area (Figure 2.1). Although this shift began by A.D. 1140, the pattern intensified after A.D. 1180, with community centers becoming most numerous and concentrated along the canyons in the western half of the study area by A.D. 1225 through A.D. 1260 (Figure 2.10).

The post-Chacoan phase had two components. One was in the A.D. 1140 to A.D. 1180 period—a time of severe drought and violence—and the other was from A.D. 1180 to A.D. 1280, the peak of ancestral Pueblo occupation in the northern San Juan region. Our data suggest the study area population was relatively stable during the period between A.D. 1140 and A.D. 1180 (Varien et al. 2007:Table 3; see also Chapter 10). There were 25 community centers occupied during this interval, and of these, 18 were established in earlier periods. Thus, seven new centers were constructed during the period of drought that coincided with the decline of the Chacoan regional system. However, the relative lack of tree-ring cutting dates for this period remains curious (see Berry and Benson 2010; Lipe and Varien 1999b). Because of the unfavorable convergence of climatic and environmental factors such as stream entrenchment, drought, and low agricultural productivity, Carla Van West and Jeffrey Dean (2000:37) suggest this period may have been a key hinge point when people were inclined to make adaptive changes in their lifestyle. Given the paucity of tree-ring dates for this period, it is possible that some of these changes resulted in reduced wood harvesting and increased reuse (see also Ryan 2010). The A.D. 1140 to A.D. 1180 hinge point may have also shaped the circumstances that provided the impetus for the marked increase in aggregation and changes in village layout and village-level organization that took hold after A.D. 1180 and dominated from A.D. 1225 until the end of the regional occupation.

Thirteen new centers were constructed between A.D. 1180 and A.D. 1225, and 27 additional centers were established from A.D. 1225 to A.D. 1260. During this final century, community centers took on a new character. Rather than being built in the linear pueblo configuration of preceding periods, centers established after A.D. 1180 tended to be more inwardly focused and increasingly arranged in relation to the topography of their canyon-rim settings (Glowacki 2012; Lipe 2010; Lipe and Ortman 2000; Ortman and Bradley 2002; Varien et al. 1996). These centers were larger and closer together than in previous periods; thus, crowding and territorial encroachment were likely of more concern (Varien 1999a; Varien et al. 2000). Aggregation also intensified after A.D. 1225. By the end of the late cycle, nearly 70 percent of the study-area population was living in community centers (Figure 14.5).

Civic-ceremonial architecture also changed significantly during the second settlement cycle. Although great kivas continued to be built at roughly one-quarter of new centers, plazas became more prevalent, particularly after A.D. 1225, when they were constructed at nearly half of the new centers (Table 14.2). The activities taking place in plazas are inherently more inclusive and more visible than those in great kivas. In addition to facilitating cooperation and village integration through the performance of rituals and feasting, plazas were settings where religious and political leaders could reinforce their authority and social conflicts could be negotiated (Chamberlin 2011). With the population increasing and becoming more aggregated, plazas would have been important loci for working through processes of acculturation and syncretism, particularly if diverse populations began living together in community centers. The latter may well have been the case, since population modeling for the VEP study area shows that there were at least low levels of immigration during the A.D. 1200s (Varien et al. 2007:Figure 5b), possibly due to emigration from the west (Glowacki 2010).

Great houses, a hallmark of the Chaco phase, were constructed at only one of the 47 community centers established after A.D. 1140 (Table 14.2). That great houses were not built at the newest and latest centers suggests that ideas about connections to traditional Chacoan practices and the Canyon itself had significantly changed (see also Glowacki 2011). Perhaps in its stead, another form of restricted-use architecture, the multiwall structure, increased in frequency. These features were primarily located at the larger centers and were not widespread; 25 percent of the centers established at this time have them (Table 14.2). Although there are several multiwalled structures found outside the region at Pueblo del Arroyo, Tohatchi Flats, and Kin Li Chee (Lekson 1983), the elaboration and proliferation of these structures is a distinctive architectural trait of the eastern northern San Juan (Glowacki 2010). Given that they appear to have been used by particular people or groups, it may be that the religious and political functions associated with multiwalled structures were not unlike those associated with great houses.

The various forms of multiwalled structures may relate to the degree of affiliation with Chacoan ideology and orthodoxy (Glowacki 2011). In the VEP area, all but one of the five circular biwalled and triwalled structures are associated with a great house, whereas nine of the ten centers with D-shaped structures lacked great houses. The latter is not surprising, since D-shaped structures are not found in Chaco or the Totah (Glowacki 2006, 2010; Lipe 2006). Thus, it may be that ancestral Pueblo people at centers with the circular biwalled and triwalled structures maintained traditional ties to Chacoan practices, whereas those living in centers with D-shaped structures had either modified existing practices or developed completely new ones. The complementary distributions of these two forms of multiwalled structure may reflect the formation of political alliances or religious factions or societies, or both, during the final decades of settlement in the northern San Juan region (see also Glowacki 2011).

After A.D. 1140, features that likely controlled access to the center—e.g., enclosing walls and towers—became increasingly prevalent at community centers (Table 14.2). The intensification of controlled access indicates a definitive change in attitudes about settlement boundaries, village identity, and defensive posturing (Kenzle 1997). Indeed, the proliferation of these features may indicate an increasingly competitive landscape and be an expression of increasing political tensions either between communities or between communities and outsiders (see Lipe 2002).

The nature of the interactions among centers in the VEP study area is a complex issue, but an understanding of their social and political networks and organization is necessary for unraveling the factors that may have prompted the depopulation of the region. From a thorough assessment of settlement rank size and the distribution of wealth, Lipe (2002) concluded that even though some inequalities in social power were evident, the large villages of central Mesa Verde were not integrated into a hierarchically organized system, but rather they were most likely competing entities. In general, our results support Lipe's conclusion. As noted, Yellow Jacket, Sand Canyon, and Goodman Point[10] Pueblos were much larger than the rest of the late centers (Figure 14.3). It therefore seems likely that they were an influential presence in the study area and were important nodes of interaction. Though not as large as these, Yucca House and Hampton Ruin were larger than most of the rest and may have also been important nodes, particularly since these centers represent the largest settlements at the southernmost and northernmost limits of the study area. Yucca House, for example, was well positioned as a gateway settlement to the Montezuma Valley and was likely a key point for people traveling north-south (Glowacki 2001). These largest villages may have functioned as primate centers that were highly connected to other settlements and centers in the region. For example, Fumiyasu Arakawa's (2006, and Chapter 12 in this book) analysis of lithic procurement practices in the late A.D. 1200s indicates that lithic assemblages from Sand Canyon and Yellow Jacket Pueblos reflect a high degree of local interaction. Another example, and perhaps the most direct evidence for interaction between community centers, is the road that connects Sand Canyon Pueblo and the Goodman Point Pueblo–Shields complex (Coffey 2010).

Analysis of the catchments associated with these largest community centers also suggests there was likely some degree of cooperation among other nearby settlements and centers. All five centers are situated on catchments with above-average agricultural potential (Figure 14.6), though Yucca House and Sand Canyon Pueblo are only slightly above average. Yet, for the three largest centers (Yellow Jacket, Sand Canyon, and Goodman Point), from 50 to 85 percent of the total land within a 2-km radius needed to be actively farmed to reliably support their peak populations (Figure 14.7). Although some of this land use might have been alleviated by increasing productivity through the use of check dams in canyon-rim settings, the need to use a high proportion of their associated catchments would have put considerable pressure on agricultural production even in the best of years, and the A.D. 1200s presented far from the best conditions. As a consequence, cooperative networks and mechanisms for food redistribution were likely essential for maintaining stability for the inhabitants of these centers. The evidence of increased suprahousehold storage at Sand Canyon Pueblo (Lipe 2002), and greater material evidence of large-scale communal gatherings and feasting—which included a rise in the number of plazas, increasing bowl sizes and exterior decoration (Hegmon 1991; Mills 1999; Ortman 2000, 2002), and the evident exchange of serving bowls or their contents (Glowacki 2006)—lend further support to the intensification of food sharing, exchange, or redistribution. Though the need for these systems is most apparent at the largest centers, it was likely necessary, perhaps to lesser degrees, at other community centers.

Our analyses and demographic model suggest that the depopulation of the study area unfolded over a 75-year period, with the main exodus taking place fairly rapidly in the last 20 to 30 years of the occupation. Only one center was vacated during the drought period from A.D. 1140 through A.D. 1180, and three more, all on above-average catchments, were abandoned before A.D. 1225. Fourteen more centers were depopulated by A.D. 1260, and, incidentally, nine of these were also established during the interval between A.D. 1225 and A.D. 1260. Notably, all

10. Shields Pueblo was treated as a separate site in our analyses. It is so close to Goodman Point, however, that it was likely part of the Goodman Point complex.

but two[11] of the community centers vacated during this period were situated on catchments with below-average productivity, suggesting this may have been a factor in their abandonment. The majority of the community centers ($n=47$) were still occupied at the beginning of the final modeling period, from A.D. 1260 through 1280.

The 1200s were, agriculturally, a very difficult century, with episodic cooling trends and a severe, prolonged drought beginning at A.D. 1276 (Kohler 2010; Van West and Dean 2000; Wright 2010). The frequent below-average agricultural productivity throughout this century (Kohler 2010; Varien et al. 2007) would have been especially problematic for those centers on less productive catchments. Against this backdrop, we see evidence of an increasingly competitive social and natural landscape (Lipe 2002; Varien et al. 2000), signaled by the marked growth in the number of towers and enclosing walls across the study area. Each community center had its own history and experienced the challenges of the time in different ways, depending on its own agricultural circumstances, village leadership, social networks, and interpersonal relationships. Although the challenges were different for each center, it seems that all community centers struggled.

If centers were somewhat dependent on each other or on nearby settlements, the disruption of regular access to necessary goods would have been disastrous. Disruptions could have included a lack of adequate surplus to continue exchange (Hegmon 1996; Kohler and Van West 1996), but they also could have derived from social tensions and conflict (Kohler, Cole, and Ciupe 2009; Kuckelman 2002, 2010a). The significant change in village organization suggests there were accompanying changes in social and religious practices. The development of new religious practices designed to relieve uncertain circumstances may have exacerbated social tensions if these new practices proved ineffectual (Glowacki 2011). The emigration that began in the A.D. 1200s, too, could have caused disruptions, for it would have required the renegotiation of social networks, particularly when entire villages and centers were depopulated. In the last decades of the occupation, this fractious social landscape was plagued by drought and intensifying violence, which further disrupted and evidently destroyed the last vestiges of the social and ceremonial networks that may have been in place. These circumstances could have provoked rapid emigration and insured the complete depopulation of the region.

The catchment and paleoproductivity analyses of the VEP have allowed for a progressively nuanced reconstruction of the history of community-center formation and abandonment in the study area. It is a reconstruction that begins to explore the potential relationships between community-center occupation and its location. In turn, our Community Center Survey and subsequent data analysis have aided the VEP not only by improving our understanding of the distribution and timing of population aggregation, but also by elucidating the histories of social institutions as reflected in architectural changes that played a role in community organization.

11. Interestingly, the two community centers on above-average catchments were the Lake View Complex and Reservoir Ruin, which are arguably among the most "Chacoan" sites in the study area. Although we would not want to make too much of this coincidence, it does make one wonder if the breakdown of the Chacoan system at large created internal problems for those villages most strongly connected to it, ideologically.

FIFTEEN

The Rise and Collapse of Villages in the Central Mesa Verde Region

Timothy A. Kohler

THE VILLAGE ECODYNAMICS PROJECT (VEP) is ambitious in scope and innovative in design. Our work continues in parallel with that of many other archaeologists; it is hard to think of this book as a conclusion. This chapter emphasizes our provisional results on two topics: (1) what caused the ebb and flow of population into and out of our study area, and (2) what caused the formation and dissolution of villages in this area. In truth, the single authorship of this chapter reflects that in many cases these are my own (possibly idiosyncratic) conclusions. I will focus on what, specifically, the models and new data presented in this book contribute to understanding these phenomena. Especially important will be results from our model-based approach, which moves back and forth between Village simulation results and the analysis of the archaeological data. I also sum up time allocation data from the previous chapters to give a sense of how hard life may have been in the closing decades of this area's occupation. Along the way I will make some observations on the concept of adaptation and its role in archaeological explanation, as well as on the fragility of early Neolithic villages around the world.

POPULATION HISTORY

> Temperature curves may well connect with general history . . . rainfall curves belong in the first instance to local or regional history.
> LE ROY LADURIE (1971:89)

To be inviting for farmers lacking domesticated animals, any area must offer, at minimum, productive soils and sufficient precipitation and length of growing season for their crops, security for their families and for their storage, the natural resources to meet daily needs for things like fuel and drinking water, and enough wild game to meet protein needs. This last requirement can be relaxed to the extent that domesticated animals provide adequate protein. Other factors being equal, areas that offer these qualities in greater per capita quantities or more dependably than other accessible areas should experience immigration; the converse will tend to encourage emigration.

Large-Scale Trends in Immigration and Emigration

In Figure 3.5 Aaron Wright extends his pollen-based (low-frequency) temperature reconstruction only back to A.D. 500, but elsewhere he demonstrates that except for a brief period in the early to middle A.D. 300s, the entire period from the bottom of his Beef Pasture core at about 100 B.C. to A.D. 600 was characterized by low-frequency temperatures that are lower than his long-term average (Wright 2006) and well below the modern baseline (Moberg 2005). This likely limited the attraction of our generally high-elevation area for the agriculturally committed Basketmaker II and III populations that occupied (albeit sporadically, in some cases) areas east, southeast, southwest, and west of our area for over a millennium before A.D. 600. This would especially have been the case if their maize varieties, or cultivation practices, were poorly suited to the mesa top soils that provide the best agricultural opportunities in our area. However, these factors together probably would not have completely precluded occupation of our area, since Basketmaker settlement clusters in the Cedar Mesa area to the west and the Animas valley to the east are at elevations similar to those in much of the VEP area. So it may also have been important that our area lies between the eastern and western Basketmaker traditions identified by R. G. Matson (e.g., 2002), as noted in Chapter 1.

The striking success of Southwestern dendroclimatologists in reconstructing precipitation has tended to detract attention from low-frequency climate variability, especially in temperature. Nevertheless, Figure 3.5 (see also Kohler 2010:Figure 5.4) shows an apparent correlation between high temperatures and immigration into the VEP area, and low temperatures and emigration. With a bit of a lag in both cases, our two cycles of population growth described in Chapter 2 map directly onto the two cycles of warm temperatures that Wright reconstructs from A.D. 600 through A.D. 800 and from A.D. 1000 through A.D. 1200.

Both this correlation and these lags have been underappreciated in discussions of the population history of the northern Southwest. One of our most interesting findings is that the peak population in the central Mesa Verde region in both cycles (the first between A.D. 840 and A.D. 880, and the second between A.D. 1225 and A.D. 1260) is achieved only *after* these stable periods of high temperatures, during periods when temperatures have been in decline for many decades. One possible explanation for this is that during warm periods farmers tended to spread both into the area we studied and north of it, and then began to recoil to the south again as temperatures declined, in each case bringing VEP population sizes above what was earlier supported under superior climatic conditions, even as local climate was deteriorating. The first warm cycle corresponds to the spread of the Fremont people into central and northern Utah (Allison 2010). It is not clear where the populations flowing into our area in the mid-A.D. 800s were coming from, but it is a good bet that it is not from the south.[1] Several researchers (e.g., Wilshusen and Van Dyke 2006) have pointed to the probability that populations leaving our area around A.D. 900 were moving southeast, eventually contributing to the rise of the Chaco phenomenon. Some data from Mesa Verde National Park, just southeast of the VEP I area, also suggest relatively high populations there during the early Pueblo II population trough in the VEP area. If this proves to be true, then it will be interesting to assess whether Mesa Verde proper was attractive during relatively cool periods because of its superior cold-

1. Wilshusen and Ortman (1999) suggest southeastern Utah as a source. Population movements can be quite complicated, and the role of slow changes in temperature in particular probably did not quickly render any areas uninhabitable. Rather they would have made for slow changes in the means around which climates varied on an annual or decadal basis. This would have caused complicated movements that were affected by both low- and high-frequency variability, some of which would have been proximately caused by the cascading effects from populations displaced from increasingly difficult areas, who then moved into adjacent areas of low population, adding incentives for the populations already there to move in turn.

air drainage, given the south-trending aspect of the cuesta.

In the second population cycle, the new populations spreading into the central Mesa Verde region—some 60 to 70 years after increasing temperatures cross the long-term mean—quite likely haled from somewhere to the south since they brought with them the trappings of Chacoan society; this flow might have included people from Mesa Verde proper. The final surge of population into the VEP area occurs some 50 years after temperatures had begun to decline, by Wright's reconstruction. In Chapter 14 we suggested that these immigrants came—in part, at least—from southeastern Utah to the west, but if expectations based on temperature trends have any merit, we might also examine the possibility of immigration from the northwest as well. Finally, the direction of emigration during the final depopulation seems to be predominately toward the northern Rio Grande (Ortman 2009, 2010).

In general, I suggest as a working hypothesis that the populations moving into our area early in a warming cycle were coming from the south, or from lower-elevation and therefore warmer areas; and that populations moving into our area after the end of a warming cycle, well after temperatures had begun to cool (early to middle A.D. 800s and early to middle A.D. 1200s), were coming from the north or from areas that—because of edaphic or local topographic conditions—were particularly susceptible to temperature decreases. I also hypothesize that in both cycles the populations leaving our area as temperatures continued to cool were moving predominately south, or at least to areas lower in elevation. This working hypothesis of course owes some inspiration to Ken Petersen's work (e.g., 1994, 1995), and Michael Berry (1982:103–126) likewise recognized some of these population movements, though he attributed them solely to precipitation variability, which certainly does contribute. Of course, the existence of this pattern—should it be borne out by future research, including ours—does nothing whatsoever to lessen the importance of understanding the social processes that accompany these dislocations (see, e.g., Glowacki 2010).

Real Versus Simulated Population Histories

Most readers will have noticed by this time that although many of our simulations provide a reasonably good fit to the growth portions of the first population cycle, after that time most runs fit the real paleodemographic sequence poorly. In particular, the model does not capture the partial depopulation that occurred at the end of the first demographic cycle or the complete depopulation at the end of the second cycle. On average, the simulation runs register only a slight pause in their otherwise steady growth during the period when the Pueblo I villages were collapsing, and during the depopulation of the area in the late A.D. 1200s the simulated households were on average *increasing* in number, though only slightly (Plate 4.1).

Assuming that there are no serious errors in our models, the two major episodes of population decline must therefore have been caused by something we did *not* model. The contributors to the edited book by Timothy Kohler, Mark Varien, and Aaron Wright (2010) consider the many possibilities for the population collapse of the A.D. 1200s in more detail than is possible here, but (as already implied) I strongly suspect that the single most important feature that the VEP model does not presently account for is maize-production variability due to low-frequency climate variability, especially in temperature.[2] If this could be properly taken into account, I suspect that our production estimates in the last half of the A.D. 800s, the A.D. 900s (especially), and the A.D. 1200s would decrease. Conversely, these estimates would probably rise from A.D. 600 to A.D. 800 and from

2. Our maize paleoproductivity reconstruction uses high-elevation bristlecone pine as a proxy for temperature (Chapter 6). These records contain more low-frequency variability than do the lower-elevation Douglas fir ring-width index series that we use for precipitation, but are still much less sensitive to centuries-long variability than is pollen.

A.D. 1000 to A.D. 1200, though the reductions are likely to be larger than the increases. As has been demonstrated repeatedly, the training period for our maize production estimates, from A.D. 1930 to 1960, falls within a low-frequency warm period that began in the late nineteenth century (e.g., Dahl-Jensen et al. 1998; Le Roy Ladurie 1971:80–128; Luterbacher et al. 2004) in the northern hemisphere at least, and probably worldwide, and that apparently continues to accelerate. Declining maize production would decrease the number of agents in the model or increase their costs of doing business—most likely, some of both.

We suspect that two linked facts provide the next-most-important reasons that our simulated paleodemography departs from the sequence we reconstruct from archaeological data. First, there is no entry or exit from the Village world except by being born into it, or by dying. The model is a demographically closed system, though we know that there was in fact immigration into and out of this area during the 700 years of occupation. Second, all decisions in the model are made at the household level. It is clear that there was some emigration from our area beginning in the middle A.D. 1200s, and the speed at which it took place suggests that the decisions were often being made at the level of whole communities, or at least by kin groups larger than the household (see also Chapter 14 and Varien 2010). Most people did not tarry to leave by dying, although clearly quite a few died before they could leave (Kuckelman 2010a). The same factors acting on a smaller scale apparently characterized the A.D. 900 depopulation.

What we have learned from the archaeological record and its comparison with the model is consistent with the view that migration decisions are based in part on rough cost-benefit calculations for making a living in one place as opposed to another, but mostly on what one's neighbors, kin, and exchange partners are doing. Such decisions, then, were highly conformist in character, especially if one's identity was largely defined by one's social group, although these decisions would also be affected by one's age—young people are more likely to emigrate—and local resource holdings (vested interests, or sunk costs), which themselves may well be bound to a social group larger than the household. This mechanism easily leads to overshoot: more people come into an area than should, based on strict economic considerations, and likewise, more leave than would probably have to by those criteria.

EMPLOYMENT OPPORTUNITIES IN PREHISTORY

We can extrapolate from our simulation results to provide rough estimates for resource costs and time budgets during the thirteenth-century population peak in the VEP. These estimates are very approximate because of all the senses in which the Village simulation is a model and not reality. The most important of these discrepancies between the model and the real world are that these costs extrapolate from the agents' distributions on the landscape, which are different from (and lower in cost than) those of the real occupants. Also, the model does not take into account the advantages, and costs, of obtaining some protein from turkey. In addition, the model assumes some degree of soil degradation, but it does not consider possible gains from such things as improvements to cultigen varieties or fuel conservation from massing architecture, nor does it allow for a wide variety of additional factors (e.g., forest fires, grasshopper plagues, human errors, mischief, and innovation) that might cause either enhanced efficiencies or disastrous production shortfalls. Finally, our agents' households average about 3.3 members (Chapter 4), whereas we suspect that the households in this area more typically had five or six members. Although real households may therefore have had nearly twice as many hands as those in our model, they also had twice the mouths to feed, exacerbating resource supply constraints over those in our model.

In sum, most, though not all, of these factors suggest that we underestimate, probably substantially, the costs even for those limited activi-

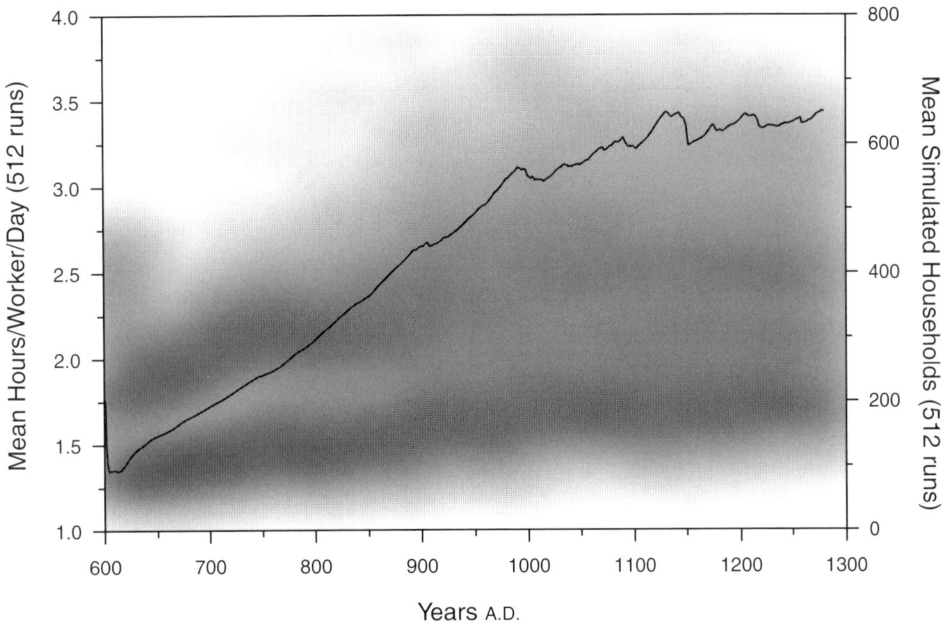

FIGURE 15.1 Mean number of simulated households through time, all 512 runs (right axis, black line), and mean hours worked / worker / day (left axis, gray cloud), through time. The gray cloud is a two-dimensional kernel density smooth (Wand 1994; Wand and Jones 1995) of the mean for all households in each run in each year.

ties that we do model. We do not calculate costs for many subsistence-related activities, such as making and maintaining ceramic, lithic, and basketry items that enable hunting and farming; building and maintaining storage features and agricultural features such as terraces; and food preparation. Food preparation, in particular was a major labor expenditure for women; corn grinding alone could easily require two hours each day (Weatherwax 1954:99). Moreover, all *nonsubsistence* activities—building and maintaining residences and civic-ceremonial structures, child care, and all the other activities of daily family and village life—are ignored entirely. Exchange, too, is costless and takes no time in the model. Finally, nobody in the model world is ever sick or injured. Data on traditional societies reviewed by Michael Gurven et al. (2000) suggest, by contrast, that on any given day somewhere between 6 and 22 percent of adults are too sick or injured to work.

Figure 15.1 shows the relationship between the average number of hours spent per day per worker on the subsistence tasks we do model through time, in each of the 512 runs reported here (in the point cloud), in relation to the number of households averaged across all those runs (black line). The average number of hours per day per worker across all the simulations remains fairly low—about two or three hours a day—although they increase gradually through time.

In Figure 15.2 we regress these mean daily hours per worker in each run on the number of households in each run at year A.D. 1131 (the average population peak for the simulations). This relationship is positive and significant, but not very strong (r^2=.07; p<.0001); the regression line splits the difference between runs with higher populations and lower time costs and those with lower populations and higher time costs. Most runs with higher populations and lower costs develop on the more productive landscapes (HARVEST_ADJUST=.75) and typically fall below the regression line, as would be expected.

If we were to extrapolate from this regression relationship to a population of 3,234

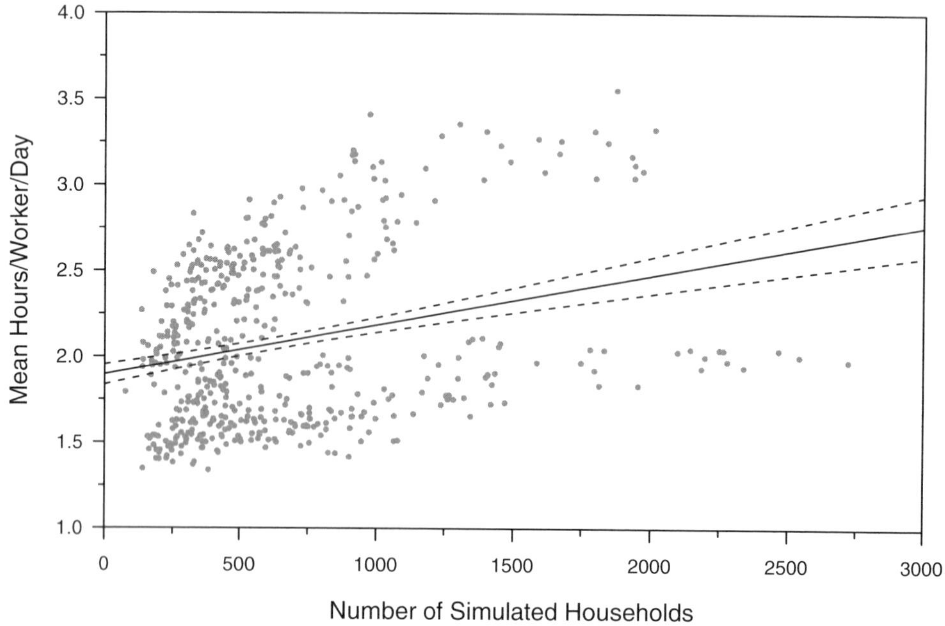

FIGURE 15.2 Regression of mean hours / worker / day on mean simulated households / run, for all 512 runs, at year 1131 ($r^2 = 0.07$, $p < 0.001$).

households—our median estimate for the number of households between A.D. 1225 and A.D. 1260 (Table 2.2)—we estimate that each worker (i.e., everyone seven years of age or older) would have been spending a little over 2.8 hours a day for the subsistence activities we model. This does not appear to indicate a time crisis in these societies, since adults in horticultural societies typically spend some four hours a day in production activities (including the foraging and farming that we model, but also commercial activities, which we do not)—though we need to keep in mind the likelihood that this represents a floor for the true time spent, and of course only for the activities we model.[3]

But remembering the small average size of the simulated households, we should also extrapolate from the number of *people* in the model to the 19,404 people notionally represented by the 3,234 households of our median estimate for the period between A.D. 1225 and A.D. 1260. Computed that way, our workers at the population peak would have been spending slightly over five hours a day on just those activities we model. This begins to look very unattractive, particularly if better options were available elsewhere.

SOCIAL (AND ANTISOCIAL) HISTORY

Maximum dispersion is the settlement pattern of the state of nature.

M. SAHLINS (1972:97)

Social history is connected to population history but is richer and less easily character-

3. This is according to data extracted by Stefani Crabtree from Ross De Forest Sackett (1996), who compiled information from disparate sources on 37 horticultural societies. Sackett (1996:126) defines horticultural societies as having "average adult farming time greater than either foraging or commercial time [with] agricultural techniques . . . mostly low-subsidy, low technology, human-powered, and land extensive systems such as swidden cultivation, arboriculture, and some rainfed upland cultivation systems." These data are far from perfect; Sackett (1996:149–150) advises that most of the cases were studied for less than one season. If, for example, this happened to be the main season for farming, then these costs represent all or most of the farming costs for the entire year, whereas all the costs we present from the model in this chapter are averaged over the entire year.

ized in a quantitative fashion. Two main aspects of this region's social history, however, overlap with factors that were examined in the model: exchange and aggregation.

Aggregation can be measured in any number of ways. In Figure 2.11C we graphed the number of households living in settlements with eight or more other households within the block survey areas. By this measure, aggregation peaks between A.D. 840 and A.D. 880, and again between A.D. 1260 and A.D. 1280, during or just before both major episodes of depopulation.

In Plate 15.1 (in color insert of this book) we use a simpler metric, but one which is easier to apply to both the simulated and the real households. For each period we graph the coefficient of variation (CV=[standard deviation × 100] / mean) in the number of household-years per cell across the simulated and real landscapes once both have been divided into matching 4-ha cells.[4] Although each of our simulations generates different numbers of households—and these are also generally different from the numbers estimated for the real sites—such scale effects are removed by this measure, since it divides by the mean household-years per period. The history of aggregation yielded by this measure—the red dots in Plate 15.1—is similar to that told by Figure 2.11C, although by the Chapter 2 measure aggregation peaked earlier in the first cycle and was more marked in the middle A.D. 1200s.

One very interesting result of the view of aggregation shown in Plate 15.1, though, is that the simulated households—whose CVs as calculated for each run, in each period, are shown by the tiny green dots—are initially (from A.D. 600–725) *more* aggregated than the real households. This rather unexpected finding puts the problem of aggregation in a new light: the slightly aggregated stance created by our agents is in fact the expected pose on this landscape, as judged by the criterion of least cost. Now, instead of just trying to explain why households are sometimes more concentrated in space than at other times, we find we should also be trying to explain why households are initially more dispersed than expected, where that expectation is derived from the model. To explain the problem of dispersion, we must consider what presumably noneconomic factors could be causing real households to live farther apart than our "rational" agents do. We should be looking for psychological predispositions or cultural habits that in times of necessity may be relaxed, with some discomfort, allowing households to live more closely together than they would prefer. The general history of group formation in our area and elsewhere in the Southwest suggests that the very strong identification of Pueblo households with a larger social group (and especially with the pueblo or town itself) was developed over the course of centuries and not given by any deep cultural or subsistence-related imperative.

During most of the first population cycle—except right at its end and during the pre-Chacoan portions of the second cycle—households are only about as aggregated as would be expected if they were independently minimizing their efforts to obtain satisfactory returns. It could still be the case that the various factors researchers have adduced to explain Pueblo I aggregation—e.g., as a means of avoiding conflict with adjacent communities for access to arable lands in conjunction with rapid rates of population growth (Orcutt et al. 1990), for efficiency of exchange (Kohler and Van West 1996), or to present strength through numbers (Wilshusen and Potter 2010)—played some role in encouraging households to overcome their evident reluctance to associate so closely with

4. For those unfamiliar with the CV, imagine a grid of nine cells. The CV for the case where one of these contains nine households, with no households in the other eight cells, is 300 (s=3, mean=1; [3 × 100] / 1=300); for the case where there is one household in each of the nine cells, the CV is zero, since the standard deviation is zero. In calculating these CVs, we first multiply the household counts in the real-site data plane by the estimated occupation span for households in small sites from Varien et al. (2007:Table 3) to make them comparable to the household-years stored in each cell by the simulation. Unsurveyed cells in the real landscape are considered missing values, not zeroes; simulated cells that are zero and also have missing (unsurveyed) values in the same location in the real-site data plane in any period are also set to missing.

others. Except for the period between A.D. 880 and A.D. 920, however, we cannot rule out the possibility that households were only behaving in an individually economically rational manner, gently nudged by competitive pressures toward greater efficiency. (Compare Plates 10.1 and 15.1.)[5] Of course, a particular settlement such as Grass Mesa Pueblo, the largest of its day, still may be viewed as anomalously large in the context of its contemporaries; the analysis in Plate 15.1 refers to mean values across the entire landscape. This is nevertheless a humbling reminder that we need better ways to build expectations about what the archaeological record "should" look like according to various processes of interest, rather than simply assuming that we can recognize a high degree of aggregation (or whatever) when we see it. Developing methods for generating such expectations is, of course, one goal of the VEP.

However, it is also clear that in the final years of the Pueblo I cycle, and throughout the Chaco and post-Chaco periods, the degree of aggregation exceeded that expected—often greatly. Most of the periods in which aggregation exceeds the expectation are also violent, reinforcing the notion that aggregation is often motivated by security concerns. The chief exception is from A.D. 1180 to A.D. 1225, in which population is considerably more concentrated than expected (Plate 15.1) but in which Sarah Cole (Chapter 13, this book) saw little osteological evidence for strife. This little-studied period is emerging from our analyses as an interesting time, since within it people succeeded in bringing the rampant violence of the middle A.D. 1100s to an end even as local populations continued to increase and climatic conditions began to deteriorate. Perhaps it is in times like this—in which structural conditions seem to point in one direction and the settlement practices in the archaeological record in another—that exercise of agency becomes most attractive as a category of explanation. Note, however, that our ability to make this suggestion at all depends on having models that provide strong structural expectations.

We can ask of our simulations what parameter settings lead to high degrees of aggregation, as indicated by high CVs in Plate 15.1. In so doing, it should be kept in mind that our agents never aggregate as much as we see throughout most of the second cycle, and that some of the factors we believe to have been important for aggregation—such as violence—play no role in the lives of our simulated households.

Nevertheless, the parameters that lead to high CVs are interesting and not intuitive. A regression model containing five parameters explains 37 percent of the variability in CVs across all runs and all periods. The five parameters that lead to high aggregation, in order of their decreasing importance as measured by their standardized slope estimates, are high values for NEED_MEAT and PROTEIN_NEED, a low value for STATE_GOOD, a high value for HARVEST_ADJUST_FACTOR, and a low value for HUNT_SRADIUS.

When NEED_MEAT is 1 rather than zero, agents are constrained to move to cells within HUNT_SRADIUS from which hunting is possible, even if not is the least-cost cell; if there are no such cells, then agents do not move at all. Since low values for HUNT_SRADIUS and high values for PROTEIN_NEED are also connected with high aggregation, such failed moves appear to greatly contribute to aggregation among our agents. This is aggregation as a "no-acceptable-place-to-move" phenomenon.

A low value for STATE_GOOD, which slows population growth, and a high value for HARVEST_ADJUST_FACTOR, which lowers overall landscape maize production, are also connected with high CVs. These two parameter settings probably work together. When production is high, cells that would not otherwise be acceptable locations

5. The fact that households in the period between A.D. 880 and A.D. 920 were decidedly more aggregated than expected under the household-level optimization hypothesis instantiated by our agents (Plate 15.1) but were also the most efficiently located in space (Plate 10.1) of those in any period implies that their choice of settlement locations was economically sensible, although the degree to which households were concentrated within them was often "excessive."

come into play, potentially decreasing CVs. High population growth rates mean that more of these less desirable, but still farmable, locations will actually be employed. Therefore, low population growth and a less productive landscape will tend to increase the CVs.

Although it is far from certain whether the specific factors that enhanced aggregation in the simulations operated in a similar manner in the lives of Pueblo people (I rather doubt it), it may be that factors in the real world creating an environment in which there was "nowhere better to go" also enhanced aggregation. This could be the case if aggregation is partly a density-dependent phenomenon. Using the insights provided by Plate 15.1, we can construct a new measure for aggregation as the difference between the CV for the realized degree of aggregation (the red dots) and the mean of the CVs for the simulations (green dots) in each period. Call this "unexpected aggregation"—i.e, the average degree to which aggregation in each period departs from that practiced by the agents. (Of course, this value can be either positive or negative.) We can then regress this measure on both the middle VEP population estimate (from Table 2.2, converted to households) and Cole's final version of the warfare index (Table 13.4). The logic for doing this is that both "population packing" and a violent environment could contribute to loss of freedom for local movement or difficulty in fissioning, either of which would promote aggregation.

As it happens, together these independent variables explain 40 percent of the variability in our new variable, unexpected aggregation ($R^2=.4$; adjusted $R^2=.29$), but given the small sample size (14 periods) the significance of the relationship is marginal ($p=.06$). Table 15.1 contains the variables used in the regression, as well as the studentized residuals from the model. As VEP-area population size and the regional level of interpersonal violence increase, this measure of aggregation (which already has the household-level economic effects removed) also increases. The largest residuals are in period 6 (A.D. 600–725), when households are markedly less aggregated than the model expects, and period 10 (A.D. 880–920), when they are markedly more aggregated than the model anticipates. In other periods the fit of the model is relatively good.

Despite a past hypothesis that one reason households aggregate is to facilitate interhousehold exchange, especially of maize (e.g., Kohler and Van West 1996), whether the COOP parameter is set to zero (no exchange) or 4 (enabling both a balanced reciprocal exchange network and a generalized reciprocity network) has almost no effect on the degree of aggregation in the model world as measured by these CVs. We will conduct more experimentation with our parameterization for exchange to determine whether increased volumes of exchange effect aggregation in our simulations, and if so, the point at which that happens.

TAKING STOCK

Villages and Their Discontents

So far in this book we have tried to stay close to the data (including those provided by our simulations). Having duly eaten our vegetables, we now all deserve some dessert. The remainder of this chapter is guided by the evidence collected by the VEP but also goes well beyond it.

If some future researcher can find a way to measure happiness in the archaeological record, we predict that the seventh-century A.D. will rank at the top of our local record. The rich and game-filled landscape open to these pioneer farmers would support a household virtually wherever one wished to dig a pit house. More than at any other time, these folks could really get away from it all—which they seem to have enjoyed.

The rapidity with which guiding settlement parameters come and go for the first 300 years (Plate 10.3) indicates that people were constantly experimenting with new ways to locate themselves on this landscape. It may also be that populations with different settlement practices are being lumped together in this figure,

TABLE 15.1
Variables for Regression of Unexpected Aggregation on Population Size and Warfare Index, and Residuals from That Model[a]

Period	CV (Real)	Mean CV (Simulated)	Difference ("Unexpected Aggregation")	Warfare Index (*W*)	Households	Student Residuals
6	231	725	–494	.1038	304	–1.8
7	616	629	–13	.0000	326	–.1
8	708	648	59	.0000	836	–.2
9	794	610	184	.0358	1,030	.0
10	1,559	579	980	.3154	370	2.3
11	791	530	261	.0000	289	.7
12	528	545	–16	.2220	653	–.7
13	585	537	48	.2353	671	–.6
14	952	544	408	.5387	1,385	–.5
15	1,172	542	631	.4065	1,940	.2
16	1,247	530	716	.8894	2,078	–.8
17	1,303	523	780	.0915	2,326	1.1
18	947	542	405	.0673	3,234	–.9
19	1,418	551	867	.4210	1,770	.9

a. Unexpected aggregation = –24.77 ($p>|t|=0.88$) + 0.112*households ($p>|t|=.14$) + 616.72*warfare index ($p>|t|=.15$).

contributing to the freedom with which the parameters seem to vary.

Nevertheless, there are at least two senses in which high-level order seems to emerge out of the chaotic-appearing settlement practices of this first population cycle. First, households move without fail in the direction of increased aggregation (especially using the metric in Table 15.1, derived from Plate 15.1), and at the same time they move in the direction of increased settlement efficiency (Plate 10.1). It is as if the gentle constraints of structure were gradually but inexorably tightening around the freedom of household activities—a good example of structuration.[6]

But what exactly were these constraints? They arise mainly, I suggest, from the competition among households—and the social groups to which they belonged—for access to key resources on the landscape. On a per capita basis these resources became scarcer as populations grew, and it is critical to remember that except for its final century, the entire occupation of the VEP area takes place in a period of unprecedented regional population growth (Kohler and Varien 2010).

Fewer resources per capita of course encouraged people to live in those areas of greatest resource productivity, especially since areas of less productivity would become depleted first. We have employed our agents—who will work for only electricity and our coding time—to discover where those places with the highest resource-renewal rates and happiest conjunction of critical resources were on the ever-changing landscape. Pueblo people increasingly mapped onto these areas throughout the first 300 years. The best areas—in general, where the Pueblo I

6. Perhaps, as Mark Varien suggested to me, these populations may have seen these new behaviors not as a loss of freedom but as ways to live better in the new circumstances.

villages were located—would have been the focus of competition among kin groups. The largest kin groups, or kin groups that could recruit other groups, would have been successful in acquiring and maintaining these locations, so long as the overhead from maintaining the larger group did not overwhelm the advantages of maintaining the more attractive territory. Part of the overhead in maintaining the larger groups was the cost of breaking through at least two social barriers.[7]

In the movement from households to Pueblo I villages, an intermediate form of organization also became prominent: these are lineages or clan segments, which I assume to be the social unit ordinarily represented by the larger hamlet-scale spatial units and the room blocks of the villages. Families in these units, having first had to learn to live with each other, then had to learn to live with other more distantly related clan segments in villages that eventually came to include 100 or so other households. This began to happen by A.D. 770 in our area. This was probably a more traumatic and delicately negotiated transition than the household-to-hamlet transition, since it appears that later, cooperative relations between lineages (or clan segments) often collapsed, whereas apparently those within lineages generally endured.

So far we have imagined households as somewhat passive participants in a process that included both aggregation and construction of new social relations allowing cooperation with increasingly distantly related people. Now I would like to suggest an interpretation of Plate 15.1 in which the distance of the red circles above the central tendency for the green dots is a measure of the average increasing returns to aggregation through social coordination across the landscape. This distance (or "unexpected aggregation") begins to be positive in the early A.D. 800s and remains so until the early A.D. 1000s, when it briefly became essentially zero (Table 15.1). After that it turns positive again throughout the remainder of the occupation—strongly so, in most periods.

What could these returns be, that are more easily achieved in large groups than by individual households or hamlet-sized groups? We have already mentioned the advantages larger groups have in claiming and maintaining desirable territories (see Shennan 2007:153 for a similar argument for Neolithic Europe). All other factors equal, when two groups are in conflict, the larger prevails; this has been formalized as Lanchester's Square Law (Lanchester 1916). One bit of evidence that prime agricultural territories were sought out is in Figure 14.8, which demonstrates that except between A.D. 1140 and A.D. 1260, catchments of large sites were more productive than the landscape as a whole. I think superior deer-hunting territories were perhaps even more important to Pueblo I peoples (Kohler and Reed 2011). Remembering the first villages were forming in the context of increasing human population size and increasing depression of deer, it would have been especially advantageous to larger groups that they could field the larger long-distance hunting parties increasingly important to hunting success. It is extremely likely, based on current studies on why men hunt, that establishing and maintaining individual prestige, and that of the lineages and clan segments, were as important to real Pueblo I households as were the grams of protein to our agents. These are advantages that villagers would have been more than aware of, and they could easily have been decisive in their decisions about where to live and with whom.[8]

7. If the pit houses of the A.D. 600s were occupied by nuclear families, one more transition was involved. In that case, these households had to come to terms with living in tight quarters with close relatives, which happened by the early A.D. 700s as hamlets of two or three households began to become more common. Eastern Pueblo peoples would never again return to the radically dispersed settlement pattern of the seventh century; the "unit Pueblos" defined by Mitchell Prudden for Pueblo II were on the same social scale as the hamlet if, as I believe, single pit house sites by Pueblo II times were typically occupied by extended families and not nuclear families.

8. Not only are villages better at this, but as the landscape increasingly took on an aggregated posture it would be difficult or impossible for the independent, single-family farmstead or hamlet to defend its land-use rights and to preempt those rights from others. There were

Conveniently, the people who could best defend their village's claims to its territories, or dislodge smaller groups from theirs, were more or less the same personnel involved in long-distance deer hunting.

A less obvious advantage of larger groups that might still have contributed to their increasing returns is their advantage in accumulating and maintaining more complex technologies and skills (Henrich 2004; Powell et al. 2009), including social technologies and skills. To the extent that these improved their quality of life, this in turn would make them a target for selective immigration from other groups (Boyd and Richerson 2009), further increasing their size. It is now well known through scaling studies that in contemporary cities, systems that increase information exchange as well as indices of wealth and creativity grow more rapidly ("superlinearly") than does city size itself (e.g., Helbing et al. 2009). Do the "public" buildings and increased storage capacities of our community centers signal something very similar, albeit on a smaller scale? There are some hints from scaling studies conducted by Kohler and Varien (2010) that this may be the case, at least in the second population cycle, and Fumiyasu Arakawa's analyses of lithic material distributions in the late Pueblo III might suggest more intercommunity movement, facilitating more exchange of information and cooperation, in the A.D. 1200s. These approaches to aggregation will be a key area for future research.

In the second population cycle, using the metric in Plate 10.1 and emphasizing the "rand2" results, settlement efficiency increased during the Chaco periods (A.D. 1060–1140) and then slowly decreased during the post-Chaco periods, trending up again slightly (or rapidly, according to the metric in Plate 10.2) during the final depopulation.

In general, the Chaco and post-Chaco populations seem to be placed on the landscape in a way that resembles that chosen by our agents. The apparent settlement efficiency in these periods would almost certainly be higher if it were not for three factors. First, our agents do not receive any of the increasing returns to aggregation (either the economies of scale, or the "superlinearities") that I suggested are necessary to understand aggregation in the record. Therefore, the real households appear, by the efficiency metric (Plate 10.1), to be more aggregated than they "ought" to be (Plate 15.1). Second, our agents seek to live within 200 m of their fields, whereas a field-house strategy in periods of aggregation relaxed that requirement for the real households. It is likely, though not certain, that if our agents could implement a field-house strategy their locations would resemble more strongly those chosen by the Pueblo II and III populations. Finally, turkey was an important source of protein for Chaco and post-Chacoan populations, whereas our VEP I agents try to get all their protein from hunting. Presumably turkey husbandry required places where maize production was high and water was handy. But was hunting no longer valued? We can see from Plate 10.3 that relatively high values for PROTEIN_NEED and NEED_MEAT typify the late periods. From these we can infer that hunting—though certainly less rewarding by this time than earlier, and no longer unique in providing high-quality protein—was still important in location of settlements.

We know from the archaeological record (Chapters 2 and 14), and can infer from the stability of the particular settlement parameters typifying the late cycle in Plate 10.3, that settlement locations were much more durable in the Chaco and post-Chacoan periods than before A.D. 1000. Since this stability and aggregation accompany each other, it is tempting to conclude that they are both caused in part by the much higher populations on the landscape, which impeded movement. The low values for SOIL_DEGRADE during this time may therefore simply be an epiphenomenon of that stability—

small sites throughout the Pueblo II–Pueblo III sequence, but they were a part of larger communities that included larger sites, and it is likely that this community membership enabled these small sites to survive (Adler 1990).

i.e, it is only by having a low value for SOIL_DEGRADE that our agents can achieve such stability. If so, these low values for SOIL_DEGRADE need not have any specific referent in the real settlement practices. However, it is also true that soil and water control devices such as check dams, terraces, and reservoirs were constructed primarily during these times (Lipe and Varien 1999a:281–282), signaling active attempts at controlling topsoil and water resources. We might speculate that turkeys foraging in maize fields also contributed to pest control and returned a little nitrogen at the same time, slightly promoting longevity of field use.

The long-term patterns of violence preserved on human bones (Chapter 13) suggest that Chacoan populations were resisted before and on their arrival and while the polity they seem to have imposed was in place. Violence peaked as this system was falling apart in the mid-A.D. 1100s, suggesting desperate measures to keep the system together, score settling on its demise, or both. Considering that a human being can endure a great deal of violence without suffering osteological consequences, violence that affected some 90 percent of the bodies found in the archaeological record sufficiently to leave marks on the bones must have been a major structuring force in these societies. Moreover, given the general tendency for populations to flee areas of violence, the continued population growth from A.D. 1120 through A.D. 1160 and on into the A.D. 1200s (when violence suddenly abates) is also remarkable. The Chaco-era reversal of the relationship between population growth and violence seen from A.D. 600 through A.D. 1000 implies that the violence from A.D. 1060 through A.D. 1140 was of an entirely different character than during the first 400 years of occupation. Likely it had become a tool for achieving compliance rather than an expression of population-resource imbalances.

The association of Chaco in our area with *both* an unexpected degree of aggregation *and* high levels of violence leads us to wonder whether the strongly built great houses that identify Chacoan centers on the landscape may have been not just places of elite residence, or community ritual, but also places of refuge for at least some community members in times of exigency. The great houses not only invoked deep thematic qualities by employing long- and widely-used San Juan architectural conventions (Lipe 2006:268–271), but one can infer by the remarkable size of their rooms and "unnecessarily" thick walls—both of which exaggerated conventions—as well as by their typical location on high ground toward the centers of communities that they could have provided places of temporary refuge. If Chacoan elites extracted goods and services in the name of a political and religious system embodied by the great houses, the high levels of violence might suggest that this extraction was not always welcomed, and that Chacoan dominance of the countryside was not complete (see also Chapter 14). If these suggestions have any merit, then it must also be true that being a follower of Chaco in the VEP area, at least, was not without its risks.[9]

Archaeologists have recently written a great deal on the last century of occupation in the VEP area and the reasons for the thirteenth-century depopulation of the northern Southwest in general (e.g., see contributions to Kohler et al., editors, 2010). There is no need to repeat those discussions here. VEP researchers, in particular, have improved the chronology for and recognized the importance of low-frequency climatic variability, especially in temperature (Chapter 3 and this chapter); defined the surge of immigration to the VEP area in the early to middle A.D. 1200s, mainly occupying small centers in its western part (Figure 2.10, top); recognized that the final depopulation, accompanied by a significant increase in violence, was already well under way by A.D. 1260 at the latest (Chapters 2,

9. Most of these great houses are relatively small structures that would not have held many community members, so if this suggestion has any merit—and I raise it as a possibility to be investigated by future research—it will need to be strengthened by lines of evidence beyond the coincidence of the Chacoan system with high levels of violence. It is always quite possible that the high levels of violence we see in our area in conjunction with Chaco were not universal or even common.

13, and 14); and have defined the time crunch, extrapolated from the simulations, for producing basic household requirements at the population peak (this chapter). Kristin Kuckelman (2010a) argues that occupants of one of the VEP area's largest centers, Sand Canyon Pueblo, had essentially reverted to a foraging lifestyle by the end of its occupation in the late A.D. 1270s. Although this position seems extreme, it could help explain the anomalous patterns in the lithic distributions discussed in Chapter 12 by Arakawa, who shows that the late Pueblo III period resembles the Basketmaker III and Pueblo I periods much more than it does the Pueblo II and early Pueblo III periods (Figure 12.6).

In general, the more we learn, the more important push factors seem to have been in the final depopulation. These definitely included a desire to move away from areas with poor returns to labor—and away from violence. One can be even more speculative and say that motivations for departure might also have included the desire to abandon a repressive social system with few opportunities for advancement in favor of a putatively more open system developing to the southeast. In my view it is no accident that many of the specific characteristics contributing to a very "high overhead" for life in the thirteenth-century central Mesa Verde region are those that are not continued in the postmigration northern Rio Grande societies (see also Lipe 2010). These obviously include the significant labor expended on the elaborate civic-ceremonial structures discussed in Chapter 14, and some of the decorated ceramics that required high labor input. More putatively it also includes the considerable demands of clan-based gerontocracies that not only controlled the best lands but also quite likely a lengthy calendar of ceremonies. If northern Rio Grande societies offered an economic and social system in which opportunities were widely available and the benefits of ceremonial participation more broadly shared, we can understand how pulls could have reinforced pushes in the final depopulation. As Eric Wolfe (1982:94) argued, "While kin ordering . . . sets upper limits to internal differentiation, under conditions of closed resources it appears more likely to produce inequalities than an egalitarian distribution of life chances."

If we look at the history of the eastern Pueblo Southwest in its entirety, a few of our earliest (Pueblo I) villages lasted about a century; most had shorter use-lives. Over 150 years later, a second cycle began in which no community center lasted for more than about two centuries. The Spanish interrupted the third wave of village life in the northern Rio Grande after it had lasted a little over two centuries, so it is hard to say how long those pueblos might have endured without that intervention. As it happened, though, quite a few contemporary pueblos have been in place for many centuries. With each of these cycles, then, village life has tended to become more stable, even as the villages themselves became larger, as if Pueblo peoples were slowly working out the practices that could make these organizations socially and environmentally sustainable.

There are some hints that the first wave of Neolithic village life may often have been somewhat short-lived in other places in the world as well, at least relative to later developments in the same areas. Nigel Goring-Morris and Anna Belfer-Cohen (2010) discuss the early Neolithic Pre-Pottery Neolithic B villages of the Levant that contained hundreds and perhaps occasionally thousands of occupants between about 8500 cal B.C. and 6400 cal B.C., before giving way to dispersed farmsteads. Given the dating resolution provided by ^{14}C, it is difficult to know how long individual villages lasted, but their duration seems to be significantly shorter than for many later towns in the same area. Goring-Morris and Belfer-Cohen attribute a large part of the demise of Pre-Pottery Neolithic B villages to zoonoses (diseases harbored by and transmitted from domestic livestock), suggesting that these early villagers had not yet learned the precautions necessary for living in close association with domesticated animals over long periods. Another possible example comes from the earliest Neolithic sites in the Balkans, which did not form the tells that were characteristic of sites from the Middle Neolithic on (Lichardus-Itten 1993).

If such experiences turn out to be fairly general, it may be because the weak social ties that maintained cooperation in early villages were unable to overcome the novel environmental problems that arose in settings that combined, for the first time, relative permanence and relatively high populations. Early villagers faced two essential problems whose solutions probably overlapped at least a little. First, they had to construct and maintain a novel social organization allowing many unrelated people to get along and coordinate their actions well enough to generate the sorts of increasing returns I suggested must have existed. Second, this had to be done while avoiding new environmental problems posed by more people in a more concentrated and sedentary situation than had ever been experienced. These two problems overlap in the development of cooperation for restraint in resource use, and in the ability for nascent leaders to reign in defection in such cooperation. The problems of the past are not always unique to the past.

Archaeologists and Their Discontents

However hard we may try, archaeologists are not computer scientists, and computer scientists are not archaeologists. Thus Village—the model—is an imperfect tool in a number of ways. Its production landscapes do not take into account the low-frequency variability in temperature I have argued to be important. Its inhabitants do not like to work too hard—they would rather take it easy than reproduce or produce as much as did the real Pueblo peoples. The households they create are probably smaller than they should be. These households are incapable of imaginative intensifications such as the domestication or management of turkey, or the building of a reservoir or a wall. They cannot band together to invent economies of scale in production or in their relations to other groups. In fact, they cannot form real groups to coordinate actions to mutual advantage, although households ordinarily have kin and exchange partners. They can neither leave our area nor come into it; if they go too far south they reappear again in the north. In effect, they live in a virtual Groundhog Day world, with the depressing difference that—with the minor exception of some learning about exchange (Chapter 11)—they cannot learn to behave better.

Still, for all their faults, our agents do a pretty good job. They generate an exact deduction from a particular model on a landscape that is changing because of both climatic variability and their own impacts. Each year of the simulation is essentially a test implication of that model. To the extent that our model and our programming of it are correct, they show us where households *ought* to be located, and in what numbers. They allow us to see how hard it would be to make a decent, if somewhat basic, living in those places. Just like real people, the agents experience trade-offs: when circumstances require that they work harder to produce their basic livelihood, they have less energy to invest in reproduction (e.g., Chapter 5). They allow us to look over their shoulders to gauge the effects of their labor on the landscape. Within the range of the parameters they have available, they make it possible for us to see what values for which parameters provide the best fits to the real settlement practices, and how those change through time. By their perfect simplicity and autonomy they throw into relief the human, social, and cultural factors that were critical in forming the real history of Pueblo peoples in the northern Southwest.

They allow us to infer that through time, in each population cycle, there was a general tendency for the efficiency of settlement location to increase, in step with population size and inferred competition among social groups. But this was not a simple monotonic trend throughout the occupation, as some evolutionists of old might have predicted; it was a complicated stop-and-start-again phenomenon that lost its power to shape practice when population size decreased and competition waned.

I assume that the optimality trends we note in settlement are echoed in other contemporaneous aspects of technology and society. Processual archaeologists (e.g., Christenson 1980) had a great interest in subsistence intensification, and I follow them in assuming that the same forces that lead to more optimal settlement location and size will at the same time encourage higher-yielding combinations of foods, techniques for achieving higher production (including simply working harder), and more efficient processing technologies. Such intensifications do indeed seem to appear in our area at the appropriate times. Subsistence intensification is just one more arena in which competition driven by population size or climate change encourages more optimal practices. Adaptation happens; to the extent that we ignore these processes, our understanding of the human past is incomplete. In particular, we will fail to understand the causes for the increases in the sizes of the unrelated groups in which we live. These increases are an important part of human history, and they set us apart from most other living creatures.

At the same time, periods such as the A.D. 600s and the century from A.D. 920 to A.D. 1020 in our area are no less interesting simply because such competition was apparently less prominent. They allow us to see how people behave in the absence of such pressures. Probably these are times when a great deal of "neutral" variability in behavior, technology and social arrangements can arise. A parallel example is, possibly, the early Neolithic in Greece, who present an "unusually high" level of intersite architectural variability (Perlès 2001:173). Despite the oft-cited pleasures and importance of sociality, especially for Pueblo peoples, these periods demonstrate that even Pueblo peoples generally disaggregate when they can.

Although I consider many of the analyses in this book to be innovative, it will not be shocking to most archaeologists that variable pressures toward optimality exist, and can be identified, in various aspects of the human past. Why should Marshall Sahlins, for example, decree that the "decisive" aspect of our human existence is the scheme of meanings that we construct for the world around us (Sahlins 1976:viii), when the ability of humans to construct and maintain ever-larger groups of unrelated individuals is equally a circumstance not shared with other organisms? In my opinion it does not have to be one or the other. The decisive aspect of humans is that we live simultaneously in a material and a social world, and a key tool in our adaptive flexibility has been that we can endow both of these with meanings that are differentially shared.

It has now been a long time since most of the world's populations lived in the sort of small villages we have been describing here—long enough for some to regard them with nostalgia (Critchfield 1981). Of course, many villages remain in the world, but most are now cogs in larger machines, victims or beneficiaries of social construction processes leading to larger groups, organized differently. Although many factors affect the structure and operation of these organizations, I hope we have made the point in this book that in the crucible of competition among groups, some organizations, and some practices, will prove to work better than others, winnowing a forest of possibilities into just a few recognizable species.

Turning this logic inward, this book has been an extended argument in favor of a model-based archaeology guided by evolutionary logic. We do not ask that it supplant other approaches to archaeology—selection is ineffective without variability—but we think its share of the intellectual marketplace is poised to grow. We now look forward to expanding the geographic and temporal scope of the VEP, and to concentrating our modeling efforts more squarely on the construction processes of creating and maintaining larger social groups, whose effects we have so far only noted.

Appendix A

Parameter Values for Each Run Reported in This Book

Run	SOIL_ DEGRADE	COOP	PROTEIN_NEED	PROTEIN_ PENALTY	HUNT_ SRADIUS	HARVEST_ ADJUST_ FACTOR	NEED_MEAT	STATE_GOOD	H_2O_TYPE
1	1	0	10	0	20	.75	0	.1	3
2	1	0	10	0	20	.75	0	.1	1
3	1	0	10	0	20	.75	0	.2	3
4	1	0	10	0	20	.75	0	.2	1
5	1	0	10	0	20	.75	1	.1	3
6	1	0	10	0	20	.75	1	.1	1
7	1	0	10	0	20	.75	1	.2	3
8	1	0	10	0	20	.75	1	.2	1
9	1	0	10	0	20	.95	0	.1	3
10	1	0	10	0	20	.95	0	.1	1
11	1	0	10	0	20	.95	0	.2	3
12	1	0	10	0	20	.95	0	.2	1
13	1	0	10	0	20	.95	1	.1	3
14	1	0	10	0	20	.95	1	.1	1
15	1	0	10	0	20	.95	1	.2	3
16	1	0	10	0	20	.95	1	.2	1
17	1	0	10	0	40	.75	0	.1	3
18	1	0	10	0	40	.75	0	.1	1
19	1	0	10	0	40	.75	0	.2	3
20	1	0	10	0	40	.75	0	.2	1
21	1	0	10	0	40	.75	1	.1	3
22	1	0	10	0	40	.75	1	.1	1
23	1	0	10	0	40	.75	1	.2	3
24	1	0	10	0	40	.75	1	.2	1

(continued)

Run	SOIL_ DEGRADE	COOP	PROTEIN_NEED	PROTEIN_ PENALTY	HUNT_ SRADIUS	HARVEST_ ADJUST_ FACTOR	NEED_MEAT	STATE_GOOD	H_2O_TYPE
25	1	0	10	0	40	.95	0	.1	3
26	1	0	10	0	40	.95	0	.1	1
27	1	0	10	0	40	.95	0	.2	3
28	1	0	10	0	40	.95	0	.2	1
29	1	0	10	0	40	.95	1	.1	3
30	1	0	10	0	40	.95	1	.1	1
31	1	0	10	0	40	.95	1	.2	3
32	1	0	10	0	40	.95	1	.2	1
33	1	0	10	1	20	.75	0	.1	3
34	1	0	10	1	20	.75	0	.1	1
35	1	0	10	1	20	.75	0	.2	3
36	1	0	10	1	20	.75	0	.2	1
37	1	0	10	1	20	.75	1	.1	3
38	1	0	10	1	20	.75	1	.1	1
39	1	0	10	1	20	.75	1	.2	3
40	1	0	10	1	20	.75	1	.2	1
41	1	0	10	1	20	.95	0	.1	3
42	1	0	10	1	20	.95	0	.1	1
43	1	0	10	1	20	.95	0	.2	3
44	1	0	10	1	20	.95	0	.2	1
45	1	0	10	1	20	.95	1	.1	3
46	1	0	10	1	20	.95	1	.1	1
47	1	0	10	1	20	.95	1	.2	3
48	1	0	10	1	20	.95	1	.2	1
49	1	0	10	1	40	.75	0	.1	3
50	1	0	10	1	40	.75	0	.1	1
51	1	0	10	1	40	.75	0	.2	3
52	1	0	10	1	40	.75	0	.2	1
53	1	0	10	1	40	.75	1	.1	3
54	1	0	10	1	40	.75	1	.1	1
55	1	0	10	1	40	.75	1	.2	3
56	1	0	10	1	40	.75	1	.2	1
57	1	0	10	1	40	.95	0	.1	3
58	1	0	10	1	40	.95	0	.1	1
59	1	0	10	1	40	.95	0	.2	3
60	1	0	10	1	40	.95	0	.2	1
61	1	0	10	1	40	.95	1	.1	3
62	1	0	10	1	40	.95	1	.1	1
63	1	0	10	1	40	.95	1	.2	3
64	1	0	10	1	40	.95	1	.2	1
65	1	0	25	0	20	.75	0	.1	3
66	1	0	25	0	20	.75	0	.1	1
67	1	0	25	0	20	.75	0	.2	3
68	1	0	25	0	20	.75	0	.2	1
69	1	0	25	0	20	.75	1	.1	3
70	1	0	25	0	20	.75	1	.1	1
71	1	0	25	0	20	.75	1	.2	3

Run	SOIL_ DEGRADE	COOP	PROTEIN_NEED	PROTEIN_ PENALTY	HUNT_ SRADIUS	HARVEST_ ADJUST_ FACTOR	NEED_MEAT	STATE_GOOD	H_2O_TYPE
72	1	0	25	0	20	.75	1	.2	1
73	1	0	25	0	20	.95	0	.1	3
74	1	0	25	0	20	.95	0	.1	1
75	1	0	25	0	20	.95	0	.2	3
76	1	0	25	0	20	.95	0	.2	1
77	1	0	25	0	20	.95	1	.1	3
78	1	0	25	0	20	.95	1	.1	1
79	1	0	25	0	20	.95	1	.2	3
80	1	0	25	0	20	.95	1	.2	1
81	1	0	25	0	40	.75	0	.1	3
82	1	0	25	0	40	.75	0	.1	1
83	1	0	25	0	40	.75	0	.2	3
84	1	0	25	0	40	.75	0	.2	1
85	1	0	25	0	40	.75	1	.1	3
86	1	0	25	0	40	.75	1	.1	1
87	1	0	25	0	40	.75	1	.2	3
88	1	0	25	0	40	.75	1	.2	1
89	1	0	25	0	40	.95	0	.1	3
90	1	0	25	0	40	.95	0	.1	1
91	1	0	25	0	40	.95	0	.2	3
92	1	0	25	0	40	.95	0	.2	1
93	1	0	25	0	40	.95	1	.1	3
94	1	0	25	0	40	.95	1	.1	1
95	1	0	25	0	40	.95	1	.2	3
96	1	0	25	0	40	.95	1	.2	1
97	1	0	25	1	20	.75	0	.1	3
98	1	0	25	1	20	.75	0	.1	1
99	1	0	25	1	20	.75	0	.2	3
100	1	0	25	1	20	.75	0	.2	1
101	1	0	25	1	20	.75	1	.1	3
102	1	0	25	1	20	.75	1	.1	1
103	1	0	25	1	20	.75	1	.2	3
104	1	0	25	1	20	.75	1	.2	1
105	1	0	25	1	20	.95	0	.1	3
106	1	0	25	1	20	.95	0	.1	1
107	1	0	25	1	20	.95	0	.2	3
108	1	0	25	1	20	.95	0	.2	1
109	1	0	25	1	20	.95	1	.1	3
110	1	0	25	1	20	.95	1	.1	1
111	1	0	25	1	20	.95	1	.2	3
112	1	0	25	1	20	.95	1	.2	1
113	1	0	25	1	40	.75	0	.1	3
114	1	0	25	1	40	.75	0	.1	1
115	1	0	25	1	40	.75	0	.2	3
116	1	0	25	1	40	.75	0	.2	1
117	1	0	25	1	40	.75	1	.1	3
118	1	0	25	1	40	.75	1	.1	1

(continued)

Run	SOIL_ DEGRADE	COOP	PROTEIN_NEED	PROTEIN_ PENALTY	HUNT_ SRADIUS	HARVEST_ ADJUST_ FACTOR	NEED_MEAT	STATE_GOOD	H_2O_TYPE
119	1	0	25	1	40	.75	1	.2	3
120	1	0	25	1	40	.75	1	.2	1
121	1	0	25	1	40	.95	0	.1	3
122	1	0	25	1	40	.95	0	.1	1
123	1	0	25	1	40	.95	0	.2	3
124	1	0	25	1	40	.95	0	.2	1
125	1	0	25	1	40	.95	1	.1	3
126	1	0	25	1	40	.95	1	.1	1
127	1	0	25	1	40	.95	1	.2	3
128	1	0	25	1	40	.95	1	.2	1
129	1	4	10	0	20	.75	0	.1	3
130	1	4	10	0	20	.75	0	.1	1
131	1	4	10	0	20	.75	0	.2	3
132	1	4	10	0	20	.75	0	.2	1
133	1	4	10	0	20	.75	1	.1	3
134	1	4	10	0	20	.75	1	.1	1
135	1	4	10	0	20	.75	1	.2	3
136	1	4	10	0	20	.75	1	.2	1
137	1	4	10	0	20	.95	0	.1	3
138	1	4	10	0	20	.95	0	.1	1
139	1	4	10	0	20	.95	0	.2	3
140	1	4	10	0	20	.95	0	.2	1
141	1	4	10	0	20	.95	1	.1	3
142	1	4	10	0	20	.95	1	.1	1
143	1	4	10	0	20	.95	1	.2	3
144	1	4	10	0	20	.95	1	.2	1
145	1	4	10	0	40	.75	0	.1	3
146	1	4	10	0	40	.75	0	.1	1
147	1	4	10	0	40	.75	0	.2	3
148	1	4	10	0	40	.75	0	.2	1
149	1	4	10	0	40	.75	1	.1	3
150	1	4	10	0	40	.75	1	.1	1
151	1	4	10	0	40	.75	1	.2	3
152	1	4	10	0	40	.75	1	.2	1
153	1	4	10	0	40	.95	0	.1	3
154	1	4	10	0	40	.95	0	.1	1
155	1	4	10	0	40	.95	0	.2	3
156	1	4	10	0	40	.95	0	.2	1
157	1	4	10	0	40	.95	1	.1	3
158	1	4	10	0	40	.95	1	.1	1
159	1	4	10	0	40	.95	1	.2	3
160	1	4	10	0	40	.95	1	.2	1
161	1	4	10	1	20	.75	0	.1	3
162	1	4	10	1	20	.75	0	.1	1
163	1	4	10	1	20	.75	0	.2	3
164	1	4	10	1	20	.75	0	.2	1
165	1	4	10	1	20	.75	1	.1	3

Run	SOIL_ DEGRADE	COOP	PROTEIN_NEED	PROTEIN_ PENALTY	HUNT_ SRADIUS	HARVEST_ ADJUST_ FACTOR	NEED_MEAT	STATE_GOOD	H_2O_TYPE
166	1	4	10	1	20	.75	1	.1	1
167	1	4	10	1	20	.75	1	.2	3
168	1	4	10	1	20	.75	1	.2	1
169	1	4	10	1	20	.95	0	.1	3
170	1	4	10	1	20	.95	0	.1	1
171	1	4	10	1	20	.95	0	.2	3
172	1	4	10	1	20	.95	0	.2	1
173	1	4	10	1	20	.95	1	.1	3
174	1	4	10	1	20	.95	1	.1	1
175	1	4	10	1	20	.95	1	.2	3
176	1	4	10	1	20	.95	1	.2	1
177	1	4	10	1	40	.75	0	.1	3
178	1	4	10	1	40	.75	0	.1	1
179	1	4	10	1	40	.75	0	.2	3
180	1	4	10	1	40	.75	0	.2	1
181	1	4	10	1	40	.75	1	.1	3
182	1	4	10	1	40	.75	1	.1	1
183	1	4	10	1	40	.75	1	.2	3
184	1	4	10	1	40	.75	1	.2	1
185	1	4	10	1	40	.95	0	.1	3
186	1	4	10	1	40	.95	0	.1	1
187	1	4	10	1	40	.95	0	.2	3
188	1	4	10	1	40	.95	0	.2	1
189	1	4	10	1	40	.95	1	.1	3
190	1	4	10	1	40	.95	1	.1	1
191	1	4	10	1	40	.95	1	.2	3
192	1	4	10	1	40	.95	1	.2	1
193	1	4	25	0	20	.75	0	.1	3
194	1	4	25	0	20	.75	0	.1	1
195	1	4	25	0	20	.75	0	.2	3
196	1	4	25	0	20	.75	0	.2	1
197	1	4	25	0	20	.75	1	.1	3
198	1	4	25	0	20	.75	1	.1	1
199	1	4	25	0	20	.75	1	.2	3
200	1	4	25	0	20	.75	1	.2	1
201	1	4	25	0	20	.95	0	.1	3
202	1	4	25	0	20	.95	0	.1	1
203	1	4	25	0	20	.95	0	.2	3
204	1	4	25	0	20	.95	0	.2	1
205	1	4	25	0	20	.95	1	.1	3
206	1	4	25	0	20	.95	1	.1	1
207	1	4	25	0	20	.95	1	.2	3
208	1	4	25	0	20	.95	1	.2	1
209	1	4	25	0	40	.75	0	.1	3
210	1	4	25	0	40	.75	0	.1	1
211	1	4	25	0	40	.75	0	.2	3
212	1	4	25	0	40	.75	0	.2	1

(continued)

Run	SOIL_ DEGRADE	COOP	PROTEIN_NEED	PROTEIN_ PENALTY	HUNT_ SRADIUS	HARVEST_ ADJUST_ FACTOR	NEED_MEAT	STATE_GOOD	H_2O_TYPE
213	1	4	25	0	40	.75	1	.1	3
214	1	4	25	0	40	.75	1	.1	1
215	1	4	25	0	40	.75	1	.2	3
216	1	4	25	0	40	.75	1	.2	1
217	1	4	25	0	40	.95	0	.1	3
218	1	4	25	0	40	.95	0	.1	1
219	1	4	25	0	40	.95	0	.2	3
220	1	4	25	0	40	.95	0	.2	1
221	1	4	25	0	40	.95	1	.1	3
222	1	4	25	0	40	.95	1	.1	1
223	1	4	25	0	40	.95	1	.2	3
224	1	4	25	0	40	.95	1	.2	1
225	1	4	25	1	20	.75	0	.1	3
226	1	4	25	1	20	.75	0	.1	1
227	1	4	25	1	20	.75	0	.2	3
228	1	4	25	1	20	.75	0	.2	1
229	1	4	25	1	20	.75	1	.1	3
230	1	4	25	1	20	.75	1	.1	1
231	1	4	25	1	20	.75	1	.2	3
232	1	4	25	1	20	.75	1	.2	1
233	1	4	25	1	20	.95	0	.1	3
234	1	4	25	1	20	.95	0	.1	1
235	1	4	25	1	20	.95	0	.2	3
236	1	4	25	1	20	.95	0	.2	1
237	1	4	25	1	20	.95	1	.1	3
238	1	4	25	1	20	.95	1	.1	1
239	1	4	25	1	20	.95	1	.2	3
240	1	4	25	1	20	.95	1	.2	1
241	1	4	25	1	40	.75	0	.1	3
242	1	4	25	1	40	.75	0	.1	1
243	1	4	25	1	40	.75	0	.2	3
244	1	4	25	1	40	.75	0	.2	1
245	1	4	25	1	40	.75	1	.1	3
246	1	4	25	1	40	.75	1	.1	1
247	1	4	25	1	40	.75	1	.2	3
248	1	4	25	1	40	.75	1	.2	1
249	1	4	25	1	40	.95	0	.1	3
250	1	4	25	1	40	.95	0	.1	1
251	1	4	25	1	40	.95	0	.2	3
252	1	4	25	1	40	.95	0	.2	1
253	1	4	25	1	40	.95	1	.1	3
254	1	4	25	1	40	.95	1	.1	1
255	1	4	25	1	40	.95	1	.2	3
256	1	4	25	1	40	.95	1	.2	1
257	2	0	10	0	20	.75	0	.1	3
258	2	0	10	0	20	.75	0	.1	1
259	2	0	10	0	20	.75	0	.2	3

Run	SOIL_ DEGRADE	COOP	PROTEIN_NEED	PROTEIN_ PENALTY	HUNT_ SRADIUS	HARVEST_ ADJUST_ FACTOR	NEED_MEAT	STATE_GOOD	H_2O_TYPE
260	2	0	10	0	20	.75	0	.2	1
261	2	0	10	0	20	.75	1	.1	3
262	2	0	10	0	20	.75	1	.1	1
263	2	0	10	0	20	.75	1	.2	3
264	2	0	10	0	20	.75	1	.2	1
265	2	0	10	0	20	.95	0	.1	3
266	2	0	10	0	20	.95	0	.1	1
267	2	0	10	0	20	.95	0	.2	3
268	2	0	10	0	20	.95	0	.2	1
269	2	0	10	0	20	.95	1	.1	3
270	2	0	10	0	20	.95	1	.1	1
271	2	0	10	0	20	.95	1	.2	3
272	2	0	10	0	20	.95	1	.2	1
273	2	0	10	0	40	.75	0	.1	3
274	2	0	10	0	40	.75	0	.1	1
275	2	0	10	0	40	.75	0	.2	3
276	2	0	10	0	40	.75	0	.2	1
277	2	0	10	0	40	.75	1	.1	3
278	2	0	10	0	40	.75	1	.1	1
279	2	0	10	0	40	.75	1	.2	3
280	2	0	10	0	40	.75	1	.2	1
281	2	0	10	0	40	.95	0	.1	3
282	2	0	10	0	40	.95	0	.1	1
283	2	0	10	0	40	.95	0	.2	3
284	2	0	10	0	40	.95	0	.2	1
285	2	0	10	0	40	.95	1	.1	3
286	2	0	10	0	40	.95	1	.1	1
287	2	0	10	0	40	.95	1	.2	3
288	2	0	10	0	40	.95	1	.2	1
289	2	0	10	1	20	.75	0	.1	3
290	2	0	10	1	20	.75	0	.1	1
291	2	0	10	1	20	.75	0	.2	3
292	2	0	10	1	20	.75	0	.2	1
293	2	0	10	1	20	.75	1	.1	3
294	2	0	10	1	20	.75	1	.1	1
295	2	0	10	1	20	.75	1	.2	3
296	2	0	10	1	20	.75	1	.2	1
297	2	0	10	1	20	.95	0	.1	3
298	2	0	10	1	20	.95	0	.1	1
299	2	0	10	1	20	.95	0	.2	3
300	2	0	10	1	20	.95	0	.2	1
301	2	0	10	1	20	.95	1	.1	3
302	2	0	10	1	20	.95	1	.1	1
303	2	0	10	1	20	.95	1	.2	3
304	2	0	10	1	20	.95	1	.2	1
305	2	0	10	1	40	.75	0	.1	3
306	2	0	10	1	40	.75	0	.1	1

(continued)

Run	SOIL_ DEGRADE	COOP	PROTEIN_NEED	PROTEIN_ PENALTY	HUNT_ SRADIUS	HARVEST_ ADJUST_ FACTOR	NEED_MEAT	STATE_GOOD	H_2O_TYPE
307	2	0	10	1	40	.75	0	.2	3
308	2	0	10	1	40	.75	0	.2	1
309	2	0	10	1	40	.75	1	.1	3
310	2	0	10	1	40	.75	1	.1	1
311	2	0	10	1	40	.75	1	.2	3
312	2	0	10	1	40	.75	1	.2	1
313	2	0	10	1	40	.95	0	.1	3
314	2	0	10	1	40	.95	0	.1	1
315	2	0	10	1	40	.95	0	.2	3
316	2	0	10	1	40	.95	0	.2	1
317	2	0	10	1	40	.95	1	.1	3
318	2	0	10	1	40	.95	1	.1	1
319	2	0	10	1	40	.95	1	.2	3
320	2	0	10	1	40	.95	1	.2	1
321	2	0	25	0	20	.75	0	.1	3
322	2	0	25	0	20	.75	0	.1	1
323	2	0	25	0	20	.75	0	.2	3
324	2	0	25	0	20	.75	0	.2	1
325	2	0	25	0	20	.75	1	.1	3
326	2	0	25	0	20	.75	1	.1	1
327	2	0	25	0	20	.75	1	.2	3
328	2	0	25	0	20	.75	1	.2	1
329	2	0	25	0	20	.95	0	.1	3
330	2	0	25	0	20	.95	0	.1	1
331	2	0	25	0	20	.95	0	.2	3
332	2	0	25	0	20	.95	0	.2	1
333	2	0	25	0	20	.95	1	.1	3
334	2	0	25	0	20	.95	1	.1	1
335	2	0	25	0	20	.95	1	.2	3
336	2	0	25	0	20	.95	1	.2	1
337	2	0	25	0	40	.75	0	.1	3
338	2	0	25	0	40	.75	0	.1	1
339	2	0	25	0	40	.75	0	.2	3
340	2	0	25	0	40	.75	0	.2	1
341	2	0	25	0	40	.75	1	.1	3
342	2	0	25	0	40	.75	1	.1	1
343	2	0	25	0	40	.75	1	.2	3
344	2	0	25	0	40	.75	1	.2	1
345	2	0	25	0	40	.95	0	.1	3
346	2	0	25	0	40	.95	0	.1	1
347	2	0	25	0	40	.95	0	.2	3
348	2	0	25	0	40	.95	0	.2	1
349	2	0	25	0	40	.95	1	.1	3
350	2	0	25	0	40	.95	1	.1	1
351	2	0	25	0	40	.95	1	.2	3
352	2	0	25	0	40	.95	1	.2	1
353	2	0	25	1	20	.75	0	.1	3

Run	SOIL_ DEGRADE	COOP	PROTEIN_NEED	PROTEIN_ PENALTY	HUNT_ SRADIUS	HARVEST_ ADJUST_ FACTOR	NEED_MEAT	STATE_GOOD	H_2O_TYPE
354	2	0	25	1	20	.75	0	.1	1
355	2	0	25	1	20	.75	0	.2	3
356	2	0	25	1	20	.75	0	.2	1
357	2	0	25	1	20	.75	1	.1	3
358	2	0	25	1	20	.75	1	.1	1
359	2	0	25	1	20	.75	1	.2	3
360	2	0	25	1	20	.75	1	.2	1
361	2	0	25	1	20	.95	0	.1	3
362	2	0	25	1	20	.95	0	.1	1
363	2	0	25	1	20	.95	0	.2	3
364	2	0	25	1	20	.95	0	.2	1
365	2	0	25	1	20	.95	1	.1	3
366	2	0	25	1	20	.95	1	.1	1
367	2	0	25	1	20	.95	1	.2	3
368	2	0	25	1	20	.95	1	.2	1
369	2	0	25	1	40	.75	0	.1	3
370	2	0	25	1	40	.75	0	.1	1
371	2	0	25	1	40	.75	0	.2	3
372	2	0	25	1	40	.75	0	.2	1
373	2	0	25	1	40	.75	1	.1	3
374	2	0	25	1	40	.75	1	.1	1
375	2	0	25	1	40	.75	1	.2	3
376	2	0	25	1	40	.75	1	.2	1
377	2	0	25	1	40	.95	0	.1	3
378	2	0	25	1	40	.95	0	.1	1
379	2	0	25	1	40	.95	0	.2	3
380	2	0	25	1	40	.95	0	.2	1
381	2	0	25	1	40	.95	1	.1	3
382	2	0	25	1	40	.95	1	.1	1
383	2	0	25	1	40	.95	1	.2	3
384	2	0	25	1	40	.95	1	.2	1
385	2	4	10	0	20	.75	0	.1	3
386	2	4	10	0	20	.75	0	.1	1
387	2	4	10	0	20	.75	0	.2	3
388	2	4	10	0	20	.75	0	.2	1
389	2	4	10	0	20	.75	1	.1	3
390	2	4	10	0	20	.75	1	.1	1
391	2	4	10	0	20	.75	1	.2	3
392	2	4	10	0	20	.75	1	.2	1
393	2	4	10	0	20	.95	0	.1	3
394	2	4	10	0	20	.95	0	.1	1
395	2	4	10	0	20	.95	0	.2	3
396	2	4	10	0	20	.95	0	.2	1
397	2	4	10	0	20	.95	1	.1	3
398	2	4	10	0	20	.95	1	.1	1
399	2	4	10	0	20	.95	1	.2	3
400	2	4	10	0	20	.95	1	.2	1

(continued)

Run	SOIL_ DEGRADE	COOP	PROTEIN_NEED	PROTEIN_ PENALTY	HUNT_ SRADIUS	HARVEST_ ADJUST_ FACTOR	NEED_MEAT	STATE_GOOD	H_2O_TYPE
401	2	4	10	0	40	.75	0	.1	3
402	2	4	10	0	40	.75	0	.1	1
403	2	4	10	0	40	.75	0	.2	3
404	2	4	10	0	40	.75	0	.2	1
405	2	4	10	0	40	.75	1	.1	3
406	2	4	10	0	40	.75	1	.1	1
407	2	4	10	0	40	.75	1	.2	3
408	2	4	10	0	40	.75	1	.2	1
409	2	4	10	0	40	.95	0	.1	3
410	2	4	10	0	40	.95	0	.1	1
411	2	4	10	0	40	.95	0	.2	3
412	2	4	10	0	40	.95	0	.2	1
413	2	4	10	0	40	.95	1	.1	3
414	2	4	10	0	40	.95	1	.1	1
415	2	4	10	0	40	.95	1	.2	3
416	2	4	10	0	40	.95	1	.2	1
417	2	4	10	1	20	.75	0	.1	3
418	2	4	10	1	20	.75	0	.1	1
419	2	4	10	1	20	.75	0	.2	3
420	2	4	10	1	20	.75	0	.2	1
421	2	4	10	1	20	.75	1	.1	3
422	2	4	10	1	20	.75	1	.1	1
423	2	4	10	1	20	.75	1	.2	3
424	2	4	10	1	20	.75	1	.2	1
425	2	4	10	1	20	.95	0	.1	3
426	2	4	10	1	20	.95	0	.1	1
427	2	4	10	1	20	.95	0	.2	3
428	2	4	10	1	20	.95	0	.2	1
429	2	4	10	1	20	.95	1	.1	3
430	2	4	10	1	20	.95	1	.1	1
431	2	4	10	1	20	.95	1	.2	3
432	2	4	10	1	20	.95	1	.2	1
433	2	4	10	1	40	.75	0	.1	3
434	2	4	10	1	40	.75	0	.1	1
435	2	4	10	1	40	.75	0	.2	3
436	2	4	10	1	40	.75	0	.2	1
437	2	4	10	1	40	.75	1	.1	3
438	2	4	10	1	40	.75	1	.1	1
439	2	4	10	1	40	.75	1	.2	3
440	2	4	10	1	40	.75	1	.2	1
441	2	4	10	1	40	.95	0	.1	3
442	2	4	10	1	40	.95	0	.1	1
443	2	4	10	1	40	.95	0	.2	3
444	2	4	10	1	40	.95	0	.2	1
445	2	4	10	1	40	.95	1	.1	3
446	2	4	10	1	40	.95	1	.1	1
447	2	4	10	1	40	.95	1	.2	3

Run	SOIL_ DEGRADE	COOP	PROTEIN_NEED	PROTEIN_ PENALTY	HUNT_ SRADIUS	HARVEST_ ADJUST_ FACTOR	NEED_MEAT	STATE_GOOD	H_2O_TYPE
448	2	4	10	1	40	.95	1	.2	1
449	2	4	25	0	20	.75	0	.1	3
450	2	4	25	0	20	.75	0	.1	1
451	2	4	25	0	20	.75	0	.2	3
452	2	4	25	0	20	.75	0	.2	1
453	2	4	25	0	20	.75	1	.1	3
454	2	4	25	0	20	.75	1	.1	1
455	2	4	25	0	20	.75	1	.2	3
456	2	4	25	0	20	.75	1	.2	1
457	2	4	25	0	20	.95	0	.1	3
458	2	4	25	0	20	.95	0	.1	1
459	2	4	25	0	20	.95	0	.2	3
460	2	4	25	0	20	.95	0	.2	1
461	2	4	25	0	20	.95	1	.1	3
462	2	4	25	0	20	.95	1	.1	1
463	2	4	25	0	20	.95	1	.2	3
464	2	4	25	0	20	.95	1	.2	1
465	2	4	25	0	40	.75	0	.1	3
466	2	4	25	0	40	.75	0	.1	1
467	2	4	25	0	40	.75	0	.2	3
468	2	4	25	0	40	.75	0	.2	1
469	2	4	25	0	40	.75	1	.1	3
470	2	4	25	0	40	.75	1	.1	1
471	2	4	25	0	40	.75	1	.2	3
472	2	4	25	0	40	.75	1	.2	1
473	2	4	25	0	40	.95	0	.1	3
474	2	4	25	0	40	.95	0	.1	1
475	2	4	25	0	40	.95	0	.2	3
476	2	4	25	0	40	.95	0	.2	1
477	2	4	25	0	40	.95	1	.1	3
478	2	4	25	0	40	.95	1	.1	1
479	2	4	25	0	40	.95	1	.2	3
480	2	4	25	0	40	.95	1	.2	1
481	2	4	25	1	20	.75	0	.1	3
481	2	4	25	1	20	.75	0	.1	1
483	2	4	25	1	20	.75	0	.2	3
484	2	4	25	1	20	.75	0	.2	1
485	2	4	25	1	20	.75	1	.1	3
486	2	4	25	1	20	.75	1	.1	1
487	2	4	25	1	20	.75	1	.2	3
488	2	4	25	1	20	.75	1	.2	1
489	2	4	25	1	20	.95	0	.1	3
490	2	4	25	1	20	.95	0	.1	1
491	2	4	25	1	20	.95	0	.2	3
492	2	4	25	1	20	.95	0	.2	1
493	2	4	25	1	20	.95	1	.1	3
494	2	4	25	1	20	.95	1	.1	1

(continued)

Run	SOIL_ DEGRADE	COOP	PROTEIN_NEED	PROTEIN_ PENALTY	HUNT_ SRADIUS	HARVEST_ ADJUST_ FACTOR	NEED_MEAT	STATE_GOOD	H_2O_TYPE
495	2	4	25	1	20	.95	1	.2	3
496	2	4	25	1	20	.95	1	.2	1
497	2	4	25	1	40	.75	0	.1	3
498	2	4	25	1	40	.75	0	.1	1
499	2	4	25	1	40	.75	0	.2	3
500	2	4	25	1	40	.75	0	.2	1
501	2	4	25	1	40	.75	1	.1	3
502	2	4	25	1	40	.75	1	.1	1
503	2	4	25	1	40	.75	1	.2	3
504	2	4	25	1	40	.75	1	.2	1
505	2	4	25	1	40	.95	0	.1	3
506	2	4	25	1	40	.95	0	.1	1
507	2	4	25	1	40	.95	0	.2	3
508	2	4	25	1	40	.95	0	.2	1
509	2	4	25	1	40	.95	1	.1	3
510	2	4	25	1	40	.95	1	.1	1
511	2	4	25	1	40	.95	1	.2	3
512	2	4	25	1	40	.95	1	.2	1

APPENDIX B

Community Centers

Center Number	Center Name	USGS Quad	Site Number(s)	600–725	725–800	800–840	840–880	880–920	920–980	980–1020	1020–1060	1060–1100	1100–1140
1	Carol's Ruin	Ruin Canyon	5MT10581A, B					2	9				
2	Kristie's Site	Ruin Canyon	5MT4421A, B						6	5			
3	Mud Springs	Mud Creek	5MT4466A, B								1		
4	Turkey House Complex	Ruin Canyon	5MT13314										
5	Lower Squaw Village	Papoose Canyon	5DL488, 5DL717										
6	Upper Squaw Village	Papoose Canyon	5DL860, 5DL861, 5DL863		1						1		
7	Mitchell Springs	Cortez	5MT10991A-T, 5MT12598	5	7	2	2	3	9	9	3	12	8
8	McPhee Village	Trimble Point	(a)	1	3	7	29	35	10				
9	5MT7414	Arriola	5MT7414	1								1	1
10	Sand Canyon Pueblo	Woods Canyon	5MT765										
11	Goodman Point Pueblo	Arriola	5MT604										
13	Hampton (Squaw Point) Ruin	Ruin Canyon	5DL859									34	36
14	Seven Towers Pueblo	Negro Canyon	5MT1000										
15	Ruin Canyon Rim Pueblo	Ruin Canyon	5MT10438										
16	Great Pithouse	Mud Creek	5MT10647	1	1								
17	Jackson's Hovenweep Castle	Negro Canyon	5MT10853										

1140–1180	1180–1225	1225–1260	1260–1280	Great kiva	Plaza	Oversized pit structure	Great house	Multiwall structure	Enclosing wall	Towers	Upland	Canyon head	Rim rock	Bench	Talus	Canyon Bottom	Spring	Mean Catchment Productivity
				X												X		249.47
3	3	7	3	X						X						X		262.06
1	20	23	21					X		X	X	X	X				X	329.17
		9	1		X				X	X		X	X		X		X	355.35
5	8	10	7			X			X	X	X							
4	4	7							X	X	X							226.57
7	8	7	6	X			X	X		X	X		X				X	338.58
				X	X	X					X			X				454.86
1				X							X							274.62
			83	X	X			X	X	X		X					X	365.48
			79	X	X			X	X	X		X					X	470.2
40	44	48	41	X					X	X		X					X	445.7
		29	30	X	X			X	X	X		X					X	257.02
		12	11						X	X		X	X		X		X	385.66
				X							X							269.59
		13		X	X					X						X		225.58

(continued)

Center Number	Center Name	USGS Quad	Site Number(s)	600–725	725–800	800–840	840–880	880–920	920–980	980–1020	1020–1060	1060–1100	1100–1140
18	Stix and Leaves Pueblo	Arriola	5MT11555						11		7		
19	Woods Canyon Pueblo	Woods Canyon	5MT11842										
20	Rohn 84 (Farmer Pueblo)	Arriola	5MT121										
21	Albert Porter Pueblo	Woods Canyon	5MT123									18	18
22	Millard Site	Arriola	5MT12936			3	3	3	4				
23	Maxwell Community	Cortez	5MT13041							3	4	3	3
24	Bass Site Complex	Woods Canyon	5MT136									16	17
25	5MT1541	Negro Canyon	5MT1541										
27	Gardner Pueblo	Ruin Canyon	5MT1647										
28	Hovenweep Mesa Top	Ruin Canyon	5MT1648										
29	Thompson Site	Negro Canyon	5MT1655										
30	Rohn 150	Woods Canyon	5MT207										
31	Escalante Ruin	Dolores West	5MT2149									1	1
32	Rio Vista Village	Trimble Point	5MT2182		5	5	8	7					
33	House Creek Village	Trimble Point	5MT2320			8	9	8					
34	Herren Site	Ruin Canyon	5MT2516										
35	5MT2681	Dolores West	5MT2681							3		5	4

1140–1180	1180–1225	1225–1260	1260–1280	Great kiva	Plaza	Oversized pit structure	Great house	Multiwall structure	Enclosing wall	Towers	Upland	Canyon head	Rim rock	Bench	Talus	Canyon Bottom	Spring	Mean Catchment Productivity
				X		X					X							442.92
16	22	32	30		X			X	X	X		X	X	X	X	X	X	257.02
		14	11						X	X		X	X		X			329.28
19	19	20				X	X			X	X						X	309.78
				X							X							340.32
3	3	3	2				X							X		X	X	333.87
18	21	19	17				X				X							245.15
	5	4	4							X	X							269.02
		29	25						X	X		X	X		X		X	266.19
		8	7			X				X	X							274.67
	7	8	7		X				X	X		X	X		X		X	236.23
		6	5						X	X			X		X			383.03
							X				X							392.42
					X									X				481.94
					X									X				471.67
14	17	14	12		X					X	X	X						399.85
3	3			X							X							503.28

(continued)

Center Number	Center Name	USGS Quad	Site Number(s)	600–725	725–800	800–840	840–880	880–920	920–980	980–1020	1020–1060	1060–1100	1100–1140
36	Cannonball Ruins	Bowdish Canyon	5MT338										
37	Stevenson Ruin	Arriola	5MT35										5
38	Easter Ruin	Woods Canyon	5MT3793										
39	Finley Ruins	Arriola	5MT3822		3	3	3	4	3				
40	Cirque Site	Cortez	5MT3879		5	6	5						
42	Isleta de Vaca Pueblo	Ruin Canyon	5MT4329										
43	Windy Ruin	Trimble Point	5MT4353					10					
44	Reservoir Ruin Group	Dolores West	5MT4450							4	6	5	4
45	Rich's Ruin	Woods Canyon	5MT4700										
46	Lancaster/Pharo Ruin	Pleasant View	5MT4803									25	26
47	Lower Cow Canyon	Ruin Canyon	5MT4813										
48	Yellow Jacket Pueblo	Yellow Jacket	5MT5									90	102
49	O'Brien Site	Dolores West	5MT5518			1				5	6	5	5
50	Cross Roads/ Radio Tower Site	Pleasant View	5MT6		14	17	12	12					
51	Smoots No.1	Dolores West	5MT6849		3	4	5	4	4				
52	Big Spring Ruin	Negro Canyon	5MT7088										
53	Ruin Canyon Mesa Top	Ruin Canyon	5MT717										
54	Hidden Spring	Negro Canyon	5MT7575										

1140–1180	1180–1225	1225–1260	1260–1280	Great kiva	Plaza	Oversized pit structure	Great house	Multiwall structure	Enclosing wall	Towers	Upland	Canyon head	Rim rock	Bench	Talus	Canyon Bottom	Spring	Mean Catchment Productivity
		26	25					X		X		X					X	289.9
6	6	7	6			X					X							316.18
	29	36	29					X	X	X		X	X		X		X	307.18
											X							300.4
					X	X					X							385.38
	4	6	5		X				X	X				X	X		X	189.98
					X	X					X							422.28
4	4	4		X		X	X	X		X	X	X						391.17
	5	6	5	X							X							190.86
29	30	27	26	X	X	X	X			X	X						X	400.26
		4		X	X	X			X	X				X				307.45
118	132	134	115	X	X		X	X		X	X		X		X		X	481.5
4	4			X							X							359.67
											X							495.42
											X							420.62
		25			X				X			X	X				X	274.42
		9				X					X							316.42
		8						X	X	X		X	X	X			X	228.34

(continued)

Center Number	Center Name	USGS Quad	Site Number(s)	600–725	725–800	800–840	840–880	880–920	920–980	980–1020	1020–1060	1060–1100	1100–1140
55	Hibbets Pueblo	Negro Canyon	5MT7656										
56	Cow Mesa 40	Ruin Canyon	5MT797										
57	Little Cow Canyon	Ruin Canyon	5MT835										
59	Miller Pueblo	Ruin Canyon	5MT875										
60	Hartman Draw	Dolores West	5MT8888					1	1	1	1	1	1
61	McVicker's Site	Ruin Canyon	CCS-3										
62	Bear Paw Pueblo	Ruin Canyon	CCS-2										
63	Yucca House	Mud Creek	5MT5006										
64	Cut Throat Castle	Negro Canyon	5MT603										
65	Pedro Point	Cahone Mesa 15	5MT4575										
66	May Canyon Ruin	Trimble Point	5MT6794				6	5					
67	Harris Pueblo	Negro Canyon	CCS-4										
68	Brewer Mesa Top	Ruin Canyon	5DL578										8
69	Brewer Canyon Pueblo	Ruin Canyon	CCS-9										
70	Brewer Well	Papoose Canyon	5DL506										
71	Truelson Site	Arriola	CCS-8		1	1	1	1	1				
73	Kearns' Site	Negro Canyon	5MT532										

1140–1180	1180–1225	1225–1260	1260–1280	Great kiva	Plaza	Oversized pit structure	Great house	Multiwall structure	Enclosing wall	Towers	Upland	Canyon head	Rim rock	Bench	Talus	Canyon Bottom	Spring	Mean Catchment Productivity
		12			X	X				X		X					X	234.68
	4	3	2		X				X	X			X		X			303.32
		37	35		X				X	X		X	X		X		X	410.17
		19	18					X	X	X		X	X		X		X	283.44
1	1			X							X							492.07
		15				X			X	X		X	X		X		X	270.15
		10			X				X	X		X		X	X		X	315.03
28	34	42	40	X	X	X	X	X	X							X	X	346.7
	7	9	8						X	X		X	X				X	219.99
		14	13		X				X	X		X					X	
						X					X							451.52
	12	15	12						X	X				X	X			184.1
9	11	9	7			X					X							315.88
	17	21	16	X					X	X		X					X	344.49
		16	14						X	X		X	X		X	X	X	254.48
				X												X	X	369
	7	8	7	X					X	X	X							259.91

(continued)

Center Number	Center Name	USGS Quad	Site Number(s)	600–725	725–800	800–840	840–880	880–920	920–980	980–1020	1020–1060	1060–1100	1100–1140
74	Ackmen Site 2—1938	Cahone	5MT2108-2		4	3							
75	Lake View Complex	Dolores West	5MT1905, 5MT4126, 5MT6970							3	2	4	6
77	Castle Rock Pueblo	Battle Rock	5MT1825										
78	Grass Mesa Village	Trimble Point	5MT23		32	31	46	47					
79	Lowry Complex	Ruin Canyon	5MT839, 5Mt1566, 5MT4802							2		4	13
80	Ackmen Site 1—1938	Cahone	5MT2108-1				17						
81	Wild Goose Pueblo	Papoose Canyon	5MT697										
82	Griffey Site	Woods Canyon	Rohn Y-149		4				3				
83	Fortress Spur	Cahone Mesa 15	5MT296		1	1	1	1					
84	Holder Site	Trimble Point	5MT10-12		2	3	2						
86	5MT2766	Dolores West	5MT2766									0	0
87	5MT214	Arriola	5MT214									0	
88	5MT11933	Ruin Canyon	5MT11933									0	0
89	5MT107	Arriola	5MT107									0	0
90	5DL554	Ruin Canyon	5DL554									0	0
100	5MT7984	Pleasant View	5MT7984	1						3	3	4	3
102	Cline Crest Ruin	Trimble Point	5MT2663					9					

1140–1180	1180–1225	1225–1260	1260–1280	Great kiva	Plaza	Oversized pit structure	Great house	Multiwall structure	Enclosing wall	Towers	Upland	Canyon head	Rim rock	Bench	Talus	Canyon Bottom	Spring	Mean Catchment Productivity
					X						X							410.61
6	4	3		X	X		X				X							416.2
			14		X			X	X	X			X			X		220.82
				X	X									X				421.71
7	4	9	8	X		X	X			X	X							401.22
				X		X					X							224.43
		8	7						X	X			X	X		X	X	
	19	18									X							307.49
				X							X		X					
		1	1		X					X	X							417.42
0	0			X														391.15
				X							X							273.78
0				X										X				361.35
0				X											X			299.96
				X							X							268.71
	3	3	3							X	X							275.02
					X	X					X							471.16

(continued)

Center Number	Center Name	USGS Quad	Site Number(s)	600–725	725–800	800–840	840–880	880–920	920–980	980–1020	1020–1060	1060–1100	1100–1140
104	Cottonwood Ruin	Ruin Canyon	5MT11601										
105	Singing Shelter	Trimble Point	5MT4683			1		2				1	
106	Casa Negra	Woods Canyon	5MT3925, 5MT3954									1	1
109	Yellowjacket Floodplain Site	Negro Canyon	5MT6359	1									
110	Spook Point Pueblo	Papoose Canyon	5DL492										
111	Sandstone Canyon Mesa Top	Pleasant View	5MT245									4	
112	Shields Community	Arriola	(b)		19	1					1	35	47

NOTE: (a) = 5MT4475, 5MT4477, 5MT4478, 5MT4479, 5MT4480, 5MT4621, 5MT4623, 5MT4624, 5MT4625, 5MT4684, 5MT4725, 5MT5103, 5MT5104, 5MT5106, 5MT5107, 5MT5108, 5MT5121, 5MT5388, 5MT5519.

(b) = 5MT3807, 5MT3965, 5MT16803, 5MT16805, 5MT16806, 5MT16808, 5MT16777, 5MT16778, 5MT16781, 5MT16783, 5MT16784, 5MT16785, 5MT16789.

Prepared by Scott Ortman and Donna Glowacki.

1140–1180	1180–1225	1225–1260	1260–1280	Great kiva	Plaza	Oversized pit structure	Great house	Multiwall structure	Enclosing wall	Towers	Upland	Canyon head	Rim rock	Bench	Talus	Canyon Bottom	Spring	Mean Catchment Productivity
	6	8	7							X						X	X	303.83
				X									X				X	467.46
1	1	1	1	X		X	X				X							426.04
		3	2							X	X							193.24
		5							X		X	X						
		5	4			X			X	X	X							261.59
47	70	59	18	X	X		X			X	X							490.9

BIBLIOGRAPHY

Adams, E. Charles

1989 Changing Form and Function in Western Pueblo Ceremonial Architecture from A.D. 1000 to A.D. 1500. In *The Architecture of Social Integration in Prehistoric Pueblos*, edited by W. D. Lipe and M. Hegmon, pp. 155–160. Occasional Papers, no. 1. Crow Canyon Archaeological Center, Cortez, CO.

Adams, Karen R., and Vandy E. Bowyer

2002 Sustainable Landscape: Thirteenth-Century Food and Fuel Use in the Sand Canyon Locality. In *Seeking the Center Place*, edited by Mark D. Varien and Richard H. Wilshusen, pp. 123–142. University of Utah Press, Salt Lake City.

Adams, Karen R., Cathryn M. Meegan, Scott G. Ortman, R. Emerson Howell, Lindsay Werth, Deborah A. Muenchrath, Michael K. O'Neill, and Candice Gardner

2006 MAÍS (Maize of American Indigenous Societies) Southwest: Ear Descriptions and Traits that Distinguish 27 Morphologically Distinct Groups of 123 Historic USDA Maize (*Zea mays* L. ssp. *mays*) Accessions, and Data Relevant to Archaeological Subsistence Models. Unpublished report submitted to the James S. McDonnell Foundation 21st Century Research Award/Studying Complex Systems, JSMF Grant no. 21002035.

Adams, Karen R., Deborah A. Muenchrath, and Dylan M. Schwindt

1999 Moisture Effects on the Morphology of Ears, Cobs and Kernels of a South-Western U.S. Maize (*Zea Mays* L.) Cultivar, and Implications for the Interpretation of Archaeological Maize. *Journal of Archaeological Science* 26:483–496.

Adams, Karen R., and Kenneth Lee Peterson

1999 Environment. In *Colorado Prehistory: A Context for the Southern Colorado Plateau*, edited by William D. Lipe, Mark D. Varien, and Richard H. Wilshusen, pp. 14–50. Colorado Council of Professional Archaeologists, Denver.

Adler, Michael A.

1990 *Communities of Soil and Stone: An Archaeological Investigation of Population Aggregation Among the Mesa Verde Anasazi, A.D. 900–1300*. Ph.D. dissertation, University of Michigan. University Microfilms, Ann Arbor.

1992 The Upland Survey. In *The Sand Canyon Archaeological Project: A Progress Report*, edited by William D. Lipe, pp. 11–23. Occasional Papers, no. 2. Crow Canyon Archaeological Center, Cortez, CO.

1994 Population Aggregation and the Anasazi Social Landscape: A View from the Four Corners. In *The Ancient Southwestern Community: Models and Methods for the Study of Prehistoric Social Organization*, edited by W. H. Wills and R. D. Leonard, pp. 85–101. University of New Mexico Press, Albuquerque.

1996 Land Tenure, Archaeology, and the Ancestral Pueblo Social Landscape. *Journal of Anthropological Archaeology* 15:337–371.

2002 The Ancestral Pueblo Community as Structure and Strategy. In *Seeking the Center*

Place: Archaeology and Ancient Communities in the Mesa Verde Region, edited by M. D. Varien and R. H. Wilshusen, pp. 25–39. University of Utah Press, Salt Lake City.

Adler, Michael A. (editor)

1996 *The Prehistoric Pueblo World,* A.D. 1150–1350. University of Arizona Press, Tucson.

Adler, Michael A., and Amber Johnson

1996 Appendix: Mapping the Puebloan Southwest. In *The Prehistoric Pueblo World,* A.D. 1150–1350, edited by Michael A. Adler. University of Arizona Press, Tucson.

Adler, Michael A., and Mark D. Varien

1994 The Changing Face of the Community in the Mesa Verde Region A.D. 1000–1300. In *Proceedings of the Anasazi Symposium, 1991,* edited by A. Hutchinson and J. E. Smith, pp. 83–97. Mesa Verde Museum Association, Mesa Verde, CO.

Ager, Alan, A., Bruce K. Johnson, John W. Kern, and John G. Kie

2003 Daily and Seasonal Movements and Habitat Use by Female Rocky Mountain Elk and Mule Deer. *Journal of Mammalogy* 84:1076–1088.

Alldredge, A. William, James F. Lipscomb, and F. Ward Whicker

1974 Forage Intake Rates of Mule Deer Estimated with Fallout Cesium-137. *Journal of Wildlife Management* 38:508–516.

Allison, James R.

2010 The End of Farming in the "Northern Periphery" of the Southwest. In *Leaving Mesa Verde: Peril and Change in the Thirteenth-Century Southwest,* edited by T. A. Kohler, M. D. Varien, and A. M. Wright, pp. 75–101. University of Arizona Press, Tucson.

Amick, Daniel S., Raymond P. Mauldin, and Steven A. Tomka

1988 An Evaluation of Debitage Produced by Experimental Bifacial Core Reduction of a Georgetown Chert Nodule. *Lithic Technology* 17:26–36.

Ammerman, Albert J., and William Andrefsky, Jr.

1982 Reduction Sequences and the Exchange of Obsidian in Neolithic Calabria. In *Contexts for Prehistoric Exchange,* edited by J. Ericson and T. Earle, pp. 149–72. Academic, New York.

Andrefsky, William, Jr.

1998 *Lithics: Macroscopic Approaches to Analysis.* Cambridge University Press, Cambridge.

Andrews, John T., Paul E. Carrara, Francis B. King, and Robert Stuckenrath

1975 Holocene Environmental Changes in the Alpine Zone, Northern San Juan Mountains, Colorado: Evidence from Bog Stratigraphy and Palynology. *Quaternary Research* 5:171–197.

Angel, J. L.

1974 Patterns of Fractures from Neolithic to Modern Times. *Anthropologiai Kozlemenyek* 18:571–588.

Arakawa, Fumiyasu

2000 Lithic Analysis of Yellow Jacket Pueblo as a Tool for Understanding and Visualizing Women's Roles in Procuring, Utilizing, and Making Stone Tools. Unpublished Master's thesis, Department of Sociology, Anthropology, and Criminal Justice, University of Idaho, Moscow.

2006 Lithic Raw Material Procurement and the Social Landscape in the Central Mesa Verde Region, A.D. 600–1300. Unpublished Ph.D. dissertation, Department of Anthropology, Washington State University, Pullman.

Arakawa, Fumiyasu, and Andrew I. Duff

2002 Transitions in Mesa Verde Region Communities and Lithic Procurement Strategies. Paper presented at the 67th Annual Meeting of the Society for American Archaeology, Denver, CO.

Arakawa, Fumiyasu, and Kimberlee M. Gerhardt

2007 Toolstone Procurement Patterns on Wetherill Mesa, A.D. 600 to 1280. *Kiva* 73:67–84.

2009 Geoarchaeological Investigation of Lithic Resources in the Central Mesa Verde Region, Colorado, USA. *Geoarchaeology* 24:204–223.

Arno, Stephen F.

1984 *Timberline: Mountain and Arctic Forest Frontiers.* Mountaineers, Seattle, WA.

Arnold, Dean E.

1985 *Ceramic Theory and Cultural Process.* Cambridge University Press, Cambridge.

Arrhenius, Gustaf, and Enrico Bonatti

1965 The Mesa Verde Loess. In *Contributions of the Wetherill Mesa Archeological Project,* Assembled by D. Osborne, pp. 92–100. Memoirs of the Society for American Archaeology, no. 19, Salt Lake City, UT.

ASTM International

2002 Standard Guide for Conceptualization and Characterization of Groundwater Systems. Active Standards D5979-96. *ASTM Book of Standards,* vol. 04.09, pp. 482–489. West Conshohaken, PA.

Axtell, Robert L., Joshua M. Epstein, Jeffrey S. Dean, George J. Gumerman, Alan C. Swedlund, Jason Harburger, Shubha

Chakravarty, Ross Hammond, Jon Parker, and Miles Parker
2002 Population Growth and Collapse in a Multiagent Model of the Kayenta Anasazi in Long House Valley. *Proceedings of the National Academy of Sciences* 99:7275–7279.

Badenhorst, Shaw and Jonathan C. Driver
2009 Faunal Changes in Farming Communities from Basketmaker II to Pueblo III (AD 1–1300) in the San Juan Basin of the American Southwest. *Journal of Archaeological Science* 36:1832–1841.

Baker, Shane A.
1990 Rattlesnake Ruin (42Sa 18434): A Case of Violent Death and Perimortem Mutilation in the Anasazi Culture of San Juan County, Utah. Unpublished Master's thesis, Department of Anthropology, Brigham Young University, Provo, UT.

Bandy, Matt
2008 Global Patterns of Early Village Development. In *The Neolithic Demographic Transition and Its Consequences*, edited by Jean-Pierre Bocquet-Appel and Ofer Bar-Yosef, pp. 333–357. Springer, Berlin.

Barkley, Fred A.
1934 The Statistical Theory of Pollen Analysis. *Ecology* 15(4):283–289.

Bellorado, Benjamin
2007 Breaking Down the Models: Reconstructing Prehistoric Subsistence Agriculture in the Durango District of Southwestern Colorado. Unpublished Master's thesis, Department of Anthropology, Northern Arizona University, Flagstaff.

Benson, Larry V.
2011a Factors Controlling Pre-Columbian and Early Historic Maize Productivity in the American Southwest, Part 1: The Southern Colorado Plateau and Rio Grande Regions. *Journal of Archaeological Method and Theory* 18:1–60.
2011b Factors Controlling Pre-Columbian and Early Historic Maize Productivity in the American Southwest, Part 2: The Chaco Halo, Mesa Verde, Pajarito Plateau/ Bandelier, and Zuni Archaeological Regions. *Journal of Archaeological Method and Theory* 18:61–109.

Benson, Larry V., Kenneth Petersen, and John Stein
2007 Anasazi (Pre-Columbian Native-American) Migrations During the Middle-12th and Late-13th Centuries—Were They Drought Induced? *Climate Change* 83:187–213.

Berlin, G. Lennis, J. Richard Ambler, Richard H. Hevly, and Gerald G. Shaber
1977 Identification of a Sinagua Agricultural Field by Aerial Thermography, Soil Chemistry, Pollen/Plant Analysis, and Archaeology. *American Antiquity* 42:588–600.

Bernardini, Wesley
2005 Reconsidering Spatial and Temporal Aspects of Prehistoric Cultural Identity: A Case Study from the American Southwest. *American Antiquity* 70:31–54.

Berry, Michael S.
1982 *Time, Space, and Transition in Anasazi Prehistory*. University of Utah Press, Salt Lake City.

Berry, Michael S., and Larry V. Benson
2010 Tree-Ring Dates and Demographic Change in the Southern Colorado Plateau and Rio Grande Regions. In *Leaving Mesa Verde: Peril and Change in the Thirteenth-Century Southwest*, edited by Timothy A. Kohler, Mark D. Varien, and Aaron M. Wright, pp. 53–74. University of Arizona Press, Tucson.

Bettinger, Robert L.
1982 Aboriginal Exchange and Territoriality in Owens Valley, California. In *Contexts for Prehistoric Exchange*, edited by J. E. Ericson and T. K. Earle, pp. 103–126. Academic, New York.
1989 *The Archaeology of Pinyon House, Two Eagles, and Crater Middens: Three Residential Sites in Owens Valley, Eastern California*. American Museum of Natural History, New York.

Billman, Brian R.
1997 Population Pressure and the Origins of Warfare in the Moche Valley, Peru. In *Integrating Archaeological Demography: Multidisciplinary Approaches to Prehistoric Population*, edited by Richard R. Paine, pp. 285–310. Center for Archaeological Investigations. Occasional Papers, no. 24. Southern Illinois University, Carbondale.
2003 *The Puebloan Occupation of the Ute Mountain Piedmont*, vol. 7, *Synthesis and Conclusions*. Soil Systems Publications in Archaeology, no. 22. Phoenix, Arizona.

Billman, Brian R., Patricia M. Lambert, and Banks L. Leonard
2000 Cannibalism, Warfare, and Drought in the Mesa Verde Region During the Twelfth Century A.D. *American Antiquity* 65:145–178.

Binford, Lewis R.
2001 *Constructing Frames of Reference: An Analytical Method for Archaeological Theory Building Using Ethnographic and*

Environmental Data Sets. University of California Press, Berkeley.

Bird, Douglas W., and James F. O'Connell

2006 Behavioral Ecology and Archaeology. *Journal of Archaeological Research* 14:143–188.

Birkedal, Terje G.

1976 Basketmaker III Residence Units: A Study of Prehistoric Social Organization in the Mesa Verde Archaeological District. Unpublished Ph.D. dissertation, Department of Philosophy and Anthropology, University of Colorado, Boulder.

Blake, Michael

2006 Dating the Initial Spread of *Zea mays*. In *Histories of Maize: Multidisciplinary Approaches to the Prehistory, Linguistics, Biogeography, Domestication, and Evolution of Maize*, edited by J. Staller, R. Tykot, and B. Benz, pp. 55–72. Elsevier, Amsterdam.

Bliege Bird, Rebecca, and Eric Alden Smith

2005 Signaling Theory, Strategic Interaction, and Symbolic Capital. *Current Anthropology* 46:221–248.

Blinman, Eric

1989 Potluck in the Protokiva: Ceramics and Ceremonialism in Pueblo I Villages. In *The Architecture of Social Integration in Prehistoric Pueblos*, edited by William D. Lipe and Michelle Hegmon, pp. 113–124. Occasional Papers, no. 1. Crow Canyon Archaeological Center, Cortez, CO.

Bocinsky, R. Kyle

2011 Is a Bird in Hand Really Worth Two in the Bush? Models of Turkey Domestication on the Colorado Plateau. Unpublished Master's thesis, Department of Anthropology, Washington State University, Pullman.

Bocquet-Appel, Jean-Pierre

2002 Paleoanthropological Traces of a Neolithic Demographic Transition. *Current Anthropology* 43:637–650.

2008 Explaining the Neolithic Demographic Transition. In *The Neolithic Demographic Transition and Its Consequences*, edited by Jean-Pierre Bocquet-Appel and Ofer Bar-Yosef, pp. 35–55. Springer, Berlin.

Boone, James L.

2002 Subsistence Strategies and Early Human Population History: An Evolutionary Ecological Perspective. *World Archaeology* 34:6–25.

Boserup, Ester

1965 *The Conditions of Agricultural Growth: The Economics of Agricultural Change Under Population*. Aldine, Chicago, IL.

Bowles, Samuel

2006 Group Competition, Reproductive Leveling, and the Evolution of Human Altruism. *Science* 314:1569–1572.

Boyd, Robert, and Peter J. Richerson

2009 Voting with Your Feet: Payoff Biased Migration and the Evolution of Group Beneficial Behavior. *Journal of Theoretical Biology* 25:331–339.

Bradfield, Maitland

1971 *The Changing Pattern of Hopi Agriculture*. Occasional Papers, no. 30. Royal Anthropological Institute of Great Britain and Ireland, London.

Bradley, Bruce A.

1988 Wallace Ruin Interim Report. *Southwestern Lore* 54(2):8–33.

1992 Excavations at Sand Canyon Pueblo. In *The Sand Canyon Archaeological Project: A Progress Report*, edited by William D. Lipe, pp. 79–97. Occasional Papers, no. 2. Crow Canyon Archaeological Center, Cortez, CO.

1993 Planning, Growth, and Functional Differentiation at a Prehistoric Pueblo: A Case Study from SW Colorado. *Journal of Field Archaeology* 20:23–42.

1996 Pitchers to Mugs: Chacoan Revival at Sand Canyon Pueblo. *Kiva* 61:241–255.

Bradley, Raymond S.

2000 1000 Years of Climate Change. *Science* 288(5470):1353–1355.

Brengle, K. G.

1992 *Principles and Practices of Dryland Farming*. Colorado Associated University Press, Boulder.

Breternitz, David A., Christine K. Robinson, and G. Timothy Gross (compilers)

1986 *Dolores Archaeological Program: Final Synthetic Report*. U.S. Department of the Interior, Bureau of Reclamation, Engineering and Research Center, Denver, CO.

Brew, John O.

1946 *Archaeology of Alkali Ridge, Southeastern Utah*. Papers of the Peabody Museum of American Archaeology and Ethnology, vol. 21. Harvard University Press, Cambridge, MA.

Brondeau, Florence

2001 Évolution de la Filière Bois Énergie et Dynamique des Formations Ligneuses Autour de l'Office du Niger. *Bois et Forêts des Tropiques* 270(4):16–34.

Brown, D. E. (editor)

1982 Biotic Communities of the American Southwest—United States and Mexico. *Desert Plants* 4(1–4).

Brown, Gary, Thomas C. Windes, and Peter J. McKenna
2008 Animas Anamnesis: Aztec Ruins or Anasazi Capital? In *Chaco's Northern Prodigies: Salmon, Aztec, and the Ascendancy of the Middle San Juan Region after* A.D. 1100, edited by Paul F. Reed, pp. 231–250. University of Utah Press, Salt Lake City.

Brown, Peter, and Aaron Podolefsky
1976 Population Density, Agricultural Intensity, Land Tenure, and Group Size in New Guinea Highlands. *Ethnology* 15:211–238.

Buck, Caitlin E., William G. Cavanaugh, and Clifford D. Linton
1996 *Bayesian Approach to Interpreting Archaeological Data.* Wiley, New York.

Burns, Barney T.
1983 *Simulated Anasazi Storage Behavior Using Crop Yields Reconstructed from Tree Rings:* A.D. 652–1968. 2 vols. Ph.D. dissertation, Department of Anthropology, University of Arizona. University Microfilms, Ann Arbor, MI.

Burt, W. H.
1943 Territoriality and Home Range Concepts as Applied to Mammals. *Journal of Mammalogy* 24:346–352.

Byers, David A., and Jack M. Broughton
2004 Holocene Environmental Change, Artiodactyl Abundances, and Human Hunting Strategies in the Great Basin. *American Antiquity* 69:235–255.

Cameron, Catherine M.
1990 Pit Structure Abandonment in the Four Corners Region of the American Southwest: Late Basketmaker III and Pueblo I. *Journal of Field Archaeology* 17:27–37.
1995 Migration and the Movement of Southwestern Peoples. *Journal of Anthropological Archaeology* 14:104–124.
2005 Exploring Archaeological Cultures in the Northern Southwest: What Were Chaco and Mesa Verde? *Kiva* 70:227–253.

Carneiro, Robert
1972 From Autonomous Villages to the State, a Numerical Estimation. In *Population Growth: Anthropological Implications,* edited by Brian Spooner, pp. 64–77. MIT Press, Cambridge, MA.

Cashdan, Elizabeth A.
1983 Territoriality Among Human Foragers: Ecological Models and an Application to Four Bushman Groups. *Current Anthropology* 24:47–66.

Cater, John D., and Mark L. Chenault
1988 Kiva Use Reinterpreted. *Southwestern Lore* 54(3):19–32.

Cattanach, George S., Jr.
1980 *Long House, Mesa Verde National Park, Colorado.* Publications in Archeology 7H, Wetherill Mesa Studies. U.S. Department of the Interior, National Park Service, Washington, DC.

Chagnon, Napoleon A.
1983 *Yanomamö: The Fierce People.* 3rd ed. Holt, Rinehart, and Winston, New York.
1988 Reproductive and Somatic Conflicts of Interest in the Genesis of Violence and Warfare Among Tribesmen. In *The Anthropology of War,* edited by Jonathan Haas, pp. 77–104. Cambridge University Press, Cambridge.

Chamberlain, Andrew
2009 Archaeological Demography. *Human Biology* 81:275–286.

Chamberlin, Matthew
2011 Plazas, Performance, and Symbolic Power in Ancestral Pueblo Religion. In *Gathering Ancestors: Religious Transformation in the Pueblo World,* edited by Donna M. Glowacki and Scott Van Keuren, pp. 130–154. University of Arizona Press, Tucson.

Chapman, Joseph A., and Gale R. Willner
1986 Lagomorphs. In *Inventory and Monitoring of Wildlife Habitat,* edited by Allen Y. Cooperrider, Raymond J. Boyd, and Hanson R. Stuart, pp. 453–473. U.S. Department of Interior, Bureau of Land Management, Denver, CO.

Charles, Mona C., and Sally J. Cole
2006 Chronology and Cultural Variation in Basketmaker II. *Kiva* 72:167–216.

Chenault, Mark L.
1983 Excavations at Tres Chapulines Pueblo (Site 5MT4725), a Pueblo I Habitation Site. *Dolores Archaeological Program Technical Reports* DAP-102. Submitted to the U.S. Bureau of Reclamation, Upper Colorado Region, Salt Lake City, UT, in compliance with Contract No. 8-07-40-S0562.
2002 Archaeological Investigations at Dead Dog Hamlet (Site 5MT11861): A Basketmaker III Habitation Site. In *From Dancing Man to Hummingbird: Long-Term Prehistoric Change in Southwest Colorado,* vol. 2, Part 2: *Basketmaker III Sites,* edited by Mark L. Chenault, Cynthia Manseau, and Todd Kohler, pp. 5-1–5-198. Cultural Resource Report 2003-16. SWCA Environmental

Consultants. Prepared for the U.S. Bureau of Reclamation, Durango, CO.

Chenault, Mark L., Cynthia Manseau, and Todd Kohler (editors)

2002 *From Dancing Man to Hummingbird: Long-Term Prehistoric Change in Southwest Colorado.* Cultural Resource Report 2003-16. SWCA Environmental Consultants. Prepared for the U.S. Bureau of Reclamation, Durango, CO.

Chernoff, Miriam

1982 Empirical Bayes Estimation of Ceramic Proportions at Teotihuacan. Paper presented at the 47th Annual Meeting of the Society for American Archaeology, Minneapolis, MN.

Chojnacky, David C.

1984 *Volume and Biomass for Curlleaf Cercocarpus in Nevada.* U.S. Department of Agriculture, Forest Service Research Paper INT-332. Intermountain Forest and Range Experiment Station, Ogden, UT.

Christenson, Andrew L.

1980 Change in the Human Food Niche in Response to Population Growth. In *Modeling Change in Prehistoric Subsistence Economies,* edited by T. K. Earle and A. L. Christenson, pp. 31–71. Academic, New York.

1981 The Evolution of Subsistence in the Prehistoric Midwestern United States. Unpublished Ph.D. dissertation, University of California, Los Angeles.

Church, Richard L., and Thomas L. Bell

1988 An Analysis of Ancient Egyptian Settlement Patterns Using Location-Allocation Covering Models. *Annals of the Association of American Geographers* 78:701–714.

Churchill, Melissa J. (editor)

2002 *The Archaeology of Woods Canyon Pueblo: A Canyon-Rim Village in Southwestern Colorado.* Available at www.crowcanyon.org/woodscanyon. Accessed August 2, 2009.

Churchill, Melissa J., Kristin A. Kuckelman, and Mark D. Varien

1998 Public Architecture in the Mesa Verde Region, A.D. 900 to 1300. Paper presented at the 63rd Annual Meeting of the Society for American Archaeology, Seattle, WA.

Clark, Jeffrey J.

2001 *Tracking Prehistoric Migrations: Pueblo Settlers Among the Tonto Basin Hohokam.* Anthropological Papers of the University of Arizona, no. 65. University of Arizona Press, Tucson.

Coffey, Grant

2010 Landscape and Social Scale at Boodman Point, Hovenweep National Monument, Southwestern Colorado. *Kiva* 76(1):55–82.

Cole, Sarah M.

2007 Population Dynamics and Sociopolitical Instability in the Central Mesa Verde Region, A.D. 600–1280. Unpublished Master's thesis, Department of Anthropology, Washington State University, Pullman.

Colorado Natural Areas Program

1996 *Best Management Practices for Wetlands Within Colorado State Parks,* vol. 2, *Caring for the Land.* Colorado State Parks, Colorado Department of Natural Resources, Denver.

Coltrain, Joan Brenner, Joel C. Janetski, and Shawn W. Carlyle

2006 The Stable and Radio-Isotope Chemistry of Eastern Basketmaker and Pueblo Groups in the Four Corners Region of the American Southwest. In *Histories of Maize: Multidisciplinary Approaches to the Prehistory, Linguistics, Biogeography, Domestication, and Evolution of Maize,* edited by J. Staller, R. Tykot, and B. Benz, pp. 275–287. Elsevier, Amsterdam.

2007 The Stable- and Radio-Isotope Chemistry of Western Basketmaker Burials: Implications for Early Puebloan Diets and Origins. *American Antiquity* 72:301–321.

Contreras-Hinojosa, José, R., Victor Volke-Haller, José Luis Oropeza-Mota, Carlos Rodríguez-Franco, Tomás Martínez-Saldaña, and Ángel Martínez-Garza

2003 Disponibilidad y Uso de Leña en El Municipio de Yanhuitlán. *Terra* 21:437–445.

Cook, Edward R., Keith R. Griffa, David M. Meko, Donald S. Graybill, and Gary Funkhouser

1995 The "Segment Length Curse" in Long Tree-Ring Chronology Development for Paleoclimatic Studies. *Holocene* 5:229–237.

Cook, Edward R., Richard Seager, Mark A. Cane, and David W. Stahle

2007 North American Drought: Reconstructions, Causes, and Consequences. *Earth Science Reviews* 81:93–134.

Cook, Edward R., Connie A. Woodhouse, C. Mark Eakin, David M. Meko, and David W. Stahle

2004 Long-Term Aridity Changes in the Western United States. *Science* 306:1015–1018.

Cooper, David J., and Lee H. MacDonald

2000 Restoring the Vegetation of Mined Peatlands in the Southern Rocky Mountains of Colorado, U.S.A. *Restoration Ecology* 8:103–111.

Cordy, Don
1976 *The Protein Book*. Naturegraph, Healdsburg, CA.

Cowgill, George L.
1968 Computer Analysis of Archeological Data from Teotihuacan, Mexico. In *New Perspectives in Archaeology*, edited by S. Binford and L. Binford, pp. 143–150. Aldine, Chicago, IL.
1974 Quantitative Studies of Urbanization in Teotihuacan. In *Mesoamerican Archaeology: New Approaches*, edited by N. Hammond, pp. 363–396. Duckworth, London.
1975 On Causes and Consequences of Ancient and Modern Population Changes. *American Anthropologist* 77:505–525.
1990 Why Pearson's *r* Is Not a Good Similarity Coefficient for Comparing Collections. *American Antiquity* 55:512–521.
1994 How Much Is Enough? Coping with Few or Small Surface Collections. Paper presented in the Symposium "Albert Spaulding and the Continuation of Archaeological Science" at the 59th Annual Meeting of the Society for American Archaeology, Anaheim, CA.

Cowgill, George L., Jeffrey H. Altschul, and R. S. Sload
1984 Spatial Analysis of Teotihuacan: A Mesoamerican Metropolis. In *Intrasite Spatial Analysis in Archaeology*, edited by H. J. Hietala, pp. 154–195. Cambridge University Press, Cambridge.

Crabtree, Keith
1968 Pollen Analysis. *Science Progress* 56(221):83–101.

Creel, Darrell, and Roger Anyon
2003 New Interpretations of Mimbres Public Architecture and Space: Implications for Cultural Change. *American Antiquity* 68:67–92.

Critchfield, Richard
1981 *Villages*. Doubleday, New York.

Cronk, Lee
1999 *That Complex Whole: Culture and the Evolution of Human Behavior*. Westview, Boulder, CO.

Crow Canyon Archaeological Center
2001 *The Castle Rock Pueblo Database*. Available at www.crowcanyon.org/castlerockdatabase. Accessed January 16, 2006.
2003 *The Crow Canyon Archaeological Center Research Database*. Available at www.crowcanyon.org/researchdatabase. Accessed January 16, 2006.
2004 *The Sand Canyon Pueblo Database*. Available at www.crowcanyon.org/sandcanyondatabase. Accessed January 16, 2006.

Crown, Patricia L., and Wirt H. Wills
1995 The Origins of Southwestern Ceramic Containers: Women's Time Allocation and Economic Intensification. *Journal of Anthropological Research* 51:173–186.

Dahl-Jensen, D., K. Mosegaard, N. Gundestrup, G. D. Clow, S. J. Johnsen, A. W. Hansen, and N. Balling
1998 Past Temperatures Directly from the Greenland Ice Sheet. *Science* 282:268–271.

Darling, J. Andrew
1999 Mass Inhumation and the Execution of Witches in the American Southwest. *American Anthropologist* 100:732–752.

Darsie, Richard Floyd III
1983 Prehistoric Agricultural Potential and Site Location in the Dolores River Valley, Southwest Colorado. Unpublished Master's thesis, Department of Anthropology, Washington State University, Pullman.

Daubenmire, Rexford F.
1954 Alpine Timberlines in the Americas and Their Interpretation. *Butler University Botanical Studies* 11:119–136.

Davis, Edward H., and E. Yale Dawson
1945 The Savage Seris of Sonora–I. *Scientific Monthly* 60:193–202.

Davis, William E.
1985 *Anasazi Subsistence and Settlement on White Mesa, San Juan County, Utah*. University Press of America, Lanham, MD.

Dean, Jeffrey S.
1988 Dendrochronology and Paleoenvironmental Reconstruction on the Colorado Plateaus. In *The Anasazi in a Changing Environment*, edited by George J. Gumerman, pp. 119–167. Cambridge University Press, Cambridge.
1995 Keynote Lecture: Dendrochronology and the Study of Human Behavior. In *Tree Rings, Environment and Humanity: Proceedings of the International Conference, Tucson, Arizona, May 17–21, 1994*, edited by Jeffrey S. Dean, David M. Meko, and Thomas W. Swetnam, pp. 461–470. Radiocarbon 1996, Department of Geosciences, University of Arizona, Tucson.
1996 Demography, Environment, and Subsistence Stress. In *Evolving Complexity and Environmental Risk in the Preshistoric Southwest*, edited by Joseph A. Tainter and Bonnie Bagley Tainter, pp. 25–56. Santa Fe Institute

Studies in the Sciences of Complexity, Proceedings, vol. 24. Addison-Wesley Publishing Company, Reading, MA.
2007 Zuni-Area Paleoenvironment. In *Zuni Origins: Toward a New Synthesis of Southwestern Archaeology*, edited by D. A. Gregory and D. R. Wilcox, pp. 77–94. University of Arizona Press, Tucson.
2010 Environmental, Demographic, and Behavioral Context of the Thirteenth Century Depopulation of the Northern Southwest. In *Time of Peril, Time of Change: Explaining Thirteenth Century Pueblo Migrations*, edited by Timothy A. Kohler, Mark D. Varien, and Aaron M. Wright. University of Arizona Press, Tucson.

Dean, Jeffrey S., Robert E. Euler, George J. Gumerman, Fred Plog, Richard H. Hevly, and Thor N. V. Karlstrom
1985 Human Behavior, Demography, and Paleoenvironment on the Colorado Plateaus. *American Antiquity* 50:537–554.

Dean, Jeffrey S., George J. Gumerman, Joshua M. Epstein, Robert L. Axtell, Alan C. Swedlund, Miles T. Parker, and Stephen McCarroll
2000 Understanding Anasazi Culture Change Through Agent-Based Modeling. In *Dynamics in Human and Primate Societies: Agent-Based Modeling of Social and Spatial Processes*, edited by T. A. Kohler and G. J. Gumerman, pp. 170–205. Santa Fe Institute and Oxford University Press, New York.

Dean, Jeffrey S., and William S. Robinson
1978 *Expanded Tree-Ring Chronologies for the Southwestern United States.* Chronology Series 3. Laboratory of Tree-Ring Research, University of Arizona, Tucson.

Dean, Jeffrey S., and Carla R. Van West
2002 Environment-Behavior Relationships in Southwestern Colorado. In *Seeking the Center Place: Archaeology and Ancient Communities in the Mesa Verde Region*, edited by Mark D. Varien and Richard H. Wilshusen, pp. 81–100. University of Utah Press, Salt Lake City.

DeBenedetti, Steven H., and David J. Parsons
1984 Postfire Succession in a Sierran Subalpine Meadow. *American Midland Naturalist* 111:118–125.

Decker, K. W., and L. L. Tiezen
1989 Isotopic Reconstruction of Mesa Verde Diet from Basketmaker III to Pueblo III. *Kiva* 55:33–46.

Del Grosso, Stephen, William Parton, Thomas Stohlgren, Daolan Zheng, Dominique Bachelet, Stephen Prince, Kathy Hibbard, and Richard Olson
2008 Global Potential Net Primary Production Predicted from Vegetation Class, Precipitation, and Temperature. *Ecology* 89:2117–2126.

Dennett, Daniel
1995 *Darwin's Dangerous Idea: Evolution and the Meanings of Life.* Simon and Shuster, New York.

Desruisseaux, Danielle, Thomas Motsinger, and Terri McBride
2004 Archaeological Investigations at Dancing Man Hamlet (Site 5MT9343): A Stockaded Basketmaker III Hamlet: Feature Descriptions. In *From Dancing Man to Hummingbird: Long-Term Prehistoric Change in Southwest Colorado, Vol. 2, Part 1: Basketmaker III Sites*, edited by Mark L. Chenault, Cynthia Manseau, and Todd Kohler, pp. 3-1–3-23. Cultural Resource Report 2003-16. SWCA Environmental Consultants. Prepared for the U.S. Bureau of Reclamation, Durango, Colorado.

Dewar, Robert E.
1984 Environmental Productivity, Population Regulation, and Carrying Capacity. *American Anthropologist* 86:601–614.

Dice, Michael H.
1993 *Disarticulated Human Remains from Reach III of the Towaoc Canal, Ute Mountain Ute Reservation, Montezuma County, Colorado.* Report no. 22. Complete Archaeological Service Associates, Cortez, CO.

Dittert, A. E., Jr., J. J. Hester, and F. W. Eddy
1961 *An Archaeological Survey of the Navajo Reservoir District, Northwestern New Mexico.* School of American Research and the Museum of New Mexico, Monograph 23, Santa Fe.

Dix, Ralph L., and Jon D. Richards
1976 Possible Changes in Spectra Structure of the Subalpine Forest Induced by Increased Snowpack. In *Ecological Impacts of Snowpack Augmentation in the San Juan Mountains, Colorado: San Juan Ecology Project, Final Report*, edited by Harold W. Steinhoff and Jack D. Ives, pp. 311–322. Colorado State University Publications, Fort Collins.

Douglass, Andrew E.
1929 The Secret of the Southwest Solved by Talkative Tree Rings. *National Geographic* 56:737–770.

Drew, Linda G. (editor)
1972 *Tree-Ring Chronologies of Western America III. California and Nevada.* Chronology

Series 1. Laboratory of Tree-Ring Research, University of Arizona, Tucson.

Driver, Jonathan C.

1996 Social Complexity and Hunting Systems in Southwestern Colorado. In *Debating Complexity: Proceedings of the Twenty-Sixth Annual Chacmool Conference*, edited by D. A. Meyer, P. C. Dawson, and D. T. Hanna, pp. 364–374. Archaeological Association of the University of Calgary, Calgary, Alberta.

2002 Faunal Variation and Change in the Northern San Juan Region. In *Seeking the Center Place*, edited by Mark D. Varien and Richard H. Wilshusen, pp. 143–160. University of Utah Press, Salt Lake City.

Duff, Andrew I., Karen R. Adams, and Susan C. Ryan

2010 The Impact of Long-Term Residential Occupation of Community Centers on Local Plant and Animal Resources. In *Leaving Mesa Verde: Peril and Change in the 13th Century Southwest*, edited by Timothy A. Kohler, Mark D. Varien, and Aaron Wright, pp. 156–179. Amerind Foundation and the University of Arizona Press, Tucson.

Duff, Andrew I., and Richard H. Wilshusen

2000 Prehistoric Population Dynamics in the Northern San Juan Region, A.D. 950–1300. *Kiva* 66:167–190.

Dyakowska, Jadwiga

1937 Research on the Rapidity of the Falling Down of Pollen of Some Forest Trees. *Bulletin de l'Academie Polonaise des Sciences—Series des Sciences Biologiques* 7:11–16.

Dyson-Hudson, Rada, and Eric A. Smith

1978 Human Territoriality: An Ecological Reassessment. *American Anthropologist* 80:21–41.

Eddy, Frank W.

1966 Prehistory in the Navajo Reservoir District, Northwestern New Mexico. Museum of New Mexico Papers in Anthropology, no. 15, Santa Fe.

Ehrlich, Paul R., and Anne H. Ehrlich

1970 *Population, Resources, Environment: Issues in Human Ecology*. W.H. Freeman, San Francisco, CA.

Ekren, E. B., and F. N. Houser

1965 Geology and Petrology of the Ute Mountains Area Colorado. *Geological Survey Professional Paper 481*. Government Printing Office, Washington, DC.

Ellis, Andrea

1991 Towers of the Gallina Area and Greater Southwest. In *Puebloan Past and Present*, edited by M. S. Duran and D. T. Kirkpatrick, pp. 57–70. Archaeological Society of New Mexico, Albuquerque.

Ember, Carol R., and Melvin Ember

1992 Resource Unpredictability, Mistrust, and War. *Journal of Conflict Resolution* 36:242–262.

Ember, Melvin

1982 Statistical Evidence for an Ecological Explanation for Warfare. *American Anthropologist* 84:645–649.

Emlen, John M.

1987 Evolutionary Ecology and the Optimality Assumption. In *The Latest on the Best: Essays on Evolution and Optimality*, edited by John Dupré, pp. 163–177. MIT Press, Cambridge, MA.

Encyclopaedia Britannica

2005 *Diseases of Nutritional Deficiency*. Available at www.search.eb.com/eb/article?tocid=63264, Accessed April 15, 2005.

2006 Kwashiorkor. Available at www.search.eb.com/eb/article-9046555. Accessed January 6, 2006.

Epstein, Joshua M., and Robert Axtell

1996 *Growing Artificial Societies: Social Science from the Bottom Up*. Brookings Institution Press and MIT Press, Washington, DC, and Cambridge, MA.

Errickson, Mary

1993 *Archaeological Investigations on Reach III of the Towaoc Canal*. Four Corners Archaeological Project, Report no. 21. Complete Archaeological Service Associates, Cortez, CO.

Esper, Jan, Edward R. Cook, and Fritz H. Schweingruber

2002 Low-Frequency Signals in Long-Term Tree-Ring Chronologies for Reconstructing Past Temperature Variability. *Science* 295:2250–2253.

Euler, Robert E., George J. Gumerman, Thor N. V. Karlstrom, Jeffrey S. Dean, and Richard H. Hevley

1979 The Colorado Plateaus: Cultural Dynamics and Paleoenvironment. *Science* 205:1089–1101.

Faegri, Knut, and Johs Iversen

1988 *Textbook of Pollen Analysis*. 4th ed. Hafner, New York.

Fall, Patricia L.

1983 Holocene Dynamics of the Subalpine Forest in Central Colorado. *American Association of Stratigraphic Palynologists Contribution Series* 16:31–46. American Association of Stratigraphic Palynologists Foundation, Dallas, TX.

1988 Vegetation Dynamics in the Southern Rocky Mountains: Late Pleistocene and Holocene Timberline Fluctuations. Unpublished Ph.D. dissertation, Department of Geosciences, University of Arizona, Tucson.

1997 Timberline Fluctuations and Late Quaternary Paleoclimates in the Southern Rocky Mountains, Colorado. *Geological Society of America Bulletin* 29:53–62.

Farmer, James D.

1997 Iconographic Evidence of Basketmaker Warfare and Human Sacrifice: A Contextual Approach to Early Anasazi Art. *Kiva* 62:391–420.

Feinman, Gary M., Kent G. Lightfoot, Steadman Upham

2000 Political Hierarchies and Organizational Strategies in the Puebloan Southwest. *American Antiquity* 65:449–470.

Ferguson, R. Brian

1984 Introduction: Studying War. In *Warfare, Culture, and Environment*, edited by Brian R. Ferguson, pp. 1–81. Academic, Orlando, FL.

1990 Explaining Warfare. In *The Anthropology of War*, edited by Jonathan Haas, pp. 26–55. Cambridge University Press, Cambridge.

1995 *Yanomami Warfare: A Political History*. School of American Research Press, Santa Fe, NM.

Fetterman, Jerry, and Linda Honeycutt

1987 *The Mockingbird Mesa Survey, Southwestern Colorado*. Cultural Resource Series, no. 22. Bureau of Land Management, Denver, CO.

1990 *In the Fremont-Anasazi Transition Zone: Excavations in Verdure Canyon*. Woods Canyon Archaeological Consultants, Yellow Jacket, CO.

1994 *Interim Report on the Archaeological Survey of Indian Camp Ranch*. Southwestern Colorado, Indian Camp Report, no. 3. Woods Canyon Archaeological Consultants, Cortez, CO.

Fetterman, Jerry, Linda Honeycutt, and Kristin Kuckelman

1988 *Salvage Excavations of 42SA12209, A Pueblo I Habitation Site in Cottonwood Canyon, Manti-Lasal National Forest, Southeastern Utah*. Woods Canyon Archaeological Consultants, Yellow Jacket, CO.

Fewkes, J. Walter

1909 *Antiquities of the Mesa Verde National Park, Spruce Tree House*. Bureau of American Ethnology, Bulletin 41, Government Printing Office, Washington, DC.

1911 *Antiquities of the Mesa Verde National Park, Cliff Palace*. Bureau of American Ethnology, Bulletin 51, Government Printing Office, Washington, DC.

1916 The Relation of Sun Temple, a New Type of Ruin Lately Excavated in the Mesa Verde National Park, to Prehistoric "Towers." *Journal of the Washington Academy of Sciences* 6:341–346.

1919 *Prehistoric Villages, Castles, and Towers of Southwestern Colorado*. Bureau of American Ethnology, Bulletin 70, Government Printing Office, Washington, DC.

1999 [1909, 1911] *Mesa Verde Ancient Architecture*. Awanyu.

Findlow, Frank J., and Marisa Bolognese

1984 Economic Aspects of Prehistoric Quarry Use: A Case Study in the American Southwest. In *Prehistoric Quarries and Lithic Production*, edited by J. E. Ericson and B. A. Purdy, pp. 77–82. Cambridge University Press, Cambridge.

Fink, T. Michael

1989 *Analysis of Human Remains from Site 42SA3724 Southeastern Utah*. Soil Systems Technical Reports 89-04. Soil Systems, Phoenix, Arizona. Woods Canyon Archaeological Consultants, Yellow Jacket, CO.

Firor, James, Rand A. Greubel, and Alan D. Reed

1998 Archaeological Data Recovery at Four Anasazi Sites on White Mesa Along US Highway 191 San Juan County, Utah. Alpine Archaeological Consultants, Montrose, CO.

Foley, Robert

1985 Optimality Theory in Anthropology. *Man, New Series*, 20:222–242.

Forde, C. Daryll

1931 Hopi Agriculture and Land Ownership. *Journal of the Royal Anthropological Institute of Great Britain and Ireland* 61:357–405.

France, Diane L.

1988 *A Human Burial from Dolores County, Colorado*. Cultural Resource Series, no. 24. Bureau of Land Management, Colorado State Office, Denver.

French, Norman R., Ray McBride, and Jack Detmer

1965 Fertility and Population Density of the Black-Tailed Jackrabbit. *Journal of Wildlife Management* 29:14–26.

Fritts, Harold C.

1991 *Reconstructing Large-Scale Climatic Patterns from Tree-Ring Data: A Diagnostic Analysis*. University of Arizona Press, Tucson.

Fritts, Harold C., David G. Smith, and Marvin A. Stokes

1965 The Biological Model for Paleoclimatic Interpretation of the Mesa Verde Tree-Ring

Series. In *Contributions of the Wetherill Mesa Archeological Project*, assembled by D. Osbourne, pp. 101–121. Memoirs of the Society for American Archaeology, no. 19. Salt Lake City, UT.

Fuller, Steven L., and James N. Morris
1991 Excavations at Knobby Knee Stockade (Site 5MT2525), a Basketmaker III-Pueblo III Habitation. In *Archaeological Excavations on the Hovenweep Laterals*, by James N. Morris, pp. 59–325. Four Corners Archaeological Project, Report no. 16. Complete Archaeological Service Associates, Cortez, CO.

Gendel, Peter A.
1984 *Mesolithic Social Territories in Northwestern Europe*. British Archaeological Reports, Oxford.

Gigerenzer, G., P. M. Todd, and the ABC Research Group
1999 *Simple Heuristics That Make Us Smart*. Oxford University Press, New York.

Gilman, Patricia A.
1987 Architecture as Artifact: Pit Structures and Pueblos in the American Southwest. *American Antiquity* 52(3):538–564.

Giuliano, W. M., C. L. Elliott, and J. D. Sole
1994 Significance of Tall Fescue in the Diet of the Eastern Cottontail. *Prairie Naturalist* 26:53–60.

Gleick, Peter H.
2000 *The World's Water 2000–2001: The Biennial Report on Freshwater Resources*. Island, Washington, DC.

Glowacki, Donna M.
2001 Yucca House (5MT5006) Mapping Project 2000. Manuscript on file at Mesa Verde National Park and Crow Canyon Archaeological Center, CO.
2006 The Social Landscape of Depopulation: The Northern San Juan, A.D. 1150–1300. Unpublished Ph.D. dissertation, Department of Anthropology, Arizona State University, Tempe.
2007 Mesa Verde Village Assessment Project. Manuscript in possession of the author.
2010 The Social and Cultural Contexts of the Central Mesa Verde Region During the Thirteenth-Century Migrations. In *Leaving Mesa Verde: Peril and Change in the Thirteenth-Century Southwest*, edited by Timothy A. Kohler, Mark D. Varien, and Aaron M. Wright, pp. 200–221. University of Arizona Press, Tucson.
2011 The Role of Religion in the in the Depopulation of the Central Mesa Verde Region. In *Religious Transformation in the Late Pre-Hispanic Pueblo World*, edited by Donna M. Glowacki and Scott Van Keuren, pp. 66–83. University of Arizona Press, Tucson.

Glowacki, Donna M., Hector Neff, and Michael D. Glascock
1995 Characterization of Mesa Verde Black-on-White Ceramics from Southwestern Colorado Using NAA. *Journal of Radioanalytical and Nuclear Chemistry, Articles* 196:215–222.
1998 An Initial Assessment of the Production and Movement of 13th Century Ceramic Vessels in the Northern San Juan Region. *Kiva* 63:217–240.

Glowacki, Donna M., and Mark D. Varien
2002 Community Centers: Community Center Survey 2002. Report on file at Crow Canyon Archaeological Center, Cortez, CO and the Department of Anthropology, Washington State University, Pullman.
2003 Community Center Survey 2003: Report and Synthesis. Report on file at Crow Canyon Archaeological Center, Cortez, CO, and the Department of Anthropology, Washington State University, Pullman.

Glowacki, Donna M., Mark D. Varien, and C. David Johnson
2003 Community Centers: Cycles of Aggregation in the Mesa Verde Region. Paper presented at the 68th Annual Meeting of the Society for American Archaeology, Milwaukee, WI.

Golub, Gene, and Charles Van Loan
1983 *Matrix Computations*. Johns Hopkins University Press, Baltimore, MD.

Godfrey-Smith, Peter
2001 Three Kinds of Adaptationism. In *Adaptationism and Optimality*, edited by S. H. Orzack and E. Sober, pp. 335–357. Cambridge University Press, Cambridge.

Goring-Morris, Nigel, and Anna Belfer-Cohen
2010 "Great Expectations," or the Inevitable Collapse of the Early Neolithic in the Near East. In *Becoming Villagers: Comparing Early Village Societies*, edited by Matthew S. Bandy and Jake R. Fox, pp. 62–80. University of Arizona Press, Tucson.

Gould, Ronald R.
1982 The Mustoe Site: The Application of Neutron Activation Analysis in the Interpretation of a Multi-component Archaeological Site. Unpublished Ph.D. dissertation, Department of Anthropology, University of Texas, Austin.

Gould, S. J., and R. C. Lewontin
1979 The Spandrels of San Marco and the Panglossian Paradigm: A Critique of the Adaptationist Programme. *Proceedings of the Royal Society of London, B Biological Science* 205:581–598.
Graber, Robert B.
1997 A Rigorous Approach to Population Pressure's Contribution to Cultural Evolution. In *Integrating Archaeological Demography: Multidisciplinary Approaches to Prehistoric Population,* edited by Richard R. Paine, pp. 263–284. Center for Archaeological Investigations. Occasional Paper, no. 24. Southern Illinois University, Carbondale.
Grayson, Donald K.
1984 *Quantitative Zooarchaeology.* Academic Press, Orlando, FL.
2001 The Archaeological Record of Human Impacts on Animal Populations. *Journal of World Prehistory* 15:1–68.
Green, Richard C.
1962 The Hormigas Site of the Largo-Gallina Phase. *El Palacio* 63:142–157.
Gregory, David A., and Fred L. Nials
2007 The Environmental Context of Linguistic Differentiation and Other Cultural Developments in the Prehistoric Southwest. In *Zuni Origins: Toward a New Synthesis of Southwestern Archaeology,* edited by D. A. Gregory and D. R. Wilcox, pp. 49–76. University of Arizona Press, Tucson.
Gross, J. E., K. C. Stoddart, and F. H. Wagner
1974 Demographic analysis of a northern Utah jackrabbit population. *Wildlife Monographs* No. 40.
Gross, Timothy G., and Allen E. Kane (compilers)
1983 *Dolores Archaeological Program: Aceramic and Late Occupations at Dolores.* U.S. Department of the Interior, Bureau of Reclamation, Engineering and Research Center, Denver, CO.
Grove, Jean M.
1988 *The Little Ice Age.* Cambridge University Press, Cambridge.
Gurney, W. S. C., and R. M. Nisbet
1998 *Ecological Dynamics.* Oxford University Press, New York.
Gurven, Michael, Wesley Allen-Arave, Kim Hill, and Magdalena Hurtado
2000 "It's a Wonderful Life": Signaling Generosity Among the Ache of Paraguay. *Evolution and Human Behavior* 21(4):263–282.
Haas, Jonathan
1982 The Evolution of Kayenta Anasazi. In *Tse Yaa Kin: Houses Beneath the Rock.* Exploration Series. School of American Research Press, Santa Fe, NM.
1989 The Evolution of the Kayenta Regional System. In *The Sociopolitical Structure of Prehistoric Southwestern Societies,* edited by S. Upham, K. G. Lightfoot, and R. A. Jewett, pp. 491–508. Westview, Boulder, CO.
1990 Warfare and the Evolution of Tribal Polities in the Prehistoric Southwest. In *The Anthropology of War,* edited by Jonathan Haas, pp. 171–189. Cambridge University Press, New York.
Haas, Jonathan, and Winifred Creamer
1993 Stress and Warfare Among the Prehistoric Kayenta Anasazi of the Thirteenth Century. *Fieldiana: Anthropology 21.* Field Museum of Natural History, Chicago, IL.
1996 The Role of Warfare in the Pueblo III Period. In *The Prehistoric Pueblo World, A.D. 1150–1350,* edited by M. A. Adler, pp. 205–213. University of Arizona Press, Tucson.
Haase, William R.
1985 Domestic Water Conservation Among the Northern San Juan Anasazi. *Southwestern Lore* 51(2):15–27.
Hack, John T.
1942 *The Changing Physical Environment of the Hopi Indians of Arizona.* Papers of the Peabody Museum of American Archaeology and Ethnology, vol. 35, no. 1. Harvard University Museum, Cambridge, MA.
Hanley, Thomas A., and Kathleen A. Hanley
1982 Food Resource Partitioning by Sympatric Ungulates on Great Basin Rangeland. *Journal of Range Management* 35:152–158.
Harriman, Ray G., and James N. Morris
1991 Excavations at Dripping Springs Stockade (Site 5MT2527): A Multiple-Occupation, Pueblo I–Pueblo II Anasazi Habitation. In *Archaeological Excavations on the Hovenweep Laterals,* by James N. Morris, pp. 327–399. Four Corners Archaeological Project, Report no. 16. Complete Archaeological Service Associates, Cortez, CO.
Harris, Marvin
1977 *Cannibals and Kings: The Origins of Culture.* Vintage, New York.
1984 A Cultural Materialist Theory of Band and Village Warfare: The Yanomamo Test. In *Warfare, Culture, and Environment,* edited by

Brian Ferguson, pp. 111–140. Academic, New York.

Harro, Douglas R.

1997 Patterns of Lithic Raw Material Procurement on the Pajarito Plateau, New Mexico. Unpublished Master's thesis, Department of Anthropology, Washington State University, Pullman.

Haskell, Horace S., and Hudson G. Reynolds

1947 Growth, Developmental Food Requirements, and Breeding Activity of the California Jack Rabbit. *Journal of Mammalogy* 28:129–136.

Hawkes, Kristen

1993 Why Hunter-Gatherers Work: An Ancient Version of the Problem of Public Goods. *Current Anthropology* 34:341–361.

Hawley, Florence, Michel Pijoan, and C. A. Elkin

1943 An Inquiry into Food Economy and Body Economy in Zia Pueblo. *American Anthropologist* 45:547–556.

Hayes, Alden C.

1964 *The Archaeological Survey of Wetherill Mesa, Mesa Verde National Park–Colorado.* Archaeological Research Series, no. 7A. National Park Service, U.S. Department of the Interior, Washington, DC.

1998 *Two Raven House, Mesa Verde National Park, Colorado.* Archaeological Research Series, no. 7-E. Mesa Verde Museum Association, Mesa Verde National Park, CO.

Hayes, Alden C., and James A. Lancaster

1975 *Badger House Community, Mesa Verde National Park.* Publications in Archaeology 7E, Wetherill Mesa Studies. National Park Service, Washington, DC.

Hegmon, Michelle

1991 Six Easy Steps to Dating Pueblo III Ceramic Assemblages: Working Draft. Manuscript on file at Crow Canyon Archaeological Center, Cortez, CO.

1996 Variability in Food Production, Strategies of Storage and Sharing, and the Pithouse-to-Pueblo Transition in the Northern Southwest. In *Evolving Complexity and Environmental Risk in the Prehistoric Southwest,* edited by Joseph A. Tainter and Bonnie Bagley Tainter, pp. 223–241. Santa Fe Institute Studies in the Sciences of Complexity, vol. 24. Addison-Wesley, Boston, MA.

Helbing, D., C. Kühnert, S. Lämmer, Anders Johansson, B. Gehlsen, H. Ammoser, and G. B. West

2009 Power Laws in Urban Supply Networks, Social Systems, and Dense Pedestrian Crowds. In *Complexity Perspectives in Innovation and Social Change,* edited by D. Lane, S. van der Leeuw, D. Pumain, and G. West, pp. 11–41. Springer, Dordrecht.

Helms, Mary W.

1994 Chiefdom Rivalries, Control, and External Contacts in Lower Central America. In *Factional Competition and Political Development in the New World,* edited by Elizabeth M. Brumfiel and John W. Fox, pp. 55–60. Cambridge University Press, New York.

Henrich, Joseph

2004 Demography and Cultural Evolution: How Adaptive Cultural Processes Can Produce Maladaptive Losses—The Tasmanian Case. *American Antiquity* 69:197–214.

Herman, Frederick J.

1970 *Manual of the Carices of the Rocky Mountains and Colorado Basin.* Agricultural Handbook no. 374. U.S. Department of Agriculture, U.S. Government Printing Office, Washington, DC.

Herweijer, Celine, Richard Seager, Edward R. Cook, and Julien Emile-Geay

2007 North American Droughts of the Last Millennium from a Gridded Network of Tree-Ring Data. *Journal of Climate* 20:1353–1376.

Hewett, Edgar L.

1908 *Les Communautés Anciennes Dans le Désert Américain: Recherches Archéologiques sur la Distribution et Organisation Sociale des Anciennes Populations au Sud-Oest des États-Unis et au Nord du Mexique.* Librairie Kundig, Geneva.

Hibben, Frank C.

1948 The Gallina Architectural Forms. *American Antiquity* 16:32–36.

Higgs, Eric S., and M. R. Jarman

1970 The Origins of Agriculture: A Reconsideration. *Antiquity* 43:31–41.

Hill, David V.

1985 Pottery Making at the Ewing Site (5MT927). *Southwestern Lore* 51(1):19–31.

Hillel, Daniel J.

1991 *Out of the Earth.* Free Press, New York.

Hirth, Kenneth

1984 Catchment Analysis and Formative Settlement in the Valley of Mexico. *American Anthropologist* 86:136–143.

Hobbs, N. Thompson and David M. Swift
1985 Estimates of Habitat Carrying Capacity Incorporating Explicit Nutritional Constraints. *Journal of Wildlife Management* 49:814–822.
Holland, John H.
1992 *Adaptation in Natural and Artificial Systems.* MIT Press, Cambridge, MA.
Holmes, William Henry
1878 Report on the Ancient Ruins of Southwestern Colorado, Examined During the Summers of 1875 and 1876. *United States Geological and Geographical Survey of the Territories.* 10th Annual Report, Washington, DC.
Homburg, Jeffrey A.
2000 Anthropogenic Influences on American Indian Soils of the Southwestern United States. Unpublished Ph.D. dissertation, Department of Soil Science, Iowa State University, Ames.
Honeycutt, Linda
1995 Dryland Gardening in Southwest Colorado: Past and Present. In *Soil, Water, Biology, and Belief in Prehistoric and Traditional Southwestern Agriculture,* edited by H. W. Toll, pp. 369–373. New Mexico Archaeological Council Special Publication, no. 2. Albuquerque.
Hovezak, T. D., L. M. Sesler, J. E. Fetterman, and L. Honeycutt
2003 The 2002 Canyon of the Ancients National Monument Survey, Montezuma County, Colorado. Woods Canyon Archaeological Consultants, Yellow Jacket, CO.
Hovezak, T. D., L. M. Sesler, M. D. Varien, and C. Goetze
2004 An Archaeological Survey of the Goodman Point Unit of Hovenweep National Monument, Montezuma County, Colorado. Manuscript on file at Crow Canyon Archaeological Center, Cortez, CO.
Howell, Joseph
1941 Pinon and Juniper Woodlands of the Southwest. *Journal of Forestry* 39:542–545.
Huckleberry, Gary A., and Brian R. Billman
1998 Floodwater Farming, Discontinuous Ephemeral Streams, and Puebloan Abandonment in Southwestern Colorado. *American Antiquity* 63:595–616.
Hughes, Malcolm K., and Henry F. Diaz (editors)
1994 *The Medieval Warm Period.* Kluwer Academic, Dordrecht.
Hunt, L. A., W. Yan, and Gregory S. McMaster
2001 Simulating Response to Temperature. In *Modeling Temperature Response in Wheat and Maize,* edited by Jeffrey W. White, pp. 23–29. NRG-GIS series 03-01. CIMMYT, Mexico City.
Hunter-Anderson, Rosalind D.
1979 Explaining Residential Aggregation in the Northern Rio Grande: A Competition Reduction Model. In *Archaeological Investigations in Cochiti Reservoir, New Mexico,* vol. 4, *Adaptive Change in the Northern Rio Grande Valley,* edited by J. V. Biella and R. C. Chapman, pp. 56–67. Office of Contract Archaeology, University of New Mexico, Albuquerque.
Hurst, C. T., and V. F. Lotrich
1932 An Unusual Mug from Yellow Jacket Canyon. *El Palacio* 33(21–22):195–198.
Hurst, Winston, and Jonathan D. Till
2002 Some Observations Regarding the "Chaco Phenomenon" in the Northwestern San Juan Provinces. Paper presented at the 67th Annual Meeting of the Society for American Archaeology, Denver, CO.
Hurst, Winston B. and Christy G. Turner II
1993 Rediscovering the "Great Discovery": Wetherill's First Cave 7 and Its Record of Basketmaker Violence. In *Anasazi Basketmaker: Papers from the 1990 Wetherill–Grand Gulch Symposium,* edited by Victoria M. Atkins, pp. 143–191. Cultural Resource Series, no. 24. Bureau of Land Management, Salt Lake City, UT.
Hussain, K., D. B. Wijetunge, S. Grubnic, and I. T. Jackson
1994 A Comprehensive Analysis of Craniofacial Trauma. *Journal of Trauma* 24:40–44.
Ingles, Lloyd G.
1941 Natural History Observations on the Audubon Cottontail. *Journal of Mammalogy* 22(3):226–250.
Israel, Debra
2002 Fuel Choice in Developing Countries: Evidence from Bolivia. *Economic Development and Cultural Change* 50:865–890.
Iversen, Gudmund R.
1984 *Bayesian Statistical Inference.* Quantitative Applications in the Social Sciences 43. Sage, Newbury Park, CA.
Jackson, William H.
1876a Ancient Ruins in Southwestern Colorado. In *Annual Report of the United States Geological and Geographical Survey of the Territories, Embracing Colorado and Parts of Adjacent Territories; Being a Report of Progress of the Exploration for the Year 1874, by F. V. Hayden,*

pp. 367–381. Government Printing Office, Washington, DC.
1876b *A Notice of the Ancient Ruins in Arizona and Utah Lying About the Rio San Juan.* Bulletin of the Geological and Geographical Survey of the Territories II(I):25–45. Washington, DC.

Jackson, William H., and William H. Holmes
1981 [1876, 1878] *Mesa Verde and the Four Corners.* Bear Creek Publishing, Ouray, CO.

James, Steven R.
2004 Hunting, Fishing, and Resource Depression in Prehistoric Southwest North America. In *The Archaeology of Global Change,* edited by Charles L. Redman, Steven R. James, Paul R. Fish, and J. Daniel Rogers, pp. 28–62. Smithsonian Books, Washington, DC.

Janssen, C. Roel
1966 Recent Pollen Spectra from the Deciduous and Coniferous-Deciduous Forests of Northeastern Minnesota: A Study in Pollen Dispersal. *Ecology* 47:804–825.

Janssen, Marco A., Timothy A. Kohler, and Marten Scheffer
2003 Sunk-Cost Effects and Vulnerability to Collapse in Ancient Societies. *Current Anthropology* 44:722–728.

Jarman, M. R., G. N. Bailey, and H. N. Jarman
1982 Territories and Mobility. In *Early European Agriculture: Its Foundations and Development: Written in Honour of Eric Higgs,* edited by M. R. Jarman, G. N. Bailey, and H. N. Jarman, pp. 26–48. Cambridge University Press, New York.

Jeske, Robert J.
1992 Energetic Efficiency and Lithic Technology—An Upper Mississippian Example. *American Antiquity* 57:467–481.

Jochim, Michael
1989 Optimization and Stone Tool Studies: Problems and Potential. In *Time, Energy and Stone Tools,* edited by R. Torrence, pp. 106–111. Cambridge University Press, Cambridge.

Johnson, Allen W., and Timothy Earle
2000 *The Evolution of Human Societies: From Foraging Group to Agrarian State.* 2nd ed. Stanford University Press, Stanford, CA.

Johnson, C. David
2003 Mesa Verde Region Towers: A View from Above. *Kiva* 68:323–340.
2006 Critical Natural Resources in the Mesa Verde Region, A.D. 600–1300: Distribution, Use, and Influence on Puebloan Settlement. Unpublished Ph.D. dissertation, Department of Anthropology, Washington State University, Pullman.

Johnson, C. David, Timothy A. Kohler, and Jason Cowan
2005 Modeling Historical Ecology: Thinking About Contemporary Systems. *American Anthropologist* 107:96–108.

Jonassen, Hagbard
1950 Recent Pollen Sedimentation and Jutland Heath Diagrams. *Dansk Botanisk Arkiv* 13:1–168.

Jones, Kevin T., and David B. Madsen
1989 Calculating the Cost of Resource Transportation: A Great Basin Example. *Current Anthropology* 30:529–534.

Judge, W. James
1989 Chaco Canyon-San Juan Basin. In *Dynamics of Southwest Prehistory,* edited by L. S. Cordell and G. J. Gumerman, pp. 209–261. Smithsonian Institution Press, Washington, DC.

Kane, Allen E.
1986 Prehistory of the Dolores River Valley. In *Dolores Archaeological Program: Final Synthetic Report,* edited by D. A. Breternitz, C. K. Robinson, and G. T. Gross, pp. 353–435. U.S. Department of the Interior, Bureau of Reclamation, Engineering and Research Center, Denver, CO.
1989 Did the Sheep Look Up? Sociopolitical Complexity in Ninth Century Dolores Society. In *The Sociopolitical Structure of Prehistoric Southwestern Societies,* edited by S. Upham, K. G. Lightfoot, and R. A. Jewett, pp. 307–361. Westview, Boulder, CO.

Kane, Allen E., and Mark Chenault
1982 Excavations at Horsefly Hamlet (Site 5MT2236), An Archaic Camp/Anasazi Habitation Site. *Dolores Archaeological Program Technical Reports* DAP-038. Submitted to the U.S. Bureau of Reclamation, Upper Colorado Region, Salt Lake City, UT in compliance with Contract No. 8-07-40-S0562.

Kane, Allen E., and G. Timothy Gross (compilers)
1986 *Dolores Archaeological Program: Anasazi Communities at Dolores: Early Anasazi Sites in the Sagehen Flats Area.* U.S. Department of the Interior, Bureau of Reclamation, Engineering and Research Center, Denver, CO.

Kane, Allen E., Timothy A. Kohler, and William D. Lipe (compilers)
1986 *Dolores Archaeological Program: Anasazi Communities at Dolores: Early Small Settlements in the Dolores River Canyon and*

Western Sagehen Flats Area. U.S. Department of the Interior, Bureau of Reclamation, Engineering and Research Center, Denver, CO.

Kane, Allen E., and Christine K. Robinson

1988 *Dolores Archaeological Program: Anasazi Communities at Dolores: McPhee Village.* U.S. Department of the Interior, Bureau of Reclamation, Engineering and Research Center, Denver, CO.

Kane, Allen E., and Christine K. Robinson (compilers)

1986 *Dolores Archaeological Program: Anasazi Communities at Dolores: Middle Canyon Area.* U.S. Department of the Interior, Bureau of Reclamation, Engineering and Research Center, Denver, CO.

Kang, Bong W.

2000 A Reconsideration of Population Pressure and Warfare: A Protohistoric Korean Case. *Current Anthropology* 41:873–881.

Kantner, John

1999a Survival Cannibalism or Sociopolitical Intimidation? Explaining Perimortem Mutilation in the American Southwest. *Human Nature* 10:1–50.

1999b Anasazi Mutilation and Cannibalism in the American Southwest. In *The Anthropology of Cannibalism,* edited by L. R. Goldman, pp. 75–104. Bergin and Garvey, Westport, CT.

Kaplan, Hillard, and Kim Hill

1985 Hunting Ability and Reproductive Success Among Male Ache Foragers: Preliminary Results. *Current Anthropology* 26:131–133.

Keeley, Lawrence, H.

1996 *War Before Civilization.* Oxford University Press, New York.

Kelly, Robert L.

1995 *The Foraging Spectrum: Diversity in Hunter-Gatherer Lifeways.* Smithsonian Institution Scholarly Press, Washington, DC.

Kendrick, James W. and W. James Judge

2000 Household Economic Autonomy and Great House Development in the Lowry Area. In *Great House Communities Across the Chacoan Landscape,* edited by John Kantner and Nancy M. Mahoney, pp. 113–129. Anthropological Papers of the University of Arizona, no. 64, University of Arizona Press, Tucson.

Kent, Susan

1989 Cross-Cultural Perceptions of Farmers as Hunters and the Value of Meat. In *Farmers as Hunters: The Implications of Sedentism,* edited by Susan Kent, pp. 1–17. Cambridge University Press, Cambridge.

Kenzle, Susan C.

1993 Enclosing Walls: A Study of Architectural Function in the American Southwest. Unpublished Master's thesis, Department of Archaeology, University of Calgary, Alberta.

1997 Enclosing Walls in the Northern San Juan: Sociophysical Boundaries and Defensive Fortifications in the American Southwest. *Journal of Field Archaeology* 24:195–210.

Khan, Abdul Qayyum, Bashir Ahmad Wani, Hakim Shah, and S. Irfanullah

2001 Assessment of Rural Energy Needs in Hilkot Watershed Pakistan. *Pakistan Journal of Forestry* 51(2):1–14.

Khattree, Ravindra, and Dayanand Naik

2000 *Multivariate Data Reduction and Discrimination with SAS Software.* SAS Institute, Cary, NC.

Kidder, Alfred V.

1924 *An Introduction to the Study of Southwestern Archaeology.* Phillips Academy, Andover, MA.

Kirch, Patrick V.

1980 The Archaeological Study of Adaptation: Theoretical and Methodological Issues. *Advances in Archaeological Method and Theory* 3:101–156.

2010 *How Chiefs Became Kings: Divine Kingship and the Rise of Archaic States in Ancient Hawai'i.* University of California Press, Berkeley and Los Angeles.

Kleidon, Jim, Shanna Diederichs, and Donna M. Glowacki

2007 Long Mesa 2002 Fire Burned Area Archeological Inventory, Mesa Verde National Park, Colorado: Puebloan Settlement, Demographics, and Community Structure, A.D. 500–1300 on Chapin and Park Mesas. Division of Research and Natural Resources, Mesa Verde National Park, CO.

Knick, Steven T., and Deanna L. Dyer

1997 Distribution of Black-Tailed Jackrabbit Habitat Determined by GIS in Southwestern Idaho. *Journal of Wildlife Management* 61:75–85.

Kobti, Ziad

2004 Learning in Dynamic Hierarchical Network Structures in Complex Systems. Unpublished Ph.D. dissertation, Department of Computer Science, Wayne State University, Detroit, MI.

Kobti, Ziad, Robert G. Reynolds, and Timothy A. Kohler

2003 A Multi-agent Simulation Using Cultural Algorithms: The Effect of Culture on the

Resilience of Social Systems. Paper delivered in the Institute of Electrical and Electronics Engineers Congress on Evolutionary Computation, December 5–12, Canberra.

Kohler, Timothy A.

1988a Predictive Locational Modeling: History and Current Practice. In *Quantifying the Past and Predicting the Present: Theory, Method, and Application of Archaeological Predictive Modeling*, edited by W. J. Judge and L. Sebastian, pp. 19–59. U.S. Department of the Interior, Bureau of Land Management, Denver, CO.

1988b The Probability Sample at Grass Mesa Village. In *Dolores Archaeological Program: Anasazi Communities at Dolores: Grass Mesa Village*, edited by W. D. Lipe, J. N. Morris, and T. A. Kohler, pp. 51–74. U.S. Department of the Interior, Bureau of Reclamation, Engineering and Research Center, Denver, CO.

1989 Introduction. In *Bandelier Archaeological Project: Research Design and Summer 1988 Sampling*, edited by T. A. Kohler, pp. 1–12. Washington State University Department of Anthropology Reports of Investigations, no. 61. Pullman.

1992a Field Houses, Villages, and the Tragedy of the Commons in the Early Northern Anasazi Southwest. *American Antiquity* 57:617–734.

1992b Prehistoric Human Impact on the Environment in the Upland North American Southwest. *Population and Environment: A Journal of Interdisciplinary Studies* 13:255–268.

2000 The Final 400 Years of Prehispanic Agricultural Society in the Mesa Verde Region. *Kiva* 66:191–204.

2004 Pre-Hispanic Human Impact on Upland North American Southwestern Environments: Evolutionary Ecological Perspectives. In *The Archaeology of Global Change: The Impact of Humans on Their Environment*, edited by Charles L. Redman, Steven R. James, Paul R. Fish, and J. Daniel Rogers, pp. 224–242. Smithsonian Books, Washington, DC.

2010 A New Paleoproductivity Reconstruction for Southwestern Colorado, and Its Implications for Understanding Thirteenth-Century Depopulation. In *Leaving Mesa Verde: Peril and Change in the Thirteenth-Century Southwest*, edited by Timothy A. Kohler, Mark D. Varien, and Aaron M. Wright, pp. 102–127. University of Arizona Press, Tucson.

Kohler, Timothy A., and Eric Blinman

1987 Solving Mixture Problems in Archaeology: Analysis of Ceramic Materials for Dating and Demographic Reconstruction. *Journal of Anthropological Archaeology* 6:1–28.

Kohler, Timothy A., and Eric Carr

1997 Swarm-Based Modeling of Prehistoric Settlement Systems in Southwestern North America. In *Proceedings of Colloquium 2, UISPP, 23rd Congress, Forli, Italy, September 1996*, edited by I. Johnson and M. North. Sydney University Archaeological Methods Series 5. Archaeology (P&H), Sydney.

Kohler, Timothy A., Sarah Cole, and Stanca Ciupe

2009 Population and Warfare: A Test of the Turchin Model in Puebloan Societies. In *Pattern and Process in Cultural Evolution*, edited by Stephen Shennan, pp. 297–295. University of California Press, Berkeley.

Kohler, Timothy A., Matt Pier Glaude, Jean-Pierre Bocquet-Appel, and Brian M. Kemp

2008 The Neolithic Demographic Transition in the U.S. Southwest. *American Antiquity* 73:645–659.

Kohler, Timothy A., and George Gumerman (editors)

2000 *Dynamics in Human and Primate Societies: Agent-Based Modeling of Social and Spatial Processes*. Santa Fe Institute and Oxford University Press, New York.

Kohler, Timothy A., C. David Johnson, Mark D. Varien, Scott G. Ortman, Robert Reynolds, Ziad Kobti, Jason Cowan, Kenneth Kolm, Schaun Smith, and Lorene Yap

2007 Settlement Ecodynamics in the Prehispanic Central Mesa Verde Region. In *The Model-Based Archaeology of Socionatural Systems*, edited by Timothy A. Kohler and Sander van der Leeuw, pp. 61–104. School of American Research Press, Santa Fe, NM.

Kohler, Timothy A., James Kresl, Carla Van West, Eric Carr, and Richard H. Wilshusen

2000 Be There Then: A Modeling Approach to Settlement Determinants and Spatial Efficiency Among Late Ancestral Pueblo Populations of the Mesa Verde Region, U.S. Southwest. In *Dynamics in Human and Primate Societies: Agent-Based Modeling of Social and Spatial Processes*, edited by T. A.

Kohler and G. J. Gumerman, pp. 145–178. Santa Fe Institute and Oxford University Press, New York.

Kohler, Timothy A., William D. Lipe, and Allen E. Kane (compilers)
1985 *Dolores Archaeological Program: Anasazi Communities at Dolores: Early Small Settlements in the Dolores River Canyon and Western Sagehen Flats Area.* U.S. Department of the Interior, Bureau of Reclamation, Engineering and Research Center, Denver, CO.

Kohler, Timothy A., and Meredith Matthews
1988 Long-Term Anasazi Land-Use Patterns and Forest Reduction: A Case Study from Southwest Colorado. *American Antiquity* 53:537–564.

Kohler, Timothy A., Janet D. Orcutt, Kenneth L. Petersen, and Eric Blinman
1986 Anasazi Spreadsheets: The Cost of Doing Agricultural Business in Prehistoric Dolores. In *Dolores Archaeological Program: Final Synthetic Report,* compiled by D. A. Breternitz, C. K. Robinson, and G. T. Gross, pp. 525–538. U.S. Department of the Interior, Bureau of Reclamation, Engineering and Research Center, Denver, CO.

Kohler, Timothy A., Robert P. Powers, and Janet D. Orcutt
2004 Bandelier from Hamlets to Towns. In *Archaeology of Bandelier National Monument: Village Formation on the Pajarito Plateau, New Mexico,* edited by T. A. Kohler, pp. 293–303. University of New Mexico Press, Albuquerque.

Kohler, Timothy A., and Charles Reed
2011 Explaining the Structure and Timing of Formation of Pueblo I Villages. In *Sustainable Lifeways: Cultural Persistence in an Ever-Changing Environment,* edited by Naomi F. Miller, Katherine M. Moore, and Kathleen Ryan, pp. 150–179. University of Pennsylvania Museum of Archaeology and Anthropology, Philadelphia.

Kohler, Timothy A., and Kathryn Kramer Turner
2006 Raiding for Women in the Prehispanic Northern Pueblo Southwest? A Pilot Examination. *Current Anthropology* 47:1035–1045.

Kohler, Timothy A., and Sander van der Leeuw
2007 Historical Socionatural Systems and Models. In *The Model-Based Archaeology of Socionatural Systems,* edited by T. A. Kohler and S. van der Leeuw, pp. 1–12. School for Advanced Research Press, Santa Fe, NM.

Kohler, Timothy A., Matthew Van Pelt, and Lorene Y. L. Yap
2000 Reciprocity and Its Limits: Considerations for a Study of the Prehispanic Pueblo World. In *Alternative Leadership Strategies in the Prehispanic Southwest,* edited by B. J. Mills, pp. 180–206. University of Arizona Press, Tucson.

Kohler, Timothy A., and Carla R. Van West
1996 The Calculus of Self-Interest in the Development of Cooperation: Sociopolitical Development and Risk Among the Northern Anasazi. In *Evolving Complexity and Environmental Risk in the Prehistoric Southwest,* edited by Joseph Tainter and Bonnie Bagley Tainter, pp. 169–196. Santa Fe Institute Studies in the Sciences of Complexity, vol. 24. Addison-Wesley, Boston, MA.

Kohler, Timothy A., Carla R. Van West, Eric P. Carr, and Christopher G. Langton
1996 Agent-Based Modeling of Prehistoric Settlement Systems in the Northern American Southwest. In *Proceedings of the Third International Conference Integrating GIS and Environmental Modeling,* Santa Fe, NM, January. National Center for Geographic Information and Analysis, Santa Barbara, CA.

Kohler, Timothy A., and Mark D. Varien
2010 A Scale Model of Seven Hundred Years of Farming Settlements in Southwestern Colorado. In *Becoming Villagers: Comparing Early Village Societies,* edited by Matthew S. Bandy and Jake R. Fox, pp. 37–61. University of Arizona Press, Tucson.

Kohler, Timothy A., Mark D. Varien, and Aaron M. Wright (editors)
2010 *Leaving Mesa Verde: Peril and Change in the Thirteenth-Century Southwest.* University of Arizona Press, Tucson.

Kohler, Timothy A., Mark D. Varien, Aaron M. Wright, and Kristin A. Kuckelman
2009 Mesa Verde Migrations: New Archaeological Research and Computer Simulation Suggest Why Ancestral Puebloans Deserted the Northern Southwest United States. *American Scientist* 96:146–153.

Kohler, Timothy A., and Lorene Yap
2003 Modeling Reciprocal Exchange in Southwestern Societies. Paper presented at the 68th Annual Meeting of the Society for American Archaeology, Milwaukee, WI.

Kolb, Michael J., and James E. Snead
1997 It's a Small World After All: Comparative Analyses of Community Organization in

Archaeology. *American Antiquity* 62:609–628.

Kolm, Kenneth E.

1996 Conceptualization and Characterization of Groundwater Systems Using Geoscientific Information Systems. In *Advanced Techniques for Solving Groundwater Contaminant Problems,* edited by J. May and K. E. Kolm. *Engineering Geology* 42(2–3):111–118.

Krausman, Paul R., Amy J. Kuenzi, Richard C. Etchberger, Kurt R. Rautenstrauch, Leonard L. Ordway, and John J. Hervert

1997 Diets of Desert Mule Deer. *Journal of Range Management* 50:513–522.

Krech, Shepard

1999 *The Ecological Indian: Myth and History.* Norton, New York.

Kuckelman, Kristin A.

1984a Excavations at Golondrinas Oriental (Site 5MT5108), a Pueblo I Habitation. *Dolores Archaeological Program Technical Reports* DAP-132. Submitted to the U.S. Bureau of Reclamation, Upper Colorado Region, Salt Lake City, UT in compliance with Contract No. 8-07-40-S0562.

1984b Excavations at Masa Negra Pueblo (Site 5MT4477), a Pueblo I/Pueblo II Habitation. *Dolores Archaeological Program Technical Reports* DAP-161. Submitted to the U.S. Bureau of Reclamation, Upper Colorado Region, Salt Lake City, UT in compliance with Contract No. 8-07-40-S0562.

1988 Excavations at Dobbins Stockade (Site 5MT8827), a Pueblo II Habitation. In *Archaeological Investigations on South Canal, Vol. I,* edited by K. A. Kuckelman and J. N. Morris, pp. 46–107. Four Corners Archaeological Project Report, no. 11. Complete Archaeological Service Associates, Cortez, CO.

2002 Thirteenth-Century Warfare in the Central Mesa Verde Region. In *Seeking the Center Place: Archaeology and Ancient Communities in the Mesa Verde Region,* edited by Mark D. Varien and Richard H. Wilshusen, pp. 233–253. University of Utah Press, Salt Lake City.

2010a Catalysts of the Thirteenth-Century Depopulation of Sand Canyon Pueblo and the Central Mesa Verde Region. In *Leaving Mesa Verde: Peril and Change in the Thirteenth-Century Southwest,* edited by Timothy A. Kohler, Mark D. Varien, and Aaron M. Wright, pp. 180–199. University of Arizona Press, Tucson.

2010b The Depopulation of Sand Canyon Pueblo, a Large Ancestral Pueblo Village in Southwestern Colorado. *American Antiquity* 75(3):497–526.

Kuckelman, Kristin A. (editor)

2000 *The Archaeology of Castle Rock Pueblo: A Thirteenth-Century Village in Southwestern Colorado.* Available at www.crowcanyon.org/castlerock. Accessed August 1, 2009.

2003 *The Archaeology of Yellow Jacket Pueblo (Site 5MT5): Excavations at a Large Community Center in Southwestern Colorado.* Available at www.crowcanyon.org/yellowjacket. Accessed June 23, 2008.

2007 A Descriptive and Interpretive Summary of Excavations, by Architectural Block. In *The Archaeology of Sand Canyon Pueblo: Intensive Excavations at a Late-Thirteenth-Century Village in Southwestern Colorado.* Available at www.crowcanyon.org/. Accessed January 29, 2009.

Kuckelman, Kristin A., Grant D. Coffey, and Steve R. Copeland

2009 *Interim Descriptive Report of Research at Goodman Point Pueblo (5MT604), Montezuma County, Colorado, 2005–2008.* Available at http://crowcanyon.org/ResearchReports/GoodmanPoint/interim_reports/2005_2008/GPP_interim_report_2005_2008.pdf.

Kuckelman, Kristin A., and Raymond G. Harriman

1984 Excavations at Rabbitbrush Pueblo (Site 5MT4480), Pueblo I Habitation. *Dolores Archaeological Program Technical Reports* DAP-163. Submitted to the U.S. Bureau of Reclamation, Upper Colorado Region, Salt Lake City, UT, in compliance with Contract No. 8-07-40-S0562.

Kuckelman, Kristin A., Ricky R. Lightfoot, and Debra L. Martin

2000 Changing Patterns of Violence in the Northern San Juan Region. *Kiva* 66:147–165.

2002 The Bioarchaeology and Taphonomy of Violence at Castle Rock and Sand Canyon Pueblos, Southwestern Colorado. *American Antiquity* 67:486–513.

Kuckelman, Kristin A., and James N. Morris (compilers)

1988 *Archaeological Investigations on South Canal.* Four Corners Archaeological Project, Report no. 11. Complete Archaeological Service Associates, Cortez, CO.

Lahren, C. H., and H. E. Berryman

1984 Fracture Patterns and Status at Chucalissa (40SY1): A Biocultural Approach. *Tennessee Anthropologist* 9:15–21.

LaMarche, Valmore C., Jr.

1973 Holocene Climatic Variations Inferred from Treeline Fluctuations in the White Mountains, California. *Quaternary Research* 2:632–660.

Lamb, Hubert H.

1977 *Climate: Past, Present, and Future.* Methuen, London.

1995 *Climate, History, and the Modern World.* 2nd ed. Routledge, London.

Lambert, Patricia M.

1999 Human Skeletal Remains Recovered During the 1995–1996 Seasons. In *The Puebloan Occupation of the Ute Mountain Piedmont,* vol. 5, *Environmental and Bioarchaeological Studies,* edited by Brian R. Billman, pp. 203–236. Soil Systems Publications in Archaeology, no. 22. Phoenix, AZ.

2002 The Archaeology of War: A North American Perspective. *Journal of Archaeological Research* 10:207–241.

Lancaster, James A., and Jean M. Pinkley

1954 Excavation at Site 16. In *Archaeological Excavations in Mesa Verde National Park, Colorado, 1950,* edited by James A. Lancaster, Jean M. Pinkley, P. Van Cleave, and Don Watson, pp. 23–86. Archeological Research Series, no. 2. National Park Service, U.S. Department of the Interior, Washington, DC.

Lancaster, James A., Jean M. Pinkley, Philip F. Van Cleave, and Don Watson

1954 *Archaeological Excavations in Mesa Verde National Park, Colorado, 1950.* Archaeological Research Series, no. 2. National Park Service, U.S. Department of the Interior, Washington, DC.

Lanchester, F. W.

1916 *Aircraft in Warfare: The Dawn of the Fourth Arm.* Constable, London.

Langdon, Christopher A.

2001 A Comparison of White-Tailed Deer Population Estimation Methods in West Virginia. Master's thesis, West Virginia University, Morgantown.

Lange, Frederick, Nancy Mahaney, Joe Ben Wheat, and Mark L. Chenault

1986 *Yellow Jacket: A Four Corners Ceremonial Center.* Johnson, Boulder, CO.

Langer, W. H., and K. E. Kolm

2001 A Systems Approach to Characterize and Evaluate Environmental Impacts from Aggregate Mining. In *Aggregate 2001—Environment and Economy,* edited by P. Kuula-Väisänen and R, Uusinoka, pp. 401–406. Tampere University of Technology, Publication no. 51, Finland.

LeBlanc, Steven A.

1999 *Prehistoric Warfare in the American Southwest.* University of Utah Press, Salt Lake City.

2006 Warfare and the Development of Social Complexity: Some Demographic and Environmental Factors. In *The Archaeology of Warfare: Prehistories of Raiding and Conquest,* edited by Elizabeth N. Arkush and Mark W. Allen, pp. 437–468. University Press of Florida, Gainesville.

Lechleitner, R. R.

1958 Movements, Density, and Mortality in a Black-Tailed Jack Rabbit Population. *Journal of Wildlife Management* 22:371–384.

Lekson, Stephen H.

1983 Dating the Hubbard Tri-wall and Other Tri-wall Structures. *Southwest Lore* 49(4):15–23.

2002 War in the Southwest, War in the World. *American Antiquity* 67:607–624.

Lekson, Stephen H., and Catherine M. Cameron

1995 The Abandonment of Chaco Canyon, the Mesa Verde Migrations, and the Reorganization of the Pueblo World. *Journal of Anthropological Archaeology* 14:184–202.

Lekson, Stephen H., Thomas C. Windes, and Peter J. McKenna

2006 Architecture. In *The Archaeology of Chaco Canyon: An Eleventh-Century Pueblo Regional Center,* edited by Stephen H. Lekson, pp. 67–116. School of American Research Press, Santa Fe, NM.

Le Roy Ladurie, Emmanuel

1971 *Times of Feast, Times of Famine: A History of Climate Since the Year 1000.* Translated by Barbara Bray. Doubleday, Garden City, NY.

Lichardus-Itten, M.

1993 Zum Beginn des Neolithikums im Tal der Struma (Südwest-Bulgarien). *Anatolica* 19:99–116.

Lightfoot, Kent

1979 Food Redistribution Among Prehistoric Pueblo Groups. *Kiva* 44:319–339.

Lightfoot, Ricky R.

1992 Architecture and Tree-Ring Dating at the Duckfoot Site in Southwestern Colorado. *Kiva* 57:213–236.

1994 The Duckfoot Site, vol. 2, *Archaeology of the House and Household.* Occasional Papers, no. 4. Crow Canyon Archaeological Center, Cortez, CO.

Lightfoot, Ricky R., Alice M. Emerson, and Eric Blinman
1988 Excavations in Area 5. In *Dolores Archaeological Program: Anasazi Communities at Dolores: Grass Mesa Village*, edited by William D. Lipe, James N. Morris, and Timothy A. Kohler, pp. 561–766. U.S. Department of the Interior, Bureau of Reclamation, Engineering and Research Center, Denver, CO.

Lightfoot, Ricky R., and Mary C. Etzkorn (editors)
1992 *The Duckfoot Site*, vol. 1, *Descriptive Archaeology*. Occasional Papers, no. 3. Crow Canyon Archaeological Center, Cortez, CO.

Lightfoot, Ricky R., and Kristin A. Kuckelman
2001 A Case of Warfare in the Mesa Verde Region. In *Deadly Landscapes: Case Studies in Prehistoric Southwestern Warfare*, edited by G. E. Rice and S. A. LeBlanc, pp. 51–64. University of Utah Press, Salt Lake City.

Lindsay, J. H.
1971 Annual Cycle of Leaf Water Potential in *Picea engelmannii* and *Abies lasiocarpa* at Timberline in Wyoming. *Arctic and Alpine Research* 2:131–138.

Lipe, William D.
1970 Anasazi Communities in the Red Rock Plateau, Southeastern Utah. In *Reconstructing Prehistoric Pueblo Societies*, edited by W. A. Longacre, pp. 84–139. University of New Mexico Press, Albuquerque.
1989 Social Scale of Mesa Verde Anasazi Kivas. In *The Architecture of Social Integration in Prehistoric Pueblos*, edited by W. D. Lipe and M. Hegmon, pp. 53–71. Occasional Papers, no. 1. Crow Canyon Archaeological Center, Cortez, CO.
1992 Introduction. In *The Sand Canyon Archaeological Project: A Progress Report*, edited by W. D. Lipe, pp. 1–10. Occasional Papers, no. 2. Crow Canyon Archaeological Center, Cortez, CO.
1995 The Depopulation of the Northern San Juan: Conditions in the Turbulent 1200s. *Journal of Anthropological Archaeology* 14:143–169.
1999 History of Archaeology. In *Colorado Prehistory: A Context for the Southern Colorado River Basin*, edited by William D. Lipe, Mark D. Varien, and Richard H. Wilshusen, pp. 51–94. Colorado Council of Professional Archaeologists, Denver.
2002 Social Power in the Central Mesa Verde Region, A.D. 1150–1290. In *Seeking the Center Place: Archaeology and Ancient Communities in the Mesa Verde Region*, edited by M. D. Varien and R. H. Wilshusen, pp. 203–232. University of Utah Press, Salt Lake City.
2006 Notes from the North. In *The Archaeology of Chaco Canyon: An Eleventh-Century Pueblo Regional Center*, edited by Stephen H. Lekson, pp. 261–313. School of American Research Press, Santa Fe, NM.
2010 Lost in Transit: The Central Mesa Verde Archaeological Complex. In *Leaving Mesa Verde: Peril and Change in the Thirteenth Century Southwest*, edited by Timothy A. Kohler, Mark D. Varien, and Aaron Wright, pp. 262–284. University of Arizona Press, Tucson.

Lipe, William D. (editor)
1992 *The Sand Canyon Archaeological Project: A Progress Report*. Occasional Papers, no. 2. Crow Canyon Archaeological Center, Cortez, CO.

Lipe, William D., and Scott G. Ortman
2000 Spatial Patterning in Northern San Juan Villages, A.D. 1050–1300. *Kiva* 66(1):91–122.

Lipe, William D., and Carla Van West
1992 Modeling Prehistoric Climate and Agriculture in Southwestern Colorado. In *The Sand Canyon Archaeological Project: A Progress Report*, edited by William D. Lipe, pp. 105–119. Occasional Papers, no. 2. Crow Canyon Archaeological Center, Cortez, CO.

Lipe, William D., and Mark D. Varien
1999a Pueblo II (A.D. 900–1150). In *Colorado Prehistory: A Context for the Southern Colorado River Basin*, edited by William D. Lipe, Mark D. Varien, and Richard H. Wilshusen, pp. 242–289. Colorado Council of Professional Archaeologists, Denver.
1999b Pueblo III (A.D. 1150–1300). In *Colorado Prehistory: A Context for the Southern Colorado River Basin*, edited by William D. Lipe, Mark D. Varien and Richard H. Wilshusen, pp. 290–352. Colorado Council of Professional Archaeologists, Denver.

Lipe, William D., Mark D. Varien, and Richard H. Wilshusen (editors)
1999 *Colorado Prehistory: A Context for the Southern Colorado River Basin*. Colorado Council of Professional Archaeologists, Denver.

Lister, Robert H. (editor)
1964 *Contributions to Mesa Verde Archaeology*, vol. 1, *Site 499, Mesa Verde National Park, Colorado*. University of Colorado Studies, Series in Anthropology, no. 9. University of Colorado Press, Boulder.

1965 *Contributions to Mesa Verde Archaeology*, vol. 2, *Site 875, Mesa Verde National Park, Colorado*. University of Colorado Studies, Series in Anthropology, no. 11. University of Colorado Press, Boulder.

1966 *Contributions to Mesa Verde Archaeology*, vol. 3, *Site 866, and the Cultural Sequence at Four Villages in the Far View Group, Mesa Verde National Park, Colorado*. University of Colorado Studies, Series in Anthropology, no. 12. University of Colorado Press, Boulder.

1968 *Contributions to Mesa Verde Archaeology*, vol. 5, *Emergency Archaeology in Mesa Verde National Park, Colorado, 1948–1966*. Studies in Anthropology, no. 15. University of Colorado, Boulder.

Lotka, Alfred

1925 *Elements of Physical Biology*. Williams and Wilkins, Baltimore, MD.

Luebben, Ralph A.

1983 The Grinnell Site: A Small Ceremonial Center Near Yucca House, Colorado. *Journal of Intermountain Archaeology* 2:1–26.

Luebben, Ralph A., and Paul R. Nickens

1982 A Mass Interment in an Early Pueblo III Kiva in Southwestern Colorado. *Journal of Intermountain Archaeology* 1:66–79.

Luterbacher, Jurg, Daniel Deitrich, Elena Xoplaki, Martin Grosjean, and Heinz Wanner

2004 European Seasonal and Annual Variability, Trends, and Extremes Since 1500. *Science* 303:1499–1503.

Lyman, R. Lee

1994 *Vertebrate Taphonomy*. Cambridge University Press, Cambridge.

Lyons, P. D.

2003 *Ancestral Hopi Migrations*. Anthropological Papers of the University of Arizona, no. 68. University of Arizona Press, Tucson.

Mackey, James, and Richard C. Green

1979 Largo-Gallina Towers: An Explanation. *American Antiquity* 44:144–154.

Mackie, Richard J.

1994 Populations and Habitat. In *Deer*, edited by Duane Gerlach, Sally Atwater, and Judith Schnell, pp. 322–329. Stackpole, Mechanicsburg, PA.

Magne, Martin P., and David Pokotylo

1981 A Pilot Study in Bifacial Lithic Reduction Sequences. *Lithic Technology* 10:34–47.

Maher, Louis J., Jr.

1961 *Pollen Analysis and Postglacial Vegetation History in the Animas Valley Region, Southern San Juan Mountains, Colorado*. Ph.D. dissertation, University of Minnesota, Minneapolis. University Microfilms, Ann Arbor, MI.

1972a Nomograms for Computing 0.95 Confidence Limits of Pollen Data. *Review of Palaeobotany and Palynology* 13:85–93.

1972b Absolute Pollen Diagram of Redrock Lake, Boulder County, Colorado. *Quaternary Research* 2:531–553.

Mahoney, Nancy M., Michael A. Adler, and James W. Kendrick

2000 The Changing Scale and Configuration of Mesa Verde Communities. *Kiva* 66(1):67–90.

Malthus, Thomas R.

1803 *An Essay on the Principle of Population*. Johnson, London.

Malville, Nancy J.

1989 Two Fragmented Human Bone Assemblages from Yellow Jacket, Southwestern Colorado. *Kiva* 55:3–22.

Mann, M. E., R. S. Bradley, and M. K. Hughes

1998 Global-Scale Temperature Patterns and Climate Forcing over the Past Six Centuries. *Nature* 392:779–787.

Manolescu, Kathleen

1995 Hopi Corn Production (unpublished report). Bureau of Indian Affairs, Phoenix Area Office, Phoenix, AZ.

Markgraf, Vera, and Linda Scott

1981 Lower Timberline in Central Colorado During the Past 15,000 Years. *Geology* 9:231–235.

Marr, John W., and R. E. Marr

1973 Environment and Phenology in the Forest-Tundra Ecotone, Front Range, Colorado (summary). *Arctic and Alpine Research* 5(3, pt. 2):A65–A66.

Martin, Debra L., and Nancy J. Akins

1994 Patterns of Violence Against Women in the Prehistoric Southwest. Paper presented at the Southwest Symposium. Tempe, AZ.

Martin, Debra L., Nancy J. Akins, and A. Goodman

1993 Health Profile for the La Plata Highway Project. Paper presented at the Fifth Occasional Anasazi Symposium. Farmington, NM.

Martin, Paul S.

1936 *Lowry Ruin in Southwestern Colorado*. Anthropological Series, Field Museum of Natural History, vol. 21, no. 1. Chicago, IL.

1938 *Archaeological Work in the Ackmen-Lowry Area Southwestern Colorado, 1937*. Field Museum of Natural History, vol. 23, no. 2, publication 419. Chicago, IL.

Martin, Paul S., and William Byers
1965 Pollen and Archaeology at Wetherill Mesa. In *Contributions of the Wetherill Mesa Archaeological Project*, assembled by Douglas Osborne, pp. 122–135. Society of American Archaeology Memoirs, no. 19. Salt Lake City, UT.

Maschner, Herbert D. G.
1997 The Evolution of Northwest Coast Warfare. In *Troubled Times: Violence and Warfare in the Past*, edited by Debra L. Martin and David W. Frayer, pp. 267–302. Gordon and Breach, Amsterdam.

Matson, R. G.
2002 The Spread of Maize Agriculture into the U.S. Southwest. In *Examining the Farming / Language Dispersal Hypothesis*, edited by Peter Bellwood and Colin Renfrew, pp. 341–356. McDonald Institute for Archaeological Research, University of Cambridge, Cambridge.

Matson, R. G., and Brian Chisholm
1991 Basketmaker II Subsistence: Carbon Isotopes and Other Dietary Indicators from Cedar Mesa, Utah. *American Antiquity* 56:444–459.

Matthews, John A., and Keith R. Briffa
2005 "Little Ice Age": Re-evaluation of an Evolving Concept. *Geografiska Annaler* 87(1):17–36.

Mauldin, Raymond P., and Daniel S. Amick
1989 Investigating Patterning in Debitage from Experimental Bifacial Core Reduction. In *Experiments in Lithic Technology*, edited by D. S. Amick and R. P. Mauldin, pp. 67–88. British Archaeological Reports International Series, no. 528. Oxford.

Maynard Smith, John
1978 Optimization Theory in Evolution. *Annual Review of Ecology and Systematics* 9:31–56.

Mayr, Ernst
1983 How to Carry Out the Adaptationist Program? *American Naturalist* 121:324–334.

McAllister, James W.
2002 Historical and Structural Approaches in the Natural and Human Sciences. In *The Future of the Sciences and Humanities*, edited by P. Tindemans, A. Verrijn-Stuart, and R. Visser, pp. 19–54. Amsterdam University Press, Amsterdam.

McCullough, Dale R.
1997 Irruptive Behaviour in Ungulates. In *The Science of Overabundance*, edited by William J. McShea, H. Brian Underwood, and John H. Rappole, pp. 69–98. Smithsonian Institution Scholarly Press, Washington, DC.

McDonald, M. G., and A. W. Harbaugh
1988 A Modular Three-Dimensional Finite-Difference Ground-Water Flow Model. *Techniques of Water-Resources Investigations of the United States Geological Survey*, Book 6, Chapter A1. U.S. Geological Survey, Denver, CO.

McGuire, Kelly R., and William R. Hildebrandt
2005 Re-Thinking Great Basin Foragers: Prestige Hunting and Costly Signaling During the Middle Archaic Period. *American Antiquity* 70:695–712.

McKenna, Peter J., and H. Wolcott Toll
1992 Regional Patterns of Great House Development Among the Totah Anasazi, New Mexico. In *Anasazi Regional Organization and the Chaco System*, edited by David E. Doyel, pp. 133–143. Anthropological Papers of the Maxwell Museum of Anthropology, no. 5. University of New Mexico, Albuquerque.

Medin, Dean E., and Allen E. Anderson
1979 Modeling the Dynamics of a Colorado Mule Deer Population. *Wildlife Monographs* 68:3–67.

Meggitt, Mervyn J.
1977 *Blood Is Their Argument: Warfare Among the Mae Enga Tribesman of the New Guinea Highlands*. Mayfield, Palo Alto, CA.

Miles, James S.
1975 *Orthopedic Problems of the Wetherill Mesa Populations, Mesa Verde National Park, Colorado*. Publications in Archaeology 7G, Wetherill Mesa Studies. National Park Service, U.S. Department of the Interior, Washington, DC.

Mills, Barbara J.
1999 Ceramics and the Social Contexts of Food Consumption in the Northern Southwest. In *Pots and People: A Dynamic Interaction*, edited by James Skibo and Gary Feinman, pp. 99–114. University of Utah Press, Salt Lake City.

Moberg, Anders, Dmitry M. Sonechkin, Karin Holmgren, Nina M. Datsenko, and Wibjörn Karlén
2005 Highly Variable Northern Hemisphere Temperatures Reconstructed from Low- and High-Resolution Proxy Data. *Nature* 433:613–617.

Morgan, Lewis Henry
1965 *Houses and Houselife of the American Aborigines*. Reprinted. University of Chicago

Press, Chicago. Originally published 1881, Contributions to North American Ethnology, vol. 12, U.S. Government Printing Office, Washington, DC.

Morley, Sylvanus G.

1908 The Excavation of the Cannonball Ruins in Southwestern Colorado. *American Anthropologist* 10:596–610.

Morley, Sylvanus G., and Alfred V. Kidder

1917 The Archaeology of McElmo Canyon, Colorado. *El Palacio* 4(4):41–70.

Morris, James N.

1984 Excavations at Weasel Pueblo (Site 5MT5106), an Anasazi Habitation. *Dolores Archaeological Program Technical Reports* DAP-131. Submitted to the U.S. Bureau of Reclamation, Upper Colorado Region, Salt Lake City, UT, in compliance with Contract No. 8-07-40-S0562.

1986 *Monitoring and Excavation at Aulston Pueblo (5MT2433), a Pueblo II Habitation Site.* Four Corners Archaeological Project, Report no. 6. Complete Archaeological Service Associates, Cortez, CO.

1988 Excavations at Casa Bisecada (Site 5MT8829), a Pueblo I Field House/Pueblo II Habitation. In *Archaeological Investigations on South Canal,* vol. 1, compiled by Kristin A. Kuckelman and James N. Morris, pp. 108–151. Four Corners Archaeological Project, Report no. 11. Complete Archaeological Service Associates, Cortez, CO.

1991 *Archaeological Excavations on the Hovenweep Laterals.* Four Corners Archaeological Project, Report no. 16. Complete Archaeological Service Associates, Cortez, CO.

Morrow, Carol A., and Richard W. Jefferies

1989 Trade or Embedded Procurement? A Test Case from Southern Illinois. In *Time, Energy and Stone Tools,* edited by R. Torrence, pp. 27–33. Cambridge University Press, Cambridge.

Muenchrath, Deborah A., Maya Kuratomi, Jonathan A. Sandor, and Jeffrey A. Homburg

2002 Observational Study of Maize Production Systems of Zuni Farmers in Semiarid New Mexico. *Journal of Ethnobiology* 22(1):1–33.

Muir, Robert J.

1999 Zooarchaeology of Sand Canyon Pueblo, Colorado. Unpublished Ph.D. dissertation, Department of Archaeology, Simon Fraser University, Burnaby, British Columbia.

Muir, Robert J., and Jonathan C. Driver

2002 Scale of Analysis and Zooarchaeological Interpretation: Pueblo III Faunal Variation in the Northern San Juan Region. *Journal of Anthropological Archaeology* 21:165–199.

2004 Identifying Ritual Use of Animals in the Northern American Southwest. In *Behaviour Behind Bones: The Zooarchaeology of Ritual, Religion, Status and Identity,* edited by S. J. O'Day, W. Van Neer, and A. Ervynck, pp. 128–143. Proceeding of the 9th Conference of the International Council of Archaeozoology, Durham, August 2002. Oxbow, Oxford.

Murdock, George P.

1949 *Social Structure.* Macmillan, New York.

Murdock, George, and S. Wilson

1972 Settlement Patterns and Community Organization: Cross-Cultural Codes. *Ethnology* 11:254–295.

Myers, K.

1964 Influence of Density on Fecundity, Growth Rates, and Mortality in the Wild Rabbit. CSIRO *Wildland Resources* 9:134–137.

Naranjo, Tessie

1995 Thoughts on Migration by Santa Clara Pueblo. *Journal of Anthropological Archaeology* 14:247–250.

National Climate Data Center

2010 http://lwf.ncdc.noaa.gov/oa/ncdc.html.

Neily, Robert B.

1983 *The Prehistoric Community on the Colorado Plateau: An Approach to the Study of Change and Survival in the Northern San Juan Area of the American Southwest.* Ph.D. dissertation, Southern Illinois University, Carbondale. University Microfilms, Ann Arbor, MI.

Nelson, G. Charles

1985 Excavations at Standing Pipe Hamlet (Site 5MT5985), a Possible Basketmaker III–Pueblo I Habitation. *Dolores Archaeological Program Technical Reports* DAP-137. Submitted to the U.S. Bureau of Reclamation, Upper Colorado Region, Salt Lake City, UT, in compliance with Contract No. 8-07-40-S0562.

Nelson, Margaret C., and Karen Gus Schollmeyer

2003 Game Resources, Social Interaction, and the Ecological Footprint in Southwest New Mexico. *Journal of Archaeological Method and Theory* 10:69–110.

Netting, Robert M.

1969 Ecosystems in Process: A Comparative Study of Change in Two West African Societies. In *Contributions to Anthropology: Ecological Essays,* edited by D. Damas, pp. 102–112. Queen's Printer, Ottawa.

1982 Territory, Property, and Tenure. In *Behavioral and Social Science Research: A National Resource*, edited by R. M. Adams, N. J. Smelser, and D. J. Treimain, pp. 446–502. National Academy Press, Washington, DC.

Nickens, Paul R., and John M. Mabry
2000 Site 5MT10967, an Isolated Prehistoric Human Burial from the Lower Montezuma Valley, Southwestern Colorado. *Southwestern Lore* 66(1):1–18.

Nordby, Larry V.
2001 *Prelude to Tapestries in Stone: Understanding Cliff Palace Architecture*. Mesa Verde National Park Archeological Research Series, Architectural Studies, no. 4. Mesa Verde National Park Division of Research and Resource Management. Mesa Verde National Park, CO.

Nordenskiold, Gustav
1973 [1893] *The Cliff Dwellers of the Mesa Verde, Southwestern Colorado: Their Pottery and Implements*. Translated by D. Lloyd Morgan. AMS, New York.
1990 [1893] *The Cliff Dwellers of the Mesa Verde*. Mesa Verde Museum Association, Mesa Verde, CO.

Norton, Jay B., Jonathon A. Sandor, and Roman R. Pawluk
2000 Farmer-Scientist Collaboration for Research and Agricultural Development on the Zuni Indian Reservation, New Mexico, USA. In *Sustainability of Agricultural Systems in Transition*, edited by William A. Payne, Dennis R. Keeney, and Srinivas C. Rao, pp. 107–120. ASA Special Publication, no. 64, American Society of Agronomy, Madison, WI.

Norton, Jay B., Jonathon A. Sandor, and C. S. White
2003 Hillslope Soils and Organic Matter Dynamic Within a Native American Agroecosystem on the Colorado Plateau. *Soil Science Society of America Journal* 67:225–234.

Oberg, Kalervo
1955 Types of Social Structure Among the Lowland Tribes of South and Central America. *American Anthropologist* 57:472–487.

O'Bryan, Deric
1950 *Excavations in Mesa Verde National Park, 1947–1948*. Medallion Papers, no. 39. Lancaster Press, Lancaster, PA.

O'Connell, James F., and Kristen Hawkes
1988 Hadza Hunting, Butchering, and Bone Transport and Their Archaeological Implications. *Journal of Anthropological Research* 44:113–161.

Odum, Eugene P.
1971 *Fundamentals of Ecology*. 3rd edition. Saunders, Philadelphia, PA.

O'Neil, Brian
1984 *Test Excavations at 5DL975. Cultural Resource Management Report, Dolores County, Colorado*. Mobil Oil Corporation. Denver, CO.

Orcutt, Janet D.
1987 Modeling Prehistoric Agricultural Ecology in the Dolores Area. In *Dolores Archaeological Program: Supporting Studies: Settlement and Environment*, compiled by Kenneth Lee Petersen and Janet D. Orcutt, pp. 649–681. U.S. Department of the Interior, Bureau of Reclamation, Engineering and Research Center, Denver, CO.

Orcutt, Janet D., Eric Blinman, and Timothy A. Kohler
1990 Explanations of Population Aggregation in the Mesa Verde Region Prior to A.D. 900. In *Perspectives on Southwestern Prehistory*, edited by Paul E. Minnis and Charles L. Redman, pp. 196–212. Westview, Boulder, CO.

Orians, G. H., and N. E. Pearson
1979 On the Theory of Central Place Foraging. In *Analyses of Ecological Systems*, edited by D. J. Horn, R. D. Mitchell, and G. R. Stairs, pp. 154–177. Ohio State University Press, Columbus.

Ortman, Scott G.
1998 Corn Grinding and Community Organization in the Pueblo Southwest, A.D. 1150–1550. In *Migration and Reorganization: The Pueblo IV Period in the American Southwest*, edited by K. A. Spielmann, pp. 165–192. Anthropological Research Papers, no. 51. Arizona State University, Tempe.
2000 Artifacts. In *The Archaeology of Castle Rock Pueblo: A Thirteenth-Century Village in Southwestern Colorado*, edited by K. A. Kuckelman. Available at www.crowcanyon.org/castlerock. Accessed May 3, 2002.
2002 Artifacts. In *The Archaeology of Woods Canyon Pueblo: A Canyon-Rim Village in Southwestern Colorado*, edited by Melissa J. Churchill. Available at www.crowcanyon.org/woodscanyon. Accessed April 11, 2005.
2003 Artifacts. In *The Archaeology of Yellow Jacket Pueblo (Site 5MT5): Excavations at a Large Community Center in Southwestern Colorado*, edited by Kristin A. Kuckelman. Available at www.crowcanyon.org/yellowjacket.
2008 Action, Place and Space in the Castle Rock Community. In *The Social Construction of*

Communities: Agency, Structure, and Identity in the Prehispanic Southwest, edited by M. D. Varien and J. M. Potter, pp. 125–154. Rowman & Littlefield Publishers, Lanham, MD.

2009 *Genes, Language and Culture in Tewa Ethnogenesis*, A.D. 1150–1400. Ph.D. dissertation, Arizona State University, Tempe.

2010 Evidence of a Mesa Verde Homeland for the Tewa Pueblos. In *Leaving Mesa Verde: Peril and Change in the Thirteenth Century Southwest*, edited by T. A. Kohler, M. D. Varien and A. Wright, pp. 222–261. University of Arizona Press, Tucson.

Ortman, Scott G., and Bruce A. Bradley

2002 Sand Canyon Pueblo: The Container in the Center. In *Seeking the Center Place: Archaeology and Ancient Communities in the Mesa Verde Region*, edited by M. D. Varien and R. H. Wilshusen, pp. 41–78. University of Utah Press, Salt Lake City.

Ortman, Scott G., and Catherine Cameron

2011 A Framework for Controlled Comparisons of Ancient Southwestern Movement. In *Movement, Connectivity and Landscape Change in the Ancient Southwest*, edited by M. Nelson and C. Strawhacker, pp. 233–252. University Press of Colorado, Boulder.

Ortman, Scott G., Donna M. Glowacki, Melissa J. Churchill, and Kristin A. Kuckelman

2000 Pattern and Variation in Northern San Juan Village Histories. *Kiva* 66:123–146.

Ortman, Scott G., and Mark D. Varien

2007 McElmo Dome-Sand Canyon Locality Settlement Patterns. In *The Archaeology of Sand Canyon Pueblo: Intensive Excavations at a Late-Thirteenth-Century Village in Southwestern Colorado*, edited by Kristin A. Kuckelman. Available at www.crowcanyon.org.

Ortman, Scott G., Mark D. Varien, and T. Lee Gripp

2007 Empirical Bayesian Methods for Archaeological Survey Data: An Application from the Mesa Verde Region. *American Antiquity* 72:241–272.

Osborne, Carolyn Miles

2004 *The Wetherill Collections and Perishable Items from Mesa Verde*. Available at www.lulu.com, Los Alamitos, CA.

Palmer, W. C.

1965 Meteorological Drought. Research Papers, no. 45. U.S. Weather Bureau, Washington, DC.

Pannell, James B.

2003 *Soil Survey of Animas-Dolores Area, Colorado, Parts of Archuleta, Dolores, Hinsdale, La Plata, Montezuma, San Juan, and San Miguel Counties*. U.S. Department of the Interior, Natural Resource Conservation Service. Available at http://soildatamart.nrcs.usda.gov/manuscripts/CO672/0/A-D%20Part%201.pdf.

Parry, William J., and Robert L. Kelly

1987 Expedient Core Technology and Sedentism. In *The Organization of Core Technology*, edited by J. K. Johnson and C. A. Morrow, pp. 285–304. Westview, Boulder, CO.

Parsons, Jeffrey R.

1972 Archaeological Settlement Patterns. *Annual Reviews in Anthropology* 1:127–150.

Patton, John Q.

2000 Reciprocal Altruism and Warfare: A Case from the Ecuadorian Amazon. In *Human Behavior and Adaptation: An Anthropological Perspective*, edited by N. Chagnon, L. Cronk, and W. Irons, pp. 417–436. Aldine de Gruyter, Hawthorne, NY.

Pearson, Gustaf A.

1931 *Forest Types in the Southwest as Determined by Climate and Soil*. Technical Bulletin no. 247. U.S. Department of Agriculture. U.S. Government Printing Office, Washington, DC.

Perlès, Catherine

2001 *The Early Neolithic in Greece*. Cambridge University Press, Cambridge.

Petersen, Ken L.

1988 Climate and the Dolores River Anasazi: A Paleoenvironmental Reconstruction from a 10,000-Year Pollen Record, La Plata Mountains, Southwestern Colorado. Anthropological Papers of the University of Utah, no. 113. University of Utah Press, Salt Lake City.

1994 A Warm and Wet Little Climatic Optimum and a Cold and Dry Little Ice Age in the Southern Rocky Mountains, U.S.A. *Climate Change* 26:243–269.

1995 Interaction of Prehistoric Climate, Ecology, and Cultures: An Example from the Dolores River Region of Southwest Colorado. In *Proceedings of the Workshop Climate Change in the Four Corners and Adjacent Regions: Implications for Environmental Restoration and Land-use Planning*, edited by W. J. Waugh, pp. 105–124. National Technical Information Service, Springfield, VA.

Petersen, Ken L., and Peter J. Mehringer, Jr.

1976 Postglacial Timberline Fluctuations, La Plata Mountains, Southwestern Colorado. *Arctic and Alpine Research* 8:275–288.

Petersen, Kenneth Lee, and Vickie L. Clay

1987 Characteristics and Archaeological Implications of Cold Air Drainage in the Dolores Project Area. In *Dolores Archaeological Program: Supporting Studies: Settlement and Environment*, compiled by Kenneth Lee Petersen and Janet D. Orcutt, pp. 187–216. U.S. Department of the Interior, Bureau of Reclamation Engineering and Research Center, Denver, CO.

Phillips, David A., and Mark L. Chenault
2004 Archaeological Investigations at the Rabbit Site (Site 5MT9168): A Basketmaker III Habitation Site: Feature Descriptions. In *From Dancing Man to Hummingbird: Long-Term Prehistoric Change in Southwest Colorado, Vol. 2, Part 1: Basketmaker III Sites*, edited by Mark L. Chenault, Cynthia Manseau, and Todd Kohler, pp. 2-1–2-29. Cultural Resource Report 2003-16. SWCA Environmental Consultants. Prepared for the U.S. Bureau of Reclamation, Durango, CO.

Pierce, Christopher D., Donna M. Glowacki, and Margaret M. Thurs
2002 Measuring Community Interaction: Pueblo III Pottery Production and Distribution in the Central Mesa Verde Region. In *Seeking the Center Place: Archaeology and Ancient Communities in the Mesa Verde Region*, edited by Mark D. Varien and Richard H. Wilshusen, pp. 185–202. University of Utah Press, Salt Lake City.

Piperno, Dolores R., and Kent V. Flannery
2001 The Earliest Archaeological Maize (*Zea mays* L.) from Highland Mexico: New Accelerator Mass Spectrometry Dates and Their Implications. *Proceedings of the National Academy of Sciences* 98:2101–2103.

Piperno, Dolores R., Anthony J. Ranere, Irene Holst, Jose Iriarte, and Ruth Dickau
2009 Starch Grain and Phytolith Evidence for Early Ninth Millennium B.P. Maize from the Central Balsas River Valley, Mexico. *Proceedings of the National Academy of Sciences* 106:5019–5024.

Potter, James M.
1997 Communal Ritual and Faunal Remains: An Example from the Dolores Anasazi. *Journal of Field Archaeology* 24:353–364.
2000 Pots, Parties, and Politics: Communal Feasting in the American Southwest. *American Antiquity* 65:471–492.
2004 The Creation of Person, the Creation of Place: Hunting Landscapes in the American Southwest. *American Antiquity* 69:322–338.

Potter, James M., and Jason Chuipka
2007 Early Pueblo Communities and Cultural Diversity in the Durango Area. *Kiva* 72:407–429.

Potter, James M., and Scott G. Ortman
2004 Community and Cuisine in the Prehispanic American Southwest. In *Identity, Feasting, and the Archaeology of the Greater Southwest*, edited by Barbara J. Mills, pp. 173–191. University Press of Colorado, Boulder.

Powell, Adam, Stephen Shennan, and Mark G. Thomas
2009 Late Pleistocene Demography and the Appearance of Modern Human Behavior. *Science* 324:1298–1301.

Powell, Shirley
1988 Anasazi Demographic Patterns and Organizational Responses: Assumptions and Interpretive Difficulties. In *The Anasazi in a Changing Environment*, edited by G. J. Gumerman, pp. 168–191. Cambridge University Press, Cambridge.

Powers, Robert P., William B. Gillespie, and Stephen H. Lekson
1983 *The Outlier Survey: A Regional View of Settlement in the San Juan Basin*. Reports of the Chaco Center, no. 3. Division of Cultural Research, National Park Service, U.S. Department of the Interior, Albuquerque, NM.

Price, Barbara J.
1984 Competition, Productive Intensification, and Ranked Society: Speculations from Evolutionary Theory. In *Warfare, Culture, and Environment*, edited by Brian Ferguson, pp. 209–240. Academic, New York.

Prudden, T. Mitchell
1903 The Prehistoric Ruins of the San Juan Watershed in Utah, Arizona, Colorado, and New Mexico. *American Anthropologist* 5:224–288.

Ramsey, Douglas K.
2003a Soils of Mesa Verde Country. In *Ancient Piñon-Jiniper Woodlands: A Natural History of Mesa Verde Country*, edited by M. Lisa Floyd, pp. 213–222. University Press of Colorado, Boulder.
2003b *Soil Survey of Cortez Area, Colorado, Parts of Dolores and Montezuma Counties*. U.S. Department of the Interior, Natural Resource Conservation Service.
2008 *Soil Survey of Ute Mountain Area, Colorado and New Mexico*. U.S. Department of the Interior, Natural Resource Conservation Service.

Rappaport, Roy A.
1968 *Pigs for the Ancestors: Ritual in the Ecology of a New Guinea People.* Yale University Press, New Haven, CT.

Redmond, Elsa M.
1994 External Warfare and the Internal Politics of Northern South American Tribes and Chiefdoms. In *Factional Competition and Political Development in the New World,* edited by Elizabeth M. Brumfiel and John W. Fox, pp. 44–54. Cambridge University Press, New York.

Reed, Alan D., Wayne K. Howell, Paul R. Nickens, and Jonathan Horn
1985 *Archaeological Investigations on the Johnson Canyon Road Project, Ute Mountain, Ute Tribal Lands, Colorado.* Nickens and Associates, Montrose, CO.

Reheis, M. C., R. L. Reynolds, H. Goldstein, H. M. Roberts, J. C. Yount, Y. Axford, L. S. Cummings, and N. Shearin
2005 Late Quaternary Eolian and Alluvial Response to Paleoclimate, Canyonlands, Southeastern Utah. *Geological Society of America Bulletin* 117:1051–1069.

Renfrew, Colin
1969 Trade and Culture Process in European Prehistory. *Current Anthropology* 10:151–169.
1977 Alternative Models for Exchange and Spatial Distribution. In *Exchange Systems in Prehistory,* edited by T. Earle and J. E. Ericson, pp. 71–90. Academic, New York.

Research Institute for Knowledge Systems
2009 *Map Comparison Kit 3 User Manual.* Research Institute for Knowledge Systems, Maastricht, The Netherlands.

Revilla, Eloy, and Francisco Palomares
2002 Spatial Organization, Group Living and Ecological Correlates in Low-Density Populations of Eurasian Badgers, *Meles meles. Journal of Animal Ecology* 71:497–512.

Reynolds, Robert G.
1979 An Adaptive Computer Model of the Evolution of Agriculture for Hunter-Gatherers in the Valley of Oaxaca, Mexico. Unpublished Ph.D. dissertation, University of Michigan, Ann Arbor.
2000 The Impact of Raiding on Settlement Patterns in the Northern Valley of Oaxaca: An Approach Using Decision Trees. In *Dynamics in Human and Primate Societies: Agent-Based Modeling of Social and Spatial Processes,* edited by T. A. Kohler and G. J. Gumerman, pp. 251–274. Oxford University Press, New York.

Reynolds, Robert G., Ziad Kobti, Timothy A. Kohler, and Lorene Y. L. Yap
2005 Unraveling Ancient Mysteries: Reimagining the Past Using Evolutionary Computation in a Complex Gaming Environment. *IEEE Transactions on Evolutionary Computation* 9:707–720.

Rice, Glen E.
2001 Warfare and Massing in the Salt and Gila Basins of Central Arizona. In *Deadly Landscapes: Case Studies in Prehistoric Southwestern Warfare,* edited by Glen E. Rice and Steven A. LeBlanc, pp. 289–329. University of Utah Press, Salt Lake City.

Richerson, Peter J., Robert Boyd, and Robert L. Bettinger
2001 Was Agriculture Impossible During the Pleistocene but Mandatory During the Holocene? A Climate Change Hypothesis. *American Antiquity* 66:387–411.

Robertson, Ian G.
1999 Spatial and Multivariate Analysis, Random Sampling Error, and Analytical Noise: Empirical Bayesian Methods at Teotihuacan, Mexico. *American Antiquity* 64:137–152.

Robinson, Hugh H.
2005 Feasting, Exterior Bowl Design and Public Space in the Northern San Juan, A.D. 1240–1300. Unpublished Master's thesis, Department of Anthropology, Washington State University, Pullman.

Rocek, Thomas R.
1995 Sedentarization and Agricultural Dependence: Perspectives from the Pithouse-to-Pueblo Transition in the American Southwest. *American Antiquity* 60:218–239.

Rohn, Arthur H.
1963 Prehistoric Soil and Water Conservation on Chapin Mesa, Southwestern Colorado. *American Antiquity* 28:441–455.
1971 *Mug House: Mesa Verde National Park—Colorado.* Archaeological Research Series, no. 7D. National Park Service. U.S. Department of the Interior, Washington, DC.
1974 Payne Site Investigations. *Southwestern Lore* 40(3&4):50–51.
1975 A Stockaded Basketmaker III Village at Yellow Jacket, Colorado. *Kiva* 40:113–120.
1977 *Cultural Change and Continuity on Chapin Mesa.* Regents Press of Kansas, Lawrence.
1989 Northern San Juan Prehistory. In *Dynamics of Southwestern Prehistory,* edited by Linda S. Cordell and George J. Gumerman, pp. 149–178. Smithsonian Institution Scholarly Press, Washington, DC.

Rosen, Robert
1967 *Optimality Principles in Biology*. Plenum, New York.
Ross, M.
1983 Political Decision Making and Conflict: Additional Cross-Cultural Codes and Scales. *Ethnology* 22:169–192.
Roth, Barbara J., and Andrea Freeman
2008 The Middle Archaic Period and the Transition to Agriculture in the Sonoran Desert of Southern Arizona. *Kiva* 73:321–354.
Rummel, R. J.
1970 *Applied Factor Analysis*. Northwestern University Press, Evanston, IL.
Ryan, Susan C.
2008 Constructing Community and Transforming Identity at Albert Porter Pueblo. In *The Social Construction of Communities*, edited by Mark D. Varien and James M. Potter, pp. 69–86. Rowman & Littlefield Publishers, Lanham, MD.
2010 Environmental Change, Population Movement, and the Post-Chaco Transition at Albert Porter Pueblo. *Kiva* 75:303–325.
Sackett, Ross De Forest
1996 Time, Energy and the Indolent Savage: A Quantitative Cross-Cultural Test of the Primitive Affluence Hypothesis. Unpublished Ph.D. dissertation, Department of Anthropology, University of California, Los Angeles.
Sahlins, Marshall
1972 *Stone Age Economics*. Aldine Atherton, Chicago, IL.
1976 *Culture and Practical Reason*. University of Chicago Press, Chicago, IL.
Salomonsson, Marcus
2010 Group Selection: The Quest for Social Preferences. *Journal of Theoretical Biology* 264:737–746.
Salzer, Matthew W.
2000a Temperature Variability and the Northern Anasazi. *Kiva* 65:295–318.
2000b *Dendroclimatology in the San Francisco Peaks Region of Northern Arizona, USA*. Ph.D. dissertation, Department of Geosciences, University of Arizona. University Microfilms, Ann Arbor, MI.
Salzer, Matthew W., and Kurt F. Kipfmueller
2005 Reconstructed Temperature and Precipitation on a Millennial Timescale from Tree-Rings in the Southern Colorado Plateau, U.S.A. *Climate Change* 70:465–487.
Sanders, William T.
1997 Prehistoric Population Studies and Anthropology's Future. In *Integrating Archaeological Demography: Multidisciplinary Approaches to Prehistoric Population*, edited by Richard R. Paine, pp. 381–385. Center for Archaeological Investigations. Occasional Papers, no. 24. Southern Illinois University, Carbondale.
Sandor, Jonathon A., P. L. Gersper, and J. W. Hawley
1990 Prehistoric Agricultural Terraces and Soils in the Mimbres Area, New Mexico. *World Archaeology* 22:70–86.
Schachner, Gregson
2001 Ritual Control and Transformation in Middle-Range Societies: An Example from the American Southwest. *Journal of Anthropological Archaeology* 20:168–194.
Schlanger, Sarah H.
1985 Prehistoric Population Dynamics in the Dolores Area, Southwestern Colorado. Unpublished Ph.D. dissertation, Department of Anthropology, Washington State University, Pullman.
1988 Patterns in the Population Movement and Long-Term Population Growth in Southwestern Colorado. *American Antiquity* 53:773–793.
Schlanger, Sarah H., and Richard H. Wilshusen
1993 Local Abandonments and Regional Conditions in the North American Southwest. In *Abandonment of Settlements and Regions: Ethnoarchaeological and Archaeological Approaches*, edited by Catherine M. Cameron and Steve A. Tomka, pp. 85–98. Cambridge University Press, Cambridge.
Scuderi, Louis A.
1987 Late Holocene Upper Timberline Variation in the Southern Sierra Nevada. *Nature* 325:242–244.
Sept, Jeanne
2001 Modeling the Edible Landscape. In *Meat-Eating and Human Evolution*, edited by Craig B. Stanford and Henry T. Bunn, pp. 73–98. Oxford University Press, Oxford.
Shankman, Paul
1991 Culture Contact, Cultural Ecology, and Dani Warfare. *Man* 26:299–321.
Shanks, Michael, and Christopher Tilley
1987 *Reconstructing Archaeology: Theory and Practice*. Cambridge University Press, Cambridge.
Sharrock, Floyd W., Keith M. Anderson, Don D. Fowler, and David S. Dibble
1961 1960 Excavations, Glen Canyon Area. Glen Canyon Series, no. 14, Anthropological

Papers, no. 2. Department of Anthropology, University of Utah, Salt Lake City.

Shelley, Steven D.

1993 Analysis of the Faunal Remains from the Wallace Ruin: A Chacoan Outlier Near Cortez, Colorado. Unpublished Ph.D. dissertation, Department of Anthropology, Washington State University, Pullman.

Shennan, Stephen

1988 *Quantifying Archaeology*. Edinburgh University Press, Edinburgh.

2002 *Genes, Memes and Human History: Darwinian Archaeology and Cultural Evolution*. Thames and Hudson, London.

2007 The Spread of Farming into Central Europe and Its Consequences: Evolutionary Models. In *The Model-Based Archaeology of Socionatural Systems*, edited by T. A. Kohler and S. van der Leeuw, pp. 141–155. School for Advanced Research Press, Santa Fe, NM.

Shermis, S.

1982/1984 Domestic Violence in Two Skeletal Populations. *Ossa* 9–11:143–151.

Short, Henry L., Wain Evans, and Erwin L. Boeker

1977 The Use of Natural and Modified Pinyon Pine-Juniper Woodlands by Deer and Elk. *Journal of Wildlife Management* 41:543–559.

Short, Susan K.

1985 Palynology of Holocene Sediments, Colorado Front Range: Vegetation and Treeline Changes in the Subalpine Forest. *American Association of Stratigraphic Palynologists Contribution Series* 16:7–30. American Association of Stratigraphic Palynologists Foundation, Dallas, TX.

Shott, Michael J.

1994 Size and Form in the Analysis of Flake Debris: Review and Recent Approaches. *Journal of Archaeological Method and Theory* 1:69–110.

Simms, Steven R.

1987 *Behavioral Ecology and Hunter-Gatherer Foraging: An Example from the Great Basin*. British Archaeological Reports International Series, no. 381. Oxford.

Simon, Herbert A.

1991 *Models of My Life*. Basic, New York.

Smiley, Terah L.

1961 Evidence of Climatic Fluctuations in Southwestern Prehistory. In *Solar Variations, Climatic Change, and Related Geophysical Problems*, edited by Rhodes W. Fairbridge, pp. 697–704. Annals of the New York Academy of Sciences, no. 95 (Art. 1). New York.

Smith, Eric A.

1979 Human Adaptation and Energy Efficiency. *Human Ecology* 7:53–54.

1988 Risk and Uncertainty in the "Original Affluent Society": Evolutionary Ecology of Resource Sharing and Land Tenure. In *Hunters and Gatherers 1: History, Evolution, and Social Change*, edited by T. Ingold, D. Riches, and J. Woodburn, pp. 222–252. Berg, Oxford.

1992 Human Behavioral Ecology: II. *Evolutionary Anthropology* 1:50–55.

Smith, Eric A., and Robert Boyd

1990 Risk and Reciprocity: Hunter-Gatherer Socioecology and the Problem of Collective Action. In *Risk and Uncertainty in Tribal and Peasant Economies*, edited by E. Cashdan, pp. 167–191. Westview, Boulder, CO.

Smith, Eric A., and Bruce Winterhalder

1992 Natural Selection and Decision Making: Some Fundamental Principles. In *Evolutionary Ecology and Human Behavior*, edited by L. Cronk, N. Chagnon, and W. Irons, pp. 25–60. Aldine de Gruyter, Hawthorne, NY.

Smith, Graham W.

1990 Home Range and Activity Patterns of Black-Tailed Jackrabbits. *Great Basin Naturalist* 50:249–256.

Smith, Schaun M.

2008 Paleohydrogeologic Modeling of Groundwater Resources from A.D. 600 to 1300 and the Relationship to Ancestral Pueblo Settlement Population Dynamics, Central Mesa Verde Region, Southwest Colorado. Unpublished Ph.D. dissertation, Department of Geology and Geological Engineering, Colorado School of Mines, Golden.

Solometo, Julie P.

2004 The Conduct and Consequences of War: Dimensions of Conflict in East-Central Arizona. Unpublished Ph.D. dissertation, Department of Anthropology, University of Michigan, Ann Arbor.

2006 The Dimensions of War: Conflict and Culture Change in Central Arizona. In *The Archaeology of Warfare: Prehistories of Raiding and Conquest*, edited by Elizabeth N. Arkush and Mark W. Allen, pp. 23–65. University Press of Florida, Gainesville.

Speth, John D., and Susan L. Scott

1989 Horticulture and Large-Mammal Hunting: The Role of Resource Depletion and the Constraints of Time and Labor. In *Farmers as Hunters: The Implications of Sedentism*, edited by S. Kent, pp. 71–79. Cambridge University Press, Cambridge.

Spielmann, Katherine A.

1994 Clustered Confederacies: Sociopolitical Organization in the Protohistoric Rio Grande. In *The Ancient Southwestern Community: Models and Methods for the Study of Prehistoric Social Organization*, edited by W. H. Wills and R. D. Leonard, pp. 55–66. University of New Mexico Press, Albuquerque.

Spielmann, Katherine A., and Eric A. Angstadt-Leto

1996 Hunting, Gathering, and Health in the Prehistoric Southwest. In *Evolving Complexity and Environmental Risk in the Prehistoric Southwest*, edited by Joseph A. Tainter and Bonnie Bagley Tainter, pp. 79–106. Addison-Wesley, Reading, MA.

Stahle, David W., Edward R. Cook, Malcolm K. Cleaveland, Matthew D. Therrell, David M. Meko, Henri D. Grissino-Mayer, Emma Watson, and Brian H. Luckman

2000 Tree-Ring Data Document 16th Century Megadrought over North America. *EOS, Transactions of the American Geophysical Union* 81(12):121, 125.

Stephens, D. W., and J. R. Krebs

1986 *Foraging Theory*. Princeton University Press, Princeton, NJ.

Steponaitis, Vincas P.

1981 Settlement Hierarchies and Political Complexity in Nonmarket Societies: The Formative Period of the Valley of Mexico. *American Anthropologist* 83:320–363.

1984 Some Further Remarks on Catchments, Nonproducers, and Tribute Flow in the Valley of Mexico. *American Anthropologist* 86:143–148.

Steward, Julian H.

1937 Ecological Aspects of Southwestern Society. *Anthropos* 32:87–104.

Stiner, Mary C., Natalie D. Munro, and Todd A. Surovell

2000 The Tortoise and the Hare: Small-Game Use, the Broad-Spectrum Revolution, and Paleolithic Demography. *Current Anthropology* 41:39–73.

Stodder, Ann W.

1987 The Physical Anthropology and Mortuary Practice of the Dolores Anasazi: An Early Pueblo Population in Local and Regional Context. In *Dolores Archaeological Program: Supporting Studies: Settlement and Environment*, compiled by Kenneth Lee Petersen and Janet D. Orcutt, pp. 339–504. U.S. Department of the Interior, Bureau of Reclamation, Engineering and Research Center, Denver, CO.

Stone, Glenn D.

1992 Social Distance, Spatial Relations, and Agricultural Production Among the Kofyar of Namu District, Plateau State, Nigeria. *Journal of Anthropological Archaeology* 11:152–172.

Stone, Glenn D., and Christian E. Downum

1999 Non-Boserupian Ecology and Agricultural Risk: Ethnic Politics and Land Control in the Arid Southwest. *American Anthropologist* 101:113–128.

Stone, Tammy

2003 Social identity and Ethnic Interaction in the Western Pueblos of the American Southwest. *Journal of Archaeological Method and Theory* 10:31–67.

Stuiver, Minze, and Paula J. Reimer

1993 Extended ^{14}C Database and Revised CALIB Radiocarbon Calibration Program. *Radiocarbon* 35:215–230.

Stuiver, Minze, Paula J. Reimer, and Ron W. Reimer

2005 *CALIB Radiocarbon Calibration*. Available at http://radiocarbon.pa.qub.ac.uk/calib/. Accessed May 13, 2006.

Swannack, Jervis D., Jr.

1969 *Big Juniper House: Mesa Verde National Park—Colorado*. Archaeological Research Series, no. 7C. U.S. Department of the Interior, National Park Service, Washington, DC.

Swedlund, Charles A.

1969 Human Skeletal Material from the Yellowjacket Canyon Area, Southwestern Colorado. Unpublished Master's thesis, Department of Anthropology, University of Boulder, CO.

Swetnam, Thomas W., Craig D. Allen, and Julio L. Betancourt

1999 Applied Historical Ecology: Using the Past to Manage the Future. *Ecological Applications* 9:1189–1206.

Tainter, J. A., and B. B. Tainter

1991 The Towers of Hovenweep. Paper presented at the 56th Annual Meeting of the Society for American Archaeology, New Orleans, LA.

Thomas, Julian

1999 *Understanding the Neolithic*. Cambridge University Press, Cambridge.

Thomson, Robert L., Jukka T. Forsman, Francesc Sardà-Palomera, and Mikko Mönkkönen

2006 Fear Factor: Prey Habitat Selection and Its Consequences in a Predation Risk Landscape. *Ecography* 29:507–514.

Toll, H. Wolcott
1993 The Role of the Totah in Regions and Regional Definitions. Paper presented at the 5th Occasional Anasazi Symposium, San Juan College, Farmington, NM.
Torrence, Robin (editor)
1989 *Time, Energy, and Stone Tools*. Cambridge University Press, Cambridge.
Tucker, Andrea J.
1983 Excavations at Chindi Hamlet (Site 5MT4684), a Basketmaker III Habitation Site. *Dolores Archaeological Program Technical Reports* DAP-106. Submitted to the U.S. Bureau of Reclamation, Upper Colorado Region, Salt Lake City, UT, in compliance with Contract No. 8-07-40-S0562.
Turchin, Peter
2003 *Historical Dynamics: Why States Rise and Fall*. Princeton University Press, Princeton, NJ.
Turchin, Peter, and Andrey Korotayev
2006 Population Dynamics and Internal Warfare: A Reconsideration. *Social Evolution and History* 5(2):112–147.
Turner, Christy G., II, and Jacqueline A. Turner
1992 On Peter Y. Bullock's "A Reappraisal of Anasazi Cannibalism." *Kiva* 58:189–201.
1999 *Man Corn: Cannibalism and Violence in the Prehistoric American Southwest*. University of Utah Press, Salt Lake City.
Underwood, Sali A.
2004 Appendix B3: Osteological Analysis of Human Remains Recovered from Dancing Man Hamlet: Site 5MT9343. In *Archaeological Investigations Along the Towaoc Lateral and Lone Pine Lateral Components of the Dolores River Reclamation Project, Montezuma County, Colorado*, vol. 2, *Basketmaker III Sites*, edited by Mark L. Chenault, Cynthia Manseau, and Todd Kohler. Cultural Resource Report 2003-16. SWCA Environmental Consultants. Prepared for the U.S. Bureau of Reclamation, Durango, CO.
Unger, Paul W., Tom V. Sneed, Wayne R. Jordon, and Ric Jensen (editors)
1988 *Challenges in Dryland Agriculture: A Global Perspective*. Proceedings of the International Conference on Dryland Farming, August 15–19, 1988. Amarillo/Bushland, TX.
Upham, Steadman, and Lori S. Reed
1989 Regional Systems in the Central and Northern Southwest: Demography, Economy, and Sociopolitics Preceding Contact. In *Columbian Consequences*, vol. 1, *Archaeological and Historical Perspectives on the Spanish Borderlands West*, edited by D. H. Thomas, pp. 57–76. Smithsonian Institution Scholarly Press, Washington, DC.
U.S. Department of Agriculture
2005 *USDA National Nutrient Database for Standard Reference, Release 18*. Available at www.nal.usda.gov/fnic/foodcomp/cgibin/list_nut_edit.pl. Accessed February 27, 2006.
van der Leeuw, Sander, and Charles L. Redman
2002 Placing Archaeology at the Center of Socio-Natural Studies. *American Antiquity* 67:597–605.
Van Dyke, Ruth M., and Anthony G. King
2010 Connecting Worlds: PIII Towers in the Northern San Juan. *Kiva* 75:351-375.
Van West, Carla R.
1994 *Modeling Prehistoric Agricultural Productivity in Southwestern Colorado: A GIS Approach*. Reports of Investigations, no. 67. Department of Anthropology, Washington State University, Pullman, and Crow Canyon Archaeological Center, Cortez, CO.
Van West, Carla R., and Jeffrey S. Dean
2000 Environmental Characteristics of the A.D. 900–1300 Period in the Central Mesa Verde Region. *Kiva* 66:19–44.
Van West, Carla R., and Timothy A. Kohler
1996 A Time to Rend, a Time to Sew: New Perspectives on Northern Anasazi Sociopolitical Development in Later Prehistory. In *Anthropology, Space, and Geographic Information Systems*, edited by M. Aldenderfer and H. D. G. Maschner, pp. 107–131. Oxford University Press, New York.
Varien, Mark D.
1997 New Perspectives on Settlement Patterns: Sedentism and Mobility in a Social Landscape. Unpublished Ph.D. dissertation, Department of Anthropology, Arizona State University, Tempe.
1999a *Sedentism and Mobility in a Social Landscape: Mesa Verde and Beyond*. University of Arizona Press, Tucson.
1999b Regional Context: Architecture, Settlement Patterns, and Abandonment. In *The Sand Canyon Archaeological Project: Site Testing*, edited by M. D. Varien, chapter 21. Available at www.crowcanyon.org/sitetesting. Accessed April 25, 2006.
2000 Introduction. *Kiva* 66(1):5–18.
2001 We Have Learned a Lot, but We Still Have More to Learn. In *Chaco Society and Polity: Papers from the 1999 Conference*, edited by Linda S. Cordell, W. James Judge, and

June-el Piper, pp. 47–61, New Mexico Archaeological Council Special Publication, no. 4, Albuquerque.
2010 Depopulation of the Northern San Juan Region: Historical Review and Archaeological Context. In *Leaving Mesa Verde: Peril and Change in the Thirteenth-Century Southwest*, edited by T. A. Kohler, M. D. Varien, and A. M. Wright, pp. 1–33. University of Arizona Press, Tucson.

Varien, Mark D. (editor)
1999 *The Sand Canyon Archaeological Project: Site Testing*, ver. 1.0. Available at www.crowcanyon.org/ResearchReports/SiteTesting/Text/Report.htm. Accessed December 17, 2005.

Varien, Mark D., William D. Lipe, Michael A. Adler, Ian M. Thompson, and Bruce A. Bradley
1996 Southwestern Colorado and Southeastern Utah Settlement Patterns, A.D. 1100 to 1300. In *The Prehistoric Pueblo World, A.D. 1150–1350*, edited by M. A. Adler, pp. 86–113. University of Arizona Press, Tucson.

Varien, Mark D., and Barbara J. Mills
1997 Accumulations Research: Problems and Prospects for Estimating Site Occupation Span. *Journal of Archaeological Method and Theory* 4:141–191.

Varien, Mark D., and Scott G. Ortman
2005 Accumulations Research in the Southwest United States: Middle Range Theory for Big-Picture Problems. *World Archaeology* 37(1):132–155.

Varien, Mark D., Scott G. Ortman, Timothy A. Kohler, Donna M. Glowacki, and C. David Johnson
2007 Historical Ecology in the Mesa Verde Region: Results from the Village Project. *American Antiquity* 72:273–300.

Varien, Mark D., Scott G. Ortman, Susan C. Ryan, and Kristin A. Kuckelman
2008 Population Dynamics Among Salmon's Northern Neighbors in the Central Mesa Verde Region. In *Chaco's Northern Prodigies: Salmon, Aztec, and the Ascendancy of the Middle San Juan Region After* AD 1100, edited by Paul F. Reed, pp. 351–365, University of Utah Press, Salt Lake City.

Varien, Mark D. and James M. Potter
1997 Unpacking the Discard Equation: Simulating the Accumulation of Artifacts in the Archaeological Record. *American Antiquity* 62:194–213.

Varien, Mark D., Carla Van West, and G. Stuart Patterson
2000 Competition, Cooperation, and Conflict: Agricultural Production and Community Catchments in the Central Mesa Verde Region. *Kiva* 66:45–64.

Varien, Mark D., and Richard Wilshusen (editors)
2002 *Seeking the Center Place: Archaeology and Ancient Communities in the Mesa Verde Region*. University of Utah Press, Salt Lake City.

Vayda, Andrew P.
1974 Warfare in Ecological Perspective. *Annual Review of Ecology and Systematics* 5:183–193.

Vierling, Lee A.
1998 Palynological Evidence for Late and Postglacial Environmental Change in Central Colorado. *Quaternary Research* 49:222–232.

Vierra, Bradley J. and Richard I. Ford
2007 Foragers and Farmers in the Northern Rio Grande Valley, New Mexico. *Kiva* 73:117–130.

Volterra, Vito
1926 Fluctuations in the Abundance of a Species Considered Mathematically. *Nature* 118:558–600.

Voth, H.R.
1903 The Oraibi Summer Snake Ceremony. *Field Museum of Natural History, Anthropological Series* 4:262–358.

Waguespack, Nicole M., and Todd A. Surovell
2003 Clovis Hunting Strategies: How to Make Out on Plentiful Resources. *American Antiquity* 68:333–352.

Walker, Phillip L.
1997 Wife Beating, Boxing and Broken Noses: Skeletal Evidence for the Cultural Patterning of Violence. In *Troubled Times: Violence and Warfare in the Past*, edited by Debra L. Martin and David W. Frayer, pp. 145–179. Gordon and Breach, Amsterdam.

Walker, William H.
1998 Where Are the Witches of Prehistory? *Journal of Archaeological Method and Theory* 5(3):245–308.

Walsh, Michael R.
1998 Lines in the Sand: Competition and Stone Selection on the Pajarito Plateau, New Mexico. *American Antiquity* 63:573–593.

Walt, Henry J.
1985 The Function of Stockade Structures. In *The Excavation of the Cortez CO_2 Pipeline Project Sites*, 1982–1983, edited by M. P. Marshall, pp. 165–170. Office of Contract Archaeology, University of New Mexico, Albuquerque.

Wand, M. P.
1994 Fast Computation of Multivariate Kernel Estimators. *Journal of Computational and Graphical Statistics* 3:433–445.
Wand, M. P., and Jones, M. C.
1995 *Kernel Smoothing*. Chapman and Hall, London.
Wardle, Peter
1968 Engelmann Spruce (*Picea engelmannii* Engel.) at Its Upper Limits on the Front Range, Colorado. *Ecology* 49:483–495.
Warner, Ted J.
1995 *The Dominguez-Escalante Journal: Their Expedition Through Colorado, Utah, Arizona, and New Mexico in 1776*. Translated by Fray Angelico Chavez. University of Utah Press, Salt Lake City.
Weatherwax, Paul
1954 *Indian Corn in Old America*. Macmillan, New York.
Webster, David
1975 Warfare and the Evolution of the State: A Reconsideration. *American Antiquity* 40:464–470.
Weiss, Kenneth M.
1973 *Demographic Models for Anthropology*. Memoirs of the Society for American Archaeology, no. 27. Salt Lake City, UT.
Wetterstrom, Wilma
1986 *Food, Diet, and Population at Prehistoric Arroyo Hondo Pueblo, New Mexico*. School of American Research Press, Santa Fe, NM.
White, Theodore W.
1953 A Method of Calculating the Dietary Percentage of Various Food Animals Utilized by Aboriginal Peoples. *American Antiquity* 19:160–164.
White, Tim D.
1991 *Prehistoric Cannibalism at Mancos 5MTUMR-2346*. Princeton University Press, Princeton, NJ.
Wilcox, David R., and Jonathan Haas
1994 The Scream of the Butterfly: Competition and Conflict in the Prehistoric Southwest. In *Themes in Southwest Prehistory*, edited by George J. Gumerman, pp. 211–238. School of American Research, Santa Fe, NM.
Wilkinson, R. G., and K. M. Van Wagenen
1993 Violence Against Women: Prehistoric Skeletal Evidence from Michigan. *Midcontinental Journal of Archaeology* 18:190–216.
Willey, Gordon R.
1953 Prehistoric Settlement Patterns in the Virú Valley, Peru. Bureau of American Ethnology, Smithsonian Institution Bulletin, no. 155. U.S. Government Press, Washington, DC.
Willey, Gordon R., and Philip Phillips
1958 *Method and Theory in American Archaeology*. University of Chicago Press, Chicago, IL.
Williams, G. C.
1966 *Adaptation and Natural Selection: A Critique of Some Current Evolutionary Thought*. Princeton University Press, Princeton, NJ.
Wilshusen, R. H., and E. Blinman
1992 Pueblo I Village Formation: A Reevaluation of Sites Recorded by Earl Morris on Ute Mountain Ute Tribal Lands. *Kiva* 57:251–269.
Wilshusen, Richard H.
1986a Excavations at Rio Vista Village (Site 5MT2182), a Multicomponent Pueblo I Village. In *Dolores Archaeological Program: Anasazi Communities at Dolores: Middle Canyon Area*, edited by A. E. Kane and C. K. Robinson, pp. 210–658. U.S. Department of the Interior, Bureau of Reclamation, Engineering and Research Center, Denver, CO.
1986b The Relationship Between Abandonment Mode and Ritual Use in Pueblo I Anasazi Protokivas. *Journal of Field Archaeology* 13:245–254.
1989 Unstuffing the Estufa: Ritual Floor Features in Anasazi Pit Structures and Pueblo Kivas. In *The Architecture of Social Integration in Prehistoric Pueblos*, edited by William D. Lipe and Michelle Hegmon, pp. 89–111. Occasional Papers, no. 1. Crow Canyon Archaeological Center, Cortez, CO.
1991 Early Villages in the American Southwest: Cross-Cultural and Archaeological Perspectives. Unpublished Ph.D. dissertation, Department of Anthropology, University of Colorado, Boulder.
1999 Pueblo I (A.D. 750–900). In *Colorado Prehistory: A Context for the Southern Colorado River Basin*, edited by William D. Lipe, Mark D. Varien, and Richard H. Wilshusen, pp. 196–241. Colorado Council of Professional Archaeologists, Denver.
2002 Estimating Population in the Central Mesa Verde Region. In *Seeking the Center Place: Archaeology and Ancient Communities in the Mesa Verde Region*, edited by M. D. Varien and R. H. Wilshusen, pp. 101–120. University of Utah Press, Salt Lake City.
Wilshusen, Richard H., Melissa J. Churchill, and James M. Potter

1997 Prehistoric Reservoirs and Water Basins in the Mesa Verde Region: Intensification of Water Collection Strategies During the Great Pueblo Period. *American Antiquity* 62:664–681.

Wilshusen, Richard H., and Scott G. Ortman
1999 Rethinking the Pueblo I Period in the San Juan Drainage: Aggregation, Migration, and Cultural Diversity. *Kiva* 64:369–399.

Wilshusen, Richard, Scott G. Ortman, Shanna Diederichs, Donna M. Glowacki, and Grant Coffey
2011 Heartland of the Early Pueblos: The Central Mesa Verde. In *Crucible of Pueblos: The Early Pueblo Period in the Northern Southwest*, edited by Richard Wilshusen, Gregson Schachner, and James Allison. Cotsen Institute of Archaeology, University of California, Los Angeles.

Wilshusen, Richard H., and James M. Potter
2010 The Emergence of Early Villages in the American Southwest: Cultural Issues and Historical Perspectives. In *Becoming Villagers: Comparing Early Village Societies*, edited by Matthew S. Bandy and Jake R. Fox, pp. 165–183. University of Arizona Press, Tucson.

Wilshusen, Richard H., and Ruth M. Van Dyke
2006 Chaco's Beginnings: The Collapse of Pueblo I Villages and the Origins of the Chaco System. In *The Archaeology of Chaco Canyon: An Eleventh-Century Pueblo Regional Center*, edited by Stephen H. Lekson, pp. 211–259. School of American Research Press, Santa Fe, NM.

Wilshusen, Richard H., and C. Dean Wilson
1995 Reformatting the Social Landscape in the Late Pueblo I–Early Pueblo II Period: The Cedar Hill Data in Regional Context. In *The Cedar Hill Special Treatment Project: Late Pueblo I, Early Navajo, and Historic Occupations in Northwestern New Mexico*, compiled by Richard H. Wilshusen, pp. 43–80. La Plata Archaeological Consultants, Research Papers, no. 1.

Wilson, A. M., and D. D. Briske
1979 Seminal and Adventitious Root Growth of Blue Grama Seedlings on the Central Plains. *Journal of Range Management* 32:209–213.

Windes, Thomas C.
2004 The Rise of Early Chacoan Great Houses. In *In Search of Chaco: New Approaches to an Archaeological Enigma*, edited by David Grant Noble, pp. 14–21. School of American Research, Santa Fe, NM.

2007 *Early Puebloan Occupations in the Chaco Region: Excavations and Survey of Basketmaker III and Pueblo I Sites, Chaco Canyon, New Mexico*, vol. 1. Anthropology Projects, Cultural Resources Management, Intermountain Region Support Office, National Park Service, Department of the Interior, Santa Fe, NM.

Wing, Elizabeth S, and Antoinette B. Brown
1979 *Paleonutrition*. Academic, New York.

Winterhalder, Bruce
1996 Social Foraging and the Behavioral Ecology of Intragroup Resource Transfers. *Evolutionary Anthropology* 5:46–57.

Winterhalder, Bruce, and Eric A. Smith
1992 Evolutionary Ecology and the Social Sciences. In *Evolutionary Ecology and Human Behavior*, edited by E. A. Smith and B. Winterhalder, pp. 3–23. Aldine de Gruyter, Hawthorne, NY.

2000 Analyzing Adaptive Strategies: Human Behavioral Ecology at Twenty-Five Years. *Evolutionary Anthropology* 9:51–72.

Wolfe, Eric R.
1982 *Europe and the People Without History*. University of California Press, Berkeley.

Woodbury, Richard W.
1961 Climatic Changes and Prehistoric Agriculture in the Southwestern United States. In *Solar Variations, Climatic Change, and Related Geophysical Problems*, edited by Rhodes W. Fairbridge, pp. 705–709. Annals of the New York Academy of Sciences, no. 95 (Art. 1). New York.

Wood-Jones, F.
1910 Fractured Bones and Dislocations. In *The Archaeology of Nubia Report for 1907–1908*, vol. 2, *Report on the Human Remains*, edited by G. Elliot-Smith and F. Wood-Jones, pp. 293–342. National Printing Department, Cairo.

Wooster, L. D.
1935 Notes on the Effects of Drought on Animal Populations in Western Kansas. *Transactions of the Kansas Academy of Science* 38:351–352.

World Data Center for Paleoclimatology
2003 http://hurricane.ncdc.noaa.gov/pls/paleo. Accessed March 3, 2003.

Wright, Aaron M.
2006 A Low-Frequency Paleoclimatic Reconstruction from the La Plata Mountains, Colorado and Its Implications for Agricultural Paleoproductivity in the Mesa Verde Region. Master's thesis, Department of Anthropology, Washington State University, Pullman.

2010 The Climate of the Depopulation of the Northern Southwest. In *Leaving Mesa Verde: Peril and Change in the Thirteenth-Century Southwest*, edited by Timothy A. Kohler, Mark D. Varien, and Aaron M. Wright, pp. 75–101. University of Arizona Press, Tucson.

Wright, Henry T., and Gregory A. Johnson

1975 Population, Exchange, and Early State Formation in Southwestern Iran. *American Anthropologist* 77:267–289.

Wright, Jonathan W.

1952 *Pollen Dispersion of Some Forest Trees.* Northeastern Forest Experiment Station Paper No. 46. U.S. Department of Agriculture, Forest Service, Upper Darby, PA.

1953 Pollen Dispersal Studies: Some Practical Applications. *Journal of Forestry* 51:114–118.

Yarnell, Richard W.

1982 Excavations at Pheasant View Hamlet (Site 5MT2192), a Pueblo I Habitation Site. *Dolores Archaeological Program Technical Reports* DAP-027. Submitted to the U.S. Bureau of Reclamation, Upper Colorado Region, Salt Lake City, UT, in compliance with Contract No. 8-07-40-S0562.

1983 Excavations at Areas 2, 3, 4, and 7, Periman Hamlet (Site 5MT4671), a Pueblo I–Pueblo II Habitation. *Dolores Archaeological Program Technical Reports* DAP-091. Submitted to the U.S. Bureau of Reclamation, Upper Colorado Region, Salt Lake City, UT, in compliance with Contract No. 8-07-40-S0562.

Yunker, Brian E.

2001 The Yellow Jacket Burials: An Analysis of Burial Assemblages from Two Basketmaker III Through Pueblo III Mesa Verde Area Sites. Unpublished Master's thesis, Department of Anthropology, University of Colorado, Boulder.

NOTES ON CONTRIBUTORS

FUMIYASU (FUMI) ARAKAWA is an Assistant Professor of Anthropology at New Mexico State University and an affiliated researcher at the Crow Canyon Archaeological Center, Cortez, CO. He has been conducting lithic research for the Village Ecodynamics Project since 2003. His recent publications include "Geoarchaeological Investigation of Lithic Resources in the Central Mesa Verde Region" (first author, with Kim Gerhardt, 2009) in *Geoarchaeology: An International Journal.*

R. KYLE BOCINSKY is a Ph.D. student in the Department of Anthropology, Washington State University; a trainee in the IGERT Program in Evolutionary Modeling; and a National Science Foundation Graduate Research Fellow. His M.A. thesis—titled "Is a Bird in Hand Really Worth Two in the Bush? Models of Turkey Domestication on the Colorado Plateau"—integrated turkey use as a protein resource into the VEP simulation framework. Kyle's current research concerns faunal exploitation, domestication, and optimal foraging in the ancestral Puebloan Southwest.

SARAH M. COLE received her M.A. in Anthropology from Washington State University in 2007. She is President and co-owner of Red River Archaeology, LLC, a cultural resource management firm located in Dallas, Texas. Other recent publications include "Population and Warfare: A Test of the Turchin Model in Pueblo Societies" (second author, with Timothy Kohler and Stanca Ciupe) in *Pattern and Process in Cultural Evolution* (edited by Stephen Shennan, University of California Press, 2009).

KEVIN COOPER is an information systems manager and adjunct faculty member in the Mathematics Department at Washington State University. His training is in differential equations and numerical analysis, and his recent work has concerned meso-scale schemes for large systems of Newtonian particles.

JASON A. COWAN is a Project Archaeologist for Cultural Resource Consultants Inc. of Bainbridge Island, WA. He is working on a multiyear project along the mid-Columbia River. He received his M.A. in anthropology from Washington State University in 2006.

STEFANI CRABTREE is a Ph.D. student in the Department of Anthropology, Washington State University; a National Science Foundation Graduate Research Fellow; and a Watson Fellowship alumna. Her research in the Village Ecodynamics Project focuses on the evolution of exchange and cooperation.

BEN FORD holds a B.S. in Computer Science from Washington State University. He is largely responsible for maintaining the Village Web site http://village.anth.wsu.edu/, providing technical support and coding assistance and building custom tools to process the abundance of data created by the Village Ecodynamics Project.

DONNA M. GLOWACKI is the John Cardinal O'Hara CSC Assistant Professor of Anthropology at the University of Notre Dame, a senior researcher on the Village Ecodynamics Project, and a long-time research associate with Crow Canyon Archaeological Center. She has conducted fieldwork at 63 of the largest sites in the central Mesa Verde region and is currently engaged in fieldwork at Mesa Verde National Park. Other recent publications include *Religious Transformation in the Late Pre-Hispanic Pueblo World* (coedited with Scott Van Keuren, University of Arizona Press, 2012).

C. DAVID (DAVE) JOHNSON is archaeologist and tribal liaison for the Bureau of Land Management in Arcata, CA, managing the agency's regional cultural resource program in cooperation with Humboldt State University faculty, staff, and students, as well as many tribal partners. Dave earned a Ph.D. at Washington State University as a research assistant in Village Ecodynamics Project I, and maintains an active interest in long-term human ecodynamics and the ongoing VEP. He has been a coauthor on numerous VEP publications and is sole author of "Investigating the Consequences of Long-Term Human Predation of R-Selected Species" in *Movement, Connectivity, and Landscape Change in the Ancient Southwest* (edited by Margaret Nelson and Colleen Strawhacker, University Press of Colorado, 2011).

ZIAD KOBTI is an Associate Professor with the School of Computer Science at the University of Windsor and adjunct researcher in Anthropology at Washington State University. He is one of the coinvestigators on Village Ecodynamics Project II and a lead computer programmer in VEP I. He is the past program co-chair of the Twentieth Conference of the Canadian Society for Computational Studies of Intelligence, Montreal, 2007 and first editor of its proceedings.

TIMOTHY (TIM) A. KOHLER is a Regents Professor in Anthropology at Washington State University; an external professor at the Santa Fe Institute; and a research associate at the Crow Canyon Archaeological Center, Cortez, CO. He is the senior principal investigator of the Village Ecodynamics Project and a past editor of *American Antiquity*. Recent publications include *Leaving Mesa Verde: Peril and Change in the Thirteenth-Century Southwest* (first editor, with Mark Varien and Aaron Wright, University of Arizona Press, 2010). He delivered WSU's Distinguished Faculty Address in 2011 on the "Prehistory of Pueblo Peoples—How We Learn and What We Know."

KENNETH (KEN) E. KOLM is a Senior Science Advisor at Integral Consultants, Inc. and an Emeritus Professor at the Colorado School of Mines, Golden, CO. He is one of the coprincipal investigators of the National Science Foundation Village Ecodynamics Project. Dr. Kolm has more than 35 years of professional experience in the fields of hydrogeology and hydrological and environmental systems analysis, with an emphasis on groundwater and surface water science and engineering, wetlands environments, expedited environmental site characterization and remediation, and watershed and ecosystem characterization, management, and restoration.

SCOTT G. ORTMAN is an Omidyar Fellow at the Santa Fe Institute and the Lightfoot Fellow at the Crow Canyon Archaeological Center, Cortez, CO. He has been a senior researcher on the Village Ecodynamics Project since 2003 and received his Ph.D. from Arizona State University in 2010. He is also the author of *Winds from the North: Tewa Origins and Historical Anthropology* (University of Utah Press, 2012); several reports on the artifacts from ancestral Pueblo archaeological sites; and a number of papers in *American Antiquity, World Archaeology, Kiva*, and edited volumes.

SCHAUN M. SMITH is a Staff Scientist in the Environmental Unit of the Chevron Energy Technology Company. As a member of the Village Ecodynamics Project, he performed research culminating in a Ph.D. dissertation at the Colorado School of Mines, titled "Paleohydrological Modeling of Groundwater Resources and Sustainability Related to Ancestral Puebloan Settlement Populations from A.D. 600 to A.D. 1300 Central Mesa Verde Region, Southwest Colorado." This research developed, for the first time, conceptual and mathematical models of prehistoric spring flow and related it to population dynamics of the region.

MARK D. VARIEN is the Research and Education Chair at the Crow Canyon Archaeological Center, where he works with the Center's staff to fulfill Crow Canyon's mission of collaborating with American Indians on archaeological research that is conducted in the context of public education programs. He is the author of *Sedentism and Mobility in a Social Landscape* and the coeditor of several books, including, most recently, *Leaving Mesa Verde: Peril and Change in the Thirteenth-Century Southwest* (both from the University of Arizona Press).

AARON M. WRIGHT defended his Ph.D. dissertation on "Hohokam Rock Art, Ritual Practice, and Social Transformation in the Phoenix Basin" in 2011 at Washington State University and is currently a Preservation Fellow at the Center for Desert Archaeology, Tucson, AZ. He coedited the 2010 University of Arizona Press volume, *Leaving Mesa Verde: Peril and Change in the Thirteenth-Century Southwest.*

INDEX

Citations ending with a "t" indicate a table; those with an "f" indicate a figure.

Composition: Westchester Book Group
Text: 9.5/13 Scala
Display: Scala Sans
Printer and Binder: Thomson-Shore